Genetic Variation

A LABORATORY MANUAL

Genetic Variation

A LABORATORY MANUAL

EDITED BY

Michael P. Weiner
RainDance Technologies

Stacey B. Gabriel
The Broad Institute of MIT and Harvard

J. Claiborne Stephens
Motif BioSciences Inc.

http://www.cshprotocols.org

COLD SPRING HARBOR LABORATORY PRESS
Cold Spring Harbor, New York • http://www.cshlpress.com

Genetic Variation
A LABORATORY MANUAL

Publisher	John Inglis
Acquisition Editor and Managing Editor	Kaaren Janssen
Development, Marketing, & Sales Director	Jan Argentine
Developmental Editors	Irene Pech, Martin Winer, and Laurie Goodman
Project Coordinator	Inez Sialiano
Production Manager	Denise Weiss
Production Editor	Patricia Barker
Desktop Editor	Lauren Heller
Sales Account Manager	Elizabeth Powers
Cover Designer	Lewis Agrell

Front cover artwork: (*Top left*) Confocal imaging of cells in mitosis (tubulin stained green, chromosome stained orange), from P. Waadsworth, in *Live cell imaging: A laboratory manual* (ed. R.D. Goldman and D. Spector); (*top center*) diagram redrawn from Chapter 3 by D. Evans and S. Purcell; (*top right*) *Mus musculus*, Gary Henderson; (*middle left*) J.D. Watson lecture note #1 entitled "What is Life," courtesy James D. Watson Collection, Cold Spring Harbor Laboratory Archives; (*middle center*) double helix by Michael P. Weiner; (*middle right*) phylogeny, Michael P. Weiner; (*bottom left*) data from an experiment showing the expression of thousands of genes on a single GeneChip® probe array, image courtesy of Affymetrix; (*bottom center*) Bonobo chimpanzee ©2005 William H. Calvin, University of Washington; (*bottom right*) maize photographs, courtesy Cold Spring Harbor Laboratory Archives.

Library of Congress Cataloging-in-Publication Data

Genetic variation : a laboratory manual / edited by Michael P. Weiner,
Stacey Gabriel, J. Claiborne Stephens.
    p. cm.
Includes bibliographical references and index.
 ISBN 978-0-87969-780-8 (pbk. : alk. paper) -- ISBN 978-0-87969-779-2
(printed hardcover : alk. paper)
 1. Molecular genetics--Laboratory manuals. 2. Variation
(Biology)--Laboratory manuals. I. Weiner, Michael P., 1958- II. Gabriel,
Stacey. III. Stephens, J. Claiborne. IV. Title.

  QH440.5.G455 2007
  572.8078--dc22

        2007023448

10 9 8 7 6 5 4 3 2 1

*Our collective gratitude goes out to the investigators who strive
to alleviate human suffering through an increased
understanding of the biological principles of the world around us*

Contents

SECTION 1: STUDY DESIGN

Section 1 covers issues that deal with the importance of study design in addressing the sheer volume of data to be evaluated. Chapter 2 highlights that, for human genetic analyses, certain populations may have more favorable characteristics for associating traits with genes or precise genomic regions (e.g., disease-gene mapping). The size of the sample from a given population is critical for ensuring that a study is adequately powered to find a genetic association when it exists. Chapter 3 considers sample size and all other relevant factors for determining the power of genetic studies. Chapter 4 reviews the analytical methodology available for human genetic studies. In particular, strengths and weaknesses of linkage mapping, where pedigrees play an essential role, are contrasted with genetic association analyses, in which study subjects are typically unrelated. Both types of studies have their place, and the choice of method is strongly dependent on the type of disease or trait being addressed. Chapter 5 gives detailed insights into one of the most important resources available: the dbSNP database, a resource that maintains the current inventory of known variable sites in the human genome. Chapter 5 is an introduction to accessing the multitude of data cataloged in dbSNP. Section 1 closes with Chapter 6, which provides a user's guide to the suite of tools for manipulating HapMap data. The HapMap data set is a unique resource in that it provides an enormous, consistent set of genotypic data (3.9 million SNPs) for 270 individuals from four world populations. These data give very high-resolution detail of the genetic architecture of these four populations and whet our appetite for more samples in these and additional populations!

SECTION 2: LABORATORY PROTOCOLS

The raw material of genetic research is DNA; individual-to-individual variation is the prism through which it is analyzed. This section deals with two critical aspects in the study of DNA variation: sample preparation and methods of common variation analysis. DNA of sufficient quality and quantity for the current generation of large-scale experimentation is often limiting. To address this, methods of DNA extraction, amplification, and cell culture are described. Parts 2 and 3 of this section turn to the generation of genotype data—beginning with the simplest and most abundant form of sequence variation, single-nucleotide polymorphism, and conclude with more complex DNA variation and the emerging study of DNA copy number variation. The specific topics are not meant to be an exhaustive description of all options for sample preparation and genotyping. Rather, the intent is to sample the wide range of sample type (cell culture and amplified DNA) and experimental scale (small-scale "home brew" genotyping vs. genome-wide standard products) that contribute to current genetic studies.

SECTION 3: DATA ANALYSIS

Following on the generation of multiple types of genetic data, Section 3 covers multiple approaches to data analysis. With rich reference data sets available, such as the Human HapMap, there are continually evolving approaches for data analysis to extract maximal information from genome variation. These include strategies for SNP selection for maximizing the "return on investment"' of SNP genotyping by pinpointing the most informative SNPs to study; approaches to data visualization to simplify complex genotype data sets; methods of analysis to study association of genotype with phenotype; the use of genetic data sets to study evolutionary phenomena such as selection. While the density and breadth of genetic data that are routinely at hand for the investigator continue to expand, the specific approaches described here represent a set of issues and challenges amenable to the current generation of genome-wide SNP data for use in genetic association analysis.

SECTION 4: VARIATION STUDIES IN MODEL ORGANISMS

Section 4 includes an overview of eight different model species: three plant and five animal. The plant chapters include *Arabidopsis*, a traditional model for plant geneticists and biochemists, and two important crop plants (corn and rice). The animal chapters include rodent (mouse and rat), canine (domestic dog), feline (domestic cat), and primate (chimpanzee) species. For each species, the authors have included a description of the genome sequence as it currently exists (at the time of this writing), historical origin of the species, phenotypic variations, association studies, and Internet resources. Several of the chapters include information about the phylogeny of related strains and related species.

SECTION 5: INSIGHTS INTO HUMAN VARIATION

Section 5 contains two diverse applications of genomic research to human beings. Chapter 33 illustrates the utility of sex-specific markers, those from mtDNA and the Y chromosome, in the elucidation of human migrations through Polynesia. Chapter 34 gives a thorough overview of the requirements and application of DNA testing to forensics. The section concludes with a third chapter that looks ahead to the future of genome research in humans and other species and to possible societal implications.

Foreword

The study of genetic variation has never been more exciting than it is at present, because of recent successes in linking specific variants with phenotypes and common diseases. In the last decade of the 20th century, the human genetics community generated linkage maps and applied these successfully to identify mutations in over 1000 Mendelian disorders. The applications to common diseases such as diabetes, asthma, inflammatory and infectious diseases, and cancer remained elusive until the advent of the human genome sequence that appeared quickly in the first years of the new millennium. This complete sequence made possible the large-scale characterization of human genome variation, culminating in an international consortium that generated the HapMap and new technologies that ranged from extensive analyses of single variants in hundreds of samples to parallel testing of hundreds of thousands of polymorphisms in thousands of samples. Parallel efforts in other organisms, from microbes to plants and animals, are proving just as exciting in linking biological variation with inherited variation. A high-level view across species reveals a richness of genetic diversity within species that is also diverse in its appearance: single-nucleotide differences (i.e., SNPs), nucleotide insertions/deletions, copy number variation, chromosomal rearrangements, and other structural variations.

In preparing a laboratory manual on genetic variation, the editors have assembled a remarkable team and generated a solid foundation for novices and experts alike. A striking feature of the manual is its breadth of scope, with only one of the five sections being focused on the traditional aspects of a laboratory manual, i.e., how to generate data (Section 2), which is rich in practical methods to be used in tomorrow's experiments. Whereas this section contains key information to be used by laboratory staff and trainees, Sections 1 and 3 provide global knowledge on major aspects of genetic research, spanning study design, utilization of relevant databases, bioinformatics tools, statistical analyses, and interpretation. Special consideration was provided to research communities interested in three plant species (*Arabidopsis*, corn, and rice) and five mammalian species (mouse, rat, cat, dog, and chimpanzee).

The tremendous growth of databases of human genetic variation in worldwide populations is boosting the field of population genetics and its use of new data in inferring the evolutionary history of man. Section 3 includes new methods that have been developed to shed light on where and how natural selection acted during human evolution to shape current human genome diversity and, in many cases, has resulted in common variants that affect intermediate biological pathways associated with disease susceptibility.

I challenge students and other readers to take the time to read the introduction, background, key concepts, and messages that are found in all sections of the manual, and not just to open the book to a specific protocol or reference on a particular analysis method. To understand genetic variation, one needs to merge biology, technology, informatics, mathematics, statistics, and evolution. Although today's techniques will surely continue to evolve as quickly as they have in recent years, the founding principles that are outlined by world-class specialists will have lasting value and will help connect new facts and data sets into rational explanations.

Congratulations to the editors and authors for an outstanding work!

THOMAS J. HUDSON, M.D.
Ontario Institute for Cancer Research

Preface

Nearly two years ago, when we first embarked on this lab manual on genotyping, our first thought was, How long will such a manual be relevant? Would the technology being developed in the universities and commercial entities for the $1000 genome sequence make genotyping, as we knew it then in late 2005, obsolete? With input from both our scientific colleagues and our editors at Cold Spring Harbor Laboratory Press, we decided to broaden the scope of the manual and place genotyping into the greater context of a complete genome variation study. Because of Cold Spring Harbor Laboratory's illustrious past, we wanted to be sure to include information that would be of interest to researchers in both the agricultural and animal fields.

In planning the sections of this manual, we thought about the kinds of information that might be useful to scientists at all levels of expertise. Therefore, we divided the manual into five sections. The first section describes study design and the tools the reader would need to understand before getting started. The sheer capacity for generating genomic variation data has an addictive effect: The more you see, the more you want. However, sheer volume of data cannot overcome poorly planned studies. In fact, good genotyping data in a bad study design, accompanied by inadequate attention to statistical issues, such as that of multiple comparisons, has the potential to yield seemingly valid results. The contributors to Section 1 have spelled out these issues and more.

Although it is easy to think that any genome study begins with choosing a method for genotyping, we hope that Section 1 shows the reader that a proper genetic study begins with the proper choice of a population and the questions to be addressed. Section 2 of this manual deals with the generation of genotyping and genome variation data. Divided into three parts, this section provides the traditional "recipe and protocol" section that one expects from the CSHL series of laboratory manuals.

In Section 2, Part 1, methods are described for the isolation and preparation of nucleic acids, including both RNA and DNA, from various samples. Chapters 7 and 8 focus on preparation of genetic materials from plants. RNA isolation is included because in many cases it is easier to deal with the less complex transcriptome. Chapter 9 includes 12 protocols for DNA isolation from mammalian sources and includes two procedures for whole-genome amplification, one using the polymerase chain reaction, and the second, rolling circle amplification. Many of these methods are available as kits from commercial suppliers. Where they can, the authors have included home recipes for the kit components and a troubleshooting guide.

Section 2, Part 2, contains methods for low- and intermediate-throughput SNP genotyping. This is the type of genotyping (hundreds to less than several thousand genotypes per day) that one would set up in an average-sized laboratory contemplating one or two medium-sized studies. The last decade has seen an explosion in the technology used for this throughput type of genotyping. Most of these methods can be thought of as having two components: assay and read-out. Example assays include single-base chain extension, TaqMan, oligonucleotide ligation, and allele-specific amplification. Read-out can be performed using gel and capillary electrophoresis, fluorescence polarization, flow cytometry, etc.

Like DNA sequencing and oligodeoxynucleotide primer synthesis, genotyping is fungible. It seems likely that in the future much of the average laboratory-based genotyping will be performed by outsourcing to separate core facilities, but an understanding of how the genotyping is generated is important in assessing its reliability.

An important aspect of genome variation that is becoming more appreciated as more information is revealed concerns the changes in copy number. The ability to discover and measure changes in chromosome copy number is becoming critical to our understanding of the differences between normal and disease states. We are fortunate to have several chapters included on this important topic.

Section 3 deals with analysis of the data generated and the tools that can be used to help with that analysis, including strategies for SNP selection, as well as genetic association analysis. In Section 4, we tried to give the reader an understanding of the current "state of the organism" for several plant and animal species that have their respective genomes sequenced and analyzed at various levels of completeness. We end with Section 5, describing some insights into human variation.

What's missing? We purposely did not include extensive information about ultra high-throughput (UHT) genotyping. These initiatives are expensive, and the technology is rapidly changing and beyond the scope of this current effort. At one point early in our discussions, we considered the inclusion of a section on one or more of the excellent examples where genetic studies have been used to increase our understanding of a genetic disease, such as cystic fibrosis or diabetes. We opted instead for the section on model organisms, which we felt would be of a broader interest. We debated extensively about including more (and different) model organisms, such as *Drosophila*, porcine, and bovine species; but in the end, we decided that the group we included was enough to give the reader an understanding of the kinds of information available.

We thank those at Cold Spring Harbor Laboratory Press who made this project possible. We appreciate the efforts of the editorial staff at CSHL Press: Inez Sialiano, our project coordinator who helped us to cope with—even avert—natural disasters; the freelance developmental editors Irene Pech, Laurie Goodman, and Martin Winer; and, especially, Kaaren Janssen, for her origination, developmental editing, and project management. We also thank Denise Weiss, production manager; Pat Barker, production editor; and Lauren Heller, desktop editor; for their expertise in production of this work, and John Inglis, Executive Director of the Press, and Jan Argentine, Book Development, Marketing and Sales Director, for their roles in supporting the project.

M.P.W. thanks some of the people who either directly or indirectly helped to inspire him and this initiative, including especially Camille Solbrig, Evan and Alan Weiner, Albert Wexler, Michael Fleming, Ron Yasbin, Stan Zahler, Ross McIntyre, Ed Buss, Ray Wu, Harold Scheraga, Joe Sorge Jr, Jay Short, Mike Luther, Dan Burns, Lee Babiss, Allen Roses, Phil Whitcome, and Jonathan Rothberg.

J.C.S. takes this opportunity to thank all his friends and colleagues who have encouraged and mentored him throughout his career. Because that is a very long list and spans 30 years, he wants to single out those mentors who strove to provide the "big picture," especially as it incorporates population genetics: Professors Wyatt W. Anderson, John Avise, Michael Clegg, Bruce S. Weir, Masatoshi Nei, Wen-Hsiung Li, Ranajit Chakraborty, Frank Ruddle, Ken Kidd, and Stephen J. O'Brien.

S.B.G. thanks family, friends, and colleagues at the Broad Institute for their continual encouragement and support during all endeavors.

Finally, we owe a huge debt of gratitude to our authors. Their timely responsiveness to our comments was greatly appreciated. Their dedication to this effort was probably best exemplified by Dr. H. Leung, who managed to send us his excellent chapter on Rice (on time!) while in the midst of a typhoon that left his Institute bereft of electricity for a week. It was a pleasure to work with all of them.

From Mendel's peas to Darwin's finches to Benzer's bacteriophage to the deciphering of the human genome, there has been a continuum to the study of genetics. Both the understanding of our genetic heritage and the technology used to elicit that understanding have allowed a common lexicon to develop across all sciences. Since the unraveling of its structure more than 50 years ago, the double helix has become much more than a biological set of instructions; it has become a campfire around which all scientists can converse. We hope this manual allows this conversation to continue.

MICHAEL P. WEINER
STACEY B. GABRIEL
J. CLAIBORNE STEPHENS

Contributors

Arnold, Kevin, *Motif BioSciences Inc., New York*

Bamshad, Michael, *Departments of Pediatrics and Genome Sciences, Division of Genetics and Developmental Medicine, University of Washington School of Medicine, Seattle, Washington*

Barbazuk, W. Brad, *Donald Danforth Plant Science Center, St. Louis, Missouri*

Barrett, Jeffrey C., *Wellcome Trust Centre for Human Genetics, University of Oxford, United Kingdom*

Billings, Nicholas, *Department of Population Health and Reproduction, School of Veterinary Medicine, University of California, Davis*

Borevitz, Justin O., *Department of Ecology and Evolution, and Committee on Genetics, University of Chicago, Illinois*

Brooks, Andrew I., *Rutgers University Cell and DNA Repository, Department of Genetics; and Bionomics Research and Technology Center, Environmental and Occupational Health Sciences Institute, University of Medicine and Dentistry of New Jersey–Robert Wood Johnson Medical School, Piscataway, New Jersey*

Brown, Clyde, *Sigma-Aldrich, St. Louis, Missouri*

Brueck, Chad, *Sigma-Aldrich, St. Louis, Missouri*

Burtt, Noël P., *Program in Medical and Population Genetics, Broad Institute of MIT and Harvard, Cambridge, Massachusetts*

Butler, John M., *Biochemical Science Division, National Institute of Standards and Technology, Gaithersburg, Maryland*

Carlson, Chris, *Fred Hutchinson Cancer Research Center, Seattle, Washington*

Chen, Hsin D., *Department of Agronomy, Iowa State University, Ames*

Cuppen, Edwin, *Hubrecht Laboratory, Utrecht, The Netherlands*

Daly, Mark J., *Center for Human Genetic Research, Massachusetts General Hospital, Boston, and Broad Institute of MIT and Harvard, Cambridge, Massachusetts*

Davis, Gary, *Sigma-Aldrich, St. Louis, Missouri*

de Bakker, Paul I. W., *Program in Population and Medical Genetics, Broad Institute of MIT and Harvard, Cambridge, Massachusetts*

Emrich, Scott, *Bioinformatics and Computational Biology Graduate Program, Department of Electrical and Computer Engineering, Iowa State University, Ames*

Evans, David M., *Wellcome Trust Centre for Human Genetics, University of Oxford, United Kingdom*

Faham, Malek, *Affymetrix, South San Francisco, California*

Feolo, Michael L., *National Center for Biotechnology Information, National Library of Medicine, National Institutes of Health, Bethesda, Maryland*

Forman, Jonathan, *Affymetrix, South San Francisco, California*

Fu, Yan, *Donald Danforth Plant Science Center, St. Louis, Missouri*

Gulcher, Jeffrey, *deCODE Genetics, Reykjavik, Iceland, and Woodridge, Illinois*

Ha, Connie, *Cardiovascular Research Institute and Center for Human Genetics, University of California, San Francisco*

Holcomb, Ilona N., *Human Biology Division, Fred Hutchinson Cancer Research Center, Department of Genome Sciences, University of Washington, Seattle*

Hsia, An-Ping, *Department of Agronomy, Iowa State University, Ames*

Hübner, Norbert, *Max-Delbrück-Center for Molecular Medicine, Berlin-Buch, Germany*

Hudson, Thomas J., *Ontario Institute for Cancer Research, Toronto, Canada*

Jacob, Howard J., *Medical College of Wisconsin, Milwaukee*

Jacobs, Sharoni, *Affymetrix, Santa Clara, California*

Karlin-Neumann, George, *Affymetrix, South San Francisco, California*

Kayser, Manfred, *Department of Forensic Molecular Biology, Erasmus University Medical Centre, Rotterdam, The Netherlands*

King, Irena B., *Molecular Diagnostics Program, Fred Hutchinson Cancer Research Center, Seattle, Washington*

Kwitek, Anne E., *Medical College of Wisconsin, Milwaukee*

Kwok, Pui-Yan, *Department of Dermatology, University of California, San Francisco*

Leung, Hei, *Plant Breeding, Genetics, and Biotechnology Division, International Rice Research Institute, Manila, Philippines*

Li, Yan, *Department of Ecology and Evolution, University of Chicago, Chicago, Illinois*

Lin, Steven, *Affymetrix, South San Francisco, California*

Lindblad-Toh, Kerstin, *Broad Institute of MIT and Harvard, Cambridge, Massachusetts*

Lipinski, Monika J., *Department of Population Health and Reproduction, School of Veterinary Medicine, University of California, Davis*

Lucito, Rob, *Cold Spring Harbor Laboratory, Cold Spring Harbor, New York*

Lyons, Leslie A., *Department of Population Health and Reproduction, School of Veterinary Medicine, University of California, Davis*

Macdonald, Stuart J., *Department of Ecology and Evolutionary Biology and Department of Molecular Biosciences, University of Kansas, Lawrence*

Mackill, David, *Plant Breeding, Genetics, and Biotechnology Division, International Rice Research Institute, Manila, Philippines*

McNally, Kenneth L., *T.T. Chang Genetic Resources Center, International Rice Research Institute, Manila, Philippines*

Michalik, Steve, *Sigma-Aldrich, St. Louis, Missouri*

Mikkelsen, Tarjei S., *Broad Institute of MIT and Harvard, Cambridge, Massachusetts*

Moorhead, Martin, *Affymetrix, South San Francisco, California*

Mueller, Ernie, *Sigma-Aldrich, St. Louis, Missouri*

Ohtsu, Kazuhiro, *Department of Agronomy, Iowa State University, Ames*

Ostrander, Elaine A., *National Human Genome Research Institute, National Institutes of Health, Bethesda, Maryland*

Purcell, Shaun, *Center for Human Genetic Research, Massachusetts General Hospital, Boston, Massachusetts*

Sahota, Amrik, *Rutgers University Cell and DNA Repository, Department of Genetics; and Bionomics Research and Technology Center, Environmental and Occupational Health Sciences Institute, University of Medicine and Dentistry of New Jersey–Robert Wood Johnson Medical School, Piscataway, New Jersey*

Sapolsky, Ronald, *Affymetrix, South San Francisco, California*

Schnable, Patrick S., *Department of Agronomy, and Department of Genetics, Development, and Cell Biology, and Center for Plant Genomics, Iowa State University, Ames*

Sedova, Marina, *Affymetrix, South San Francisco, California*

Sherry, Stephen T., *National Center for Biotechnology Information, National Library of Medicine, National Institutes of Health, Bethesda, Maryland*

Smith, Albert Vernon, *Genthor ehf., Reykjavik, Iceland, and Icelandic Heart Association, Kopavogur, Iceland*

Springer, Nathan M., *Department of Plant Biology, University of Minnesota, St. Paul*

Stoneking, Mark, *Max Planck Institute for Evolutionary Anthropology, Leipzig, Germany*

Tischfield, Jay A., *Rutgers University Cell and DNA Repository, Department of Genetics; and Bionomics Research and Technology Center, Piscataway, New Jersey*

Trask, Barbara J., *Human Biology Division, Fred Hutchinson Cancer Research Center, Department of Genome Sciences, University of Washington, Seattle*

van der Walt, Joelle, *Motif BioSciences Inc., New York*

Vassar-Nieto, Deborah, *Sigma-Aldrich, St. Louis, Missouri*

Wade, Claire M., *Center for Human Genetic Research, Massachusetts General Hospital, Boston, and Broad Institute of MIT and Harvard, Cambridge, Massachusetts*

Wang, Yuker, *Affymetrix, South San Francisco, California*

Zody, Michael C., *Broad Institute of MIT and Harvard, Cambridge, Massachusetts*

1 Ethical Issues in Human Genetic Research: The Global Experience

Kevin Arnold and Joelle van der Walt

Motif BioSciences, New York, New York 10017

INTRODUCTION

The rapid pace of change in the potential of human genetic research has produced two powerful, but conflicting, social reactions. On the one hand, there is very strong public support for breakthroughs promising better medical diagnosis and treatments. On the other, there are heightened anxieties about the increased loss of privacy, the widening scope of forensic databases (Simoncelli 2006; Smith 2006), and the potential for genetic discrimination, as well as concerns about the capacity to regulate genetic science in the best public interest (ALRC 2003; Manaouil et al. 2005; Swartling et al. 2007).

In general, there is agreement that, in order to protect the rights of human sub-

jects involved in genetic research, three fundamental tenets should be followed for ethical genetic research—respect for persons, beneficence, and justice (WHO/CIOMS 2002). There remain, however, important ethical questions that must be addressed before conducting genetic research. These ethical questions were simply distilled in the ALRC report (ALRC 2003):

Who should be permitted to collect, use, or disclose genetic information about a person or persons?

- From whom and to whom?
- For what purposes?
- With whose consent?
- In what manner?
- On what conditions?

The remainder of this chapter summarizes the experience from international organizations and a range of countries actively engaged in human genetic research. These groups have defined the key areas in ethics of genetic research and through their work have elaborated on the issues associated with each of these ethical questions. This discussion is intended to be a distillate of current thinking on regulation of human genetic research and a source pointing to far more authoritative resources. This field is continually evolving. For those who have the time, the 1100-page Australian report (ALRC 2003) is highly recommended reading for instant immersion in the complexities of defining what is ethical in obtaining, using, and storing human genetic material and derived genetic information.

BRIEF HISTORY OF BIOMEDICAL ETHICS

Most of the perceived ethical issues of conducting genetic research are neither new nor unique within the larger context of biomedical ethics. The development of ethical frameworks for genetic research has been continuously under discussion as yet another facet in advancing research in human biology and medicine. In fact, these ethical issues have been debated actively over the last 60 years, since the "medical" experiments that occurred during World War II underscored the need for such a code of ethics.

Table 1-1 lists some historically important documents dealing with the ethics of biomedical research. It is not meant to be a comprehensive list, but rather to demonstrate that ethics in medical research is not a new field. Although most of these documents were not developed for genetic research, the concepts proposed are applicable to, and form the basis of, ethical thinking in genetic research.

TABLE 1-1. Important Documents in the Development of a Code of Ethics for Biomedical Research

Document	Date	Comment
Nuremberg Codes	1947	The 1st internationally recognized proposed code of ethics for medical research, written as a result of World War II. http://ohsr.od.nih.gov/guidelines/nuremberg.html
Declaration of Helsinki	1964 (revised in 1975, 1983, and 1989)	World Medical Association (WMA) developed own ethical guidelines, initially based on an expansion of Nuremberg thinking. http://www.wma.net/e/policy/b3.htm
Belmont Report	1979	Culmination of a revision and expansion of regulations for protection of human subjects (45 CFR part 46,) published by the U.S. government. http://www.hhs.gov/ohrp/humansubjects/guidance/belmont.htm
International Ethical Guidelines for Biomedical Research Involving Human Subjects	1982 (revised in 1993 and 2002)	World Health Organization (WHO) in collaboration with Council for International Organizations of Medical Sciences (CIOMS) publishes and routinely updates these guidelines. http://www.cioms.ch/frame_guidelines_nov_2002.htm
Statement on Principled Conduct of Genetic Research	1996	Human Genome Organization (HUGO) ethical, legal, and social implications (ELSI) committee published as part of Human Genome Project. http://www.csu.edu.au/learning/eubios/HUGO.html
Universal Declaration on the Human Genome and Human Rights	1997	United Nations Educational, Scientific and Cultural Organization (UNESCO) document attempting to embody the rights of an individual in genetic research. http://portal.unesco.org/shs/en/ev.php-URL_ID=1881&URL_DO=DO_TOPIC&URL_SECTION=201.html
Essentially Yours: The Protection of Human Genetic Information in Australia, ALRC 96	2003	Comprehensive document on bioethics serves as basis for Australian policy on obtaining and using genetic material and information in Australia. http://www.austlii.edu.au/au/other/alrc/publications/reports/96/
Pharmacogenetics ethical issues	2003	Nuffield Council on Bioethics document attempting to stimulate constructive thinking on the ethical dilemmas raised by increased use of pharmacogenetics. http://www.nuffieldbioethics.org/
Human Tissue Act	2004	U.K. framework for use of organs and tissue—introduces the offense of "DNA theft" in U.K. law. http://www.opsi.gov.uk/acts/acts2004/20040030.htm
Human Biological Material: Recommendations for Collection, Use, and Storage in Research	2005	Report by the Irish Council for Bioethics with recommendations to facilitate ethical conduct of human research. http://www.bioethics.ie/publications/index.html

KEY COMPONENTS OF ETHICAL RESEARCH

Informed Consent: General

It is commonly and universally agreed that voluntary informed consent should be obtained from each individual prior to recruitment into research projects involving genetic studies. For every project, the study organizers are responsible for protecting the privacy and confidentiality, and for respecting all rights, of the individual and the family that is approached to participate in the study.

Each of the many government agencies and international organizations concerned with these issues has its own form for informed consent. The U.S. guidelines of necessary elements for informed consent are specified by two agencies—the Department of Health and Human Services (DHHS), in its document DHHS 45 CFR 46 part 116/117, and the Federal Drug Administration (FDA), in its document FDA 21 CFR 50 part 25. There are minor differences between these U.S. guidelines and the international guidelines developed by the World Health Organization in collaboration with the Council for International Organizations of Medical Sciences (WHO/CIOMS); however, the core elements of all informed consent forms are the same. We describe and compare below some representative requirements from various organizations and institutions.

The Belmont Report, published by the U.S. government, clearly defines the minimum three requirements for informed consent: information, comprehension, and voluntariness. The Center for Disease Control maintains that informed consent should contain and convey the relevant information in a comprehensible manner to each potential subject. The issues derived from the CDC guidelines are described in Box 1-1.

Box 1-2 presents a comprehensive list of basic elements that should be included in informed consent, as practiced at the Massachusetts General Hospital (MGH) and the Partners Healthcare network. These requirements are derived from the MGH instructions on preparing informed consent forms.

Informed Consent: Child Consent

The participation of children is obviously essential for research into childhood diseases; however, the justification of genetic research involving children is far more limited. U.S. federal regulations were modified to further protect children as a special population, but an underlying issue remains that is immutable. Simply put, a child's participation in a research study is a choice made by the parent (or guardian); at some point in the future, the individual concerned will have the legal status to evaluate that decision. In the ideal case, the child may have provided assent to be enrolled in the

---

BOX 1-1. CDC GUIDELINES: ELEMENTS FOR INFORMED CONSENT

(see http://www.cdc.gov/genomics/population/publications/consent.htm)

- Study title, and description of why study is being done.
- What is involved in being part of this study, and what rights does a participant have, including process for withdrawal?
- How is information kept confidential, and what happens to the samples when the study is over?
- What are the risks/benefits of the study?
- Are there any costs involved for the participants?
- Are the results returned to participants?
- Whom do participants contact with questions, problems, and to withdraw from the study?
- Voluntary consent is required from each individual to participate in the study.

BOX 1-2. MGH GUIDELINES: ELEMENTS FOR INFORMED CONSENT

(see http://healthcare.partners.org/phsirb/consfrm.htm)

- A statement that the study involves research and an explanation of the purpose of the research, including expected duration of the individual's participation.

- A description of the procedures that will be used and notification of which, if any, procedures are experimental.

- A description of any reasonably foreseeable risks or discomforts to individuals.

- A description of any benefits to the subject or to others, which may be reasonably expected from the research.

- A disclosure of appropriate alternative procedures or courses of treatment, if any, that may be advantageous to individuals.

- A statement describing the extent, if any, to which confidentiality of records identifying individuals will be maintained.

- For research involving more than minimal risk, an explanation of whether compensation and explanation of medical treatments are available if injury occurs, what they consist of, and where further information can be obtained.

- An explanation of whom to contact for answers to relevant questions about the research and the individual's rights in the research study and whom to contact in event of a research-related injury.

- A statement that participation is voluntary and that refusal to participate will not alter any health benefits to which the individual is entitled. A follow-on statement that the individual may also withdraw from the study at any time without incurring any penalty or change in benefits to which individual is entitled.

And, Where Appropriate, One or More of the Following Elements Should Be Provided:

- A statement that the treatment or procedures may involve risks that are currently unforeseeable.

- Anticipated circumstances under which an individual's participation in the study may be terminated by the investigator without regard to the individual's consent.

- Additional costs, if any, to the individual that may result from participating in the study.

- A statement on the consequences of an individual's decision to withdraw from the research, including procedures for orderly withdrawal of participation.

- A statement that any findings developed as a result of the research and that may relate to an individual's willingness to continue participation will be provided to the subject.

- The approximate number of individuals that are involved in the study.

study, based on an age-appropriate informed consent. Yet, as for all informed consent, comprehension of the material and future implications of taking part in the study are highly subjective. Therefore, ensuring that assent was given without parental pressure is difficult (Fisher 2006a,b; Paulson 2006).

Without knowledge of how genetic information may be used in the future or how having genetic information may infringe on the child's rights to know and rights to privacy, it is unclear what may be deemed ethically sound in any future legal ruling. A cautious approach would lower the potential for misinterpretation of an individual's rights within the study. Such an approach would include, for example, the ability for an individual to withdraw once the age of consent is reached, and stating a clear position on genetic findings—either automatic disclosure or no disclosure.

It is now widely agreed that the sponsor of any new therapeutic or diagnostic, likely to be indicated for use in children, should evaluate its safety and efficacy in children before it is released for general distribution. Before undertaking research involving children, however, the investigator must ensure that:

- the research may not be carried out equally well with adults
- the purpose of the research is to obtain knowledge relevant to the health needs of children
- a parent or legal representative of each child has given permission
- the assent of each child has been obtained, depending on the extent of the child's capabilities
- a child's refusal to participate or continue in the research will be respected

A parent or guardian who gives permission for a child to participate in research should be given the opportunity, to a reasonable extent, to observe the research as it proceeds, so as to be able to withdraw the child if the parent or guardian decides it is in the child's best interests to do so (WHO/CIOMS 2002).

International and National Regulatory Standards and Gatekeepers

In recent years, national regulatory bodies and other independent entities in the U.S., the U.K., Australia, Europe, and the Middle East have considered how best to balance the ethical and scientific questions raised by genetic research. These considerations have enabled these agencies and organizations to develop frameworks for genetic research.

A number of documents have been developed that provide guidelines for conducting human genetic research, all reflecting the rapid changes in the use and potential of genetic data. The primary consensus of all the national and international guidelines is that responsibility for review, oversight, and approval of genetic studies should reside with local institutional review boards (IRBs), research ethics committees (RECs), or ethical review committees (ERCs). These boards and committees serve effectively as the gatekeepers of this research and how it is conducted.

Individual Country and Multi-country Examples of Ethical Review Committees in Practice

United Kingdom

In the U.K., the Human Tissue Act of 2004 makes it illegal to store or analyze any human tissue containing human cells without previous proper consent from the individual. When the European Union (EU) issued a directive for the operation of ethics committees in clinical trial research, the U.K. implemented a set of regulations, the U.K. Medicines for Human Use (Clinical Trials) Regulations 2004. These regulations make it illegal to start, recruit for, or conduct a clinical trial of an investigational medicinal product until there is a favorable opinion from a recognized REC and authorization from the licensing authority—the Medicines and Healthcare Products Regulatory Agency (MHRA).

U.K. biomedical research must be approved by authorized RECs, and there are over 100 RECs within the National Health Service (NHS). These committees operate according to standard operating procedures, written by the Central Office for Research Ethics Committees (COREC, http://www.corec.org.uk/index.htm), which is responsible for daily REC operations. COREC is under the control of U.K. Ethics Committee Authority (UKECA), a political body that must recognize an ethics committee before it is allowed to assess research trials.

The U.K. Biobank set up its own ethics and governance council in collaboration with the Wellcome Trust, Medical Research Council, and the Department of Health, producing their Ethics and Governance Framework (July 2006) (http://www.wellcome.ac.uk/doc_WTD003284.html). This framework states: "The core scientific protocol and operation procedures of the U.K. Biobank resource, as well as proposed uses of it, will have approval from appropriate ethics committees in accordance with guidance from relevant bodies and with relevant provisions. Participants will be told that such independent ethics approval will be obtained. Consent will be based on an explanation and understanding of, among other things, the assurance that only research uses that have been approved by both U.K. Biobank and a relevant ethics committee will be allowed, and that data and samples will be anonymized before being provided to research users."

The U.K. also maintains the Nuffield Council on Bioethics (http://www.nuffieldbioethics.org/), an independent body whose sole function is to consider ethical issues raised by new developments in biology and medicine. The council has published a paper considering the question of research performed in countries without formal ethical review procedures. Its conclusion stated that it is appropriate for a comparable body in another country to perform this function, although preferable that at least part of the protocol be reviewed by a local committee, for relevance to local health needs.

The Wellcome Trust, also U.K.-based, funds a Biomedical Ethics Program to build the ethics knowledge base and develop research capacity in both the U.K. and developing countries. As a funding agency, the Wellcome Trust produced a position statement and guidance notes for applicants. This document states their position on human participants in research: that the responsibility for ethical review of a submitted protocol is through an ethics review committee (http://www.wellcome.ac.uk/node5240.html). However, the Wellcome Trust does accept that if no independent and properly constituted ethics review committee exists in the country where research is to be carried out, an alternative source for review is acceptable. In these cases, the Wellcome Trust reserves the right to send any application to its Standing Advisory Group on Ethics (SAGE) for advice.

Australia

In Australia, a comprehensive inquiry initiated in 2001 resulted in the publication of the comprehensive 1100-page report in 2003 — "Essentially Yours: The Protection of Genetic Information in Australia"(ALRC 2003). This report accepted that a new legislative framework was not needed, as genetic information was covered within the existing Privacy Act of 1998. It did, however, lead to a review of the National Statement on Ethical Conduct in Research Involving Humans by the Australian National Health Medical Research Council (NHMRC) and the formation of a new statutory body to give advice to the government on current and emerging issues in human genetics. But, in line with international opinion, the oversight for any particular genetic research project in Australia is through a REC.

The Australian report also acknowledged that international guidelines already exist and should be considered, such as the UNESCO 2003 Universal Declaration on the Human Genome and Human Rights (www.unesco.org/ethics).

As in the other countries, the National Statement issued by the Australian National Health and Medical Research Council stipulated that research proposals involving human participants must be reviewed and approved by a Human Research Ethics Committee (HREC): "The primary function of an HREC is to protect the welfare and rights of participants in research. An often overlooked secondary purpose of the National Statement and thus of HREC's is to facilitate research that is or will be of benefit to the researcher's community or to humankind."

United States

In the U.S., the FDA effectively regulates genetic studies in a fashion similar to that in the U.K., requiring any study involving human subjects to be approved by an IRB (or equivalent body). The FDA has also published guidelines for IRBs; for example, functionality of IRBs is covered in 21 CFR Part 56 (http://www.fda.gov/oc/ohrt/irbs/appendixc.html). All biomedical research in the U.S. involving human subjects requires IRB approval with the IRBs under the purview of the FDA and the DHHS. IRBs basically follow the ethical principles laid out in the Belmont Report.

The National Institutes of Health (NIH), which administers and funds the large federal research budget in biomedical sciences, stipulates that: "No grant award will be made without IRB approval." Similarly, the FDA requires human genetic data collected and generated by pharmaceutical companies to have been approved by IRBs.

The FDA recognized that pharmaceutical companies would be reluctant to conduct FDA-regulated clinical trials with a pharmacogenetic component without guidance on how the FDA would use the genetic data in the approval process. Therefore, in March 2005, the FDA published Guidance for Industry: Pharmacogenomic Data Submissions, allowing voluntary genomic data submissions (VGDS) (http://www.fda.gov/cber/gdlns/pharmdtasub.htm) with extremely limited scenarios where the data could be used to influence an approval decision. The FDA does note that if any invasive test, including phlebotomy, is to be used for pharmacogenetic testing, this must be noted both in the protocol and in the informed consent. This approach is an effort by FDA to stimulate the use of genetic studies, to evaluate whether a large proportion of an individual's response to a drug can be readily genetically determined.

Sweden

Sweden, which follows all EU directives related to ethical research, has a history in development of ethical frameworks. Enacted law on Ethical Review of Research involving humans came into effect in 2004, updating which procedures required review, specifying a new review board structure with central and regional review boards, and defining the composition of regional review boards.

Within Sweden, the Swedish Research Council has broad-based responsibility for ethics guidelines and, in collaboration with the Centre for Bioethics at the Karolinska Institute and Uppsala University, runs CODEX (http://www.codex.vr.se/codex_eng/codex/index.html), an Internet gateway with comprehensive information on ethics in research. The government, in reviewing the 2003 Ethics Review Act, has recommended that the Swedish Research Council be responsible for appointing members of all review boards, except the Central Review Board, which is assigned by the government.

International HapMap Consortium

Within the last 5 years, one of the most complex research projects has succeeded in resolving the ethics and science of genetic research satisfactorily on an international level. The International HapMap Consortium is a multi-center, multi-national consortium involving scientists and funding agencies in Canada, China, Japan, Nigeria, U.K., and U.S. Established to develop a public resource that will help researchers find genes associated with human health and disease, the HapMap Consortium was successfully governed by a network of local IRBs in the participating countries (International HapMap Consortium 2004).

CURRENT THINKING ON ETHICALLY ACCEPTABLE STUDIES IN GENETICS

The ethics of genetic research is a continually evolving field, and the current thinking, as actively practiced by IRBs and RECs, appears to be moving away from acceptance of open-ended, undefined projects toward projects that are strongly defined in all aspects. Currently, virtually all projects with a genetic component are defined by disease of interest, genes to be examined and/or genetic information to be collected with whole-genome analysis, and with a set limit on the time the samples are stored.

This evolution has not been driven by individual review boards, but rather appears to be the result of considerations of ethical, legal, and social implications of genetic research by national and international organizations. These ideas are primarily based on the potential of both publicly and commercially funded large-scale biobanks, and the uncertainty with regard to future uses and potential of both the samples and the genetic information already generated. In the U.S., the National Human Genome Research Institute's (NHGRI) Ethical, Legal, and Social Implications (ELSI) Research Program (http://www.genome.gov/PolicyEthics/) was established in 1990, as an integral part of the Human Genome Project to foster research into the ethical, legal, and social implications of genetic and genomic research.

LIMITS ON INFORMED CONSENT

Many guidelines have been written by UNESCO, WHO, COREC, Australian NHMRC, EU, HUGO-ELSI, and others. The current literature, however, also describes standards on bioethics of consent and disclosure of results (Clayton 2005; Deschenes and Sallee 2005; Hoeyer et al. 2005; Knoppers 2005; Rothstein 2005; Corrigan and William-Jones 2006; Hansson et al. 2006; Shickle 2006). As these guidelines and standards are updated, it is likely that the revised documents will include recommendations on limits of informed consent and/or tiered informed consent, the need for re-consent, a new requirement for disclosure of results to enrolled individuals, and a strong push for capacity development in support of ethical review in developing countries (Hyder et al. 2004; Bhat and Hegde 2006; Gbadegsin and Wendler 2006).

Blanket consent forms allow researchers to keep samples indefinitely with unlimited scope for current genetic research, with no requirement for renewed consent for research in the future, and with limited or no ability to withdraw from the study. Even with the use of anonymous samples, such constraints on consent would be deemed acceptable only in exceptional circumstances; for example, for projects such as HapMap and the U.K. Biobank. IRBs now expect that consent forms and projects are clearly defined and limited in scope—not to limit the researcher, but to provide much better definition to individuals recruited into the study. Obviously, the change from a more comprehensive open-ended project design with similar open-ended consent coverage to a defined disease-focused project with defined genetic boundaries and time-limited storage allows much clearer oversight of projects by IRBs.

Clearly, for a return to the open-ended, undefined project designs, there would need to be a major shift away from individual rights to a more community-centric approach. This necessity seems to have been somewhat mitigated by a move to anonymization; that is, making samples anonymous at a certain point in the study. However, this move destroys (or severely limits) an individual's ability to withdraw from a study, and at the time of this writing, review boards have been far more accepting of expanded genetic testing on an anonymized sample set. In the future, however, IRBs may demand that an individual retain the ability to withdraw from the anonymized sample set. The concern is that these withdrawals may limit the scope of genetic testing on the entire anonymized sample set. Given current advances in molecular genetic technology, however, withdrawal of an individual sample is certainly possible. An individual wishing to withdraw from an anonymized data set would need only send a second sample, with a statement to withdraw all data that match this sample.

SUMMARY

In conclusion, the ethical issues surrounding genetic research are neither novel nor unique. The field of ethical considerations in human biological research continues to expand and evolve; however, there remains broad consensus that certain key concepts are required for ethical research. Any research project using human subjects should be reviewed and approved by a research ethics committee (REC, or an equivalent entity, IRB, ERC, etc.) before the study is initiated. Although the membership and the number of RECs vary widely among countries, national and international guidelines consistently defer review and oversight to these agencies.

Finally, from the participant perspective, the various guidelines are forthright that individual informed consent should be obtained from all participants prior to the study, and that the informed consent should be voluntary, informative, and comprehended by each person asked to participate.

REFERENCES

Australian Law Reform Commission (ALRC). 2003. *Essentially yours: The protection of human genetic information in Australia,* report 96. Australian Law Reform Commission, Sydney. http://www.austlii.edu.au/au/other/alrc/publications/reports/96/

Belmont Report. 1979. *Ethical principles and guidelines for the protection of human subjects of research.* The National Commission for the Protection of Human Subjects of Biomedical and Behavioral Research. U.S. Department of Health, Education, and Welfare, Washington, D.C. http://www.hhs.gov/ohrp/humansubjects/guidance/belmont.htm

Bhat S.B. and Hegde T.T. 2006. Ethical international research on human subjects research in the absence of local institutional review boards. *J. Med. Ethics* **32**: 535–536.

Clayton E.W. 2005. Informed consent and biobanks. *J. Law Med. Ethics* **33**: 15–21.

Corrigan O.P. and William-Jones B. 2006. Pharmacogenetics: The bioethical problem of DNA investment banking. *Stud. Hist. Philos. Biol. Biomed. Sci.* **37**: 550–565.

Deschenes M. and Sallee C. 2005. Accountability in population biobanking: Comparative approaches. *J. Law Med. Ethics* **33**: 40–53.

Fisher C.B. 2006a. Privacy and ethics in pediatric environmental health research—Part II: Protecting families and communities. *Environ. Health Perspect.* **114**: 1622–1625.

———. 2006b. Privacy and ethics in pediatric environmental health research—Part I: Genetic and prenatal testing. *Environ. Health Perspect.* **114**: 1617–1621.

Gbadegesin S. and Wendler D. 2006. Protecting communities in health research from exploitation. *Bioethics* **20**: 248–253.

Hansson M.G., Dillner J., Bartram C.R., Carlson J.A., and Helgesson G. 2006. Should donors be allowed to give broad consent to future biobank research? *Lancet Oncol.* **7**: 266–269.

Hoeyer K., Olofsson B.O., Mjorndal T., and Lynoe N. 2005. The ethics of research using biobanks: Reason to question the importance attributed to informed consent. *Arch. Intern. Med.* **165**: 97–100.

Hyder A.A., Wali S.A., Khan A.N., Teoh N.B., Kass N.E., and Dawson L. 2004. Ethical review of health research: A perspective from developing country researchers. *J. Med. Ethics* **30**: 68–72.

International HapMap Consortium 2004. Integrating ethics and science in the International HapMap project. *Nat. Rev. Genet.* **5**: 467–475.

Knoppers B.M. 2005. Biobanking: International norms. *J. Law Med. Ethics* **33**: 7–14.

Manaouil C., Graser M., Chatelain D., and Jarde O. 2005. The examination of genetic characteristics since the adoption of the French law on bioethics. *Med. Law* **24**: 783–789.

Paulson J.A. 2006. An exploration of ethical issues in research in children's health and the environment. *Environ. Health Perspect.* **114**: 1603–1608.

Rothstein M.A. 2005. Expanding the ethical analysis of biobanks. *J. Law Med. Ethics* **33**: 89–101.

Shickle D. 2006. The consent problem within DNA biobanks. *Stud. Hist. Philos. Biol. Biomed. Sci.* **37**: 503–519.

Simoncelli T. 2006. Dangerous excursions: The case against expanding the forensic DNA databases to innocent persons. *J. Law Med. Ethics* **34**: 390–397.

Smith M.E. 2006. Let's make the DNA identification database as inclusive as possible. *J. Law Med. Ethics* **34**: 385–389.

Swartling U., Eriksson S., Ludvigsson J., and Helgesson G. 2007. Concern, pressure and lack of knowledge affect choice of not wanting to know high-risk status. *Eur. J. Hum. Genet.* **15**: 556–562.

WHO/CIOMS 2002. *International ethical guidelines for biomedical research involving human subjects.* Council for International organizations of Medical Sciences (CIOMS) in collaboration with the World Health Organization (WHO), Geneva. http://www.cioms.ch/frame_guidelines_nov_2002.htm

WWW RESOURCES

http://healthcare.partners.org/phsirb/consfrm.htm Consent forms. Partners HealthCare System, Partners Human Research Committee.

http://www.cdc.gov/genomics/population/publications/consent.htm Informed consent template for population-based research involving genetics. National Office of Public Health Genomics, Department of Health and Human Services, Centers for Disease Control, Atlanta, Georgia.

http://www.codex.vr.se/codex_eng/codex/index.html CODEX rules and guidelines for research homepage. CODEX is The Swedish Research Council's gateway to various research ethics guidelines. In collaboration with The Centre for Bioethics at Karolinska Institute and Uppsala University.

http://www.corec.org.uk/index.htm Central Office for Research Ethics Committees (COREC) homepage. National Patient Safety Agency, London, United Kingdom.

http://www.fda.gov/oc/ohrt/irbs/appendixc.html Information Sheets. Guidance for Institutional Review Boards and Clinical Investigators 1988 Update. U.S. Food and Drug Administration, Rockville, Maryland.

http://www.fda.gov/cber/gdlns/pharmdtasub.htm Guidance for Industry page. Pharmacogenomic Data Submissions. U.S. Food and Drug Administration, Rockville, Maryland.

http://www.genome.gov/PolicyEthics/ Policy & Ethics homepage. National Human Genome Research Institute, National Institutes of Health, Bethesda, Maryland.

http://www.wellcome.ac.uk/doc_WTD003284.html U.K. Biobank Ethics and Governance Framework homepage. Wellcome Trust, London, United Kingdom.

http://www.wellcome.ac.uk/node5240.html Biomedical ethics. Reports on meetings and other activities organized by the Wellcome Trust Biomedical Ethics Programme, London, United Kingdom.

www.unesco.org/ethics

2 Population Choice as a Consideration for Genetic Analysis Study Design

J. Claiborne Stephens[1] and Michael Bamshad[2]

[1]Motif BioSciences Inc., New York, New York 10017; [2]Departments of Pediatrics and Genome Sciences, Division of Genetics and Developmental Medicine, University of Washington School of Medicine, Seattle, Washington 98195

INTRODUCTION

Genetic association and gene mapping studies have five main components: study design, phenotyping, sample collection and processing, genotyping, and statistical analysis. Several sections of this book are devoted to the last three of these components, and phenotyping is largely beyond our scope. With respect to study design, the main considerations include whether to focus on families or unrelated individuals, the sample size required for adequate statistical power, and whether particular popula-

tions offer scientific advantages. The focus of this chapter is the latter, since it has now become clear that population characteristics play a major role in the tractability of genetic studies (Jorde 1995; de la Chapelle and Wright 1998; Wright et al. 1999).

Several types of populations offer special advantages for genetic studies. There is a long, rich history in human genetics of studies that focus on isolated or "founder" populations. This is due, in part, to the notion that identifying disease genes might be easier in founder populations because the prevalence of many disorders is either substantially higher or lower than in the general population. These differences in prevalence exist because the founder population typically went through a severe constriction in size, or "bottleneck," and carried with it only a subset of the genetic variation in the original population. Thus, disease genes in this small founder population either quickly became much more common or were eliminated.

Populations that have certain nonrandom mating patterns may also be more amenable to genetic studies than are more general populations. Chief among these are "consanguineous" populations, in which the preferred pattern of marriage is among relatives (such as between first cousins). Because of the consanguinity, otherwise rare autosomal recessive disorders are exposed. Homozygosity or "autozygosity" mapping is a special strategy for disease gene mapping that makes use of the increased frequency of long autozygous tracts of DNA that occur within the genomes

11

of children from consanguineous marriages. Autozygosity is essentially the homozygosity of the DNA produced when both copies of a given genomic region derive from the same recent common ancestor. Another term in common usage for this is "identity-by-descent" mapping.

Mapping by admixture disequilibrium or simply "admixture mapping" has been used recently to map genes of considerable medical importance. Admixed populations are those produced by fusion or re-contact between populations that have been genetically isolated historically. The rationale for using admixed populations for mapping studies is that any genetic trait with a large frequency difference between the original parental populations can be mapped relatively easily in the admixed population.

In this chapter, we discuss founder populations, homozygosity mapping, and admixed populations, and consider the advantages of each in mapping human disease genes.

FOUNDER POPULATIONS

The notion of a founder population and founder effects is well ingrained in population genetics. As applied to humans, a founder population is one that is initiated by a very small number of members from some larger population. Over time, the new population remains isolated from other related populations and may therefore, by chance, have allele frequencies that diverge dramatically from other populations. Many island populations are founder populations due to historical geographical isolation. Other sources of genetic isolation arise from barriers to intermarriage, which may stem from religious, linguistic, social, or cultural reasons.

Classically, human founder populations have been of tremendous value in elucidating the genetic basis of many Mendelian disorders. For instance, many rare autosomal recessive disorders are exposed when carriers of the pathogenic allele are, by chance, more common in the founder population than in the source population (see also Homozygosity Mapping, below). This circumstance leads to more families with an otherwise rare disorder. Furthermore, such families will typically have an identical genetic basis for the disorder, which makes the genetic basis far easier to elucidate.

More recently, there has been widespread recognition that the restricted genetic variability of founder populations also makes the genetic study of common, complex diseases more tractable (Service et al. 2006; Freimer and Sabatti 2007). In addition to the more homogeneous genetic background, individuals in a founder population quite often share a more homogeneous environment. If a common, complex disease has dozens of genetic and environmental factors contributing to individual susceptibility in a large, diverse, continental population, it is easy to imagine that a founder population will lack many of these factors, both genetic and environmental. Hence, if a common, complex disease has a comparable frequency between a continental population and a founder population, the hope is that the underlying variation leading to disease in the founder population will have a much more restricted genetic basis. If so, the likelihood of one or a few genes accounting for a significant fraction of disease susceptibility in the founder population increases dramatically. Identifying one or a few genes with large effects is a much easier task than identifying many genes with small effects.

More subtly, the genetic architecture of founder populations is also conducive to disease gene identification and mapping. The attribute of genetic architecture that is of fundamental importance is the level and pattern of linkage disequilibrium (LD). LD is the tendency, observed among individuals within a population, for genetic markers to co-occur. It is this lack of independence that allows us to take one set of markers, for instance a set of single-nucleotide polymorphisms (SNPs) in a microarray, and make inferences about the entire genome as to which regions are, or are not, associated with a given phenotype. It has now been shown, in a variety of founder populations, that levels of LD are typically higher, and that fewer gaps exist in the overall pattern of LD (Service et al. 2006), than in the general population.

FOUNDER POPULATIONS: PRACTICAL APPLICATIONS

There have been a very large number of excellent studies in isolated populations that have elucidated the genetic basis of important Mendelian disorders. Several populations stand out in terms of the community organization, participation, and the amount of genetic work carried out. The Ashkenazi Jewish population (Risch et al. 1995; for recent reviews, see Charrow 2004; Weinstein 2007) has many well-organized Web sites and genetic screening services for at least nine diseases (Tay-Sachs disease, Canavan, Gaucher, Niemann-Pick, familial dysautonomia, Bloom syndrome, Type 1A glycogen storage disorder, Fanconi anemia, Mucolipidosis IV) that have unusually high prevalence in this population. Furthermore, common diseases like breast cancer and cystic fibrosis have high-frequency variants that are found exclusively or predominantly in the Ashkenazi population (Quint et al. 2005; Chen et al. 2006; Weinstein 2007).

Another population that has had a disproportionate amount of genetic research given its size is the Finnish population (de la Chapelle and Wright 1998; Peltonen et al. 1999; Peltonen 2000; Kere 2001). Specific autosomal dominant and recessive diseases occur at a higher frequency in Finland, presumably due to a founder effect (see Table 1 in Kere 2001). Because of the rich registries and active research into the genetic basis of diseases, Finnish genetic research has provided insights and guidance into the process of disease gene identification as well as insight into the basis of numerous genetic diseases.

The study of communities within the U.S. that have isolated themselves because of their religious beliefs has had a major impact on gene mapping research. Victor McKusick's work in the Old Order Amish was some of the earliest, and it clearly demonstrated the potential for establishing that certain diseases had a genetic basis (McKusick et al. 1964; McKusick 1973). Furthermore, the large, extended kindreds allowed evaluation of the mode of inheritance (e.g., dominant, recessive) as well. Much of this work was done well in advance of the availability of genetic markers for mapping the disease genes to specific chromosomes. Once markers were available, the stage was set for mapping many of the diseases peculiar to the Old Order Amish to specific chromosomes (Velinov et al. 1993; Polymeropoulos et al. 1996).

Studies of asthma in the Hutterites (Chan et al. 2006; Kurz et al. 2006) have recently indicated the existence of risk variants in multiple regions of the genome. The Mennonite community of southeastern Pennsylvania has been the subject of genetic studies of diseases that are rare elsewhere, such as Hirschsprung disease and maple syrup urine disease (Puffenberger 2003). The frequency of the latter is over 1000 times higher in Mennonites than in outbred European populations.

One especially important case study of genetic work in an isolated community is that of Lake Maracaibo, Venezuela, and Huntington disease (HD). Although HD is one of the early success stories of gene mapping, having been mapped to the short arm of chromosome 4 in the mid-1980s (Gusella et al.1983), it took nearly another decade to identify the actual gene underlying HD (MacDonald et al. 1992). The Venezuelan cohort, having nearly 100 related individuals with HD, was vital in the mapping and ultimate identification of the gene.

A very intriguing example of a founder population having a more tractable genetic architecture comes from a recent study by Bonnen et al. (2006). They evaluated the potential for a whole-genome association study on the Micronesian island of Kosrae. With genome-wide SNP data (113,240 SNPs typed upon 30 trios), they found increased LD and decreased allelic diversity compared to those same SNPs genotyped on the HapMap (International HapMap Consortium 2005) panels. Furthermore, they demonstrated that the extent of the LD was longer than that found in the HapMap panels. The authors claimed that these attributes of the Kosraen population should result in improved efficiency for genetic studies relative to other populations.

HOMOZYGOSITY MAPPING

In a landmark paper in 1987, Eric Lander and David Botstein proposed that excess homozygosity in the DNA of affected children could be used to indicate the genomic location of genes for autosomal recessive diseases (Lander and Botstein 1987). Specifically, a small number of affected children from consanguineous marriages, together with a relatively small genomic marker set (very small by current standards), would be sufficient to map a disease or trait that is inherited in an autosomal recessive pattern. The principle relies on the idea that affected, closely related children from a consanguineous population will all be homozygous for the same ancestral trait-causing allele, and this trait-causing allele will be embedded in a very large tract of homozygosity in each child. Each child would be expected to have multiple homozygous tracts in his genome, but only the tract shared by all affected children would be the presumptive location of the trait-causing allele. Furthermore, the actual location would be refined to the overlap of homozygous tracts from affected children and would exclude any tract that was also homozygous (for the same set of alleles) in their unaffected siblings.

Homozygosity mapping has now been used to map dozens of autosomal recessive diseases in numerous populations (Sheffield et al. 1998). Many of these populations commonly prefer first-cousin marriages or other consanguineous marriages and hence have high frequencies of otherwise rare disorders (Teebi and El-Shanti 2006). Other populations may avoid consanguineous marriages, but may have increased consanguinity due to founder effects or geographical or other isolation (Strauss et al. 2005).

ADMIXED POPULATIONS

More than 50 years ago, well before the availability of the modern plethora of hundreds of thousands of well-documented genetic markers, an insightful paper by Rife (1954) suggested the notion that "admixed" populations could be valuable for mapping disease genes. Individuals in admixed populations are those whose ancestry comes from two or more source or parental populations that had previously been isolated from one another. Thirty-four years later, Chakraborty and Weiss (1988) extended and revitalized this idea in a theoretical paper describing a strategy for mapping disease genes. Under this strategy, any disease or trait that has a prevalence difference between the source populations should show LD to linked markers that also have frequency differences between the source populations. A considerable advantage of this strategy is that the markers linked to the trait need not be extremely close to the trait locus itself. Hence, the number of markers necessary for "mapping by admixture LD," or more simply "admixture mapping," is considerably smaller than the number required in more conventional studies.

There are several assumptions underlying this strategy. First and foremost, the trait in question should have a genetic basis, and the prevalence difference among source populations should be due to differences in the frequencies of these genetic factors. Second, the time elapsed since the initial formation of the admixed population should be sufficient to erode any LD between the trait and unlinked markers. This does not preclude the possibility of ongoing admixture with either source population, but ongoing admixture will increase the noise created by apparent linkage to unlinked markers. Note that the time since population formation, the relative contribution of parental populations, and the dynamics of the admixture process are just a few of the important considerations that determine the feasibility of mapping disease genes in a given population by using this strategy.

In the mid- to late 1990s, a spate of papers emerged that addressed additional practical issues for the application of this mapping approach (Stephens et al. 1994; McKeigue 1997, 1998; Shriver et al. 1997). Shortly thereafter, genome-wide resources for admixture mapping became available (Dean et al. 1994; Smith et al. 2001, 2004; Collins-Schramm et al. 2002), culminating in a mapped set of 4222 well-documented, highly informative, validated SNPs currently available for application in the African-American population (Tian et al. 2006). Undoubtedly, such marker sets will increase as more markers are identified and evaluated, and as additional admixed populations are studied.

As these marker sets started to become available, several empirical studies affirmed the theoretical expectations. One such study (Lautenberger et al. 2000) demonstrated extended LD in African-Americans between the Duffy blood group locus (FY) and markers up to 30 centiMorgans (cM) away from it. Another study (Collins-Schramm et al. 2003) showed multiple megabase-scale regions of LD, again in the African-American population.

In addition to the physical resources of well-characterized sets of markers, the theoretical and analytical resources have kept pace, with the development of several algorithms specifically addressing admixture mapping: MALDSOFT (Montana and Pritchard 2004), ANCESTRYMAP (Patterson et al. 2004), and ADMIXMAP (Hoggart et al. 2004). More recently, Tang et al. (2006) have introduced a Markov–hidden Markov model (MHMM) algorithm for reconstructing the blocks of genetic ancestry in admixed individuals (SABER, http://www.fhcrc.org/science/labs/tang/).

The power of using admixture mapping for disease genes has recently been demonstrated in several studies. The first successful application was in a microsatellite scan of an African-American cohort, in which two genomic regions conferring susceptibility to hypertension were identified (Darvasi and Shifman 2005; Zhu et al. 2005). Both the number of markers (269) and the number of subjects (cases plus controls, 1310) were relatively modest. Furthermore, one of the two regions identified had been previously implicated as being linked to hypertension, which gives support to the validity of the admixture mapping approach.

A second success of admixture mapping made use of population prevalence differences of multiple sclerosis (MS). MS is thought to be highly heritable, yet no convincing candidate genes have yet been identified. Because MS is more frequent in people of northern European origin than in those of African origin, an application of admixture mapping to African-American MS cases could potentially identify MS susceptibility genes of European origin. Indeed, in a case-control study of 1648 African-Americans, a highly significant susceptibility locus was found on chromosome 1 (Reich et al. 2005).

The third success of note is application of admixture mapping of prostate cancer susceptibility in African-American men (Freedman et al. 2006). Epidemiologically, younger African-American men are at a significantly higher risk for prostate cancer than are men of the same age from other populations. This made prostate cancer an attractive target for application of the admixture mapping strategy. Linkage results had also implicated the same genomic region, 8q24, but the linkage results only partially explained the strength of the admixture signal. The authors correctly predicted that multiple loci conferring susceptibility resided in 8q24, which has recently been confirmed by several laboratories, including their own (Gudmundsson et al. 2007; Haiman et al. 2007; Yeager et al. 2007).

So far, the main emphasis in admixture mapping has been on African-American cohorts. However, it seems clear that the successes with these cohorts in multiple disease states will motivate creation of marker sets and research into other populations.

CONCLUSIONS

If the focus of genetic research is on a particular trait, it makes a great deal of sense to evaluate whether that trait might be more advantageously studied in a particular population. "Looking under the lamppost," as it were, simply because a trait is known to exist in a certain population, might not be the most efficient way to decipher the genetic basis of a complex trait. This may mean that a specific foreign population may be more amenable to genetic characterization of a disease of interest than a researcher's own population or country. Genetic research is becoming a much more global enterprise as we recognize that, with rare exception, we all have the same set of genes, so that genetic discoveries made anywhere on the planet can be extrapolated to all human populations. Since we all have so much to gain from the globalization of human genetic research, it is likely that we will soon see studies like the above replicated and initiated in a much wider variety of human populations.

REFERENCES

Bonnen P.E., Pe'er I., Plenge R.M., Salit J., Lowe J.K., Shapero M.H., Lifton R.P., Breslow J.L., Daly M.J., Reich D.E., et al. 2006. Evaluating potential for whole-genome studies in Kosrae, an isolated population in Micronesia. *Nat. Genet.* **38:** 214–217.

Chakraborty R. and Weiss K.M. 1988. Admixture as a tool for finding linked genes and detecting that difference from allelic association between loci. *Proc. Natl. Acad. Sci.* **85:** 9119–9123.

Chan A., Newman D.L., Shon A.M., Schneider D.H., Kuldanek S., and Ober C. 2006. Variation in the type I interferon gene cluster on 9p21 influences susceptibility to asthma and atopy. *Genes Immun.* **7:** 169–178.

Charrow J. 2004. Ashkenazi Jewish genetic disorders. *Fam. Cancer* **3:** 201–206.

Chen S., Iversen E.S., Friebel T., Finkelstein D., Weber B.L., Eisen A., Peterson L.E., Schildkraut J.M., Isaacs C., Peshkin B.N., et al. 2006. Characterization of *BRCA1* and *BRCA2* mutations in a large United States sample. *J. Clin. Oncol.* **24:** 863–871.

Collins-Schramm H.E., Chima B., Operario D.J., Criswell L.A., and Seldin M.F. 2003. Markers informative for ancestry demonstrate consistent megabase-length linkage disequilibrium in the African American population. *Hum. Genet.* **113:** 211–219.

Collins-Schramm H.E., Phillips C.M., Operario D.J., Lee J.S., Weber J.L., Hanson R.L., Knowler W.C., Cooper R., Li H., and Seldin M.F. 2002. Ethnic-difference markers for use in mapping by admixture linkage disequilibrium. *Am. J. Hum. Genet.* **70:** 737–750.

Darvasi A. and Shifman S. 2005. The beauty of admixture. *Nat. Genet.* **37:** 118–119.

Dean M., Stephens J.C., Winkler C., Lomb D.A., Ramsburg M., Boaze R., Stewart C., Charbonneau L., Goldman D., Albaugh B.J., et al. 1994. Polymorphic admixture typing in human ethnic populations. *Am. J. Hum. Genet.* **55:** 788–808.

de la Chapelle A. and Wright F.A. 1998. Linkage disequilibrium mapping in isolated populations: The example of Finland revisited. *Proc. Natl. Acad. Sci.* **95:** 12416–12423.

Freedman M.L., Haiman C.A., Patterson N., McDonald G.J., Tandon A., Waliszewska A., Penney K., Steen R.G., Ardlie K., John E.M., et al. 2006. Admixture mapping identifies 8q24 as a prostate cancer risk locus in African-American men. *Proc. Natl. Acad. Sci.* **103:** 14068–14073.

Freimer N.B. and Sabatti C. 2007. Human genetics: Variants in common diseases. *Nature* **445:** 828–830.

Gudmundsson J., Sulem P., Manolescu A., Amundadottir L.T., Gudbjartsson D., Helgason A., Rafnar T., Bergthorsson J.T., Agnarsson B.A., Baker A., et al. 2007. Genome-wide association study identifies a second prostate cancer susceptibility variant at 8q24. *Nat. Genet.* **39:** 631–637.

Gusella J.F., Wexler N.S., Conneally P.M., Naylor S.L., Anderson M.A., Tanzi R.E., Watkins P.C., Ottina K., Wallace M.R., Sakaguchi A.Y., et al. 1983. A polymorphic DNA marker genetically linked to Huntington's disease. *Nature* **306:** 234–238.

Haiman C.A., Patterson N., Freedman M.L., Myers S.R., Pike M.C., Waliszewska A., Neubauer J., Tandon A., Schirmer C., McDonald G.J., et al. 2007. Multiple regions within 8q24 independently affect risk for prostate cancer. *Nat. Genet.* **39:** 638–644.

Hoggart C.J., Shriver M.D., Kittles R.A., Clayton D.G., and McKeigue P.M. 2004. Design and analysis of admixture mapping studies. *Am. J. Hum. Genet.* **74:** 965–978.

International HapMap Consortium. 2005. A haplotype map of the human genome. *Nature* **437:** 1299–1320.

Jorde L.B. 1995. Linkage disequilibrium as a gene-mapping tool. *Am. J. Hum. Genet.* **56:** 11–14.

Kere J. 2001. Human population genetics: Lessons from Finland. *Annu. Rev. Genomics Hum. Genet.* **2:** 103–128.

Kurz T., Hoffjan S., Hayes M.G., Schneider D., Nicolae R., Heinzmann A., Jerkic S.P., Parry R., Cox N.J., Deichmann K.A., and Ober C. 2006. Fine mapping and positional candidate studies on chromosome 5p13 identify multiple asthma susceptibility loci. *J. Allergy Clin. Immunol.* **118:** 396–402.

Lander E.S. and Botstein D. 1987. Homozygosity mapping: A way to map human recessive traits with the DNA of inbred children. *Science* **236:** 1567–1570.

Lautenberger J.A., Stephens J.C., O'Brien S.J., and Smith M.W. 2000. Significant admixture linkage disequilibrium across 30 cM around the FY locus in African Americans. *Am. J. Hum. Genet.* **66:** 969–978.

MacDonald M.E., Novelletto A., Lin C., Tagle D., Barnes G., Bates G., Taylor S., Allitto B., Altherr M., Myers R., et al. 1992. The Huntington's disease candidate region exhibits many different haplotypes. *Nat. Genet.* **1:** 99–103.

McKeigue P.M. 1997. Mapping genes underlying ethnic differences in disease risk by linkage disequilibrium in recently admixed populations. *Am. J. Hum. Genet.* **60:** 188–196.

———. 1998. Mapping genes that underlie ethnic differences in disease risk: Methods for detecting linkage in admixed populations, by conditioning on parental admixture. *Am. J. Hum. Genet.* **63:** 241–251.

McKusick V.A. 1973. Genetic studies in American inbred populations with particular reference to the Old Order Amish. *Isr. J. Med. Sci.* **9:** 1276–1284.

McKusick V.A., Hostetler J.A., Egeland J.A., and Eldridge R. 1964. The distribution of certain genes in the Old Order Amish. *Cold Spring Harbor Symp. Quant. Biol.* **29:** 99–114.

Montana G. and Pritchard J.K. 2004. Statistical tests for admixture mapping with case-control and cases-only data. *Am. J. Hum. Genet.* **75:** 771–789.

Patterson N., Hattangadi N., Lane B., Lohmueller K.E., Hafler D.A., Oksenberg J.R., Hauser S.L., Smith M.W., O'Brien S.J., Altshuler D., et al. 2004. Methods for high-density admixture mapping of disease genes. *Am. J. Hum. Genet.* **74:** 979–1000.

Peltonen L. 2000. Positional cloning of disease genes: Advantages of genetic isolates. *Hum. Hered.* **50:** 66–75.

Peltonen L., Jalanko A., and Varilo T. 1999. Molecular genetics of the Finnish disease heritage. *Hum. Mol. Genet.* **8:** 1913–1923.

Polymeropoulos M.H., Ide S.E., Wright M., Goodship J., Weissenbach J., Pyeritz R.E., Da Silva E.O., Ortiz De Luna R.I., and Francomano C.A. 1996. The gene for the Ellis-van Creveld syndrome is located on chromosome 4p16. *Genomics* **35:** 1–5.

Puffenberger E.G. 2003. Genetic heritage of the Old Order Mennonites of southeastern Pennsylvania. *Am. J. Med. Genet. C Semin. Med. Genet.* **121:** 18–31.

Quint A., Lerer I., Sagi M., and Abeliovich D. 2005. Mutation spectrum in Jewish cystic fibrosis patients in Israel: Implication to carrier screening. *Am. J. Med. Genet. A* **136:** 246–248.

Reich D., Patterson N., De Jager P.L., McDonald G.J., Waliszewska A., Tandon A., Lincoln R.R., DeLoa C., Fruhan S.A., Cabre P., et al. 2005. A whole-genome admixture scan finds a candidate locus for multiple sclerosis susceptibility. *Nat. Genet.* **37:** 1113–1118.

Rife D.C. 1954. Populations of hybrid origin as source material for the detection of linkage. *Am. J. Hum. Genet.* **6:** 26–33.

Risch N., de Leon D., Ozelius L., Kramer P., Almasy L., Singer B., Fahn S., Breakefield X., and Bressman S. 1995. Genetic analysis of idiopathic torsion dystonia in Ashkenazi Jews and their recent descent from a small founder population. *Nat. Genet.* **9:** 152–159.

Service S., DeYoung J., Karayiorgou M., Roos J.L., Pretorious H., Bedoya G., Ospina J., Ruiz-Linares A., Macedo A., Palha J.A., et al. 2006. Magnitude and distribution of linkage disequilibrium in population isolates and implications for genome-wide association studies. *Nat. Genet.* **38:** 556–560.

Sheffield V.C., Stone E.M., and Carmi R. 1998. Use of isolated inbred human populations for identification of disease genes. *Trends Genet.* **14:** 391–396.

Shriver M.D., Smith M.W., Jin L., Marcini A., Akey J.M., Deka R., and Ferrell R.E. 1997. Ethnic-affiliation estimation by use of population-specific DNA markers. *Am. J. Hum. Genet.* **60:** 957–964.

Smith M.W., Lautenberger J.A., Shin H.D., Chretien J.P., Shrestha S., Gilbert D.A., and O'Brien S.J. 2001. Markers for mapping by admixture linkage disequilibrium in African American and Hispanic populations. *Am. J. Hum. Genet.* **69:** 1080–1094.

Smith M.W., Patterson N., Lautenberger J.A., Truelove A.L., McDonald G.J., Waliszewska A., Kessing B.D., Malasky M.J., Scafe C., Le E., et al. 2004. A high-density admixture map for disease gene discovery in African Americans. *Am. J. Hum. Genet.* **74:** 1001–1013.

Stephens J.C., Briscoe D., and O'Brien SJ. 1994. Mapping by admixture linkage disequilibrium in human populations: Limits and guidelines. *Am. J. Hum. Genet.* **55:** 809–824.

Strauss K.A., Puffenberger E.G., Craig D.W., Panganiban C.B., Lee A.M., Hu-Lince D., Stephan D.A., and Morton D.H. 2005. Genome-wide SNP arrays as a diagnostic tool: Clinical description, genetic mapping, and molecular characterization of Salla disease in an Old Order Mennonite population. *Am. J. Hum. Genet.* **138:** 262–267.

Tang H., Coram M., Wang P., Zhu X., and Risch N. 2006. Reconstructing genetic ancestry blocks in admixed individuals. *Am. J. Hum. Genet.* **79:** 1–12.

Teebi A.S. and El-Shanti H.I. 2006. Consanguinity: Implications for practice, research, and policy. *Lancet* **367:** 970–971.

Tian C., Hinds D.A., Shigeta R., Kittles R., Ballinger D.G., and Seldin M.F. 2006. A genomewide single-nucleotide-polymorphism panel with high ancestry information for African American admixture mapping. *Am. J. Hum. Genet.* **79:** 640–649.

Velinov M., Sarfarazi M., Young K., Hodes M.E., Conneally P.M., Jackson C.E., and Tsipouras P. 1993. Limb-girdle muscular dystrophy is closely linked to the fibrillin locus on chromosome 15. *Connect. Tissue Res.* **29:** 13–21.

Weinstein LB. 2007. Selected genetic disorders affecting Ashkenazi Jewish families. *Fam. Comm. Health* **30:** 50–62.

Wright A.F., Carothers A.D., and Pirastu M. 1999. Population choice in mapping genes for complex diseases. *Nat. Genet.* **23:** 397–404.

Yeager M., Orr N., Hayes R.B., Jacobs K.B., Kraft P., Wacholder S., Minichiello M.J., Fearnhead P., Yu K., Chatterjee N., et al. 2007. Genome-wide association study of prostate cancer identifies a second risk locus at 8q24. *Nat. Genet.* **39:** 645–649.

Zhu X., Luke A., Cooper R.S., Quertermous T., Hanis C., Mosley T., Gu C.C., Tang H., Rao D.C., Risch N., and Weder A. 2005. Admixture mapping for hypertension loci with genome-scan markers. *Nat. Genet.* **37:** 177–181.

WWW RESOURCE

http://www.fhcrc.org/science/labs/tang/ SABER: SNP-based Ancestry Block Estimation and Reconstruction program. Fred Hutchinson Cancer Research Center, Seattle, Washington (see Tang et al. 2006).

3 | Power Calculations

David M. Evans[1] and Shaun Purcell[2]

[1]*Wellcome Trust Centre for Human Genetics, University of Oxford, Oxford, United Kingdom;* [2]*Psychiatric and Neurodevelopmental Genetics Unit, Center for Human Genetic Research, Massachusetts General Hospital, Boston, Massachusetts 02114*

INTRODUCTION

The power of a statistical test is the probability that it will yield a statistically significant result given that the null hypothesis is false. In other words, it represents the chance that the study will be successful in detecting a true effect and is dependent on a number of factors, including the magnitude of the effect, the sample size and study design, and the specified false-positive rate. Power calculations are primarily performed during the planning stages of a study, most typically in determining the sample size required. Consideration of statistical power can also sometimes shed light on the results of completed studies, particularly in the interpretation of negative results. In this chapter, we review the fundamentals of statistical power, discuss how power is calculated (using a genetic case/control study as an example), and consider the

most pertinent factors that influence power in genetic studies. Finally, we focus on power in the context of modern whole-genome association studies, where issues of coverage, multiple testing, and staged designs are paramount.

To make the concept of power concrete, consider the following example: a genetic case/control study to test association with a functional locus (i.e., a locus that actually does confer risk of disease). We sample repeatedly from some hypothetical population, phenotype and genotype each sample, calculate the association odds ratio, and then perform a significance test for whether the odds ratio was different from 1.0 (i.e., to reject the null hypothesis of "no association"). Purely due to the random sampling process, each replicate sample differs slightly in its estimated odds ratio. Because of sampling error, some proportion of tests would be significant, and some not; the exact ratio would depend on a number of factors, including the true effect size, so it is possible that all replicates might be significant. The proportion of replicates in which a significant result is obtained is an estimate of the power of the test, and it tells us how likely the study would be to succeed if it were repeated once more under exactly the same conditions. In practice, we do not calculate power by performing the experiment multiple times. Instead, we rely on analytic theory or use computer simulations to perform the repeated set of experiments. Power calculations can then be used to make studies more efficient, by indicating the smallest possible study that still has good power to detect all effects of a particular size.

Thinking about statistical power (the chance of detecting true effects) immediately brings two types of inferential error into focus: namely, not detecting true effects (type II error) and "detecting" nonexistent effects (type I error). A "false-positive" or type II error corresponds to failing to reject the null hypothesis (i.e., obtaining a nonsignificant test result) when the alternative hypothesis is true (i.e., the specific, non-null alternative hypothesis is in fact true). The probability of committing a type II error, denoted by the Greek letter β, is equal to one minus the statistical power: If the alternative hypothesis is true, we either detect the effect (at rate $1 - \beta$) or fail to detect it (at rate β).

In contrast, if the null hypothesis of no effect is in fact true, then we can either correctly fail to reject it (i.e., if the test is not significant), or incorrectly reject the null (i.e., if the test is significant, at rate α). Typically, the experimenter controls α by deciding at what threshold the test will be called significant. In this way, the sensitivity and specificity of a statistical test can be described and controlled. Naturally, there is a trade-off between these two properties: Lowering the threshold increases sensitivity (i.e., increases power, reduces type II errors) but also decreases specificity (i.e., increases type I error rate). Historically, for many small-scale studies, investigators have by convention adopted values such as $\alpha = 0.05$ and $\beta = 0.20$ (80% power) as representing a realistic and adequate trade-off.

The relationship between these quantities is displayed in Figure 3-1 (for ease of representation we assume a one-sided test of significance). Figure 3-1 shows the distribution of the test statistic X under both the null hypothesis (left-hand side) and the alternative hypothesis (right-hand side). The vertical line represents the critical value of the test so that values lying to the right of this line are declared significant, whereas values to the left are not. The area to the right of the critical value under the alternative hypothesis represents the power of the test

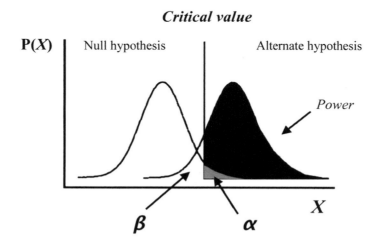

FIGURE 3-1. Distribution of the test statistic under both null (*left*) and alternative (*right*) hypotheses. The vertical line is the critical value of the test. The black and gray areas under the alternative hypothesis represent the power of the test. The shaded gray area refers to the type I error rate of the test. The unshaded area under the alternative hypothesis is the probability of committing a type II error.

(i.e., the sum of the black and gray areas), whereas the area to the left of the critical value under the alternative hypothesis represents the probability of committing a type II error. The probability of a type I error is represented by the gray area under the null hypothesis to the right of the critical value.

Besides altering the critical value, there are two other ways to increase the power of a statistical test: increasing the size of the sample, and increasing the effect size. As described below, the expected test statistic is a function of both sample size and effect size. Sample size is typically clearly under the investigator's control. Less obviously, effect size can also be increased; for example, by reducing measurement error with more accurate phenotyping, or through genotyping a region of interest more densely.

If we know the distribution of the test statistic under the null and alternative hypotheses, it is possible to calculate the expected power of the statistical test analytically (see Sham et al. 2000). For example, many tests of genetic association follow a chi-squared distribution. The shape of this distribution is determined by the "non-centrality parameter" (NCP). When the null hypothesis is true, the NCP is zero and the resulting chi-squared distribution is called a "central chi-squared distribution." Under the alternative hypothesis, the NCP will be greater than zero and the resulting distribution is called a "non-central chi-squared" distribution. This means that the distribution of the test statistic under the alternative hypothesis changes in shape and is shifted to the right by an amount quantified by the NCP. The expected test statistic equals the NCP plus the degrees of freedom. Because the shapes of the central and non-central chi-squared distributions are known, it is possible to deduce the areas under the curve in Figure 3-1 (and therefore the power) if we know the expected NCP, degrees of freedom of the test, and the critical value used.

PARAMETERIZING THE GENETIC MODEL

An investigator conducting a power analysis must decide in advance which values certain parameters will take. These include quantities that are known or under the control of the investigator (e.g., sample size, type I error rate, disease prevalence). The investigator must also make assumptions about parameters that are unknown, including the size of the genetic effect and underlying mode of inheritance, the level of linkage disequilibrium between marker and trait loci, and the allele frequencies at these loci. For example, the relative risk parameters R_{AA} and R_{Aa} quantify the effect size of the gene of interest and represent the number of times more likely an individual with genotype AA (or an individual with Aa genotype) is to be affected with disease than an individual with a lower risk aa genotype (arbitrarily assuming that aa is the low-risk, reference genotype). Given these parameters and knowledge of the disease prevalence (K), it is possible to calculate the absolute risk that an individual with the aa genotype is affected with disease (g_{aa}):

$$g_{aa} = K/(P_A^2 R_{AA} + 2p_A p_a R_{Aa} + p_a^2)$$

It is generally accepted that common complex diseases are unlikely to be the result of loci of large effect. Rather, genotypic relative risks are more likely to be in the range of 1–1.5. In the case of quantitative phenotypes, the measure of effect size is the proportion of the total phenotypic variance that is due to the quantitative trait locus. Typically, this value will be on the order of a few percent (0–5%) except perhaps in situations where there is evidence for a major locus. In such a case, a larger effect size might be assumed.

The underlying mode of inheritance is specified by the relationship between the relative risk parameters R_{AA} and R_{Aa}. For example, relative risks of $R_{AA} = 4$ and $R_{Aa} = 1$ indicate a recessive mode of inheritance for the A allele, since the Aa heterozygote contributes no more risk than the recessive aa homozygote. In contrast, relative risks of $R_{AA} = 16$ and $R_{Aa} = 4$ imply a multiplicative mode of inheritance, since the risk of the *AA* genotype is the square of the risk associated with the heterozygote.

There is also the very real possibility that the marker typed may not be the functional variant responsible for disease, but rather a marker in linkage disequilibrium with the true functional locus. The degree of linkage disequilibrium between trait and marker loci can be quantified by Lewontin's (1964) *D'* coefficient. Values of *D'* = 1 indicate an absence of ancestral recombination between marker and disease loci and thus complete disequilibrium. In contrast, a *D'* = 0 indicates independence between marker and trait loci. If one is interested in typing a small genomic region or a number of candidate genes thoroughly, one could assume that at least one marker will be in high linkage disequilibrium with a putative functional variant (assuming, of course, that a common variant actually exists in the region). In contrast, this assumption is unrealistic in genome-wide association studies, since the degree to which currently available commercial panels capture common variation in the genome is less than perfect. For example, it has recently been estimated that the Affymetrix 111K and 500K chips and the Illumina HumanHap300 panel capture around 31%, 65%, and 75%, respectively, of the common genetic variation in Western European populations through linkage disequilibrium (Barrett and Cardon 2006).

Finally, it is necessary to specify the allele frequencies at the trait and marker loci. Unless the investigator is only interested in a specific single-nucleotide polymorphism (SNP) whose frequency is known a priori, one would typically conduct power calculations across a range of possible allele frequencies. If using a commercial product that specifically tags common variants (e.g., the Illumina 300K panel of markers), one might have to assume the existence of a "common" disease allele (i.e., minor allele frequency >5%), since the tag SNPs in these panels do not capture rare variation well (Barrett and Cardon 2006). In contrast, if the focus of research is on nonsynonymous SNPs (i.e., SNPs that produce an amino-acid change in a protein), one might assume that the frequency of the disease-causing allele is less (since nonsynonymous SNPs tend to have lower minor allele frequencies).

CALCULATING POWER TO DETECT ASSOCIATION ANALYTICALLY

Given assumptions about the underlying genetic model, it is possible to calculate the expected NCP and hence the power of tests for genetic association. Closed-form expressions for the NCPs of several tests of association have been described in the literature, including the transmission disequilibrium test (TDT; McGinnis et al. 2002), case/control tests of genetic association (Schork 2002), and the quantitative trait disequilibrium test (Sham et al. 2000). With these formulae, an investigator can quickly and easily determine power under a variety of different scenarios without having to perform time-consuming data simulations. In addition, the NCP of a statistical test is a useful quantity in itself in that it increases linearly with sample size, unlike power to detect association (Witte et al. 2000). A number of utilities are available for helping investigators calculate the expected power to detect genetic association for a variety of different study designs, including QUANTO (http://hydra.usc.edu/GxE), power for association with error (PAWE; http://linkage.rockefeller.edu/pawe), and Purcell's on-line Genetic Power Calculator (http://pngu.mgh.harvard.edu/~purcell/gpc/)(Gauderman 2002; Gordon et al. 2002; Purcell et al. 2003).

To illustrate how the power of a genetic test of association can be calculated analytically, we describe a procedure for calculating the NCP for the Pearson Chi-square test of allelic association. The Pearson Chi-square test for allelic association compares the frequency of marker alleles in cases and controls:

$$\chi^2 = \sum_{i=0,1} \sum_{j=A,U} \frac{(n_{ij} - E[n_{ij}])^2}{E[n_{ij}]}$$

where the summation is over alleles (i) and affection status (j), n_{ij} is the cell containing the observed number of alleles of type i from individuals with affection status j, and

$$E[n_{ij}] = \frac{n_{i.}n_{.j}}{n_{..}}$$

is the expected cell count given the totals at the margins. Implicit in this allelic test is the assumption of a multiplicative mode of inheritance and that Hardy-Weinberg equilibrium holds in the general population (i.e., such that it is appropriate to treat an individual's two alleles as independent observations). That is, this test collapses across genotypes to form a 2 × 2 contingency table, counting alleles as opposed to individuals. If these assumptions do not hold, a genotypic test of association or the Armitage test of trend may be more appropriate (Sasieni 1997).

To illustrate the derivation of the NCP for the allelic test of association, consider a single biallelic disease-causing locus with alleles A and a in n cases and rn control individuals (so that the ratio of controls to cases is r). We denote the prevalence of disease by K, the frequency of the high-risk allele as p_A, and the genotypic relative risks for the Aa and AA genotypes as R_{Aa} and R_{AA}, respectively. Given these quantities, it is simple to calculate the risk that an individual possessing the low-risk genotype aa is affected:

$$g_{aa} = \frac{K}{R_{AA}p_A^2 + 2R_{Aa}p_A(1 - p_A) + (1 - p_A)^2}$$

and, consequently, the *absolute* risk of disease in individuals with Aa and AA genotypes:

$$g_{Aa} = R_{Aa}g_{aa}$$
$$g_{AA} = R_{AA}g_{aa}$$

From these quantities we can calculate the expected proportions of genotypes in cases using Bayes' theorem:

$$c_{aa} = \frac{g_{aa}(1 - p_A)^2}{K}$$
$$c_{Aa} = \frac{2g_{Aa}p_A(1 - p_A)}{K}$$
$$c_{AA} = \frac{g_{AA}p_A^2}{K}$$

and similarly in controls:

$$u_{aa} = \frac{(1 - g_{aa})(1 - p_A)^2}{1 - K}$$
$$u_{Aa} = \frac{2(1 - g_{Aa})p_A(1 - p_A)}{1 - K}$$
$$u_{AA} = \frac{(1 - g_{AA})p_A^2}{1 - K}$$

It is now a simple matter to calculate the expected allelic frequencies in case (c_A) and control (u_A) samples:

$$c_A = c_{AA} + \frac{c_{Aa}}{2}$$

and

$$u_A = u_{AA} + \frac{u_{Aa}}{2}$$

TABLE 3-1. Expected Allele Frequencies for Case/Control Allelic Test of Association

Allele	Case	Control
A	$2c_A n$	$2u_A rn$
a	$2(1 - c_A)n$	$2(1 - u_A)rn$
	$2n$	$2rn$

These expected allelic frequencies are summarized in the contingency table (Table 3-1).

It is now possible to use these values to calculate the NCP (denoted by the Greek letter λ) as first derived by Mitra (1958):

$$\lambda = \frac{2nr(c_A - u_A)^2(1 + r)}{(c_A + ru_A)(1 + r - c_A - ru_A)}$$

The power of the allelic test of association is found by integrating under the non-central chi-square distribution:

$$1 - \beta = \int_{\chi_\alpha'^2(v,0)}^{\infty} d\chi'^2(v, \lambda)$$

where $\chi_\alpha'^2(v, 0)$ is the $100(1 - \alpha)$ percentage point of the central χ^2 distribution with v degrees of freedom and $\chi'^2(v, \lambda)$ is a non-central χ^2 distribution with non-centrality parameter λ.

This basic procedure assumes that the investigator is lucky enough to genotype the actual trait locus. Of course, in reality this is seldom the case, except perhaps when attempting to replicate a known association. It is therefore important to extend these calculations to situations where the trait locus has not been typed, but rather a marker in linkage disequilibrium. Consider a biallelic marker locus with alleles M and m and corresponding allele frequencies p_M and p_m. We quantify the amount of linkage disequilibrium between the trait and marker locus by Lewontin's (1964) D' coefficient:

$$D' = \frac{p_{AM} - p_A p_M}{\min(p_A p_m, p_a p_M)} = \frac{\delta}{\delta_{max}} \text{ when } \delta > 0$$

and

$$D' = \frac{p_{AM} - p_A p_M}{\min(p_A p_M, p_a p_m)} = \frac{\delta}{\delta_{max}} \text{ when } \delta < 0$$

where p_{AM} is the frequency of the haplotype with A and M alleles, and δ quantifies the difference between observed and expected (under independence) haplotype frequencies. For $\delta > 0$, δ_{max} is the most positive value of δ possible given the specific allele frequencies at the marker and disease loci, whereas for $\delta < 0$, δ_{max} is the most negative possible value of δ (Devlin and Risch 1995; McGinnis et al. 2002). The haplotype frequencies at the marker and trait loci are expressed in terms of δ in Table 3-2.

To incorporate linkage disequilibrium into calculation of the NCP, we simply determine the risks associated with each marker genotype. This can be obtained by multiplying the risks at the

TABLE 3-2. Haplotype Frequencies at Marker and Disease Loci

		Trait allele	
		A	a
Marker	M	$p_A p_M + \delta$	$p_a p_M - \delta$
Allele	m	$p_A p_m - \delta$	$p_a p_m + \delta$

trait locus by the probability of the trait genotype given the marker genotype and then summing over all possible trait genotypes. For example, the probability that an individual is affected given they have an *MM* genotype at the marker locus is calculated thus:

$$P(D = Affected \mid G_M = MM) = \sum_{G_T} P(D = Affected \mid G_T)P(G_T \mid MM)$$

$$= \sum_{G_T} P(D = Affected \mid G_T)\frac{P(G_T, G_M = MM)}{P(G_M = MM)}$$

$$= \frac{g_{AA}(p_A p_M + \delta)^2 + 2g_{Aa}(p_A p_M + \delta)(p_a p_M - \delta) + g_{aa}(p_a p_M - \delta)^2}{p_M^2}$$

where G_T and G_M denote the genotypes at the trait and marker loci, respectively. The risk associated with other marker genotypes is derived in a similar fashion. The procedure for calculating the NCP of the test at the marker locus proceeds exactly as before, except now the frequencies and risks at the marker locus are used in the calculation instead of at the disease locus.

CALCULATING POWER TO DETECT ASSOCIATION THROUGH SIMULATION

As shown, closed-form expressions for the NCP of a statistical test for genetic association can be used quickly and easily to calculate expected power under a range of scenarios. However, such formulae are useful only for the simplest of study designs. In reality, data sets often involve much more complicated situations. For example, a data set may consist of a number of different types of pedigree structures, or individuals may have been ascertained on the basis of an extreme score on a quantitative trait. Web-based tools do not exist for calculating power in complicated situations such as these, and it may not be a simple matter for an investigator to derive the NCP and calculate power analytically on his own. In these situations, power calculations can still be performed with the aid of data simulation.

The basic idea behind data simulation is to generate thousands of replicate data sets under the alternative hypothesis of association for a given genetic model. In other words, each replicate data set is like drawing a random sample from the same underlying population. Data generation may involve simulating a number of different processes (e.g., transmitting alleles from parents to offspring) or perhaps ascertaining individuals on the basis of some simulated trait value. Each replicate data set is then tested for genetic association, and the proportion of replicates in which the evidence of association exceeds the critical value is an estimate of the power of the test. Simulation is therefore a very flexible strategy for estimating the power to detect genetic association. The downside is that it is computationally intensive; thousands of replicate samples need to be generated for each specific set of population parameters.

FACTORS INFLUENCING THE POWER TO DETECT ASSOCIATION

A number of factors can affect the power to detect association. These include the size of a genetic effect, its mode of inheritance, the prevalence of disease, the ratio of case to control individuals, allelic frequencies, and the extent of linkage disequilibrium between marker and trait loci. For all situations discussed below, we determined the expected power to detect association using Purcell's on-line Genetic Power Calculator (http://pngu.mgh.harvard.edu/~purcell/gpc; Purcell et al. 2003).

Mode of Inheritance

Table 3-3 displays the effects that varying the mode of inheritance, genotype relative risk, and disease allele frequency have on the power to detect association. Specifically, the table displays the number of parent–child trios (case/control pairs) required to obtain 90% power to detect association at a significance threshold of either $\alpha = .05$ or $\alpha = 5 \times 10^{-7}$. A significance level of $\alpha = .05$ might be appropriate in the analysis of a single polymorphism as in a candidate gene study, whereas a significance level of $\alpha = 5 \times 10^{-7}$ would be more typical of that required in a genome-wide association study. In both cases, we assume a relatively rare disease ($K = 0.001$), complete linkage disequilibrium between trait and marker loci (i.e., $D' = 1$), and that marker and disease loci have equal allele frequencies (which is in effect saying that we are directly genotyping and test-

TABLE 3-3. Number of (A) Case Individuals or (B) Parent–Child Trios Required to Obtain 90% Power to Detect Association at a Significance Threshold of Either $\alpha = .05$ or $\alpha = 5 \times 10^{-7}$

A. Case Individuals

		$\alpha = 0.05$				$\alpha = 5 \times 10^{-7}$			
		multiplicative	additive	recessive	dominant	multiplicative	additive	recessive	dominant
GRR_{Aa}	p_A								
4.0									
	.01	301	312	1.195×10^6	318	1137	1179	4.525×10^6	1201
	.10	40	53	1531	64	150	201	5796	242
	.50	27	60	55	195	101	225	207	736
	.80	60	153	71	2611	225	577	267	9888
2.0									
	.01	1602	1602	1.065×10^7	1656	6065	1305	4.032×10^7	6270
	.10	190	190	1.007×10^4	263	718	718	4.678×10^4	995
	.50	92	92	260	512	348	348	983	1936
	.80	175	175	233	6099	661	661	881	2.31×10^4
1.5									
	.01	5319	5179	3.462×10^7	5465	2.014×10^4	1.961×10^4	3.462×10^7	2.069×10^4
	.10	609	483	3.913×10^4	800	2306	1826	3.913×10^4	3032
	.50	260	126	691	1266	983	477	3207	4795
	.80	457	192	527	1.406×10^4	1728	725	2447	5.325×10^4

B. Parent–Child Trios

		$\alpha = 0.05$				$\alpha = 5 \times 10^{-7}$			
		multiplicative	additive	recessive	dominant	multiplicative	additive	recessive	dominant
GRR_{Aa}	p_A								
4.0									
	.01	304	315	1.197×10^6	321	1150	1192	4.534×10^6	1214
	.10	43	56	1537	67	160	212	5819	253
	.50	30	62	58	198	111	235	217	748
	.80	63	156	74	2619	235	588	278	9917
2.0									
	.01	1608	1608	1.067×10^7	1662	6089	6089	4.04×10^7	6294
	.10	193	193	1.238×10^4	266	730	730	4.688×10^4	1007
	.50	95	95	263	515	359	359	995	1950
	.80	178	178	236	6114	672	672	893	2.315×10^4
1.5									
	.01	5333	5193	3.469×10^7	5479	2.02×10^4	1.967×10^4	1.612×10^8	2.075×10^4
	.10	613	486	4.811×10^4	805	2321	1840	1.822×10^5	3048
	.50	263	129	852	1271	995	488	3223	4814
	.80	460	195	651	1.409×10^4	1741	736	2462	5.337×10^4

Results are shown for multiplicative, additive, recessive, and dominant models. An equal number of case and control individuals and a disease prevalence of $K = 0.001$ are assumed.

ing the disease locus itself). For case/control pairs, the test of association is Pearson's Chi-squared test of allelic independence (Sasieni 1997). For parent–child trios, the relevant test is Spielman's TDT (Spielman et al. 1993).

Not surprisingly, the power to detect association is greatest when the effect size is large. Both case/control and parent/offspring designs have considerable power to detect common alleles that have moderate-to-large effects (i.e., genotypic relative risks > 2). In most scenarios, only tens or hundreds of cases/trios are required. In contrast, in order to detect loci of small effect (e.g., a genotypic relative risk of about 1.5, which is probably indicative of the size of effects underlying most complex traits and diseases), sample sizes need to be much larger. This is particularly true for genome-wide association where sample sizes will need to contain thousands of individuals in order to have appreciable power to detect alleles of small effect. In this respect, it is interesting to note that although the required level of significance has changed by five orders of magnitude, the required sample size only increases around five times (i.e., the required sample is roughly proportional to the logarithm of the alpha level [Witte et al. 2000]). In general, power is greatest to detect association when the frequency of the disease allele is intermediate, although this depends to some extent on the underlying mode of inheritance. For example, in the case of a dominant disease, many more individuals are required when the frequency of the disease allele is high. The power to detect rare variants is low in general.

Table 3-3 also shows that for a broad array of disease models, when the disease is rare, the number of trios required to detect association is approximately equal to the number of case individuals for a given level of power (assuming equal numbers of case and control individuals). In other words, the total number of individuals requiring genotyping in a case/control study is around two-thirds that required for a study of trios in the case of a rare disease (Bacanu et al. 2000). This reduction is especially relevant for genome-wide association studies where the number of individuals and markers analyzed (and hence the cost of the study) is very large. Nevertheless, it must be remembered that family-based tests of association do have certain advantages over a case/control design. For example, the TDT controls for the effect of population stratification that can produce spurious associations in case/control designs and also allows for the detection of parent-of-origin effects and genotyping error.

Disease Prevalence

As the prevalence of disease increases, the power to detect association using the case/control design also increases. This is because the frequency of the high-risk allele decreases in control individuals (assuming that control individuals are selected on the basis of not having the risk allele). Hence, there is a greater divergence in the frequency of alleles between case and control groups. Thus, for common diseases, the number of individuals required is far larger for the TDT than for the case/control test of association, assuming the same underlying genetic model. However, prevalence of disease does not affect the power of the TDT, provided that both the genotypic relative risks and the mode of inheritance remain constant. This is because the TDT uses family-based controls that are not screened for the presence/absence of disease (i.e., the genetic composition of the parents is not expected to change as a function of disease prevalence).

Ratio of Case to Control Individuals

The power of the case/control test of association also changes as a function of the relative proportion of affected and unaffected individuals in the sample. In general, for a given number of individuals, the most powerful ratio is an equal number of cases and controls. However, in some situations, it may be difficult or expensive to ascertain affected individuals. In these situations, recruiting more control individuals can still increase power. McGinnis et al. (2002) showed that in the hypothetical situation of an infinite number of controls, one would only need to recruit half the number of cases to obtain the same power as a study with equal numbers of cases and controls.

However, most of this potential gain in power can be achieved by recruiting three to five times the number of control individuals. In the end, the optimal ratio of cases and controls will depend on the relative cost of acquiring each (McGinnis et al. 2002).

Linkage Disequilibrium and Marker Allele Frequency

Power to detect association is greatest when linkage disequilibrium is high (i.e., there is little ancestral recombination between marker and trait loci), and when trait and marker loci have similar allele frequencies. This poses a problem for the detection of rare alleles via genetic association, since most of the SNPs found in the dbSNP (Sachidanandam et al. 2001) and HapMap databases (Altshuler et al. 2005), and consequently those available on commercial marker panels, are common variants (i.e., minor allele frequency > 5%). Therefore, association studies that employ markers from these sources may have little power to detect rare variants (remember that the power to detect rare disease alleles is already very low). This limitation can be addressed somewhat by forming haplotypes of adjacent SNPs that might have a closer frequency to the disease allele. Such an approach is not entirely satisfactory (Lin et al. 2004).

GENOME-WIDE ASSOCIATION

Recent advances in high-throughput technology and decreased genotyping costs (Matsuzaki et al. 2004), as well as the publication of the International Haplotype Map of the human genome (Altshuler et al. 2005), make it possible to search for complex disease genes by screening thousands of individuals on hundreds of thousands of polymorphisms across the genome. The first genome-wide association studies have appeared in the literature with encouraging results (Ozaki et al. 2002; Cheung et al. 2005; Klein et al. 2005; Maraganore et al. 2005). Whereas these designs hold great promise for dissecting the genetic basis of complex traits and diseases, there are a number of statistical issues pertaining to this sort of study design. One issue that we examine here is that of using a two-stage approach to decrease the costs associated with genotyping.

To decrease the cost of genotyping thousands of subjects with hundreds of thousands of markers, a more cost-effective strategy might be to genotype subjects in stages. Initially, a proportion of individuals ($\pi_{individuals}$) would be genotyped on all markers ($N_{markers}$). Subsequently, a proportion of markers displaying the most promising results ($\pi_{markers}$) would then be typed in the remaining individuals (Sobell et al. 1993). Compared to one-stage approaches where all markers in all individuals are genotyped, two-stage designs can lead to appreciable savings in genotyping costs while still maintaining power (Satagopan et al. 2002, 2004; Satagopan and Elston 2003; Maraganore et al. 2005; Thomas et al. 2005; Skol et al. 2006).

Recently, Skol et al. (2006) examined the performance of two different approaches to analyzing the data from two-stage genome-wide association scans. In the "replication-based" strategy, association was only tested on markers and individuals genotyped in the second phase of the procedure. To maintain the genome-wide error rate below α_{genome} = 0.05, a Bonferonni corrected significance level of α_{genome} / ($N_{markers} \times \pi_{markers}$) was employed. In the "joint analysis" strategy, test statistics in stages one and two were combined and compared against an approximate significance level of α_{genome} / $N_{markers}$. The study found that despite the more stringent significance level, jointly analyzing the data from both stages almost always resulted in increased power to detect association as compared to a replication-based strategy. With appropriately chosen thresholds, the power provided by this method was comparable to that of the one-stage design. The power of the joint analysis decreased as the proportion of samples typed in stage one decreased, presumably because variants that predispose to disease were less likely to be selected for genotyping in stage two. Similarly, taking too few markers through to stage two also decreased the power of the joint analysis because of the decreased probability of taking a true risk variant through to the next stage. In contrast, power increased when fewer markers were selected for follow-up in the replication-based strategy, because fewer statistical tests were performed and hence there were fewer penalties due to multiple testing. In fact, the only situation in which the joint analysis was not more powerful than a replication-based strategy was when the strength of association was far greater in stage two than in stage one. It is therefore recommended that two-stage genome-wide association scans be analyzed using a joint analysis strategy, that a large proportion of samples be genotyped in stage one (i.e., $\pi_{individuals}$ >30%), and that a relatively large proportion of markers be selected for follow-up in stage two ($\pi_{markers}$ >1%). The Center for Statistical Genetics provides a useful Web-based tool (http://csg.sph.umich.edu) that can be used by researchers to calculate the power to detect association in two-stage designs.

CONCLUSIONS

We conclude with some basic guidelines investigators might find useful when performing power calculations in either the design or analysis phase of genetic association studies:

1. During the design phase of a study, use a Web-based utility such as Purcell's GPC (http://pngu.mgh.harvard.edu/~purcell/gpc/) to calculate the expected number of individuals who will need to be genotyped given what is known about the disease (e.g., mode of inheritance, size of effect). If one is performing a candidate gene study and is willing to assume that a functional variant lies somewhere in the region being typed (e.g., a previous linkage analysis might indicate that a disease locus lies in the region), then it might be safe to assume high levels of linkage disequilibrium between marker and disease loci. In contrast, if a genome-wide association study is being planned, then the possibility that a causal variant may not be in appreciable linkage disequilibrium with any of the markers must be considered. The ability of commercially available marker panels to capture common variation in the genome has been quantified in Barrett and Cardon (2006).

2. If employing a complicated study design, it may be necessary to perform data simulations in order to calculate the expected power.

3. If a genome-wide association study is being considered, a two-stage design is an efficient and cost-effective method of genotyping subjects while maintaining power to detect association. Generally, a large proportion of samples should be genotyped in the first stage (>30%), a relatively large proportion of markers should be selected for follow-up (>1%), and the data should be analyzed using a joint analysis strategy. A Web-based tool that can be used to calculate the power to detect association in two-stage designs is located at http://csg.sph.umich.edu.

REFERENCES

Altshuler D., Brooks L.D., Chakravarti A., Collins F.S., Daly M.J., and Donnelly P. 2005. A haplotype map of the human genome. *Nature* **437:** 1299–1320.

Bacanu S.A., Devlin B., and Roeder K. 2000. The power of genomic control. *Am. J. Hum. Genet.* **66:** 1933–1944.

Barrett J.C. and Cardon L.R. 2006. Evaluating coverage of genome-wide association studies. *Nat. Genet.* **38:** 659–662.

Cheung V.G., Spielman R.S., Ewens K.G., Weber T.M., Morley M., and Burdick J.T. 2005. Mapping determinants of human gene expression by regional and genome-wide association. *Nature* **437:** 1365–1369.

Devlin B. and Risch N. 1995. A comparison of linkage disequilibrium measures for fine-scale mapping. *Genomics* **29:** 311–322.

Gauderman W.J. 2002. Sample size requirements for association studies of gene-gene interaction. *Am. J. Epidemiol.* **155:** 478–484.

Gordon D., Finch S.J., Nothnagel M., and Ott J. 2002. Power and sample size calculations for case-control genetic association tests when errors are present: Application to single nucleotide polymorphisms. *Hum. Hered.* **54:** 22–33.

Klein R.J., Zeiss C., Chew E.Y., Tsai J.Y., Sackler R.S., Haynes C., Henning A.K., SanGiovanni J.P., Mane S.M., Mayne S.T., et al. 2005. Complement factor H polymorphism in age-related macular degeneration. *Science* **308:** 385–389.

Lewontin R.C. 1964. The interaction of selection and linkage. I. General considerations; heterotic models. *Genetics* **49:** 49–67.

Lin S., Chakravarti A., and Cutler D.J. 2004. Exhaustive allelic transmission disequilibrium tests as a new approach to genome-wide association studies. *Nat. Genet.* **36:** 1181–1188.

Maraganore D.M., de Andrade M., Lesnick T.G., Strain K.J., Farrer M.J., Rocca W.A., Pant P.V., Frazer K.A., Cox D.R., and Ballinger D.G. 2005. High-resolution whole-genome association study of Parkinson disease. *Am. J. Hum. Genet.* **77:** 685–693.

Matsuzaki H., Loi H., Dong S., Tsai Y.Y., Fang J., Law J., Di X., Liu W.M., Yang G., Liu G., et al. 2004. Parallel genotyping of over 10,000 SNPs using a one-primer assay on a high-density oligonucleotide array. *Genome Res.* **14:** 414–425.

McGinnis R., Shifman S., and Darvasi A. 2002. Power and efficiency of the TDT and case-control design for association scans. *Behav. Genet.* **32:** 135–144.

Mitra S.K. 1958. On the limiting power function of the frequency chi-square test. *Ann. Math. Stat.* **29:** 1221–1233.

Ozaki K., Ohnishi Y., Iida A., Sekine A., Yamada R., Tsunoda T., Sato H., Sato H., Hori M., Nakamura Y., and Tanaka T. 2002. Functional SNPs in the lymphotoxin-alpha gene that are associated with susceptibility to myocardial infarction. *Nat. Genet.* **32:** 650–654.

Purcell S., Cherny S.S., and Sham P.C. 2003. Genetic Power Calculator: Design of linkage and association genetic mapping studies of complex traits. *Bioinformatics* **19:** 149–150.

Sachidanandam R., Weissman D., Schmidt S.C., Kakol J.M., Stein L.D., Marth G., Sherry S., Mullikin J.C., Mortimore B.J., Willey D.L., et al. (International SNP Map Working Group). 2001. A map of human genome sequence variation containing 1.42 million single nucleotide polymorphisms. *Nature* **409:** 928–933.

Sasieni P.D. 1997. From genotypes to genes: Doubling the sample size. *Biometrics* **53:** 1253–1261.

Satagopan J.M. and Elston R.C. 2003. Optimal two-stage genotyping in population-based association studies. *Genet. Epidemiol.* **25:** 149–157.

Satagopan J.M., Venkatraman E.S., and Begg C.B. 2004. Two-stage designs for gene-disease association studies with sample size constraints. *Biometrics* **60:** 589–597.

Satagopan J.M., Verbel D.A., Venkatraman E.S., Offit K.E., and Begg C.B. 2002. Two-stage designs for gene-disease association studies. *Biometrics* **58:** 163–170.

Schork N.J. 2002. Power calculations for genetic association studies using estimated probability distributions. *Am. J. Hum. Genet.* **70:** 1480–1489.

Sham P.C., Cherny S.S., Purcell S., and Hewitt J.K. 2000. Power of linkage versus association analysis of quantitative traits, by use of variance-components models, for sibship data. *Am. J. Hum. Genet.* **66:** 1616–1630.

Skol A.D., Scott L.J., Abecasis G.R., and Boehnke M. 2006. Corrigendum: Joint analysis is more efficient than replication-based analysis for two-stage genome-wide association studies. *Nat. Genet.* **38:** 390.

Sobell J.L., Heston L.L., and Sommer S.S. 1993. Novel association approach for determining the genetic predisposition to schizophrenia: Case-control resource and testing of a candidate gene. *Am. J. Med. Genet.* **48:** 28–35.

Spielman R.S., McGinnis R.E., and Ewens W.J. 1993. Transmission test for linkage disequilibrium: The insulin gene region and insulin-dependent diabetes mellitus (IDDM). *Am. J. Hum. Genet.* **52:** 506–516.

Thomas D.C., Haile R.W., and Duggan D. 2005. Recent developments in genomewide association scans: A workshop summary and review. *Am. J. Hum. Genet.* **77:** 337–345.

Witte J.S., Elston R.C., and Cardon L.R. 2000. On the relative sample size required for multiple comparisons. *Stat. Med.* **19:** 369–372.

WWW RESOURCES

http://csg.sph.umich.edu The CaTS Power Calculator software at the Center for Statistical Genetics Web site is a simple, multiplatform interface for carrying out power calculations for large genetic association studies, including two-stage genome-wide association studies.

http://hydra.usc.edu/GxE Gene × Environment, Gene × Gene Interaction Home page

http://linkage.rockefeller.edu/pawe The Power for Association With Error (PAWE) program is designed to perform asymptotic power and sample size calculations for genetic case/control studies with a diallelic locus (for example, a SNP) in the presence of errors.

http://pngu.mgh.harvard.edu/~purcell/gpc/ The on-line Genetic Power Calculator provides automated power analysis for variance components, quantitative trait loci linkage and association tests in sibships, and other common tests.

4 Genetic Analysis: Moving between Linkage and Association

Albert Vernon Smith

Genthor ehf., 101 Reykjavik, Iceland, and Icelandic Heart Association, 201 Kopavogur, Iceland

INTRODUCTION

In the study of human disease genetics, the approaches to identifying genes and genomic regions associated with disease can be broadly grouped into two categories: linkage analysis and genetic association analysis. Family-based linkage studies have been the foundation for many successes in mapping of genes associated with Mendelian disorders, but they lack power to detect alleles conferring moderate risks that are likely to be the norm in complex disease. Linkage analysis is particularly well-suited to diseases of high penetrance that run strongly within families. In linkage analysis, researchers identify and collect individuals from pedigrees with multiple individuals affected by disease. A researcher then analyzes an extended set of markers covering the entire genome and pedigree, and looks for chromosomal regions that are shared by those affected with the disease with greater frequency (Ott 1999). A large number of disease loci have been identified in this manner, particularly traits with Mendelian inheritance. Although there have been notable successes from such methods, many of the diseases studied are rare and do not fully reflect the disease spectrum of greatest concern to medical practitioners. Linkage studies have limited power to detect situations where there are multiple genes with smaller effects, which is thought to be the case with common diseases, such as heart disease, asthma, and diabetes.

An alternative method for detecting disease genes is the genetic association study. In such an analysis, one gathers a large numbers of individuals affected by a particular trait and looks for individual marker alleles (or genotypes) that are more frequent in cases compared to controls (Cardon and Bell 2001). Association analysis can be performed on unrelated individuals and relies on linkage disequilibrium (LD) between tested loci and disease-predisposing variants at the total population level. Those alleles that predispose to disease (or are in LD with those that do) will be more frequent in cases compared to controls. Prior to the recent development of genotyping systems capable of genotyping hundreds of thousands of markers concurrently, association studies had typically been performed on a smaller number of candidate genes that were thought to be involved in a biological process related to a specific disease. This approach has been attempted in a large number of studies over many years, but it has led to the identification of relatively few genes that are consistently associated with disease across a large number of independent studies.

Although one can draw a distinction between these two different approaches, they are often heavily intertwined. Typically, a linkage study might initially identify a locus (or loci) associated with a particular disease. However, one of the limitations of genetic linkage analysis is that it has very low resolution and might only be able to identify disease-associated regions covering several megabases of DNA. Once regions associated with disease have been identified by linkage, subsequent analysis to resolve the signal is typically via genetic association, because that has much finer resolution. Following identification of a disease-associated locus, a set of candidate genes is identified within the region of interest. Markers within this locus can then be genotyped for a series of cases and controls, looking for specific marker alleles that are more frequent in the cases than in the controls.

The production of the HapMap has revolutionized approaches to the study of human disease genetics, particularly for the so-called common diseases (International HapMap Consortium 2005). The dense genetic map available from the HapMap, as well as development of commercially available, high-density genotyping products, has now facilitated a paradigm shift in genetic association studies. Given products which concurrently genotype 500,000 to 1,000,000 single-nucleotide polymorphisms (SNPs), it is now possible to do a genetic association for the whole genome rather than for a limited number of candidate genes.

Although whole-genome association analysis is newly available, and has current favor over linkage analysis in the study of common disease, it is important to be familiar with both approaches. Family-based material can be used to improve association analysis, and genetic association will certainly continue to be used to follow signals originally identified by linkage analysis. This chapter is intended as a general introduction to tools needed to perform both genetic linkage and genetic association analysis.

CREATING INPUT FILES FOR GENETIC ANALYSIS

Although there is no single standard input file format for genetic analysis packages, file formats based on that used for the LINKAGE program (Lathrop et al. 1984) are relatively predominant and can be used as input for the different genetic analysis packages described below. There are small differences in the exact formats for each, but they have the same basic design.

Describing Relationships

To input data for genetic analysis, relationships between individuals must be described. Although pedigrees are often complex, the essential information can be reduced to five items: a family identifier, an individual identifier, an identifier for each parent (if known), and the sex of the individual. Therefore, a *pedigree file* is constructed, which describes this information. For example, in a small pedigree with two siblings, their parents, and paternal grandparents, the essential information can be described as follows:

```
FAMILY    PERSON    FATHER    MOTHER    SEX
example   grandpa   unknown   unknown   m
example   grandma   unknown   unknown   f
example   father    grandpa   grandma   m
example   mother    unknown   unknown   f
example   sister    father    mother    f
example   brother   father    mother    m
```

In practice, these text identifiers are typically replaced with unique numeric values. Then, using unique integers and recoding sexes with 1 (male) and 2 (female), the example looks like this:

```
1  1  0  0  1
1  2  0  0  2
1  3  1  2  1
1  4  0  0  2
1  5  3  4  2
1  6  3  4  1
```

Describing Phenotypes and Genotypes

Often, these five standard columns are followed by phenotypes and marker genotypes. Columns are added that contain the phenotype information. For a discrete trait, the convention is typically 1 for unaffected, 2 for affected, and 0 for missing information. For quantitative traits, the actual value would be inserted. Then, marker genotypes are added as two consecutive integers, one for each allele. For SNPs, which are biallelic, 1 can be used for one allele, 2 for the other, and 0 for missing information. In the following example, information is added for a disease state, a quantitative trait, and two markers.

```
1  1  0  0  1  1     x    1 2   2 1
1  2  0  0  2  1     x    1 1   1 1
1  3  1  2  1  1     x    1 1   2 1
1  4  0  0  2  1     x    2 2   1 1
1  5  3  4  2  2   5.6    1 1   2 1
1  6  3  4  1  2   1.3    2 1   2 1
```

Typically, the pedigree file is saved with a *.ped* file name extension.

Describing the Pedigree File

Because the files described above can contain information on disease status, as well as any number of quantitative trait variables and genotyped markers, a companion data file must be constructed that describes the contents of the pedigree file. Although the exact convention used can vary slightly for different programs, the data file will contain one row per data item in the pedigree file, indicating the data type (encoded as M: marker, A: affection status, T: quantitative trait, and C: covariate) and label for each item. For the pedigree file above, which has one affection status, one quantitative trait, and two marker genotypes, the following data file would be constructed:

```
A   some_disease
T   some_trait
M   marker_one
M   marker_two
```

Typically, the data file is saved with a *.dat* file name extension. It is critical that there are the same number of rows in this *dat* file which is paired with a specific *ped* file.

Genetic Map Information

For many genetic analyses, it is critical to know the relative positions of markers. For linkage analysis programs, a genetic map position needs to be provided, because that is a component of the analysis. However, for genetic association analyses, a physical location is often provided without a genetic map position. Linkage analysis relies on the genetic positions as part of the analysis, whereas many tests of genetic association test each marker independently, and consequently, do not rely on genetic map positions. The typical map file lists the chromosome, the marker name, the genetic position, and (optionally) the physical position.

```
CHROMOSOME   MARKER       MAP    PHYSICAL
24           marker_one   14.1   12000000
24           marker_two   15.3   13500000
```

Typically, this map file is saved with a .*map* file name extension. In some cases, the map file must have the same number of rows as there are markers in the *ped* file. However, in other cases, untyped markers may be ignored, and map positions are only used for markers that are also listed in the *dat* file.

Working with the Files

Although there are slight variations in these file formats used by different genetic analysis programs, the basics are often very similar and consist of the following files: a *ped* file with the pedigree information (including information on the familial relationships, affection status, and marker genotypes); a *dat* file, which describes the columns in the pedigree file; and a *map* file, which gives information specific for the markers being typed. For a given analysis, the three files are made together as set. For different genetic analysis packages, there can be slight differences in the exact columns required, but this basic file structure is used by multiple different genetic analysis packages, including those described below.

PERFORMING LINKAGE ANALYSIS WITH MERLIN

MERLIN is a freely available program capable of carrying out analyses of pedigree data (Abecasis et al. 2002). In contrast to many other packages designed for the analysis of pedigree data, the computational algorithms used in MERLIN are well suited to dense genetic maps. MERLIN is a complete package that can analyze pedigree data in a number of ways, including calculating identity-by-descent (IBD) and kinship coefficients, performing nonparametric and variance component linkage analysis, detecting errors, and mapping information content. Following is a description of how to perform linkage analysis with MERLIN, which is a good entry point to the package. Linkage analysis tests for co-segregation of a chromosomal region and a trait of interest. To perform a basic linkage analysis:

1. Download and install the MERLIN software package, which is available at http://www.sph.umich.edu/csg/abecasis/MERLIN/. Binaries are available for Linux, Windows, Sun, and Mac. Additionally, the source code is available, and one can compile MERLIN as necessary. The programs in the package are run from the command line.

2. Create three appropriate input files: a data file (*file.dat*), a pedigree file (*file.ped*), and a map file (*file.map*). The input files are as described above, although no physical position should be included in the *map* file. (Three columns are needed in this file: chromosome, marker name, and genetic position.)

3. Once the files are created, test to ensure that the files are being run correctly. This can be done by running pedstats, a program that is part of the MERLIN package. It requires an input *dat* file (**-d** parameter) and a *ped* file (**-p** parameter). This would be run as:

```
pedstats -d file.dat -p file.ped
```

If the files are formatted properly, summary statistics about the files are obtained. Example files are provided at the distribution site, and running these test files gives good examples of the output from this program (as well as other programs from the MERLIN package).

4. If pedstats has run correctly, run MERLIN with the following command line parameters: an input *dat* file (**-d**), a *ped* file (**-p**), and a *map* file (**-m**). Additionally, specify the type of analysis. MERLIN has a number of options, but it is often best to begin with nonparametric linkage (NPL) analysis, and NPL all can be specified with the **--npl** command line options. The standard NPL analysis carried out by MERLIN uses the Kong and Cox (1997) linear model to evaluate the evidence for linkage. This would be run as:

```
merlin -p file.ped -d file.dat -m file.map --npl
```

The output starts with a summary of the options being used:

```
MERLIN 1.0.1 - (c) 2000-2005 Goncalo Abecasis

References for this version of Merlin:

    Abecasis et al (2002) Nat Gen 30:97-101          [original citation]
    Fingerlin et al (2004) AJHG 74:432-43            [case selection for association studies]
    Abecasis and Wigginton (2005) AJHG 77:754-67     [ld modeling, parametric analyses]

The following parameters are in effect:
                  Data File :        file.dat (-dname)
              Pedigree File :        file.ped (-pname)
         Missing Value Code :         -99.999 (-xname)
                   Map File :        file.map (-mname)
          Allele Frequencies : ALL INDIVIDUALS (-f[a|e|f|m|file])
                Random Seed :          123456 (-r9999)

Data Analysis Options
                    General : --error, --information, --likelihood, --model [param.tbl]
                 IBD States : --ibd, --kinship, --matrices, --extended, --select
                NPL Linkage : --npl [ON], --pairs, --qtl, --deviates, --exp
                 VC Linkage : --vc, --useCovariates, --ascertainment
                Haplotyping : --best, --sample, --all, --founders, --horizontal
              Recombination : --zero, --one, --two, --three, --singlepoint
                  Positions : --steps, --maxStep, --minStep, --grid, --start, --stop
            Marker Clusters : --clusters [], --distance, --rsq, --cfreq
                     Limits : --bits [24], --megabytes, --minutes
                Performance : --trim, --noCoupleBits, --swap, --cache []
                     Output : --quiet, --markerNames, --frequencies, --perFamily, --pdf,
                              --prefix [merlin]
                 Simulation : --simulate, --reruns, --save
```

And is followed by the analysis results:

```
Phenotype: affection [ALL] (200 families)
==================================================
        Pos    Zmean   pvalue    delta     LOD   pvalue
        min   -20.00      1.0   -0.707  -60.21      1.0
        max    20.00  0.00000    0.707   60.21  0.00000
      0.000     0.96      0.2    0.092    0.27     0.13
      5.268     1.39     0.08    0.126    0.54     0.06
     10.536     1.27     0.10    0.110    0.43     0.08
     15.804     1.43     0.08    0.128    0.56     0.05
     21.072     0.88      0.2    0.083    0.22      0.2
     26.340     1.37     0.08    0.130    0.55     0.06
```

```
31.608   1.53    0.06    0.151   0.71    0.04
36.876   2.18   0.014    0.197   1.32   0.007
42.144   2.60   0.005    0.218   1.75   0.002
47.412   3.00  0.0014    0.251   2.33  0.0005
52.680   3.43  0.0003    0.286   3.05 0.00009
...
```

In this output, the first two lines indicate the maximum possible scores for the data set, and these are followed by results for each marker location. In this example, linkage peaks at location 52.68 with a Z-score of 3.43 (asymptotic p-value of 0.0003), corresponding to a Kong and Cox LOD score of 3.05 with probability 0.00009, where the LOD score is the logarithm of the likelihood of genetic linkage for the region relative to no linkage.

The protocol listing here is only a basic introduction to a very sophisticated software package, which can perform a multitude of analyses of pedigree data. A detailed description of each of these options is beyond the scope of this chapter. Consult the excellent on-line documentation for details on how to perform additional analyses. From the output above, which lists all the available options, one can get a sense of the multiple possibilities.

Of particular note, MERLIN can perform analysis on simulated chromosome data, conditional on the input familial structure and marker spacing. This type of analysis can indicate how many peaks of similar height are expected, conditional on the phenotypes being examined and the available marker map. Additionally, MERLIN has a number of features for detecting genotyping errors. Error detection can significantly improve power in linkage analysis by removing problematic genotypes. Haplotypes can also be reconstructed by information on marker flow through the pedigrees. IBD analysis can be performed to estimate the probability that any two individuals inherited the same chromosome from founders in the pedigree. Although they are not described in detail here, the various functions contained in the MERLIN package can significantly enhance linkage analysis.

USING PLINK TO PERFORM ASSOCIATION ANALYSIS

PLINK is a recently developed, freely available package, which is specifically designed for the analysis of whole-genome association studies (Purcell et al. 2007). The software can perform a wide variety of analyses critical for the interpretation of genome-wide association studies, including analysis of both qualitative and quantitative traits. It has features to help with data management and quality control, and it can implement numerous methods for association analysis.

To perform basic association analysis with PLINK:

1. Download and install PLINK, which is available at http://pngu.mgh.harvard.edu/purcell/plink/. Binaries are available for Linux, Windows, and Mac OS X. Additionally, the source code is available, and one can compile PLINK as necessary.

2. Create appropriate input files for PLINK. Construct a *ped* file as described above. PLINK requires only one column for affection status prior to the columns containing the genotypes. (In PLINK, the default missing phenotype value is '–9'.) Construct a *map* file. PLINK requires that there be exactly one line corresponding to each marker in the *ped* file. The *map* file contains 4 columns: "chromosome," "marker name," "genetic position," and "physical position." In many analyses, the genetic position is not used, and can typically be set to 0 in the *map* file.

3. To run PLINK and check the integrity of the input files, use the following command, which will give a summary statistic:

```
plink --file hapmap1
```

(The input files listed are from an on-line PLINK tutorial and contain a simulated disease variant of rs2222162.)

In this case, the **--file** option implies the *ped* file hapmap1.ped and the *map* file hapmap1.map. In each run of PLINK, there is detailed output to a log file, which gives the options used for the run. For the example given above, the following output is observed:

```
@----------------------------------------------------------@
|       PLINK!       |     v0.99q     |    17/Jan/2007    |
|----------------------------------------------------------|
|  (C) 2007 Shaun Purcell, GNU General Public License, v2  |
|----------------------------------------------------------|
|       http://pngu.mgh.harvard.edu/purcell/plink/         |
@----------------------------------------------------------@

Web-based version check ( --noweb to skip )
Connecting to web...  OK, v0.99q is current

*** Pre-Release Version ***

Writing this text to log file [ plink.log ]
Analysis started: Mon Apr 23 12:15:31 2007

Options in effect:
        --file hapmap1

83534 (of 83534) markers to be included from [ hapmap1.map ]
89 individuals read from [ hapmap1.ped ]
89 individuals with nonmissing phenotypes
Assuming a disease phenotype (1=unaff, 2=aff, 0=miss)
Missing phenotype value is also -9
44 cases, 45 controls and 0 missing
89 males, 0 females, and 0 of unspecified sex
Before frequency and genotyping pruning, there are 83534 SNPs
Applying filters (SNP-major mode)
89 founders and 0 non-founders found
0 of 89 individuals removed for low genotyping ( MIND > 0.1 )
Total genotyping rate in remaining individuals is 0.99441
859 SNPs failed missingness test ( GENO > 0.1 )
16994 SNPs failed frequency test ( MAF < 0.01 )
After frequency and genotyping pruning, there are 65803 SNPs
```

Other Considerations for Using PLINK

1. Because whole-genome association data often have a large number of markers, it is greatly beneficial to make a more compact binary representation of the data than the starting *ped/map* files. The binary representation also loads much faster for subsequent analyses. To do this:

   ```
   plink --file hapmap1 --make-bed --out hapmap1
   ```

   To work with these files, substitute the **--file** command line option with the **--bfile** option. For all commands, use the **--out** option to specify the file name root of the output.

2. A number of different summary statistics can be generated with command line options, including information on missing rates (**--missing** flag) and allele frequency (**--freq**).

3. Perform basic association analysis by adding the **--assoc** command line flag. The basic command is then:

   ```
   plink --bfile hapmap1 --assoc --out as1
   ```

   which generates an output file as1.assoc and starts as follows:

```
CHR        SNP   A1      F_A      F_U   A2      CHISQ          P          OR
  1   rs6681049    1   0.1591   0.2667    2      3.067    0.07991     0.5203
  1   rs4074137    1  0.07955  0.07778    2   0.001919     0.9651      1.025
  1   rs1891905    1   0.4091      0.4    2    0.01527     0.9017      1.038
...
```

The columns in this output are as follows:

- CHR: Chromosome
- SNP: SNP identifier
- A1: Code for allele 1 (the minor, rare allele based on the entire sample frequencies)
- F_A: The frequency of this variant in cases
- F_U: The frequency of this variant in controls
- A2: Code for the other allele
- CHISQ: The chi-squared statistic for this test (1 df)
- P: The asymptotic significance value for this test
- OR: The odds ratio for this test

In a Unix/Linux environment, simply use the available command line tools to sort the list of association statistics and print out the top ten. For example:

```
sort --key=7 -nr as1.assoc | head
```

This gives:

```
13   rs9585021   1   0.625    0.2841    2        20.62    5.586e-06         4.2
 2   rs2222162   1   0.2841   0.6222    2        20.51    5.918e-06       0.2409
 9   rs10810856  1   0.2955   0.04444   2        20.01    7.723e-06        9.016
 2   rs4675607   1   0.1628   0.4778    2        19.93    8.05e-06        0.2125
```

In addition to providing methods for analysis of discrete traits, as listed above, PLINK can analyze association of qualitative traits. As an alternative to reading phenotypes directly from the *ped* file, PLINK can read phenotypes from an external file. This file must specify three columns, the first two being the family and individual identifiers used in the *ped* file. (The individuals do not need to be in the same order as the *ped* file.) A third column contains the phenotypic value, which in this case would be the measurement of the quantitative trait. Then, to run quantitative association analysis, very similar command line options are used as above, because PLINK automatically detects whether the trait being tested is discrete or quantitative. One could then run the analysis as:

```
plink --bfile hapmap1 --assoc --pheno qt.phe --out quant1
```

This analysis would generate output as follows (found in the file with the *.qassoc* suffix).

```
CHR        SNP    NMISS       BETA        SE        R2         T          P
  1   rs6681049      89    -0.2266    0.3626   0.004469   -0.6249     0.5336
  1   rs4074137      89    -0.2949    0.6005   0.002765   -0.4911     0.6246
  1   rs1891905      89    -0.1053    0.3165   0.001272   -0.3328     0.7401
  1   rs9729550      89     0.5402    0.4616   0.0155      1.17       0.2451
```

The fields in this file represent:

- Chromosome
- SNP identifier
- Number of non-missing individuals for this analysis
- Regression coefficient
- Standard error of the coefficient
- The regression r-squared (multiple correlation coefficient)
- t-statistic for regression of phenotype on allele count
- Asymptotic significance value for coefficient

This protocol describes only the basics of association analysis, which is a complex topic. There are numerous additional features that should be exploited as part of analyzing genome-wide association data. PLINK has options for many variations of association analysis, as well as various tests that can correct for population structure. Features exist for checking data and sample quality, as well as looking for excess relatedness.

A graphical user interface (GUI) called gPLINK is being developed which allows a simple interface of common PLINK operations. Menus and dialogs are available to create valid PLINK commands, keep records of command runs, and track input and output files. This interface is still under active development, and details on running it can be found at the PLINK homepage. For more details on how to use PLINK in the analysis of genome-wide association data, refer to the excellent documentation available on-line.

DISCUSSION AND CONCLUSIONS

Although the two genetic analysis packages described above, MERLIN and PLINK, are primarily designed for two different aspects of genetic analysis, they share a number of features. Both are comprehensive packages that are designed to address multiple issues, including error detection, summary statistics, and genetic analysis. Both are also very well documented and open source programs, which facilitates their incorporation into a genetic analysis work flow.

Until very recently, studies designed to unravel genetic contributions to a disease typically began with linkage analysis. Characterized genetic markers did not exist at sufficient density to initiate an association scan at the genome-wide level. With the advent of the HapMap, as well as a number of genotyping platforms that have markers at very high density, it is now possible to initiate genome-wide association scans for markers associated with disease. Studies utilizing this methodology are just beginning to be published. The striking successes with diabetes and prostate cancer (Gudmundsson et al. 2007; Sladek et al. 2007; Yeager et al. 2007) indicate that genome-wide association studies will be an important way to identify novel genes, each with moderate effects, associated with disease.

It is hoped that genome-wide association analysis will lead to great success in unraveling genetic contributions to complex disease. However, many issues can confound the analysis, and these must be taken into consideration during the design and implementation of an association study. Of particular note is population stratification. Since both disease rates and allele frequencies can vary across populations, testing for genetic association within admixed populations can lead to false-positive results. A number of approaches have been developed to control for this issue, which can be done, given the high density of genetic markers. Genomic control postulates that, with population structure, the results of Chi-squared tests are inflated by a constant factor, and all tests can be divided by that factor (Devlin and Roeder 1999). Another approach involves structured association, in which individuals are clustered on the basis of genotypes, and association results are conditioned upon cluster membership (Pritchard et al. 2000a,b). An additional solution is to do family-based tests of association, such as the transmission disequilibrium test (TDT), which are insensitive to population structure (Spielman et al. 1993).

Family-based designs are robust against population admixture, and they offer other advantages over case/control designs. They allow both linkage and association to be tested, and also offer benefits in multiple-hypothesis testing (for review, see Laird and Lange 2006). Despite the clear benefits, a potential drawback is that it can be difficult to collect family-based samples. A comprehensive, user-friendly package of Family-Based Association Tests (FBAT) has been developed that allows the user to test for association/linkage between disease phenotypes and haplotypes by utilizing family-based controls (Laird et al. 2000). This broad class of tests is adjusted for population admixture. The software is well-documented and freely available at http://biosun1.harvard.edu/~fbat/fbat.htm.

Although linkage and association studies rely on different designs and methods to unravel genetic disease, they can often be used in complementary ways. Since linkage studies have coarse resolution, refinement of loci identified by linkage typically relies on association analysis of genes within the loci to determine the gene associated with the trait under study. Although genome-wide association analysis can be performed on unrelated case/control samples, incorporating family-based material can offer clear benefits in the analysis and interpretation of results. Therefore, genetic analysis of human disease will continue to rely on aspects of genetic linkage analysis, as well as association analysis, for the foreseeable future.

REFERENCES

Abecasis G.R. and Wigginton J.E. 2005. Handling marker-marker linkage disequilibrium: Pedigree analysis with clustered markers. *Am. J. Hum. Genet.* **77:** 754–767.

Abecasis G.R., Cherny S.S., Cookson W.O., and Cardon L.R. 2002. Merlin-rapid analysis of dense genetic maps using sparse gene flow trees. *Nat. Genet.* **30:** 97–101.

Cardon L.R. and Bell J.I. 2001. Association study designs for complex diseases. *Nat. Rev. Genet.* **2:** 91–99.

Devlin B. and Roeder K. 1999. Genomic control for association studies. *Biometrics* **55:** 997–1004.

Fingerlin T.E., Boehnke M., and Abecasis G.R. 2004. Increasing the power and efficiency of disease-marker case-control association studies through use of allele-sharing information. *Am. J. Hum. Genet.* **74:** 432–443.

Gudmundsson J., Sulem P., Manolescu A., Amundadottir L.T., Gudbjartsson D., Helgason A., Rafnar T., Bergthorsson J.T., Agnarsson B.A., Baker A., et al. 2007. Genome-wide association study identifies a second prostate cancer susceptibility variant at 8q24. *Nat. Genet.* **39:** 631–637.

International HapMap Consortium. 2005. A haplotype map of the human genome. *Nature* **437:** 1299–1320.

Kong A. and Cox N.J. 1997. Allele-sharing models: LOD scores and accurate linkage tests. *Am J. Hum. Genet.* **61:** 1179–1188.

Laird N.M. and Lange C. 2006. Family-based designs in the age of large-scale gene-association studies. *Nat. Rev. Genet.* **7:** 385–394.

Laird N., Horvath S., and Xu X. 2000. Implementing a unified approach to family based tests of association. *Genet. Epidemiol.* (suppl. 1) **19:** S36–S42.

Lathrop G.M., Lalouel J.M., Julier C., and Ott J. 1984. Strategies for multilocus linkage analysis in humans. *Proc. Natl. Acad. Sci.* **81:** 3443–3446.

Ott J. 1999. *Analysis of human genetic linkage*, 3rd edition. Johns Hopkins University Press, Baltimore, Maryland.

Pritchard J.K., Stephens M., and Donnelly P. 2000a. Inference of population structure using multilocus genotype data. *Genetics* **155:** 945–959.

Pritchard J.K., Stephens M., Rosenberg N.A., and Donnelly P. 2000b. Association mapping in structured populations. *Am. J. Hum. Genet.* **67:** 170–181.

Purcell S., Neale B., Todd-Brown K., Thomas L., Ferreira M.A.R., Bender D., Maller J., de Bakker P.I.W., Daly M.J., and Sham P.C. 2007. PLINK: A toolset for whole-genome association and population-based linkage analysis. *Am. J. Hum. Genet.* (in press).

Sladek R., Rocheleau G., Rung J., Dina C., Shen L., Serre D., Boutin P., Vincent D., Belisle A., Hadjadj S., et al. 2007. A genome-wide association study identifies novel risk loci for type 2 diabetes. *Nature* **445:** 881–885.

Spielman R.S., McGinnis R.E., and Ewens W.J. 1993. Transmission test for linkage disequilibrium: The insulin gene region and insulin-dependent diabetes mellitus (IDDM). *Am. J. Hum. Genet.* **52:** 506–516.

Yeager M. Orr N., Hayes R.B., Jacobs K.B., Kraft P., Wacholder S., Minichiello M.J., Fearnhead P., Yu K., Chatterjee N., et al. 2007. Genome-wide association study of prostate cancer identifies a second risk locus at 8q24. *Nat. Genet.* **39:** 645–649.

WWW RESOURCES

http://biosun1.harvard.edu/~fbat/fbat.htm. FBAT software implements a broad class of Family Based Association Tests, adjusted for population admixture.

http://pngu.mgh.harvard.edu/purcell/plink-PLINK, a free, open-source whole-genome association analysis toolset, designed to perform a range of basic, large-scale analyses in a computationally efficient manner (see Purcell et al. 2007).

http://www.sph.umich.edu/csg/abecasis/MERLIN. MERLIN, a program for analyses of pedigree data. Abecasis Laboratory, Center for Statistical Genetics, University of Michigan.

5 | NCBI dbSNP Database: Content and Searching

Michael L. Feolo and Stephen T. Sherry

National Center for Biotechnology Information, National Library of Medicine, National Institutes of Health, Bethesda, Maryland 20894-3804

INTRODUCTION

The association or linkage of genomic sequence variation with heritable phenotypes is the focus of several very large and highly funded projects worldwide. The International Haplotype Map (HapMap) Project is a partnership of scientists and funding agencies from Canada, China, Japan, Nigeria, the United Kingdom, and the United States to develop a public resource that will help researchers find genes associated with human disease and response to pharmaceuticals (International HapMap Consortium 2005). SeattleSNPs is funded as part of the National Heart, Lung and Blood Institute's (NHLBI) Programs for Genomic Applications (PGA). It is focused on identifying, genotyping, and modeling the associations between single-nucleotide polymorphisms (SNPs) in candidate genes and pathways that underlie inflammatory responses in humans (http://pga.gs.washington.edu). The National Institute of Environmental Health Sciences (NIEHS) SNPs Program at the University of Washington is targeted on the systematic identification and genotyping of SNPs in environmental response genes, as part of the Environmental Genome Project (EGP; http://egp.gs.washington.edu).

The association of genomic sequence variation with heritable phenotypes is also the topic of active research in data management and statistical inference (Cardon and Bell 2001; Carlson et al. 2004; Hirschhorn and Daly 2005). SNPs are among the most common class of genetic variation. When any two copies of a human chromosome are compared, SNPs are observed to occur approximately every 500–1000 base pairs (bp). The National Center for Biotechnology Information (NCBI) dbSNP database (Sherry et al. 2001; www.ncbi.nlm.nih.gov/SNP) is the world's largest repository of variation data, classifying nucleotide sequence variations with the following types and percentage composition of the database: (1) single-nucleotide substitutions, 99.77%; (2) small insertion/deletion polymorphisms, 0.21%; (3) invariant regions of sequence, 0.02%; (4) microsatellite repeats, 0.001%; (5) named variants, <0.001%; and (6) uncharacterized heterozygous assays, <0.001%.

There is no requirement or assumption about minimum allele frequencies or functional neutrality for polymorphisms in the database, and there is no requirement for validation status. Thus, the scope of dbSNP includes disease-causing clinical mutations as well as neutral polymorphisms. In addition to the record identifiers assigned by both the submitter and NCBI, dbSNP entries record the sequence information around the polymorphism and, when available, the observed instances of the polymorphism (alleles), the specific experimental conditions necessary to perform an experiment, descriptions of the population containing the variation, and frequency information by population or individual genotype. In this chapter, you may substitute any class of nucleotide sequence variation for the term SNP.

The current level of activity in the discovery of general sequence variation suggests that relatively common (i.e., >5% minor allele frequency) SNP markers with unknown selective effects will be the majority of submitted records. The genotypes submitted to dbSNP provide validation, frequency, and linkage disequilibrium information used for the selection of SNPs in commercial arrays needed for large-scale genome-wide association studies, as well as for use by smaller projects that need to pick tag-SNP sets for candidate regions (as discussed in Chapters 19 and 20). Although the examples given in this chapter usually reference human data, the data structures and concepts have direct analogs to the nonhuman species in dbSNP, including model organisms (mouse), pets (cat, dog), pests (mosquito), as well as agriculturally significant species such as rice and cattle. The data within dbSNP are available freely in a variety of formats.

The SNP database has four main types of submissions: (1) SNP discovery, (2) allele frequency and/or genotype frequency, (3) individual genotypes, and (4) haplotypes. The formats of the various submission sections are described on-line at http://www.ncbi.nlm.nih.gov/projects/SNP/get_html.cgi?whichHtml=how_to_submit. Each submitter defines the methods in his submission as either the techniques used to assay variation or the techniques used to estimate allele frequencies. Details of the techniques are provided in a free-text description of the method.

SNP DISCOVERY

The essential component of a SNP discovery submission to dbSNP is the nucleotide sequence surrounding the polymorphism used to align the variation feature to genome assemblies and mRNA sequences. The dbSNP accepts submissions as either genomic DNA or cDNA (i.e., sequenced mRNA transcript) sequence. Sequence submissions have a minimum length requirement of 100 bp to maximize the specificity of the sequence in larger contexts, such as a reference genome sequence.

ACCESSIONING AND BUILD CYCLE

Because SNP accession numbers are used ubiquitously in general discussions and manuscripts describing genomic variation, it is worth describing the accessioning process here. Periodical releases of the database are referred to as builds and, in general, dbSNP builds are coordinated to be released with each successive build of an organism's respective genome assembly. When genome assembly activity is infrequent, dbSNP builds are released when a substantial new data set(s) is submitted between assemblies. In 2007, dbSNP moved to a more structured release cycle, with four to six builds per year.

Each new build begins with a "close of data" which defines the set of new submissions that will be aligned to genome sequence by MegaBLAST for subsequent reclustering and annotation. The details of our alignment heuristics are available on-line within the dbSNP FAQ at http://www.ncbi.nlm.nih.gov/books/bv.fcgi?rid=handbook.section.ch5.ch5-s8. The set of new data entering each build typically includes all submissions received since the close of data in the previous build.

SUBMISSION

Each submission of a sequence variation to dbSNP, or *subSNP* in dbSNP jargon, receives "submitted sequence" accession ID, i.e., ss #. These numbers are typically published as a range of numbers in polymorphism discovery articles.

---

dbSNP Builds Create a Nonredundant Set of Clustered Submissions

Each record within a SNP discovery submission is initially given a unique submitted SNP (subSNP) accession number, designated with the prefix "ss" followed by an integer. In the processing of a new build, or build cycle, these new submissions are aligned to one or more genome assemblies and clustered to existing records to create a nonredundant set of variation positions on each target assembly. A cluster is created when two or more subSNPs are determined to colocate to the same set of positions in the mapped assemblies. Each cluster of ss numbers is then given a unique reference SNP (refSNP) accession number, designated with the prefix "rs," followed by an integer.

REFERENCE SNP ACCESSION

A dbSNP rs#, or *refSNP* in dbSNP jargon, defines a unique mapping result within a dbSNP build. SNP submissions that share the exact same set of filtered alignments are clustered into a common rs# record. Although submissions with sufficient flanking sequence complexity are expected to align to a unique location in a genome, there are important exceptions to this rule. Variations in regions of low complexity or duplicated genomic sequence will have ambiguous placement, and variations in regions of alternative haplotypes (e.g., HLA DR51/52/53, or the X-chromosome pseudoautosomal region) will have multiple correct locations, but will still be scored as "unique" in the genome.

A new subSNP that does not cluster with any existing subSNP is given a new rs number. Accession numbers are never reused; however, subSNPs may be withdrawn by the submitter, and as a result, accessions may be inactivated in the next build. Withdrawn submissions may also result in the inactivation of refSNPs in those clusters that contained a single subSNP.

It is important to note that accessioning relies on the mapping of new ss and existing refSNPs, and that the mapping result (set of hits) of these records is in turn dependent on the underlying assembly sequence(s), as well as the algorithm used to do the mapping. RefSNPs can therefore "merge" in a build cycle as a result of assembly sequence changes and/or changes to the mapping algorithm. Whenever refSNPs are merged, the new cluster is designated using the rs number having the lowest integer. Query by the retired number is still supported.

Cluster "Strandedness" Is Defined by Submitted Flanking Sequence, Not the Assembly

Submitters can arbitrarily define variations on either strand of DNA sequence; therefore, the submissions in a refSNP cluster may be individually reported on the forward or reverse strand. The orientation of the refSNP and, hence, its sequence and allele string, are set by a cluster exemplar. By convention, the clustering process picks the submission with the longest sequence as the cluster exemplar. In subsequent builds, a new submission may appear with a longer flank (and thus become a candidate for cluster exemplar). Since submission flanking sequence orientation is arbitrary, there is a 50% chance the new exemplar candidate will be in reverse orientation to the current RefSNP exemplar. When this occurs, we preserve the orientation of the refSNP (in FASTA format for BLAST/BLAT alignment tasks) by using the candidate cluster exemplar flanking sequence in reverse complement.

Once established, an rs# cluster has a stable strand orientation regardless of future submissions. When the clustering process has determined the orientation of all member sequences in a cluster, it gathers a comprehensive set of "observed" alleles for a refSNP cluster normalized for strand orientation.

ANNOTATION OF ASSEMBLY SEQUENCES

The "weight" of an rs cluster is assigned by dbSNP processing. The weight of an rs cluster is a reflection of the number of locations the exemplar ss flanking sequence maps (with high quality) to a genome assembly. Specifically, a refSNP with weight 1 to the human reference assembly has a single high-quality alignment to that assembly; weight 2 is for two map locations on a single chromosome; weight 3 is for two map locations on two different chromosomes; and weight 10 is for three or more locations. The weight of a refSNP is computed separately for each assembly to which it has a set of high-quality alignments. Therefore, an rs cluster may be assigned weight 1 for the reference assembly and higher weight for an alternate assembly.

We annotate weight 1 and weight 2 refSNP variations as variation features with multiple allele qualifiers (one per allele) on NCBI RefSeq chromosomes, contiguous (contig) sequences, mRNAs, and proteins. Weight 2 features receive an additional warning note to indicate the ambiguous nature of a mapping result produced by true paralogy or imperfect sequence assembly. We do not believe that weight 3 and weight 10 variations have sufficient utility to warrant their annotation, but the mapping results for these variations are still available in dbSNP. We annotate NoVariation records on NCBI RefSeq chromosomes, contig sequences, mRNAs, and proteins as a miscellaneous feature, or misc_feat. All dbSNP annotations also include a db_Xref cross-reference pointer back to dbSNP, the refSNP ID (rs) number, and a bitmap-coded summary of the validation and link information for the rs cluster.

GenBank records can be annotated only by their original authors. Therefore, when dbSNP processing finds high-quality hits of refSNP records to the High-Throughput Genome Sequence (HTGS) and nonredundant divisions of GenBank, they are connected using an internal NCBI sequence annotation database. These can be retrieved by selecting the SNP checkbox in the GenBank record display.

FUNCTIONAL ANALYSIS

The dbSNP computes a functional context for sequence variations by inspecting the flanking sequence for gene features during the contig annotation process. Current builds include this processing on RefSeq and GenBank mRNAs. Table 5-1 defines variation functional classes. When a variation is near a transcript or in a transcript interval but not in the coding region, we define the functional class by the position of the variation relative to the structure of the aligned transcript.

TABLE 5-1. dbSNP Variation Functional Classes

Class code	Class ID	Class description
1	Locus region	Variation is within 2 Kb 5′ or 500 bp 3′ of a gene feature (on either strand), but the variation is not in the transcript for the gene. This class is indicated with an L in graphical summaries.
2	Coding	Variation is in the coding region of the gene. This class is assigned if the allele-specific class is unknown. This class is indicated with a C in graphical summaries.
3	Coding-synon	The variation allele is synonymous with the contig codon in a gene. An allele receives this class when substitution and translation of the allele into the codon makes no change to the amino acid specified by the reference sequence. A variation is a synonymous substitution if all alleles are classified as contig reference or coding-synon. This class is indicated with a C in graphical summaries.
4	Coding-nonsynon	The variation allele is nonsynonymous for the contig codon in a gene. An allele receives this class when substitution and translation of the allele into the codon changes the amino acid specified by the reference sequence. A variation is a nonsynonymous substitution if any alleles are classified as coding-nonsynon. This class is indicated with a C or N in graphical summaries.
5	mRNA-UTR	The variation is in the transcript of a gene but not in the coding region of the transcript. This class is indicated by a T in graphical summaries.
6	Intron	The variation is in the intron of a gene but not in the first two or last two bases of the intron. This class is indicated by an L in graphical summaries.
7	Splice-site	The variation is in the first two or last two bases of the intron. This class is indicated by a T in graphical summaries.
8	Contig-reference	The variation allele is identical to the contig nucleotide. Typically, one allele of a variation is the same as the reference genome. The letter used to indicate the variation is C or N, depending on the state of the alternative allele for the variation.
9	Coding-exception	The variation is in the coding region of a gene, but the precise location cannot be resolved because of an error in the alignment of the exon. The class is indicated by a C in graphical summaries.

Most gene features are defined by the location of the variation with respect to transcript exon boundaries. Variations in coding regions, however, have a functional class assigned to each allele because these classes depend on allele sequence.

In other words, a variation may be near a gene (locus region), in an untranslated region (UTR) (mRNA-utr), in an intron, or in a splice site. If the variation is in a coding region, the functional class of the variation depends on how each allele may affect the predicted translated peptide sequence. Because functional classification is assigned by positional and sequence parameters, two facts emerge: (1) If a gene has multiple transcripts because of alternative splicing, a variation may have several different functional relationships to the gene and (2) if multiple genes are densely packed within a contig region, a variation at a single location in the genome may have multiple, potentially different, relationships to its local gene neighbors.

Typically, variation is biallelic—one allele of a variation will be the same as the contig (contig reference), and the other allele will be either a synonymous change or a nonsynonymous change. In cases where more than one allele has been identified, it is possible that one allele will be a synonymous change and the other allele will be a nonsynonymous change. If any allele is a nonsynonymous change, the variation is classified as a nonsynonymous variation. Otherwise, the variation is classified as a synonymous variation. When possible, nsSNPs expressed as protein peptide variants are aligned to 3D structures for the protein.

POPULATION FREQUENCIES

Diploid organisms have two copies of each autosomal chromosome, one inherited from each parent; thus, each person contains two instances (alleles) of each autosomal SNP. If one examines an individual for a SNP with the alleles G and T, one would expect to see one of three possible genotypes: GG, GT, and TT. The alleles of a genetic variation typically exist at different frequencies in different populations. A very common allele in one population may be quite rare (or not present at all) in another population for a variety of reasons. Allelic variants can emerge as private polymorphisms when particular populations have been reproductively isolated from neighboring groups, as is the case with religious isolates or island populations. Estimates of population frequency for each allele of a SNP, as well as individual genotype data (see below), are available in dbSNP. Frequency estimates can be submitted to dbSNP as allele counts or binned frequency intervals, depending on the precision of the experimental method used to make the measurement. The dbSNP contains records of allele frequencies for specific population samples, as defined by each submitter.

Each submitter defines a population (set of samples) as the group used to discover variation, or as the group used to identify population-specific measures of allele frequencies. These samples may overlap in some experimental designs. The dbSNP assigns a population class to the sample based on its geographic origin. These broad categories provide a general framework for organizing the sample descriptions in dbSNP, but are not appropriate for rigorous population genetic analyses. Broad heritage information is specified at the individual level (see below), when possible. Similar to method descriptions, population descriptions require the submitter to provide a Population ID and a free-text description of the sample set.

INDIVIDUAL GENOTYPES

The dbSNP repository was used to provide nonredundant SNP properties and mapping information to the International HapMap Project, which genotyped more than 3.3 million SNPs from dbSNP across 270 individuals. There are, as of this writing, approximately 1.1 billion public genotypes derived from a public–private project for genome-scale genotyping project produced by the International HapMap Consortium in phase 1 and Perlegen in phase 2 (see Fig. 5-1). The other major source of human genotype data is targeted gene resequencing, such as PGA and EGP (both discussed in the Introduction), which have also contributed a substantial number of new SNP discovery records. Approximately 1.2 billion human genotypes have been deposited from all submitters. Genotype data from human samples are expected to grow exponentially in number for the foreseeable future. Publicly available genotypes are useful as known samples when new assays are evaluated and tested, and when information is compiled for the selection of tag-SNPs.

Starting in December 2006, NCBI launched the database of Genotypes and Phenotypes, dbGaP, to distribute phenotype and genotype information submitted from case/control and longitudinal whole-genome association studies, or universal control repositories. These data, unlike the contents of dbSNP mentioned to this point, have been given consent for use in medical research, but not for unrestricted distribution. Therefore, these data are available under a controlled access and distribution mechanism operated by sponsoring NIH programs. dbGaP can be browsed at http://view.ncbi.nlm.nih.gov/dbgap/. Requests for access and downloads of approved data sets are managed within dbGaP's authorized access login system http://view.ncbi.nlm.nih.gov/dbgap-controlled.

DOWNLOADING dbSNP DATA USING FILE TRANSFER PROTOCOL (FTP)

FTP Site Directory Structure

Since build 125, the design of dbSNP has been altered to a "hub and spoke" model, where the dbSNP_Main acts as the hub of a wheel, storing all of the central tables of the resource, and each spoke of the wheel is an organism-specific database that contains the latest data for a specific

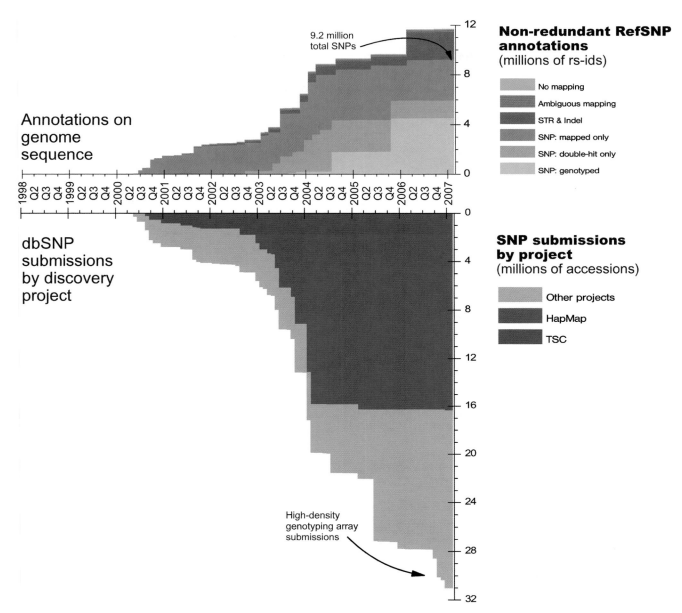

FIGURE 5-1. dbSNP growth for human, 1998 to present. The graph below the timeline shows the number of submissions (scaled as millions of ss#'s) to dbSNP_human since database inception in 1998. Two notable projects: TSC content from "The SNP Consortium," produced from 1999 to 2002, and HapMap content, derived from shotgun resequencing in 2003 and 2004, represent approximately 50% of the submitted content to dbSNP. The balance of submissions for human have been supplied by the rest of dbSNP's submitters. The upper graph shows the growth of nonredundant variations for human with 2.3 million insertion/deletion records and 9.2 million SNPs as of build 127. SNPs are shown in progressively darker shades of blue for those that are genotyped, double-hit (common but not validated), and computationally predicted, respectively.

organism. The main ftp directory for dbSNP contains a number of directories, the most useful of which are organisms/, database/, specs/. The FTP "specs/" directory contains text, .pdf, .asn, and .xsd files on everything from dbSNP submission instructions and build 125 mapping information to genotype resource information and haplotype submission .xsd files. The "organisms/" directory contains a list of the organisms for which dbSNP has data, organized by common name, and followed by its NCBI taxonomy id number. For example, the organism *Homo sapiens* is listed as human_9606. Subdirectories of human_9606 include:

FTP Subdirectory	Comment
ASN1_bin/	ASN1 binary file in doc sum format
ASN1_flat/	ASN text files in doc sum format
XML/	XML files in doc sum format
chr_rpts/	Selected fields from doc sum arranged as tab delimited file
rs_fasta/	Fasta formatted files rs cluster using ss exemplar sequence
submit_format/	Submission examples
genotype_by_gene/	Genotypes across gene regions in genoExchange format
genotype/	Chromosome files in genoExchange format
haplotypes/	Submitted haplotypes in hapExchange format
database/	Pointer to database schema directory. (see contents below)
misc/	Miscellaneous organism-specific files

Of special note is the database/ directory pointer. It contains the schema, data, and SQL statements needed to create the tables and indices for dbSNP:

The ../shared_schema subdirectory contains the schema DDL (SQL Data Definition Language) for dbSNP_main.

The ../shared_data subdirectory contains data housed in the dbSNP_main database that is shared by all organisms.

The ../organism_schema subdirectory contains links to the schema DDL for each organism-specific database.

The ../organism_data subdirectory contains links to the data housed in each organism-specific database.

The data are organized in tables, with one file per table. The file name convention is <table-name>.bcp.gz. The file name convention for the mapping table also includes the dbSNP build ID number and the NCBI genome build ID number. For example, b127_SNPContigLoc_36_2 means that during dbSNP build 127, this SNPContigLoc table has SNPs mapped to NCBI contig build 36 version 2. The data files have one line per table row. Fields of data within each file are tab delimited. dbSNP uses standard SQL DDL to create tables, views for those tables, and indexes. There are many utilities available to generate table/index creation statements from a database.

BROWSING dbSNP CONTENT

The dbSNP provides numerous options for the display and download of its content using a Web browser. Instead of simply listing the data contained in, and links to and from, each dbSNP Web page, we ask the reader to follow the examples given below to explore the range of content in dbSNP.

Searching Using SNP IDs

Using a Web browser is often the easiest way to search the dbSNP database for a specific rs number, ss number, or submitter's SNP identifier. In the following example, we are looking for the Celera variation "hCV1234567."

1. Starting at the dbSNP home page, http://www.ncbi.nlm.nih.gov/SNP/, click type hCV1234567 into the text box under the "Search by Id on All Assemblies" section.
2. Choose "Submitter's SNP ID" option from the drop-down select box.
3. Press the "Search" button to retrieve a list of records matching the query.

In this case, only one SNP will be listed. This query option allows the use of the wild card character *. For example, the query "hcv12345*" means: starts with hcv12345, followed by anything else. Using the wild card query will return 65 records.

Similar searches can be made for rs or ss numbers by choosing the appropriate option in the drop-down select box in #2 above. Entrez SNP, as described below, is an alternative way to search for an rs number. However, it is not for queries using ss or submitter identifiers.

Searching for SNPs in Gene Regions

To search for dbSNP records located in a gene region, begin the search at Entrez gene home page. The following section describes the steps to follow to obtain a list of variations in the human CYP2E1 gene, as well as the display options available using the standard dbSNP Web pages.

1. Starting at Entrez gene home page: http://www.ncbi.nlm.nih.gov/entrez/query.fcgi?db=gene, enter "CYP2E1[gene name] AND Human [orgn]" without the quotes into the text box labeled 'for', and click on the "go" button. This brings you to the summary view of the CYP2E1 gene in Entrez Gene.
2. Click on the "Links" link located on the far right of this view. This displays many resources/view options that are available from this record.
3. Click on the "GeneView in dbSNP" option. This will display a gene-centric dbSNP Web page.
   By default, coding region (cSNP) SNPs are displayed in this view. To show all of the variations in this view, click on the "In gene region" radio button, followed by the "refresh" button.

The GeneView Web page provides users with a detailed view of all the variations that map to a gene region via the NCBI mapping pipeline. The Gene Model section lists one or more refSeq mRNA to Genomic contig mapping results, each referred to here as a gene model. The highlighted gene model determines the graphical display of the gene structure and SNP locations as well as the SNP listing data table that follows. To change the gene model, click "View snp on GeneModel" link located above the data display. The SNP listing is ordered by the genomic contig order, which may be reverse to the gene order and colored by SNP function class. Coding region variations include protein residue and codon position information. A separate table following the gene model data table lists variations mapped to genomic positions flanking the transcript location.

CREATING A LOCAL COPY OF dbSNP

The dbSNP is a relational database that contains hundreds of tables. The content of dbSNP is available via FTP download at: ftp://ftp.ncbi.nih.gov/snp/, as a set of tab-delimited data files (one for each relational database table), entity relationship diagrams (ERDs), and data dictionary. Due to security concerns and vendor endorsement issues, dbSNP does not provide users with direct database dumps. This method of dbSNP data retrieval is most appropriate for large-scale genotyping centers. Core informatics groups in both academic and private sectors may need to incorporate their data with that of dbSNP, or to provide custom query options. The task of creating a local copy of dbSNP can be complicated, and should be left to an experienced programmer. Pitfalls include keeping data synchronized to current dbSNP and/or genomic build. Subsequent schema changes will require code changes and space.

Required Software

Relational Database Software

If you are planning to create a local copy of dbSNP, you must first have a relational database server, such as MySQL, Sybase, Microsoft SQL server, or Oracle. dbSNP at NCBI runs on an MSSQL server version 2000, but users have successfully created their local copy of dbSNP on Oracle, MySQL, and others.

Data-Loading Tool

Loading data from the dbSNP FTP site into a database requires a bulk data-loading tool, which usually comes with a database installation. An example of such a tool is the bcp (bulk-copy) utility that comes with Sybase, or the "bulkinsert" command in the MSSQL server.

Decompression Software

dbSNP uses winzip/gzip to compress/decompress FTP files. Complete instructions on how to uncompress *.gz and *.Z files can be found on the dbSNP FTP site.

Required Hardware

Computer Platforms/OS

Databases can be maintained on any PC, Mac, or UNIX with an Internet connection.

Disk Space

Currently, a complete copy of dbSNP that will include all organisms contained in dbSNP requires 500 GB of space. Depending on the organism you are interested in, you can simply create a local database that only includes data for that organism. Please allow room for growth.

Memory

The current sql server for dbSNP has 4GB of memory.

Internet Connection

We recommend a high-speed connection to download such large database files.

THE dbSNP PHYSICAL MODEL

Understanding the database schema is a necessary part of constructing and maintaining your own copy of dbSNP, because it is a visual representation of dbSNP that shows the logical relationship between data elements in dbSNP. All users should review the file db_DataDictionary.bcp.gz in the ..database/shared_data/ directory as well as the ERD file, erd_dbSNP.pdf, located in ..database/ directory. Auxiliary files are also included with the prefixes: "dn_" derived-content data tables, and "db_" database-specific data tables, such as db_index_sql.bcp.gz.

Outline of the Process of Creating a Local Copy of dbSNP

These instructions assume knowledge of relational databases, and are not intended for the novice user.

1. Create a database structure necessary for your needs. This may include one or more organism-specific databases with or without a shared dbSNP_main database. Tables from dbSNP_main should always be included as part of your local copy.

2. Create tables for each database without constraints or indices.

3. Bulk insert data from .bcp files to tables using database utility.

4. Create indices and constraints for tables.

If you have problems establishing a local copy of dbSNP, please contact snp-admin@ncbi.nlm.nih.gov.

STRUCTURED FILE FORMATS ASN1 AND XML

Data may be downloaded from the dbSNP FTP site from the structured file formats ASN1 binary, ASN1 text, and XML. These files are located in subdirectories of the organism level (i.e., human_9606) FTP directory. As of dbSNP build 125, the DocSum ASN1 format was directly mapped to the DocSum XML which is defined by the XML schema found at ftp://ftp.ncbi.nih.gov/snp/specs/docsum_2005.xsd. Schema versions change periodically and the newest version should be used. Graphical representation of this schema is found at ftp://ftp.ncbi.nih.gov/snp/database/b124/mssql/schema/erd_dbSNP.pdf.

Several flat file formats are also made available at each build, such as the chromosome reports, rs and ss fasta and genome reports (which are also found in subdirectories of the organism).

WEB SERVICE FOR GENOTYPES

We conclude our discussion of dbSNP with a description of how to use the Web services available for querying genotype information. As of build 126, dbSNP genotype information can be queried either directly using a Web browser and the genotype server Web pages, or programmatically by calling the service directly.

Genotype Queries Using a Web Browser

The genotype query form http://www.ncbi.nlm.nih.gov/projects/SNP/snp_gf.cgi was designed to be a simple, interactive way for users to view or download a small set (less than 20,000 genotypes) of genotype and allele frequency records. Individual genotype and allele frequency information is obtained in a three-step process, with the current step shown in bold at the top of the page, followed by one or more action buttons. The first step allows the user to select species and display orientation of the genotypes as reference genome orientation or in the orientation of the rs. The user then has the option of limiting the variations included in the results by genomic region, a list of rs numbers, or a gene region by clicking on the three tabs "Region," "RS Numbers," or "Gene." Three output formats are available for the report: html, tab-delimited text, and xml. The following scenarios demonstrate the use of many of the genotype server options.

Scenario 1: Create a genotype report in rs orientation for two specific SNPs located in the Mouse IL6 gene.

1. Select **Mouse(10090)** for **Species** and Display orientation and **RS**–default, radio button.

2. Click **RS Number** tab and type **rs8259847 rs8259848** in text area.

3. Click the **Next** button to proceed to step 2 Select Populations.

4. Expand the population tree by clicking on the boxed plus signs and check the **ROCHBIO** check box.

5. Click **Download XML** for xml output. You may have to allow the browser to display or download the file explicitly, depending on your browser settings.

Scenario 2: Create a genotype report in genomic orientation for the human ABO gene excluding intron SNPs, limited to populations of European descent.

1. Use default selections for Species **human(9606)** and Display orientation and **Genome**–default, radio button.

2. Click **Gene** tab and type **ABO** in the **Enter the Entrez Symbol or ID of a gene** text box.

3. Check **Weight 1** for **Limit Search To...**

4. Choose **intron** from the **SNP Function Class** select box and click **invert** located below the selections. Clicking **invert** inverts the selections in the select box, and clicking **all** followed by **invert** clears all selections.

5. Click the **Next** button to proceed to step 2 Select Populations.

6. Check **European**. By default, populations are grouped in terms of broad heritage. Scenario 3 demonstrates how to group populations by submitter Handle.

7. Click **Display Results button** for html output.

Scenario 3: Create a genotype report in rs orientation for human chromosome 6 between positions 31,350,000 and 31,430,000, and limit results to the HapMap YRI population.

1. Use default selections for Species **human(9606)** and Display orientation and **Genome**–default, radio button.

2. Click **Region** tab and type **6; 31,350,000;** and **31,430,000** in the **Chromosome, From, and To** text boxes, respectively.

3. Use default selections for SNP Properties section **Weight 1** checked and no selection of **SNP Function Class**.

4. Click the **Next** button to proceed to step 2 Select Populations.

5. Select **Handle...** from the **View population by** select box.

6. Click **boxed plus sign** next to **CSHL-HAPMAP** to expand available populations.

7. Click on **HapMap-YRI** checkbox.

8. Click **Download Text** for text output of the individual genotypes. You may have to allow the browser to display or download the file explicitly, depending on your browser settings.

The genotype server may also be accessed using program or script that can call and interpret http requests. Table 5-2 lists the available http parameters that can be used to modify server requests. Sample Script 1 illustrates a query of dbSNP for refSNP content using NCBI EUtils functions in Perl. Sample Script 2 provides an example of a script that may be used to retrieve genotype and allele frequency reports. Supported return report types are listed in Table 5-3.

TABLE 5-2. HTTP Parameters Recognized by the dbSNP Genotype Web Server

Name	Page	Value
pg	All	0, 1, or 2 for first, second, or third page, respectively.
species	first	The species or tax ID to be searched. This is either the tax ID in an integer form or the name of the species followed by an underscore and the tax id. For example, human_9606 and 9606 are equivalent.
tax_id	first	Same as species. Ignored if species is specified.
RSPick	first	1 if selecting SNPs by RS numbers, 2 if selecting a gene by name or ID, empty otherwise.
pickRS	first	A list of rs number delimited by spaces, commas, or newlines. This is used only if RSPick is 1.
Gene	first	The name or NCBI ID of a gene. This is used only if RSPick is 2.
chr	first	Chromosome. This can be a number or a string such as X, Y, Un, or MT. This is used only if RSPick is not 1.
rng	first	Range of SNPs to be selected, valid values are C and F. F specifies a range and C specifies that the range is centered around an RS number or base pair (bp). This is used only if RSPick is not 1. See below for more details.
rngspec	first	Values are chr_pos or rs. If chr_pos, then the range will be selected by number of bp within the chromosome; otherwise it is a range between two RS numbers or within a specified number of bp of an RS, depending on the value of rng. This is used only if RSPick is not 1.
from	first	This is the position in bp within the chromosome or an RS number for the beginning or center of the range, depending on the value of rng (see above). RS numbers may optionally be preceded with 'rs' and bp numbers may optionally end with K or M for kilo or mega bp. This is used only if RSPick is not 1.
to	first	If rng is F, this is the ending RS number or bp, using the same format as in the from parameter above. If rng is C, this is the number of bp from the centered position specified in from parameter above. This is used only if RSPick is not 1.
weight1	first	1 if selecting weight 1 SNPs only, omitted otherwise. This is used only if RSPick is not 1.
trusnp	first	1 if selecting SNPs only, omitted otherwise. This is used only if RSPick is not 1.
diall	first	1 if selecting diallelic SNPs only, omitted otherwise. This is used only if RSPick is not 1.
snpfnc	first	SNP Function. An integer representing the SNP function. Multiple values can be selected, but they must be separate name=value pairs. For example, snpfnc=2&snpfnc=3&snpfnc=4, etc. This is optional and is used only if RSPick is not 1. Values given in Table 5-1.
founders	second	1 if selecting founders only within a pedigree, omitted otherwise.
pop	second	For Human (9606): Population and type. A string consisting of the integer pop_id from the Population table followed by an exclamation point (!) and the integer population group. The population groups are: 0 - Undefined 1 - African-American 2 - Asian 3 - European 4 - Global 5 - Hispanic 6 - Native American 7 - North African/Middle Eastern 8 - Sub-Saharan African 9 - Tissue For nonhuman species, this is the integer pop_id only. Multiple name=value pairs may be used.
ind	second	For nonhuman only. Submitted_ind_id from the SubInd table. Multiple name=value pairs may be used.
type	second, third	The value is xml for downloading xml, text for tab separated text, and empty or omitted for retrieving html.
reportId	third	A report ID used for further requests until the data are ready. A request with the reportId and pg=2 will retrieve the data if they are ready, otherwise it will retrieve html with a form containing reportId again. When using a browser, this request is repeated every 5 seconds until the data are retrieved.

TABLE 5-2. (Continued.)

Name	Page	Value
api	third	When retrieving data in an automated manner, i.e., from a script, always include this parameter with a value of 1. The response will either be the data or XML with one tag, reportId with a session number to be used for subsequent requests and a single attribute, status. The values for status are: 0 - the data are not ready 1 - the data are ready −1 - the reportId is invalid or expired Subsequent requests should have the following parameters: reportId=*session number retrieved* pg=2 api=1 type=xml or type=text status=1 (optional, see below) The purpose of sending a session number is to prevent the server from timing out when a large request is sent.
status	third	When retrieving data in an automated manner, i.e., from a script, always include this parameter with a value of 1 if you only want to receive a report ID and status. This is recommended in order to avoid parsing the results to determine whether you are receiving the report or only the status.

TABLE 5-3. Efetch Report Parameters for Querying EntrezSNP Using NCBI EFetch

Rettype	Report type description
Brief	Docsum Brief (Entrez default)
FLT	Flat File
ASN1	ASN.1
XML	XML
FASTA	FASTA
RSR	RS Cluster Report
ssexemplar	SS Exemplar List
CHR	Chromosome Report
GENB	Genotype
GEN	Genotype Detail
GENXML	Genotype XML
DocSet	Summary
FREQXML	Frequency XML
uilist	UI (RS) List
MergeStatus	RS Merge Status

The **report** return type codes are available to display EntrezSNP query results. See Sample Script 1 (below) for an example coded in PERL.

Sample Script 1: Using Efetch to Query dbSNP and Return Results

```perl
#!/usr/local/bin/perl -w
# ============================================================================
#
#                              PUBLIC DOMAIN NOTICE
#                 National Center for Biotechnology Information
#
#   This software/database is a "United States Government Work" under the
#   terms of the United States Copyright Act.  It was written as part of
#   the author's official duties as a United States Government employee and
#   thus cannot be copyrighted.  This software/database is freely available
#   to the public for use. The National Library of Medicine and the U.S.
#   Government have not placed any restriction on its use or reproduction.
#
#   Although all reasonable efforts have been taken to ensure the accuracy
#   and reliability of the software and data, the NLM and the U.S.
#   Government do not and cannot warrant the performance or results that
#   may be obtained by using this software or data. The NLM and the U.S.
#   Government disclaim all warranties, express or implied, including
#   warranties of performance, merchantability or fitness for any particular
#   purpose.
#
#   Please cite the author in any work or product based on this material.
#
# ============================================================================
#
# Author:  Michael L. Feolo and Lon Phan
#          Staff Scientists
#          National Center For Biotechnology Information
#          National Library of Medicine
#          National Institutes of Health


#----------------------------------------------------------------------
# EUTIL ACCESS TO dbSNP FOR PROGRAMMATIC DATA RETRIEVAL
#----------------------------------------------------------------------
use LWP::Simple;
my $utils = "http://www.ncbi.nlm.nih.gov/entrez/eutils";
my $db       = 'snp';

#-- Three sample queries. Replace my $term to explore.
# Sample query term to retrieve nsSNPs in a human gene (OXTR)
# my $term  = 'OXTR AND human';

# sample query term to retrieve SNPs in a 2MB interval on chromosome 1
# restricted to SNPs with genotype results (any organism)
#my $term = '8810000:8830000[CHRPOS] AND 1[CHR] AND TRUE[GTYPE] AND Human[ORGAN-
ISM]';

#-- Sample query to retrieve SNPs in human ABO gene with genotypes.

my $term = 'TRUE[GTYPE] AND Human[ORGANISM] AND ABO[GENE]';

# report options uilist, XML, ASN.1, FASTA, etc.
# addition reports and eUtils help are online:
# http://www.ncbi.nlm.nih.gov/projects/SNP/SNPeutils.htm
# Refer to Table 3 to see current list of report type options.

my $report = 'XML';

#-- replace $report with the following values to explore the various types of
reports.

#my $report = 'FASTA';
#my $report = 'uilist';
#my $report = 'GENXML';


# ----------------------------------------------------------------------
# $esearch contains the PATH & parameters for the ESearch call
# $esearch_result contains the result of the ESearch call
```

```
# the results are displayed and parsed into variables
# $Count, $QueryKey, and $WebEnv for later use and then displayed.

my $esearch = "$utils/esearch.fcgi?" .
    "db=$db&retmax=1&usehistory=y&term=$term";

my $esearch_result = get($esearch);

print "\nESEARCH RESULT: $esearch_result\n";

$esearch_result =~
m|<Count>(\d+)</Count>.*<QueryKey>(\d+)</QueryKey>.*<WebEnv>(\S+)</WebEnv>|s;

my $Count    = $1;
my $QueryKey = $2;
my $WebEnv   = $3;

print "Count = $Count; QueryKey = $QueryKey; WebEnv = $WebEnv\n";

# ------------------------------------------------------------------------
# this area defines a loop which will display $retmax citation results from
# Efetch each time the Enter Key is pressed, after a prompt.

my $retstart;
my $retmax=1000;


for($retstart = 0; $retstart < $Count; $retstart += $retmax) {
  my $efetch = "$utils/efetch.fcgi?" .
              "rettype=$report&retmode=text&retstart=$retstart&retmax=$retmax&" .
              "db=$db&query_key=$QueryKey&WebEnv=$WebEnv";

  print "\nQUERY=$efetch\n\n";

  my $efetch_result = get($efetch);

  print "$efetch_result\n\n-----PRESS ENTER!!!-------\n";
  <>;
}
```

Sample Script 2: Query dbSNP Genotype Server Using HTTP Requests

```perl
#!/usr/bin/perl
# =============================================================================
#
#                          PUBLIC DOMAIN NOTICE
#              National Center for Biotechnology Information
#
#  This software/database is a "United States Government Work" under the
#  terms of the United States Copyright Act.  It was written as part of
#  the author's official duties as a United States Government employee and
#  thus cannot be copyrighted.  This software/database is freely available
#  to the public for use. The National Library of Medicine and the U.S.
#  Government have not placed any restriction on its use or reproduction.
#
#  Although all reasonable efforts have been taken to ensure the accuracy
#  and reliability of the software and data, the NLM and the U.S.
#  Government do not and cannot warrant the performance or results that
#  may be obtained by using this software or data. The NLM and the U.S.
#  Government disclaim all warranties, express or implied, including
#  warranties of performance, merchantability or fitness for any particular
#  purpose.
#
#  Please cite the author in any work or product based on this material.
#
# =============================================================================
#
# Author:   Douglas J. Hoffman
#           Contractor
#           National Center For Biotechnology Information
```

```perl
#               National Library of Medicine
#               National Institutes of Health
#
# File Description:
#
#     Example Perl script to retrieve a genotype report in XML
#

use strict 'vars';
use IO::Socket::INET;
use XML::SAX::PurePerl;

package XmlHandler;

sub new
{
  my $class = shift;
  my $self = {};
  bless ($self,$class);
  $self;
}


sub start_document
{
}

sub end_document
{
}

sub start_element
{
  my ($self,$properties) = @_;
    my $name = $properties->{'Name'};
  if($name eq "reportId")
  {
    $self->{inReportId} = 1;
    my $attr = $properties->{'Attributes'};
    my $status = $attr->{'{}status'};
    my $nStatus = $status->{'Value'};
    $self->{status} = $nStatus ? $nStatus : 0;
  }
  elsif ($name =~ m/error/i)
  {
    die("Error in server request");
  }

}

sub end_element
{
  my ($self,$properties) = @_;
    my $name = $properties->{'Name'};
  if($name eq "reportId")
  {
    $self->{inReportId} = undef;
  }
}
sub characters
{
  my ($self,$properties) = @_;
  if($self->{inReportId})
  {
    $self->{reportId} .= $properties->{'Data'};
  }
}
sub entity_reference
{
  my ($self,$properties) = @_;
  my $sChar = '&' . $properties->{'Name'} . ';';
  my $data = {'Data' => $sChar};
```

```perl
      $self->characters($data);
}

sub comment
{
}

sub processing_instruction
{
}

1;

package main;

my $sHost = "www.ncbi.nlm.nih.gov";
my $sPort = 80;
my $sScript = "/projects/SNP/snp_gf.cgi";

sub esc
{
  sub hexFunc
  {
    my $c = shift;
    my $rtn = "%" . sprintf("%02x",ord($c));
    $rtn;
  }

  my $str = shift;
  $str =~ s|[^ \w\./]|&hexFunc($&)|ge;
  $str =~ s/ /+/g;
  $str;
}

sub BuildContents
{
  my $hRequest = shift;
  my $contents = "";
  my $s;
  my $sName;
  my $sValue;
  my $bNotEmpty = undef;
  my @aValues;
  for $s (keys %$hRequest)
  {
    $sName = &esc($s);
    $sValue = $hRequest->{$s};
    if(length($sValue))
    {
      @aValues = split /\t/,$sValue;
    }
    else
    {
      @aValues = ("");
    }
    for $sValue (@aValues)
    {
      $bNotEmpty && ($contents .= '&');
      $bNotEmpty = 1;
      $contents .= $sName;
      $contents .= "=";
      $contents .= &esc($sValue);
    }
  }
  $contents;
}

sub RunRequest
{
  my $hRequest = shift;
  my $contents = &BuildContents($hRequest);
  my $nLen = length($contents);
```

```
      my $sHTTP =
        "POST ${sScript} HTTP/1.0\r\n" .
        "Host: ${sHost}:${sPort}\r\n" .
        "User-Agent: Mozilla/5.0 (Perl Script)\r\n" .
        "Connection: close\r\n" .
        "Content-Type: application/x-www-form-urlencoded\r\n" .
        "Content-Length: ${nLen}\r\n" .
        "\r\n${contents}\r\n";
      my $io = IO::Socket::INET->new
        (PeerHost => $sHost, PeerPort => $sPort, Proto =>"TCP");
      $io || (die "Could not connect to ${sHost}\n");
      $io->print($sHTTP);
      my $bDone = 0;
      my $line = 1;
      my $lines = [];
      while($line)
      {
        $line = $io->getline;
        $line ||
          (die "Did not receive entire header from ${sHost}\n");
        $line =~ s/[\r\n]//g;
        push @$lines,$line;
      }
      ($#$lines > -1) ||
        (die "Did not receive any header from ${sHost}\n");

      $line = $lines->[0];
      ($line =~ m/ 200 OK/) ||
        (die
          "Error in response header - request was as follows:\n\n${sHTTP}\n\n");
      [$io,$lines];
    }

    sub CopyHash
    {
      my $h = shift;
      my $hRtn = {};
      my $k;
      for $k (keys %$h)
      {
        $hRtn->{$k} = $h->{$k};
      }
      $hRtn;
    }

    sub Run
    {
      my $hRequestIn = shift;
      my $nHeaders = shift;
      my $done = 0;
      my $line;
      my $line2;
      my $str;
      my $rq;
      my $hRequest = &CopyHash($hRequestIn);
      my $sType = $hRequest->{type};
      $hRequest->{status} = 1;
      $hRequest->{api} = 1;
      $hRequest->{pg} = 2;
      while (!$done)
      {
        $rq = &RunRequest($hRequest);
        my $xml = new XmlHandler;
        my $sax = new XML::SAX::PurePerl->new(Handler => $xml);
        my $headerLines = $rq->[1];
        my $io = $rq->[0];
        my $contents = "";

        while($line = $io->getline)
        {
          $contents .= $line;
        }
```

```
        $io->close;
        $sax->parse(Source => { String => $contents });
        my $status = $xml->{status};
        my $reportId = $xml->{reportId};
        $hRequest =
        {
         reportId => $reportId,
         api => 1,
         keepAlive => 200,
         type => $sType,
         pg => 2
        };

        if($status > 0)
        {
          $done = 1;
          $rq = &RunRequest($hRequest);
          $io = $rq->[0];
          $headerLines = $rq->[1];
          if($nHeaders)
          {
            local $, = "\r\n";
            local $\ = "\r\n";
            print @$headerLines;
          }
          while($line = $io->getline)
          {
            print $line;
          }
        }
        elsif(!$status)
        {
          $hRequest->{status} = 1; ## explicitly get status only on next loop
          sleep 2;
        }
        else
        {
          print STDERR "Genotype report is not available\n";
          $done = 1;
        }
    }
    0;
}


#
#     three example structures for "sub Run" above
#


my $hExampleRange =
{
 type => "xml",
 species => "human_9606",
 RSPick => "",
 chr => 6,
 rng => "F",
 rng_spec => "chr_pos",
 from => "30000000",
 to =>   "30050030",
 weight1 => 1,
 pop => "904!3\t902!3\t1409!3\t1371!3", ## tab separated list

};


my $hExampleRS =
{
 type => "xml",
 species => "human_9606",
 RSPick => "1",
 RSlist => "8176742 8176740 8176739 8176721 8176720 512770",
```

```
    pop => "904!3\t902!3\t1409!3\t1371!3" ## tab separated list

};


my $hExampleGene =
{
 type => "xml",
 species => "human_9606",
 RSPick => "2",
 Gene => "brca1",
 weight1 => 1,
 pop => "904!3\t902!3\t1409!3\t1371!3" ## tab separated list

};

##
## use one of the 3 examples above
##

&Run($hExampleRange,0);
0;
```

ACKNOWLEDGMENTS

This research was supported, in part, by the Intramural Research Program of the National Institutes of Health, National Library of Medicine.

REFERENCES

Cardon L.R. and Bell J.I. 2001. Association study designs for complex diseases. *Nat. Rev. Genet.* **2:** 91–99.

Carlson C.S., Eberle M.A., Kruglyak L., and Nickerson D.A. 2004. Mapping complex disease loci in whole-genome association studies. *Nature* **429:** 446–452.

Hirschhorn J.N. and Daly M.J. 2005. Genome-wide association studies for common diseases and complex traits. *Nat. Rev. Genet.* **6:** 95–108.

International HapMap Consortium. 2005. A haplotype map of the human genome. *Nature* **437:** 1299–1320. [http://www.hapmap.org]

Sherry S.T., Ward M.H., Kholodov M., Baker J., Phan L., Smigielski E.M., and Sirotkin K. 2001. dbSNP: The NCBI database of genetic variation. *Nucleic Acids Res.* **29:** 308–311.

WWW RESOURCES

http://pga.gs.washington.edu/ The SeattleSNPs PGA is focused on identifying, genotyping, and modeling the associations between single-nucleotide polymorphisms (SNPs) in candidate genes and pathways that underlie inflammatory responses in humans. SeattleSNPs is funded as part of the National Heart, Lung, and Blood Institute's (NHLBI) Programs for Genomic Applications (PGA).

http://www.ncbi.nlm.nih.gov/projects/SNP/get_html.cgi?which Html=how_to_submit NCBI database of single nucleotide polymorphisms (SNPs). Guidelines for dbSNP submission process.

http://www.ncbi.nlm.nih.gov/books/bv.fcgi?rid=handbook.section.ch5.ch5-s8 The Single Nucleotide Polymorphism Database (dbSNP) of Nucleotide Sequence Variation.

6 Using the HapMap Web Site

Albert Vernon Smith

Cold Spring Harbor Laboratory, Cold Spring Harbor, New York 11724; Genthor ehf., 101 Reykjavik, Iceland, and Icelandic Heart Association, 201 Kopavogur, Iceland

INTRODUCTION

The primary goal of the International Haplotype Map Project (International HapMap Consortium 2005) has been to develop a haplotype map of the human genome that describes the common patterns of genetic variation, in order to accelerate the search for the genetic causes of human disease. Within the project, approximately 3.9 million distinct single-nucleotide polymorphisms (SNPs) have been genotyped in 270 individuals from four worldwide populations. The project data are available for unrestricted public use at the HapMap Web site, http://www.hapmap.org (Thorisson et al. 2005). This site, which is the primary portal to genotype data produced by the project, offers bulk downloads of the data set, as well as interactive data browsing and analysis tools that are not available elsewhere.

This chapter describes the Web site and tools that have been developed for viewing, retrieving, and analyzing the project data. Details are provided on how to perform several useful and popular tasks. Protocols include instructions for retrieving genotype and frequency data, picking tag-SNPs for use in genetic association analysis, viewing haplotypes graphically, downloading phased genotype data, and examining marker-to-marker linkage disequilibrium (LD) patterns.

Browsing HapMap Data Using the Genome Browser

Research into the genetic contributions to a human disease commonly focuses on candidate genes identified from linkage and/or association studies, as well as from pathways suspected to be involved in a particular disease process. In studying candidate genes, a researcher will want to know whether there are any common SNPs in the immediate vicinity, what those SNPs' alleles are, and the relative frequencies of the alleles in the population. The researcher will also be particularly interested in coding SNPs, whose alleles change the amino acid sequence of the gene product and therefore might represent functional variations.

METHODS

Finding and Browsing to a Region of Interest

The genome browser at the HapMap Web site provides access to small to medium sized-regions of the genome for this type of interactive exploration. This basic protocol shows how to start using the genome browser.

1. Using any modern Web browser, go to www.hapmap.org.

2. Click the "Browse Project Data" link under the "Project Data" section of the hapmap.org homepage. This will take you to a genome browser based on the GBrowse package (Fig. 6-1).

3. Locate the "Landmark or Region" search box, and enter a search term. Any of the following types of search terms will work:

   - a chromosome name (e.g., "Chr19")

   - a chromosomal position in the format Chromosome:start..stop (e.g., "Chr10: 25000..300000")

   - the name of a SNP using its dbSNP "rs" name (e.g., "rs6870660")

   - a gene using its NCBI RefSeq accession number (e.g., "NM 153254")

   - a gene using its common name (e.g., "BRCA2")

   - a chromosomal band (e.g., "5q31")

4. After entering one of these landmarks, press the "Search" button (or hit "Enter"). This will return a page showing the region surrounding the requested feature (Fig. 6-2). If multiple features match, then the page will show a graphical summary (including genomic location) of all possible features and prompt you to choose one.

   By default, the genome browser goes to the most recent release of HapMap data. Previous releases are available via this interface, and the different releases can be selected under the "Data Source" menu.

   i. At the top of the returned page is an "Overview" section that shows the cytogenetic map of the selected chromosome. A red box indicates the section of the chromosome in view.

   ii. Below this is a Region overview, displaying 2 Mb surrounding the region of interest. Again, a red box indicates the section of chromosome.

FIGURE 6-1. The initial page shown when starting to use the HapMap genome browser for the first time. Depending on your computer language settings, this page can appear in one of several languages, although this section assumes English. The page can also be reached directly at http://www.hapmap.org/cgi-perl/gbrowse/.

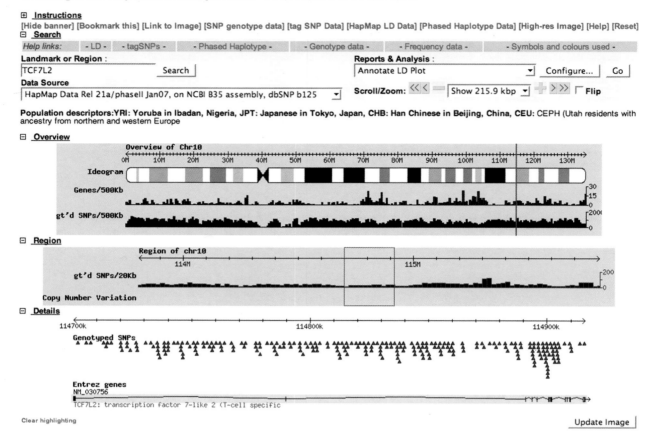

Showing 215.9 kbp from chr10, positions 114,700,200 to 114,916,057

FIGURE 6-2. The HapMap genome browser displaying a requested feature.

iii. Beneath this is a "Detail" section that has horizontal tracks showing various types of data. By default, only a small number of genomic tracks are displayed initially for the region. The two most useful tracks are the "Genotyped SNPs" track that provides information on the position, alleles, and allele frequencies of each SNP characterized by the HapMap project, and the Entrez genes track, which shows the positions and structures of human protein-coding genes.

A number of additional information tracks are available, which can particularly help with the understanding and design of association studies. A number of analyses derived from HapMap data, as well as outside data sources, are available (Table 6-1). Particularly noteworthy are a number of tracks related to structural variation in the genome, as well as links to the Reactome database (http://www.reactome.org;Vastrik et al. 2007), a curated resource of core pathways and reactions in human biology.

5. Use the controls at the top of the page to scroll left, right, or to change the magnification of the region. You can also click anywhere on the "Overview," "Region," or the scale at the top of the "Details" section in order to center the view on this position. The genotyped SNP track changes its appearance in a manner appropriate to the scale of the image:

i. At low magnifications, genotyped SNPs appear as equilateral triangles.

These colors can be customized by selecting the "Highlight SNP Properties" item in the "Reports and Analysis" menu.

TABLE 6-1. Available Tracks in the Genome Browser (as of February 2007)

Category	Track
HapMap tools	LD plot
	Phased haplotype display
	tag-SNP picker
Genes	Ensembl genes (Hubbard et al. 2007)
	Entrez genes (Wheeler et al. 2007)
Pathways	Reactome pathways (Vastrik et al. 2007)
Structural variation	Copy number variation (CNV) data sets (Iafrate et al. 2004; Sebat et al. 2004; Sharp et al. 2005; Tuzun et al. 2005)
	CNVs typed in HapMap samples (Redon et al. 2006)
	Deletions (Conrad et al. 2006; Hinds et al. 2006; McCarroll et al. 2006)
Variation	dbSNP SNPs (Wheeler et al. 2007)
	Heterozygosity/1kb
	Recombination rate (cM/Mb)
	SNP coverage/1kb genotyped SNPs
	Recombination hot spots
	Sequence tagged sites (Wheeler et al. 2007)
	Fit r^2 in genomic intervals (Smith et al. 2005)

    ii. At higher magnifications, the genotyped SNPs change to display the alleles associated with the SNP. The allele shown in blue is the allele present in the reference genomic sequence at that location, and the red allele is the other allele present in the SNP.

    iii. When zoomed in still further, the genotyped SNPs track changes to show pie charts representing the allele frequency for each genotyped population. The blue wedge of the pie chart indicates the frequency of the allele that appears in the reference genome sequence. The red wedge is the frequency of the alternative allele.

The pie chart display provides the researcher with the ability to easily distinguish SNPs that are highly polymorphic in all four of the HapMap populations and, therefore, more likely to be polymorphic in other populations as well. Alternatively, the researcher can identify SNPs that are more polymorphic in a single population and are therefore suitable as markers in population-specific genetic screens.

6. Click on the glyph for an individual SNP to see a text-based page with detailed genotype and allele counts, and assay information.

This provides the researcher with the information needed to generate an assay for the SNP, including the left and right flanking sequences needed to create PCR primers.

    i. Click on the hypertext link to dbSNP (http://www.ncbi.nlm.nih.gov/SNP; Wheeler et al. 2007) for more information about how the SNP was first discovered and any other population genetic information that may exist for it outside the HapMap project.

    ii. Click on the link to Ensembl (http://www.ensembl.org; Hubbard et al. 2007) to reach a site where the structural impact of the SNP on coding sequence, splice sites, and other features of nearby genes can be examined.

Viewing the Extent of Linkage Disequilibrium

When a researcher designs a study to detect the association between a common allelic variation of a gene and a disease of interest, knowledge of the extent of LD in the region is essential for reducing the number of SNPs that need to be genotyped across the region. If there is high LD in the region, then only a few SNPs need to be genotyped because their linkage to other SNPs in the region will serve as proxies for the genotypes of non-characterized SNPs. In contrast, a region of low LD will need to be sampled more heavily because the allelic state of a genotyped SNP will be a poor predictor of the state of non-genotyped SNPs. The determination of patterns of LD in the populations characterized by the HapMap project has been one of the major goals of this project. The International HapMap Project has precalculated patterns of LD among the genotyped SNPs. The data can be downloaded in bulk from the HapMap Web site or browsed interactively using the HapMap genome browser. The latter method allows researchers to see patterns of LD in context with the distribution of genes of interest.

7. To view available LD data precalculated from HapMap genotypes, browse to a region of interest (see Steps 1–4).

8. Select the "Annotate LD plot" plug-in from the "Reports and Analysis" menu.

9. Click the "Configure" button to bring up a configuration page that will allow you to adjust the display properties to your liking.

   Key parameters on this page are the HapMap populations to display, which measure of LD to use (choice of D', r², or LOD), whether the triangle plot should be oriented with the vertex pointing upward or downward, color scheme, and whether the box size in the plot should be proportional to genomic distance between markers or of uniform size (see Fig. 6-3).

10. Click on the "Configure" button to return to the main display. The display will now show one triangle plot for each population selected (see Fig. 6-4).

    In regions with many genotyped SNPs, the LD plug-in adds significantly to the time it takes for the Web page to load. You can turn off the LD display at any time by deselecting the appropriate checkbox in the "Tracks" section of the browser. The LD plug-in settings are stored in a browser cookie, so there is no need to visit the configuration page each time the plug-in is turned on.

11. The traditional D' and r² metrics reflect the degree of pair-wise LD between two SNPs, but differ in their sensitivity and specificity across different size scales. See Mueller (2004) for a discussion of the practical application of these measurements. The LOD metric used in the HapMap Web site display is described in Daly et al. (2001).

Picking and Viewing tag-SNPs

tag-SNPs are a reduced set of SNPs that capture much of the LD in regions; they can be used in association studies to reduce the number of SNPs needed to detect LD-based association between a trait of interest and a region of the genome. For small regions, it is possible to select tag-SNPs by hand using the graphical and numeric displays of LD generated above, but for best results, it is recommended that the researcher use an algorithm that chooses tag-SNPs by formally maximizing the number of linked SNPs captured by the tag set. There is no single set of tag-SNPs that will satisfy the diverse requirements of every association study design. Researchers may wish to select SNPs that work well with a particular genotyping system (for example, those

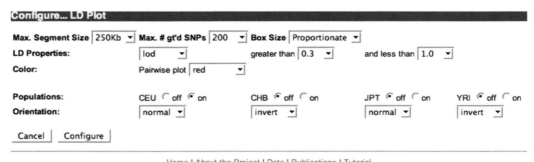

FIGURE 6-3. The configuration page of the HapMap genome browser allows the user to customize numerous style features of the data display.

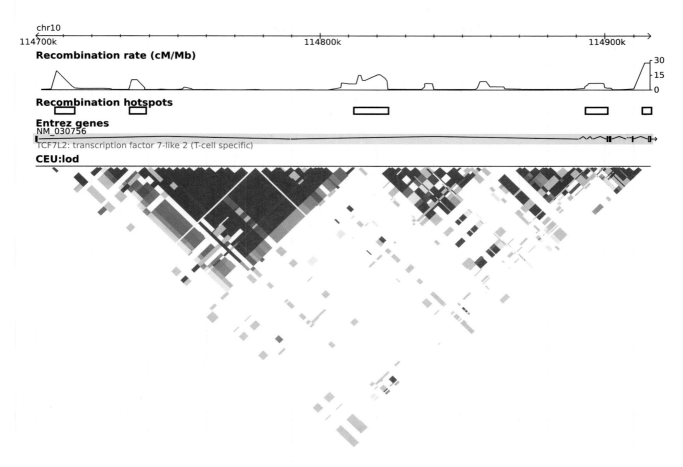

FIGURE 6-4. The HapMap genome browser displaying a triangle plot of LD values for multiple populations. A typical region of LD demonstrating "patches" of high LD separated by relatively well-defined boundaries of low LD is shown. The triangle plot is constructed by connecting every pair of SNPs along lines at 45 degrees to the horizontal track line. The color of the diamond at the position where two SNPs intersect indicates the amount of LD; more intense colors indicate higher LD. A gray diamond indicates that data are missing.

that have been included on a particular "SNP chip") and may be willing to accept different tradeoffs between the cost of genotyping a study population and the strength of the association they can detect. For this reason, the HapMap Web site does not offer a static set of preselected tag-SNPs, but instead offers researchers a tool for interactively selecting tag-SNPs based on user-provided criteria. The tag-SNP lists are generated from algorithms in the Tagger program (http://www.broad.mit.edu/mpg/tagger/; de Bakker et al. 2005).

12. Navigate to a region of interest (see Steps 1–4).

13. Under the "Reports and Analysis" menu, select the "Annotate tag SNP Picker" option.

14. Press "Configure" to select the desired options for tag-SNP selection (see Fig. 6-5). Options include:

- selecting a population and an algorithm
- uploading a list of SNP IDs to be included in the set of tag-SNPs
- uploading a list of SNP IDs to be excluded from the set of tag-SNPs
- uploading a list of design scores (priorities) for each SNP
- selecting cutoffs for minimum acceptable LD value and allele frequency for SNPs to be included in the set

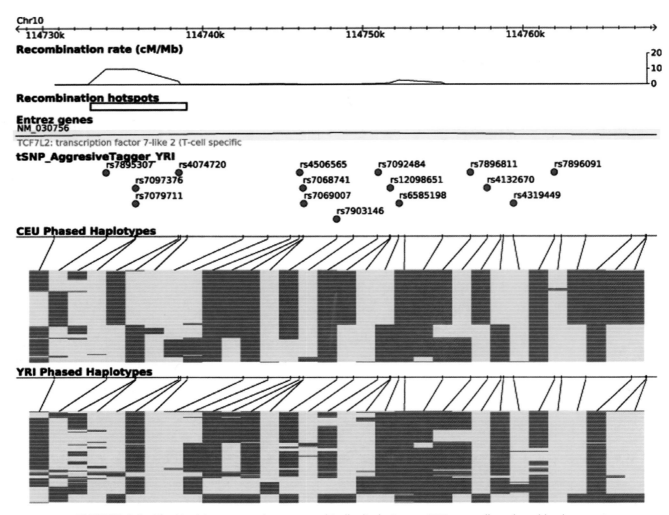

FIGURE 6-5. The HapMap genome browser graphically displaying tag-SNPs, as well as phased haplotypes.

15. Click the "Configure" button to run the analysis and return to the main display. Results are shown on a new feature track (see Fig. 6-5).

As with the LD display above, settings are stored in a browser cookie, and the plug-in track can be turned off when it is not needed.

Viewing Phased Haplotypes

A researcher may wish to correlate the tag-SNP set selected by the tag-SNP picker algorithm with the underlying haplotype structure of the region. One way to do this is to turn both the pair-wise LD and tag-SNP tracks on simultaneously (Steps 7–11 and 12–15, respectively). An alternative, however, is to activate a track that displays the phased haplotypes themselves. The phased haplotype data described in this section were generated by the International HapMap Project Consortium using the program PHASE version 2.1 (Stephens and Donnelly 2003). During phasing, each allele in a genotype is assigned to one or the other parental chromosome, using a maximum likelihood algorithm that uses trio (lineage) information in the HapMap population groups, or, if trio information is not available, by fitting the data to a model that minimizes the number of implied historical crossovers in the population. The phased haplotypes are displayed as a graphic in which each chromosome of the individuals sampled by the project is represented as a line one pixel high, and each SNP allele is arbitrarily colored blue or yellow. A region of high LD will appear as a region in which there are long runs of SNPs that share alleles across multiple chromosomes, indicating that there is little recombination among them. A region of low LD will appear as an area where the runs are shorter and more fragmentary.

16. Navigate to a region of interest (see Steps 1–4).

17. Select "Annotate Phased Haplotype Display" from the "Reports and Analysis" menu.

18. Press "Configure" to set options for Haplotype display.

The options give you the ability to select the population for which to display haplotype information.

19. After selecting the desired population(s), click the "Configure" button to return to the main display. A new feature track will appear for each population selected. Each track shows the haplotypes for that population using the two-color scheme described earlier (Fig. 6-5).

The order of chromosomes is determined by a fast hierarchical clustering methodology, which places chromosomes that share similar haplotypes together.

The advantage of this display over the pair-wise LD "triangle display" is that it is more compact and therefore better suited for the display of large regions. This makes it easy to correlate the position of long common haplotypes with SNPs chosen by the tag-SNP picker. The disadvantage of this display is that it conceals much of the fine structure of LD in the region; in particular, strong linkage disequilibrium among SNPs that are not adjacent to one another.

20. To retrieve the detailed phased genotypes, click on the track of the desired population. This will take you to a page that provides the haplotype information in tabular form. Each row of the table is an individual chromosome, and each column is an individual SNP. The background of each table entry is set to a color corresponding to that seen in the graphical track.

Generating Text Reports Using the Genome Browser

In many cases, a researcher will be interested in downloading HapMap data from a region of interest for local analysis. The Web site allows direct download of genotype, frequency, tag-SNPs, and other reports from the genome browser.

METHODS

Generating a Text Listing of Genotypes

After a researcher has browsed a region graphically and has centered his view on a candidate gene and the region surrounding it, he may want to generate a space-delimited text dump of the genotyping results across this region. These data can then be imported into an Excel spreadsheet or other data analysis tools.

1. Navigate to a region of interest (see Protocol 1, Steps 1–4).

2. Select "Download SNP genotype data" from the "Reports and Analysis" menu.

3. Click the "Configure" button.

   This will open a configuration page that allows you to select the desired HapMap population, and whether to save the data to disk or view it in the Web browser.

4. Choose the desired options. Click the "Go" button to retrieve the data and produce a report. The text-dump format consists of a set of rows containing each SNP's dbSNP ID, the two reference and alternative alleles, the position of the SNP on the genome, and the genotypes of the SNP on each of the individuals in the selected HapMap population. Because this format is identical to that used by the bulk downloadable files (see Protocol 5), it can be easily loaded into the HaploView program (http://www.broad.mit.edu/mpg/haploview/; Barrett et al. 2005) for detailed analysis on the researcher's local computer.

   The dumper configuration settings are stored in a browser cookie, so next time you can click the "Go" button on the main page and dump data directly, without having to configure the dumper first.

Generating a Text Listing of Genotype Frequencies

In a similar fashion, the researcher may wish to download a summary of the frequencies of alleles across a region of interest. The researcher can then select from this set those SNPs that meet certain criteria; for instance, those that are most highly polymorphic in a particular population of interest. This protocol describes how to create a tab-delimited summary of HapMap allele frequency data across a particular genomic region.

5. Navigate to a region of interest (see Protocol 1, Steps 1–4).

6. Select "Download SNP Genotype Frequency Data" from the "Reports and Analysis" menu.

7. Click the "Configure" button.

8. Click the "Go" button to retrieve the data and produce a report. The report generated by this section contains one row for each SNP consisting of each SNP's dbSNP ID, its genomic position, the number of times each possible genotype was seen in the selected population, and the heterozygosity of the population for that SNP.

Generating a Text Listing of Linkage Disequilibrium Values

After selecting a region of interest and visually inspecting the extent of LD across the gene or genes of interest, the researcher may wish to download a tab-delimited numeric summary of the LD values in the selected region. This information can be used to select a set of "tag" SNPs that will act as proxies for other SNPs that are in high LD with them.

9. Navigate to a region of interest (see Protocol 1, Steps 1–4).

10. Select "Download HapMap LD Data" from the "Reports and Analysis" menu.

11. Click the "Configure" button.

12. Click the "Go" button to retrieve the data and produce a report. The report generated by this protocol will show pair-wise LD between all SNPs within 250 kb of each other. Each row of the report corresponds to one pair of SNPs. The first two columns indicate the positions of the SNPs on the chromosome, the third column is the population for which the LD values were calculated, and the fourth and fifth columns indicate the dbSNP IDs for the SNP pair. These are followed by the D′, r^2, and LOD scores for LD between the two SNPs.

Generating a Text Listing of tag-SNPs

Using the interactive tag-SNP selection track described in Protocol 1, Steps 12–15, researchers can adjust selection criteria until satisfied with the characteristics of the tag set. This protocol describes how to generate a text dump of the SNPs in this set so that they can be used to create a screening set when combined with information from other HapMap-generated reports.

13. Navigate to a region of interest (see Protocol 1, Steps 1–4).

14. Select "Download tag SNP Data" from the "Reports and Analysis" menu.

15. Click the "Configure" button to set up the interactive options for tag-SNP selection as described in Protocol 1, Steps 12–15. Options include:
    - selecting a population and an algorithm
    - uploading a list of SNPs to necessarily be included as a tag-SNP
    - uploading a list of SNPs to be excluded from the tag-SNP list
    - setting cutoffs for LD and allele frequency values

16. Click the "Go" button to retrieve the data and produce a report. The generated report contains a tab-delimited list of tag-SNP names, chromosome, position, and allele frequency in the region. This is followed by a section that lists the tag-SNPs, the non-tag-SNPs that they capture, and the strength of LD between each tag-SNP and the non-tag-SNPs it captures.

Manipulating HapMap Data Using HaploView

Advanced users who wish to exercise finer control over the display of regions of high LD or who wish to experiment with new algorithms for tag-SNP picking may wish to analyze HapMap data using the HaploView program (Barrett et al. 2005). This program works well in combination with the HapMap genome browser. A big advantage of HaploView over the HapMap Web site genome browser is that it displays simultaneous high- and low-power views of regions of LD, and gives immediate feedback during scrolling and zooming operations. Additional information on using HaploView can be found in Chapter 21.

METHOD

1. To install HaploView, go to http://www.broad.mit.edu/mpg/haploview/ and follow the download link.

    HaploView requires that the Java Runtime Environment (JRE) is installed on the local computer. If the JRE is not installed, the newest version can be found at http://www.java.com.

2. Download the HaploView program appropriate for your operating system:

    i. For Windows computers, download the Windows installer file. Double clicking the installer file will create a HaploView folder on the Start Menu.

    ii. For MacOS X and Unix, download the HaploView.jar file.

3. Download genotypes from a region of interest using Protocol 2, Steps 1–4.

4. To start HaploView, open the Haploview.jar file.

    i. On Windows computers, open Haploview.jar from the HaploView folder on the Start Menu.

    ii. On other operating systems, double-click on the Haploview.jar file.

5. In the HaploView welcome window, click on the "Load HapMap Data" button to load genotypes.

6. Browse to the downloaded file containing genotypes. Open the file.

    Once the data are loaded, HaploView provides you with options to view a high-resolution "triangle plot" of LD across the region, view shared haplotypes and their recombination frequencies, and select tag-SNP sets by a number of methods.

7. Select among these analyses and visualizations by choosing the appropriately labeled tab along the top of the HaploView window.

8. Additionally, it is possible to add genome tracks from the HapMap Web site directly into the HaploView display, including Entrez gene tracks and recombination rates.

9. The "Display" and "Analysis" menus allow you to change the size and coloring scheme of the LD triangle plot, as well as to select among a variety of algorithms for defining "haplotype blocks," regions of SNPs that are in high mutual LD.

Protocol 4

Retrieving HapMap Data Using HapMart

Because of performance considerations, interactive access to HapMap data via the genome browser is limited to regions no more than 5 Mb wide. Researchers who wish to obtain data for chromosome- or genome-wide data have two choices: bulk download or HapMart access. HapMart, described in this section, allows researchers to select SNPs using diverse criteria and to display just those aspects of the data set that they are interested in.

METHOD

1. Go to the MartView interface at http://www.hapmap.org/BioMart/martview. Click the "next" button to start a new query using the default database and data set.

2. On the filter page (see Fig. 6-6), select SNPs to be retrieved on the basis of numerous criteria (either individually or in combination). You can apply filters in any order and revise existing filters by using the "next" and "back" buttons. As you apply filters, the number of SNPs selected is shown in the summary panel on the right. Available filters include:
   - lists of SNPs to be included or excluded
   - the minimum minor allele frequency of the selected SNPs
   - limit SNPs to those in Intronic Regions, mRNA/UTR regions, coding nonsynonymous SNPs, or coding synonymous SNPs
   - limit SNPs to a specific genomic region
   - limit SNPs to those that overlap specific gene ID(s)

3. After selecting and refining the appropriate filters, click the "next" button to go to the output selection page. This page allows you to select the fields you wish to be output in the report. There are many output options, organized as a series of tabs along the top of the screen. To retrieve the information of interest, select the appropriate tab at the top of the page, then select the appropriate checkbox. Output options include:
   - genotypes
   - SNP chromosome position
   - alleles
   - genotype frequencies
   - allele frequencies

4. Optionally, if the number of SNPs to be retrieved is large, you can select "gzip file compression." This will compress the report prior to sending it to your browser, and can reduce the time it takes to download the report.

5. To retrieve the results, select the "export" button.

   The reports generated by HapMart are in a tab-delimited text format suitable for importation into Excel or for loading into a relational database. The engine underlying HapMart is a generic data-mining framework named BioMart.

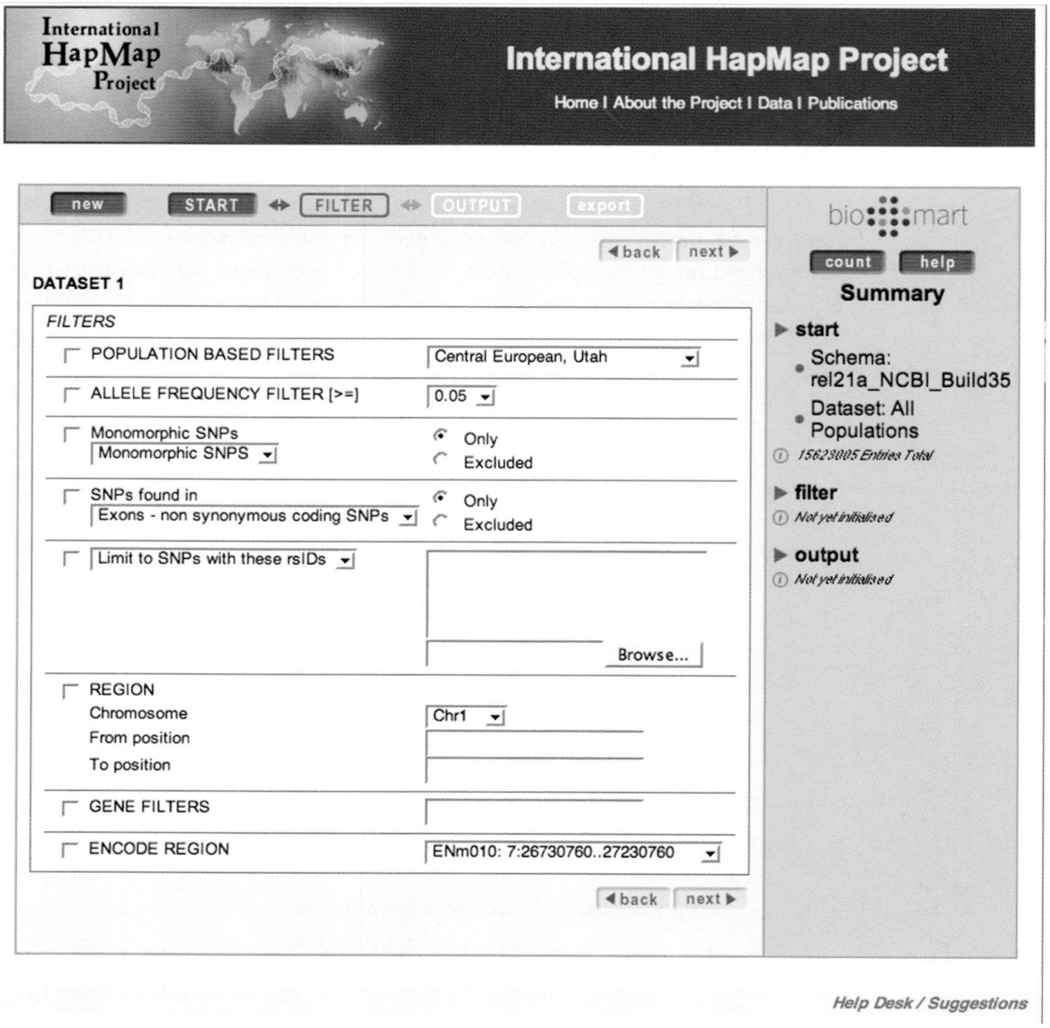

FIGURE 6-6. The HapMart interface is good for filtering HapMap data on specific criteria.

Retrieving Data Via Bulk Download

Bulk downloads of chromosome- or genome-wide data provide text dumps of the entire HapMap data set. Although complete, such downloads do not provide any filtering or selection services.

METHOD

1. To access bulk data, browse to the download page (http://www.hapmap.org/downloads/; the download repository can also be accessed via anonymous FTP at ftp://www.hapmap.org). Datafiles are split by chromosome/region and population, and come in three varieties, each within its own subdirectory:

   - non-redundant/: These data sets contain only one set of genotypes per SNP/population. All genotype sets have passed quality control checks, and multiple submissions of the same SNP (occurring because of QA exercises, corrected submissions, and planned redundancy in the project) have been removed. This is the data set most users will want.
   - redundant-filtered/: All these data sets have passed quality control checks, but redundant data have not been removed.
   - redundant-unfiltered/: This data set contains all SNPs genotyped by the project irrespective of quality control checks. This constitutes the "raw" data for users who want to look at potentially biologically interesting data that are normally filtered out by project quality control checks.

2. Select the appropriate link from the list of available bulk data downloads.

   i. Click the "Genotypes" link to go to the genotypes download directory. The "latest/" subdirectory always points to the current data freeze.

   ii. Click the "LD Data" link to go to the LD data download directory. The "latest/" directory again points to the most recent data freeze available. LD values are represented as D′, LOD, and r^2 values.

   iii. To download phased genotypes, click the "Phasing Data" link. This will take you to a directory containing data files representing the output of the PHASE program.

3. In addition to the bulk data sets listed above, a number of other bulk downloads are available. Calculated Frequency values are available for SNPs typed in the project, as well as raw signal intensity data from some of the genotypes (only from Affymetrix GeneChip 100k and 500k Mapping Arrays, at present). Downloads of recombination rate data are also available, as are protocols used in the project, as well as details on the samples genotyped in the project.

DISCUSSION

A number of public on-line resources have been developed as portals to high-volume genome-wide data sets. The UCSC Genome Browser (http://genome.ucsc.edu; Kent et al. 2002) and the EnsEMBL project (http://www.ensembl.org; Hubbard 2007) have developed multispecies genome browsers that display genomic annotations graphically and offer retrieval of the underlying data. dbSNP (Wheeler et al. 2007) is a repository for information on single-nucleotide polymorphisms, but does not yet contain extensive information on the relationships among those SNPs.

The HapMap Web site, located at http://www.hapmap.org, has a distinct focus. It aims to be a resource in the display, retrieval, and analysis of high-throughput, high-quality, genome-wide human genetic data, with an emphasis on the support of tools for facilitating disease association studies. Although the resource is still in development, it currently provides the basic tools for visualizing patterns of common polymorphism among the populations surveyed by the HapMap project, selecting tag-SNP sets based on a variety of criteria, and generating customized extracts of the data set. In the future, the HapMap Web site will evolve to provide more services to those designing and interpreting genetic association studies.

REFERENCES

Barrett J.C., Fry B., Maller J., and Daly M.J. 2005. HaploView: Analysis and visualization of LD and haplotype maps. *Bioinformatics* **21**: 263–265.

Conrad D.F., Andrews T.D., Carter N.P., Hurles M.E., and Pritchard J.K. 2006. A high-resolution survey of deletion polymorphism in the human genome. *Nat. Genet.* **38**: 75–81.

Daly M.J., Rioux J.D., Schaffner S.F., Hudson T.J., and Lander E.S. 2001. High-resolution haplotype structure in the human genome. *Nat. Genet.* **29**: 229–232.

de Bakker P.I.W., Yelensky R., Pe'er I., Gabriel S.B., Daly M.J., and Altshuler D. 2005. Efficiency and power in genetic association studies. *Nat. Genet.* **37**: 1217–1223.

Hinds D.A., Kloek A.P., Jen M., Chen X., and Frazer K.A. 2006. Common deletions and SNPs are in linkage disequilibrium in the human genome. *Nat. Genet.* **38**: 82–85.

Hubbard T.J.P., Aken B.L., Beal K., Ballester B., Caccamo M., Chen Y., Clarke L., Coates G., Cunningham F., Cutts T., et al. Ensembl 2007. 2007. *Nucleic Acids Res.* **35**: D610–D617.

Iafrate A.J., Feuk L., Rivera M.N., Listewnik M.L., Donahoe P.K., Qi Y., Scherer S.W., and Lee C. 2004. Detection of large-scale variation in the human genome. *Nat. Genet.* **36**: 949–951.

International HapMap Consortium. 2005. A haplotype map of the human genome. *Nature* **437**: 1299–1320.

Kent W.J., Sugnet C.W., Furey T.S., Roskin K.M., Pringle T.H., Zahler A.M., and Haussler D. 2002. The Human Genome Browser at UCSC. *Genome Res.* **12**: 996–1006.

McCarroll S.A., Hadnott T.N., Perry G.H., Sabeti P.C., Zody M.C., Barrett J.C., Dallaire S., Gabriel S.B., Lee C., Daly M.J., and Altshuler D.M. 2006. Common deletion polymorphisms in the human genome. *Nat. Genet.* **38**: 86–92.

Mueller J.C. 2004. Linkage disequilibrium for different scales and applications. *Brief Bioinform.* **5**: 355–364.

Redon R., Ishikawa S., Fitch K.R., Feuk L., Perry G.H., Andrews T.D., Fiegler H., Shapero M.H., Carson A.R., Chen W., et al. 2006. Global variation in copy number in the human genome. *Nature* **444**: 444–454.

Sebat J., Lakshmi B., Troge J., Alexander J., Young J., Lundin P., Maner S., Massa H., Walker M., Chi M., et al. 2004. Large-scale copy number polymorphism in the human genome. *Science* **305**: 525–528.

Sharp A.J., Locke D.P., McGrath S.D., Cheng Z., Bailey J.A., Vallente R.U., Pertz L.M., Clark R.A., Schwartz S., Segraves R., et al. 2005. Segmental duplications and copy-number variation in the human genome. *Am. J. Hum. Genet.* **77**: 78–88.

Smith A.V., Thomas D.J., Munro H.M., and Abecasis G.R. 2005. Sequence features in regions of weak and strong linkage disequilibrium. *Genome Res.* **15**: 1519–1534.

Stephens M. and Donnelly P. 2003. A comparison of bayesian methods for haplotype reconstruction from population genotype data. *Am. J. Hum. Genet.* **73**: 1162–1169.

Thorisson G.A., Smith A.V., Krishnan L., and Stein L.D. 2005. The International HapMap Project Web site. *Genome Res.* **15**: 1592–1593.

Tuzun E., Sharp A.J., Bailey J.A., Kaul R., Morrison V.A., Pertz L.M., Haugen E., Hayden H., Albertson D., Pinkel D., Olson M.V., and Eichler E.E. 2005. Fine-scale structural variation of the human genome. *Nat. Genet.* **37**: 727–732.

Vastrik I., D'Eustachio P., Schmidt E., Joshi-Tope G., Gopinath G., Croft D., de Bono B., Gillespie M., Jassal B., Lewis S., et al. 2007. Reactome: A knowledge base of biologic pathways and processes. *Genome Biol.* **8**: R39.

Wheeler D.L., Barrett T., Benson D.A., Bryant S.H., Canese K., Chetvernin V., Church D.M., DiCuccio M., Edgar R., Federhen S., et al. 2007. Database resources of the National Center for Biotechnology Information. *Nucleic Acids Res.* **35**: D5–D12.

WWW RESOURCES

http://www.broad.mit.edu/mpg/haploview/ HaploView bundles many everyday analysis tasks into one easy-to-use package. It is an open-source program written in Java and capable of running on Windows, MacOS, and UNIX platforms.

http://www.ensembl.org Ensembl produces and maintains automatic annotation on selected eukaryotic genomes. Ensembl concentrates on vertebrate genomes, but other groups have adapted the system for use with plant and fungal genomes.

http://www.hapmap.org The primary portal to genotype data produced as part of the International Haplotype Map Project. The HapMap Web site provides researchers with a number of tools that allow them to analyze the data as well as to download data for local analyses.

http://www.java.com This site has downloads of the Java Runtime Environment available for Windows, Macintosh, and UNIX.

http://www.ncbi.nlm.nih.gov/SNP dbSNP is the central standard repository for genotyping information, including HapMap as well as other targeted resequencing projects.

http://www.reactome.org Reactome is a free on-line curated resource of core pathways and reactions in human biology. In addition to curated human events, inferred orthologous events in 22 non-human species including mouse, rat, chicken, zebra fish, worm, fly, yeast, two plants, and *E. coli* are also available.

http://www.broad.mit.edu/mpg/tagger Tagger can be used for tag-SNP selection and evaluation using HapMap data.

Isolation of Plant DNA for Genotyping Analysis

Nathan M. Springer

Department of Plant Biology, University of Minnesota, St. Paul, Minnesota 55108

INTRODUCTION

There are many protocols for isolating plant DNA (see Saghai-Maroof et al. 1984; Dellaporta et al. 1985; Doyle and Doyle 1987; Oard and Dronavalli 1992; Wang et al. 1993; Richards et al. 1994; Csaikl et al. 1998). A general difficulty in isolation of DNA from plant cells is the presence of a cell wall. It is necessary to degrade plant cell walls, either physically or enzymatically, in order to effectively isolate plant DNA. Additionally, some tissues (such as

endosperm) or some species contain high levels of starches or phenolic compounds that can complicate DNA isolation. A number of plant DNA isolation protocols are designed to overcome species-specific difficulties (see, e.g., Baker et al. 1990; Weeb and Knapp 1990; Maliyakal 1992; Lohdi et al. 1994; Porebski et al. 1997).

The protocols for isolation of plant DNA can be divided into three groups. The first group of protocols was designed to isolate large quantities of DNA for Southern blots (see, e.g., Chen and Dellaporta 1994). Typically, these utilize 5- to 15-ml extraction volumes and require grinding with a mortar and pestle or using a homogenizer. The second group of protocols is designed to isolate smaller quantities of DNA for PCR (see protocol below or the Qiagen DNeasy Plant Mini Kit). In general, these protocols also require grinding with a mortar and pestle or homogenization using a mill or paint shaker. They typically yield intermediate amounts of DNA that is stable for long periods. The third group of protocols is designed to utilize commercial reagents for a one-step rapid DNA isolation (see, e.g., Manen et al. 2005 or the Extract-N-Amp Plant PCR Kit from Sigma). In many of them, an extraction reagent is added to a plant sample and the sample is simply heated to release the DNA. Although these protocols are often the most rapid, the resulting DNA can be unstable for storage or can be contaminated with plant compounds that inhibit PCR. A short Internet search for plant DNA isolation will provide a wide variety of protocols that are designed to overcome problems associated with specific species or tissues. Additionally, an Internet search can identify commercial products designed for plant DNA isolation, including products from Sigma, Qiagen, Epicentre, Invitrogen, MO BIO, Clontech, Whatman, Promega, and Bio-Nobile that often document the most appropriate tissues and species. An individual researcher will need to balance such factors as cost, quantity of DNA, quality of DNA, and existing equipment to determine the most suitable protocol.

Protocol 1

Isolation of Plant DNA for PCR and Genotyping Using Organic Extraction and CTAB

Nathan M. Springer

Department of Plant Biology, University of Minnesota, St. Paul, Minnesota 55108

The following is a relatively simple protocol that can be used for many plant species. It provides a substantial amount of high-quality DNA that is suitable for PCR procedures and is stable for long periods of time. The cost per sample is very low. In addition, this protocol is relatively robust and can be performed by individuals who have had relatively little training. A typical undergraduate student can perform about 200–300 isolations in a day using this protocol. The disadvantages are that it requires a freeze-dryer and a mill or paint-shaker-like device and that it utilizes an organic extraction step, requiring the use of a fume hood.

MATERIALS

CAUTION: See Appendix for appropriate handling of materials marked with <!>.

Reagents

Chloroform <!> (Fisher): Isoamyl alcohol <!> (Fisher) (24:1)
CTAB extraction buffer
> *Mix 100 ml of 1 M Tris (pH 7.5) (Fisher), 140 ml of 5 M NaCl (Fisher), and 20 ml of 0.5 M EDTA (Fisher) with 10 g of CTAB <!> (hexadecyltrimethylammonium bromide) (Calbio Chem). Add H_2O to a volume of 990 ml. Just prior to use, add 10 ml of β-mercaptoethanol <!> (Fisher) to the solution and mix by shaking. The final concentrations are 0.1 M Tris (pH 7.5), 0.75 M NaCl, 0.01 M EDTA, 1% CTAB, 1% β-mercaptoethanol.*

Ethanol (70%)
Isopropanol <!> (Fisher)
Plant tissue sample (e.g., 1–2-cm^2 piece of leaf tissue from maize seedling)
> *In general, younger tissues provide the best quantity and quality of DNA. The protocol has also been used for root tissue, silk tissue, and mature leaf tissue.*

Tris (1 mM, pH 8.0) (Fisher)

Equipment

Freeze-dryer (Lab Conco Freezone 12 L Freeze dry system)
Glass beads (5 mm; Fisher)
Retsch 300 matrix mill (Qiagen)
> *Modified paint shakers or jigsaws can be used in place of the matrix mill.*

Test tubes (1.5 ml, 2 ml)
Transfer pipettes (Fisher) or 1-ml pipettor

Vortex mixer
Water bath preset to 65°C

METHOD

1. Collect plant tissue into a labeled 2-ml test tube.

2. Freeze-dry tissue for 1–2 days.

   The amount of time will depend on the type of tissue and the capacity of the freeze-dryer. Once the tissue is brittle, it is ready to be milled.

3. Add two 5-mm glass beads to each tube and grind the tissue with a Retsch 300 matrix mill at 30 revolutions/second for 2 minutes to powder the tissue.

   The grinding requires about 5 minutes for a set of 24 samples.

4. Add 700 µl of CTAB extraction buffer to each tube and mix by inversion.

5. Incubate at 65°C for 30–60 minutes, mixing by inversion every 15 minutes.

6. Add 400 µl of 24:1 chloroform:isoamyl alcohol and mix by vortexing.

7. Centrifuge at ~10,000g for 5 minutes.

8. Transfer aqueous phase to a fresh 1.5-ml tube, using either transfer pipettes or 1-ml pipettor.

   Optional: For increased purification of DNA for long-term storage, repeat Steps 6–8.

9. Add 300 µl of isopropanol and mix by inversion.

10. Centrifuge at ~10,000g for 5 minutes. A whitish DNA pellet should be visible.

11. Pour off liquid, add 500 µl of 70% ethanol, and mix by vortexing.

12. Centrifuge at ~10,000g for 2 minutes, then pour off liquid.

13. Allow pellet to dry, then resuspend in 250 ml of 1 mM Tris (pH 8.0) (see Troubleshooting).

    Extraction (Steps 3–13) requires ~1.5 hours for 24 samples, but multiple sets of 24 can be processed simultaneously.

TROUBLESHOOTING

Problem (Step 13): DNA pellet cannot be resuspended.
Solution: This can be the result of starch contamination. Use less tissue for the preparation or use an alternate protocol designed for high-starch tissues.

Problem (Step 13): Little or no DNA is isolated.
Solution: Try a younger tissue, a longer incubation in extraction buffer, or an alternate protocol. Consult the literature to determine whether there are specialized protocols for the species of interest.

Problem: DNA is highly fragmented.
Solution: This can be the result of environmental stresses or poor handling of plant tissue. Put the plant tissue on ice or in the freeze-dryer within an hour of harvesting to reduce fragmentation.

Problem: DNA is recovered but PCR fails.
Solution: This could be due to contamination with chloroform or the presence of species-specific compounds that inhibit PCR. Use a column-based protocol such as the Qiagen DNeasy plant kit.

Protocol 2

DNA Extraction from Freeze-dried Plant Tissue with CTAB in a 96-Well Format

An-Ping Hsia,[1] Hsin D. Chen,[1] Kazuhiro Ohtsu,[1] and Patrick S. Schnable[1,2,3]

[1]Department of Agronomy, [2]Department of Genetics, Development, and Cell Biology, and
[3]Center for Plant Genomics, Iowa State University, Ames, Iowa 50011

This protocol is a modified version of DNA isolation (Dietrich et al. 2002) using hexadecyltrimethyl-ammonium bromide (CTAB) (Rogers and Blendich 1985) and 96-well plates. It is high-throughput, which facilitates the analysis of large mapping populations. It has been applied to maize and *Arabidopsis* leaf tissues, and the DNA yield is adequate for at least 100–500 PCR procedures.

MATERIALS

CAUTION: See Appendix for appropriate handling of materials marked with <!>.

Reagents

Chloroform (~0.75% ethanol as preservative, Technical grade, Fisher Scientific) <!>/
    Octanol <!> (24:1)
CTAB extraction buffer
    Per liter:

[Stock]	[Addition]	[Final]
1 M Tris, pH 7.5	100 ml	100 mM
5 M NaCl	140 ml	0.7 M
0.5 M EDTA	20 ml	10 mM
*BME (β-Mercaptoethanol,14 M stock) <!>	10 ml	1%
CTAB (Sigma) <!>	10 g	1%
H_2O	730 ml	—

    *Add BME just before use for optimal results.
Ethanol (70%; Aaper Alcohol)
Isopropanol <!>
TE (Optional, see Step 17)
Tissue samples from the youngest leaves on a small seedling (5–7 days after planting)

Equipment

Adhesive sheet (Qiagen)
AirPore sheet (Qiagen)
Clamps
Collection rack (96 well)

84

Freeze-dryer
Glass beads (1.7–2.5 mm, MO-Sci Corporation)
Incubator preset to 58°C
Paint shaker (Red Devil Equipment)
Paper towels (thick, KimTowels, Erie Cotton Products)
PCR plate (96-well, skirted)
Pipette tip (200 μl)
Plexiglas (10.75″ Length [L] x 8.5″ Width [W] x 3/8″ Depth [D])
Strip tubes (1.2-ml, with caps)
V-bottom plates (96-well, 0.6 ml, Costar/Fisher)
Wood planks (11.5″ L x 4.5″ W x 0.75″ D)

METHOD

1. Preload 1.2-ml strip tubes (in 96-well racks) with 1.7- to 2.5-mm glass beads prior to collection as follows:

    i. Fill a 96-well skirted PCR plate (0.2-ml tubes) with 1.7- to 2.5-mm glass beads.

    ii. Drag a 200-μl pipette tip across the rows and columns of the plate to remove beads stuck between the tubes.

    iii. Carefully invert the plate of beads over a 96-well collection rack. This assures equal volumes of beads per tube.

2. Place tissue samples in the 1.2-ml strip tubes.

3. Seal each plate of tubes with an AirPore sheet and freeze-dry for 1–3 days.

4. Remove AirPore sheets and replace with strip caps. Secure two collection plates between two pieces of wood planks (11.5″ L x 4.5″ W x 0.75″ D). Load them into a paint shaker. Break freeze-dried tissue samples with the paint shaker for 5–10 minutes. Each sample should be ground into a fine powder.

5. Gently tap the racks on benchtop and centrifuge quickly to settle the powder. Carefully remove strip caps and add 600 μl of CTAB extraction buffer. Replace strip caps and mix well by gentle inversion.

6. Incubate for 15 minutes at 58°C. Invert samples about 10 times, and then incubate for an additional 15 minutes at 58°C.

7. Remove samples from the incubator and allow to cool in a fume hood for 10 minutes.

8. Centrifuge at about 2500 rpm briefly.

9. Add 300 μl of 24:1 chloroform/octanol. Sandwich the plates between two pieces of Plexiglas (10.75″ L x 8.5″ W x 3/8″ D) and clamp. Invert gently for 5 minutes.

10. Centrifuge at 3500 rpm for 10 minutes.

11. During Step 9, add 300 μl of isopropanol to each well of 0.6-ml V-bottom 96-well plates.

12. After centrifugation, remove 250–300 μl of the aqueous (top) layer and transfer to the isopropanol plates. Cover firmly with an adhesive sheet and invert 10 times.

13. Centrifuge at 3500 rpm for 10 minutes to collect DNA.

14. Remove adhesive sheet, invert plates over sink, and shake liquid from each plate. DNA pellets should be visible in the conical bottom of each tube.

15. Wash pellets by adding 50 µl of 70% ethanol to each tube. Centrifuge at 3500 rpm for about 3–5 minutes.

16. Gently invert plates onto thick paper towels to drain the ethanol. Allow plates to dry (10–15 minutes).

17. Rehydrate pellets with about 50–100 µl of TE or H_2O. Allow DNA to dissolve for 20 minutes to overnight. Store at –20°C (long term) or 4°C (short term).

Protocol 3

Purification of *Arabidopsis* DNA in 96-Well Plates Using the PUREGENE DNA Purification Kit

Yan Li

Department of Ecology and Evolution, University of Chicago, Chicago, Illinois 60637

This protocol describes the purification of *Arabidopsis* DNA in high-throughput form using the PUREGENE DNA Purification Kit adapted to 96-well plates.

MATERIALS

CAUTION: See Appendix for appropriate handling of materials marked with <!>.

Reagents

Arabidopsis leaf tissue (2 full leaves, ~100 mg fresh weight; or full-size young rosette leaf)
Ethanol (70%)
Isopropanol (100%) <!>
PUREGENE DNA Purification Kit (Gentra Systems), including the following reagents:
    Cell Lysis Solution
    DNA Hydration Solution
    Protein Precipitation Solution
    RNase A Solution

Equipment

Bead dispenser
Benchtop paper (absorbent)
Geno/Grinder (SPEX CertiPrep)
Grinding ball (5 mm)
Hybridization oven preset to 65°C
Ice
Incubator preset to 37°C
Plates (96 deep well)
Rubber mat
Vortex mixer

METHOD

1. Collect leaf tissue in a 96-deep-well plate (use fresh or store at −80°C until needed).

2. Add one grinding ball (5 mm) to each well using a bead dispenser.

3. Add 120 µl of Cell Lysis Solution. Cover the plate using a rubber mat, and seal tightly.

4. Load the plate onto the Geno/Grinder and run at 400 strokes/minute for 45 seconds.

5. Remove the plate and centrifuge at 3000g for 1 minute.

   $rcf(g)=1.12r(rpm/1000)^2$, *where r = radius of the rotor in mm*

6. Add 280 µl of Cell Lysis Solution to the ground leaf tissue in each well.

7. Seal the plate and load onto the Geno/Grinder. Run at 400 strokes/minute for 45 seconds. Foam will appear after shaking. The cover should be tightly sealed to avoid contamination.

8. Centrifuge the plate at 3000g for 1 minute. Vortex the plate at high speed.

9. Incubate the plate in a hybridization oven at 65°C for 60 minutes. (Make sure the cover is tightly sealed.) Cool to room temperature.

10. (Optional) Add 3 µl of RNase A Solution to each well. Make sure the cover is tightly sealed and then vortex the plate at low speed. Incubate the plate at 37°C for 30 minutes (15–60 minutes).

11. Cool to room temperature by placing the plate on ice for 1 minute.

12. Add 133 µl of Protein Precipitation Solution to each well.

13. Vortex at high speed for 20 seconds or use the Geno/Grinder for 20 seconds.

14. Incubate the plate on ice for (15–)60 minutes.

15. Centrifuge the plate at 4500g for 15 minutes at 10°C to collect the proteins. If the pellet is not tight, repeat Steps 13–15.

16. During centrifugation, pipette 200 µl of 100% isopropanol into a second plate.

17. Transfer 200 µl of supernatant containing DNA into the new plate containing 200 µl of 100% isopropanol.

18. Seal the plate with a new cover and mix by inverting gently 50 times.

19. Centrifuge the plate at 4500g for 5 minutes. The DNA will be visible as a pellet.

20. Pour off the supernatant by inverting the entire 96-well plate onto clean absorbent benchtop paper. Blot 10 times to dry (watch the pellet).

21. Add 200 µl of 70% ethanol and gently invert 10 times to wash the pellet.

22. Centrifuge the plate at 4500g for 5 minutes. Pour off the supernatant by inverting the entire 96-well plate onto clean absorbent benchtop paper. Blot 10 times to dry (watch the pellet).

23. Air-dry the samples by keeping the plate inverted for 5–10 minutes.

24. Add 50–75 µl of DNA Hydration Solution to each well and seal the plate. (Can use H_2O in place of DNA Hydration Solution.)

25. Rehydrate the DNA at room temperature overnight.

26. Store the DNA at 4°C or −20°C for short-term storage, or −80°C for long-term storage.

REFERENCES

Baker S.S., Rugh C.L., and Kamalay J. 1990. RNA and DNA isolation from recalcitrant plant tissues. *Biotechniques* **9:** 268–272.

Chen J. and Dellaporta S. 1994. Urea-based plant DNA miniprep. In *The maize handbook* (ed. M. Freeling and V. Walbot), pp. 526–527. Springer-Verlag, New York.

Csaikl U.M., Bastian H., Brettschneider R., Gauch S., Meir A., Schauerte M., Scholz F., Sperisen C., Vornam B., and Ziegenhagen B. 1998. Comparative analysis of different DNA extraction protocols: A fast, universal maxi-preparation of high quality plant DNA for genetic evaluation and phylogenetic studies. *Plant Mol. Biol. Rep.* **16:** 69–86.

Dellaporta S.L., Wood J., and Hicks J.B. 1985. Maize DNA miniprep. In *Molecular biology of plants: A laboratory course manual* (ed. R. Malmberg et al.), pp. 36–37. Cold Spring Harbor Laboratory, Cold Spring Harbor, New York.

Dietrich C.R., Cui F., Packila M.L., Li J., Ashlock D.A., Nikolau B.J., and Schnable P.S. 2002. Maize *Mu* transposons are targeted to the 5′ untranslated region of the gl8 gene and sequences flanking *Mu* target-site duplications exhibit nonrandom nucleotide composition throughout the genome. *Genetics* **160:** 697–716.

Doyle J.J. and Doyle J.L. 1987. A rapid DNA isolation procedure for small quantities of fresh leaf tissue. *Phytochem. Bull.* **19:** 11–15.

Lodhi M.A., Ye G.-N., Weeden N.F., and Reisch B.I. 1994. A simple and efficient method for DNA extractions from grapevine cultivars and *Vitis* species. *Plant Mol. Biol. Rep.* **12:** 6–13.

Maliyakal E.J. 1992. An efficient method for isolation of RNA and DNA from plants containing polyphenolics. *Nucleic Acids Res.* **20:** 2381.

Manen J.F., Sinitsyna O., Aeschbach L., Markov A.V., and Sinitsyn A. 2005. A fully automatable enzymatic method for DNA extraction from plant tissues. *BMC Plant Biol.* **5:** 23.

Oard J.H. and Dronavalli S. 1992. Rapid isolation of rice and maize DNA for analysis by random-primer PCR. *Plant Mol. Biol. Rept.* **10:** 236–241.

Porebski S., Bailey L.G., and Baum B.R. 1997. Modification of a CTAB DNA extraction protocol for plants containing high polysaccharide and polyphenol components. *Plant Mol. Biol. Rep.* **15:** 8–15.

Richards E., Reichardt M., and Rogers S. 1994. Preparation of genomic DNA from plant tissue. In *Current protocols in molecular biology* (ed. F.M. Ausubel et al.), vol. 1, pp. 2.3.1–2.3.7. Wiley, New York.

Rogers S.O. and Blendich A.J. 1985. Extraction of DNA from milligram amounts of fresh herbarium and mummified plant tissues. *Plant Mol. Biol.* **5:** 69–76.

Saghai-Maroof M.A., Soliman K.M., Jorgensen R.A., and Allard R.W. 1984. Ribosomal DNA spacer-length polymorphism in barley: Mendelian inheritance, chromosomal location, and population dynamics. *Proc. Natl. Acad. Sci.* **81:** 8014–8019.

Wang H., Qi M., and Cutler A. 1993. A simple method of preparing plant samples for PCR. *Nucleic Acids Res.* **21:** 4153–4154.

Weeb D.M. and Knapp S.J. 1990. DNA extraction from a previously recalcitrant plant genus. *Plant Mol. Biol. Rep.* **8:** 180–185.

8 | Preparing RNA from Plant Tissues

An-Ping Hsia,[1] Hsin D. Chen,[1] Kazuhiro Ohtsu,[1] and Patrick S. Schnable[1,2,3]

[1]Department of Agronomy, [2]Department of Genetics, Development, and Cell Biology, and [3]Center for
Plant Genomics, Iowa State University, Ames, Iowa 50011

INTRODUCTION

The use of RNA for genotyping analysis can be advantageous in certain studies because transcriptomes are significantly smaller than genomes and typically contain far fewer repetitive sequences. Laser capture microdissection (LCM) has been used successfully to isolate sequences (especially rare transcripts) that accumulate in specific tissues. Where the quantity of isolated material is limiting, amplification can be used to increase the amount of product. Upon conversion to cDNA, the product serves as template for 454 sequencing to produce expressed sequence tags for subsequent SNP analysis and detection (see Chapter 11, Protocol 6).

Laser-assisted Microdissection of Plant Tissue Sections

Laser-assisted microdissection microscopy was developed to obtain samples from distinct tissue types (Emmert-Buck et al. 1996; Simone et al. 1998) and has been applied to plant samples (Asano et al. 2002; Kerk et al. 2003; Nakazono et al. 2003). Various laser-assisted microdissection platforms are commercially available, and it is recommended that users evaluate these platforms with their own samples. The protocol for maize shoot apical meristems (SAMs) in this chapter was adapted from Asano et al. (2002) and Kerk et al. (2003) and used on the PALM system (P.A.L.M. Microlaser Technologies, Carl Zeiss, Bernried, Germany). Laser microdissection and pressure catapulting (LMPC) technology utilizes a pulsed ultraviolet (UV-A) laser beam to cut cells from tissue sections and laser pressure catapulting to catapult isolated tissues into collection caps.

RNA Amplification

The amount of RNA collected in a standard microdissection is often insufficient for global gene expression analysis, but it can be increased via linear amplification. A procedure modified from Eberwine et al. (1992) uses an oligo(dT)-T7 chimeric primer to preferentially select polyadenylated RNA species (e.g., mRNA) through two rounds of sequential reverse transcription and RNA transcription. The amplification is reproducible (Nakazono et al. 2003) and is typically on the order of 50,000- to 500,000-fold (e.g., assuming 1% poly(A) mRNA in 10 ng of total RNA starting material, yields of 5 µg to 50 µg of amplified RNA are routinely obtained).

Maize Tissue Preparation and Extraction of RNA from Target Cells

This protocol describes the preparation of acetone-fixed and paraffin-embedded maize seedling tissue sections. Once the sections are prepared, the PALM MicroBeam System, with its laser microdissection and pressure catapulting (LMPC) technology, is used to cut out cells of interest and "catapult" isolated tissues into collection caps. This tissue can then be used for RNA extraction.

MATERIALS

CAUTION: See Appendix for appropriate handling of materials marked with <!>.

Reagents

Acetone (100%, Fisher Scientific) <!>, ice-cold and room temperature (RT)
Diethylpyrocarbonate (DEPC; Sigma) <!>
Maize seedlings, 14 days old
Paraplast chips (Paraplast +, 56°C, Oxford Labware)
RNA extraction buffer (e.g., XB from the PicoPure RNA Isolation Kit, Arcturus)
Xylene (Fisher Scientific) <!>

Equipment

Dissecting microscope
Fine paintbrush
Fisher probe-on⁺ slides
Gradient metal warming plate
> *A paraffin-embedding center can be used if one is available.*
Ice
Metal weighing dish
Microcentrifuge tubes (500 µl, with caps)
Oven preset to 60°C
PALM MicroBeam System (P.A.L.M. MicroBeam 115V Z, P.A.L.M. Microlaser Technologies, Carl Zeiss, http://www.palm-microlaser.com/dasat/index.php?cid=100106& conid=0&sid=dasat)
Paper towels
Paraffin reservoir (2.5 l, Electron Microscopy Sciences)
Petri dishes (glass)
Plastic bags
Razor blade (single-edged)
Rotary microtome (Leica RM 2135, Leica Microsystems, or equivalent)
Rotator (e.g., Ted Pella)
Scalpel
Scintillation vials (20 ml, Fisher Scientific)
Slide-warming tray (Fisher Scientific)
Tissue paper
Tissue Tek embedding rings (Electron Microscopy Sciences)

Tissue Tek metal base molds (22 x 22 x 6, Electron Microscopy Sciences)
Vacuum apparatus
Weighing spatula

METHOD

Fixation

Ice-cold 100% acetone is used as the fixative. The fixation should be carried out in a fume hood.

1. Cut off seedlings at the coleoptile node with a new single-edged razor blade and immediately place in a glass Petri dish containing ice-cold acetone. Immediately execute a second cut for collection, approximately 1 cm above the coleoptile node.

2. Trim SAM-containing tissues which are submersed in fixative to a final size of approximately 0.3 x 0.3 x 0.2 cm. Place them in a scintillation vial with 15 ml of ice-cold 100% acetone and keep on ice.

   Prepare 8–10 seedlings in this manner and place into the same vial. Work rapidly; the total preparation time for a single vial of 8–10 seedlings should not exceed 10 minutes.

3. Vacuum infiltrate the sample (on ice) by subjecting the vial to a vacuum of 400 mm Hg for 10–15 minutes. Slowly equilibrate to atmospheric pressure. Replace the fixative with fresh ice-cold 100% acetone and place the vial on a rotator for 1 hour at 4°C.

4. Empty the tissues into a Petri dish, place the dish under a dissecting microscope, and trim off extra leaf and nodal tissue using a scalpel. The resulting tissue blocks should be approximately 0.3 x 0.2 x 0.1 cm. Repeat Step 3 twice more (3 infiltrations total). Store samples overnight at 4°C on a rotator.

Dehydration/Xylene Infiltration

5. Bring the samples to RT, replace the fixative with fresh 100% acetone at RT, and rotate for 1 hour at RT.

6. Replace the fixative with a mixture of acetone: xylene (1:1) and rotate for 1.5 hours at RT.

   If the target tissue is very dense, or hidden under other tissue layers, perform slower gradual infiltration with xylene, using 3:1, 1:1, and 1:3 mixtures of acetone: xylene.

7. Perform three solution changes of pure xylene, incubating for 1 hour after each change.

Paraplast Infiltration

8. Add a small amount (usually about 10–15 chips, about 1/10 to 1/5 the volume of the vial) of Paraplast chips to each vial and incubate overnight at RT on a rotator.

9. Place the vials in an oven at 60°C to dissolve any remaining Paraplast chips. When chips are dissolved, gently invert the vials to mix xylene and Paraplast.

10. Incubate the vials for a further 1.5 hours at 60°C, add more Paraplast chips (up to half the volume of the vial), and incubate for an additional 1.5 hours at 60°C.

11. Replace half of the xylene/Paraplast mixture with pure molten paraffin from a paraffin reservoir, mix by inversion, and incubate overnight at 60°C.

12. Replace the Paraplast mixture with pure Paraplast and incubate at 60°C. Change the Paraplast at least 3 times throughout the day (every 4 hours). Incubate overnight at 60°C after the last Paraplast change.

13. Exchange the Paraplast early in the day and incubate for 4 hours at 60°C.

Embedding

14. Assemble Tissue Tek metal base molds and Tissue Tek embedding rings. Prepare a "home-made" embedding device consisting of a gradient metal warming plate which is hot on one end and RT on the opposite end, or use a paraffin-embedding center (if one is available). Dispense Paraplast from a 2.5-l reservoir into the assembled base molds plus embedding rings.

15. Pour tissue and Paraplast from the vial into a metal weighing dish which has been placed on the hot side of the warming plate. Using a weighing spatula, scoop the tissue into the assembled base mold/embedding ring combination and orient the tissue for sectioning on a rotary microtome. Cool blocks to RT and then place on ice for easy un-molding. Store blocks in plastic bags at 4°C.

Sectioning

16. Trim Paraplast blocks into a narrow trapezoidal shape with parallel horizontal cuts at the top and bottom, and slanted cuts on the sides. Section the blocks (10-μm thick) on a rotary microtome.

17. Place ribboned sections onto Fisher probe-on$^+$ slides, and float them on DEPC-treated H_2O at 40°C on a slide warming tray for approximately 5 minutes, or until the sections stretch (not to exceed 20 minutes).

18. Remove H_2O by tipping the slide onto absorptive paper towels while holding one end of the ribbon with a fine paintbrush. Wick off residual H_2O with tissue paper. The slide should not cool down during H_2O removal. Quickly place the slide back on the slide-warming tray and leave it until it is completely dry, or overnight. Store dry slides at 4°C until the PALM procedure.

PALM Procedure and Extraction of RNA

19. Warm the section slides to RT and deparaffinize the sections by washing in two changes of xylene (10 minutes each). Remove a slide from the xylene, allow it to air-dry, and place it onto the microscope stage to mark tissue regions to be collected. Leave the remaining slides in xylene until needed.

20. Fill the tube cap of a 500-μl microcentrifuge tube with about 50 μl of RNA extraction buffer (e.g., XB from the PicoPure RNA Isolation Kit, Arcturus) and mount on PALM to use for collection. (See Troubleshooting.)

21. Prior to cell collection, optimize the energy and focus of the laser for the sections that include target cells.

22. Mark areas of target cells using the PALM software.

23. Collect target cells via the "Close and Cut plus AutoLPC" method according to the vendor's manual. The focused laser cuts the outline of the target cells, and the defocused laser catapults the outlined target cells into the tube cap filled with RNA extraction buffer that is above the section.

    Test and optimize the PALM capturing settings for each tissue type.

24. After collecting a sufficient amount of tissue, extract RNA according to a protocol of choice (e.g., XB from the PicoPure RNA Isolation Kit, Arcturus).

    6–10 maize shoot apical meristems (SAMs, which contain ~ 15,000–18,000 cells) will provide about 10 ng of RNA.

TROUBLESHOOTING

Problem (Step 20): RNA extraction buffer dries out.

Solution: Some RNA extraction buffers dry out very quickly under dry weather conditions. Crystals form on the outside of the collection tube cap and eventually touch the slide with the tissue. This may cause the rest of the liquid buffer to flow onto the slide and ruin the sample. To avoid this, substitute about 40 ml of mineral oil for RNA extraction buffer (M. Scanlon, pers. comm.). Collect tissue into the mineral oil, then add RNA extraction buffer to the mineral oil and mix thoroughly. The remainder of the RNA extraction will not be affected by the presence of the oil.

Protocol 2

T7-based RNA Amplification From Maize SAM

This protocol describes the amplification of RNA from shoot apical meristem (SAM) cells. It uses an oligo(dT)-T7 chimeric primer to preferentially select polyadenylated RNA species through two rounds of sequential reverse transcription and RNA transcription. The protocol can be completed in 2–4 days.

MATERIALS

CAUTION: See Appendix for appropriate handling of materials marked with <!>.

Reagents

β-Mercaptoethanol (Acros)

β-Nicotinamide adenine dinucleotide hydrate (β-NAD+; 260 μM, min. 98% from yeast, Sigma)

Chloroform (~0.75% ethanol as preservative, Technical grade, Fisher Scientific) <!>

Diethylpyrocarbonate (DEPC) (Sigma) <!>

dNTP Set (10 mM, Intermountain Scientific)

E. coli DNA ligase (10 U/μl, New England Biolabs)

E. coli DNA polymerase I (10 U/μl, New England Biolabs)

>   *Supplied with 10x DNA polymerase buffer.*

Ethanol (Absolute, Aaper Alcohol)

MEGAscript T7 Kit (Ambion)

>   *40 reactions; includes rNTP solutions, 10x reaction buffer, T7 RNA polymerase enzyme mix, and RNase-free DNase I*

Phenol (Saturated, Fisher Scientific) <!>

>   *pH 6.6, BP1750I-400 for Step 10*
>
>   *pH 4.3, BP1751I-400 for Step 18*

QIAquick PCR Purification Kit (Qiagen)

>   *250 columns; includes Buffer PB, Buffer PE, and Buffer EB*

Random hexamer primer (1 μg/μl, Roche Diagnostics)

Ribonuclease H (RNase H; 2 U/μl, Invitrogen)

RNA extracted from laser microdissection (LM) sample

RNaseOUT Recombinant Ribonuclease Inhibitor (40 U/μl, Invitrogen)

RNeasy Mini Kit (Qiagen)

>   *50 columns; includes 1.5- and 2.0-ml collection tubes and RNase-free reagents and buffers*

Sodium acetate (100 mM, pH 5.2, certified ACS, Fisher Scientific) <!>

SuperScript II Reverse Transcriptase (200 U/μl, Invitrogen)

>   *Supplied with 5x first-strand buffer and 0.1 M DTT* <!>

T4 DNA polymerase (3 U/μl, New England Biolabs)

T4 gene 32 protein (5 ug /μl, USB Corporation)

T7-oligo(dT) primer (0.5 μg/μl, HPLC Purified, Integrated DNA Technologies)
(5′-TCTAGTCGACGGCCAGTGAATTGTAATACGACTCACTATAGGGCGTTTTTTTTTTTTTTT TTTTTT-3′)

Equipment

Concentrator/evaporator (Labconco CentriVap DNA System, Fisher Scientific)
Heating block or water bath preset to 16°C, 37°C, 42°C, 65°C, 70°C, 95°C
Ice
Microcentrifuge tubes (nuclease-free)
Vortex mixer

METHOD

All centrifugation steps are performed in a benchtop microcentrifuge at room temperature.

First-Round RNA Amplification

1. Mix the following components in a nuclease-free microcentrifuge tube:

0.5 µg/µl T7-oligo(dT) primer	1 µl
total RNA extracted from LM sample	5–100 ng
H$_2$O (DEPC-treated) to make	11 µl

2. Incubate the samples at 65°C for 10 minutes and cool on ice for 5–10 minutes.

3. Collect the sample by centrifugation and equilibrate to 42°C for 5 minutes.

4. Add 8 µl of the following mixture to each tube:

10 mM dNTP mix	1 µl
5× first-strand buffer	4 µl
0.1 M DTT	2 µl
40 U/µl RNaseOUT	0.5 µl
5 µg/µl T4 gene 32 protein	0.5 µl

5. Mix gently and add 1 µl of Superscript II (200 U/µl) to each tube.

6. Incubate for 1 hour at 42°C.

   At this point, the samples can be stored at –20°C if necessary.

7. Add 130 µl of the following mixture to each 20-µl reaction:

10× *E. coli* DNA polymerase I buffer	15 µl
10 mM dNTP mix	3 µl
260 µM β-NAD$^+$	15 µl
10 U/µl *E. coli* DNA polymerase I	4 µl
2 U/µl RNase H	1 µl
10 U/µl *E. coli* DNA ligase	1 µl
H$_2$O	91 µl

8. Mix gently and incubate for 2 hours at 16°C.

9. Add 2 µl of T4 DNA polymerase (3 U/µl) and incubate for 10 minutes at 16°C.

10. Extract the double-stranded DNA with an equal amount of 1:1 phenol (pH 6.6)/chloroform.

11. Extract with an equal volume of chloroform and transfer the aqueous layer to a new microcentrifuge tube.

12. Purify the DNA using a Qiagen QIAquick PCR Purification column as follows:

    i. Add 35 μl of 100 mM sodium acetate (pH 5.2) to each tube.

    ii. Add 500 μl of Buffer PB to each tube and mix by inverting.

    iii. Proceed as per manufacturer's instructions until elution.

    iv. Add 15 μl of H_2O to each column, allow the column to stand for 1 minute, and centrifuge at maximum speed for 1 minute. Repeat once.

13. Concentrate the sample to 8 μl in a concentrator/evaporator at 50°C.

14. Prepare the reagents from the MEGAscript T7 Kit.

    i. Thaw the rNTP solutions, mix by vortexing, collect by centrifugation, and place on ice.

    ii. Thaw 10x reaction buffer, mix until the precipitate has dissolved, and keep at room temperature (not on ice).

15. Assemble the 20-μl reaction in the following order:

cDNA	8 μl
rNTP mix (2 μl each of ATP, CTP, GTP, and UTP)	8 μl
10x reaction buffer	2 μl
T7 RNA polymerase enzyme mix	2 μl

16. Incubate the reaction mix for 5 hours at 37°C.

17. Add 1 μl of RNase-free DNase I (2 U/μl) and incubate for 15 minutes at 37°C.

18. Add 80 μl of nuclease-free H_2O to the sample and extract with an equal volume (100 μl) of 1:1 phenol (pH 4.3)/chloroform.

19. Extract with an equal volume of chloroform and transfer the aqueous layer to a new microcentrifuge tube.

20. Concentrate sample in RNeasy mini column:

    i. Add 350 μl of Buffer RLT (with 3.5 μl of β-mercaptoethanol) and mix thoroughly by inverting.

    ii. Add 250 μl of absolute ethanol and mix thoroughly by pipetting. Do not centrifuge.

    iii. Apply the entire sample (700 μl) to an RNeasy minicolumn placed in a 2-ml collection tube.

    iv. Centrifuge at 10,000 rpm for 15 seconds, and discard the flowthrough.

    v. Transfer the RNeasy column to a new 2-ml collection tube.

    vi. Pipette 500 μl of Buffer RPE onto the RNeasy column.

    vii. Centrifuge at 10,000 rpm for 15 seconds, and discard the flowthrough.

    viii. Add another 500 μl of Buffer RPE to the RNeasy column and centrifuge at 10,000 rpm for 2 minutes to dry the RNeasy silica-gel membrane. Discard the flowthrough.

    ix. To elute, transfer the RNeasy column to a new 1.5-ml collection tube and pipette 15 μl of H_2O onto the RNeasy column.

    x. Allow the column to stand for 1 minute, and centrifuge at 10,000 rpm for 1 minute.

    xi. Pipette another 15 μl of H_2O onto the RNeasy column, allow the column to stand for 1 minute, and centrifuge at 10,000 rpm for 1 minute.

21. Concentrate the amplified RNA (aRNA) sample to 10 μl in a concentrator/evaporator at 50°C. If quantification of a RNA is desired at this stage, concentrate sample to about 11 μl and use 1 μl for quantification.

Second-Round RNA Amplification

22. Assemble the first-strand reaction by mixing 1 μl of random hexamer primer (1 μg/μl) with 10 μl of aRNA and incubate for 10 minutes at 70°C. Cool on ice for 5 minutes.

23. Collect the sample by centrifugation and equilibrate the tube at room temperature for 10 minutes.

24. Add 8 μl of the following mixture to each tube:

10 mM dNTP mix	1 μl
5x first-strand buffer	4 μl
0.1 M DTT	2 μl
40 U/μl RNaseOut	0.5 μl
5 μg/μl T4 gene 32 protein	0.5 μl

25. Mix gently, add 1 μl of Superscript II (200 U/μl), and incubate for 1 hour at 37°C.

26. Add 1 μl of RNase H (2 U/μl), and incubate for 30 minutes at 37°C.

27. Heat for 2 minutes at 95°C. Cool sample on ice for 5 minutes.

28. Add 1 μl of 0.5 μg/μl T7-oligo(dT) primer and incubate for 5 minutes at 70°C.

29. Incubate for 10 minutes at 42°C, and place sample on ice for 5 minutes.

30. Add 128 μl of the following mixture to each tube:

10x $E.\ coli$ DNA polymerase I buffer	15 μl
10 mM dNTP mix	3 μl
260 μM β-NAD$^+$	15 μl
10 U/μl $E.\ coli$ DNA polymerase I	4 μl
2 U/μl RNase H	1 μl
H_2O	90 μl

31. Follow Steps 8–20, viii, of First-Round RNA Amplification.

32. To elute the aRNA, transfer the RNeasy column to a new 1.5-ml collection tube, pipette 30 μl of H_2O onto the RNeasy column, allow the column to stand for 1 minute, and centrifuge at 10,000 rpm for 1 minute.

33. Pipette another 30 μl of H_2O onto the RNeasy column, allow the column to stand for 1 minute, and centrifuge at 10,000 rpm for 1 minute.

34. Use a 1-μl aliquot for RNA quantification.

    The purified, concentrated RNA may now be used in cDNA synthesis to generate templates for 454 transcriptome sequencing runs (see Chapter 11, Protocol 6).

ACKNOWLEDGMENTS

We thank the Schnable lab personnel and alumni Mikio Nakazono, Dave Skibbe, and Marianne Smith for their contributions to the development of protocols listed in this volume. This project was supported by a competitive grant from the National Science Foundation Plant Genome Program (DBI-0321595) and Hatch Act and State of Iowa funds to P.S.S.

REFERENCES

Asano T., Masumura T., Kusano H., Kikuchi S., Kurita A., Shimada H., and Kadowaki K. 2002. Construction of a specialized cDNA library from plant cells isolated by laser capture microdissection: Toward comprehensive analysis of the genes expressed in the rice phloem. *Plant J.* **32:** 401–408.

Eberwine J., Yeh H., Miyashiro K., Cao Y., Nair S., Finnell R., Zettel M., and Coleman P. 1992. Analysis of gene expression in single live neurons. *Proc. Natl. Acad. Sci.* **89:** 3010–3014.

Emmert-Buck M.R., Bonner R.F., Smith P.D., Chuaqui R.F., Zhuang Z., Goldstein S.R., Weiss R.A., and Liotta L.A. 1996. Laser capture microdissection. *Science* **274:** 998–1001.

Kerk N.M., Ceserani T., Tausta S.L., Sussex I.M., and Nelson T.M. 2003. Laser capture microdissection of cells from plant tissues. *Plant Physiol.* **132:** 27–35.

Nakazono M., Qiu F., Borsuk L.A., and Schnable P.S. 2003. Laser-capture microdissection, a tool for the global analysis of gene expression in specific plant cell types: Identification of genes expressed differentially in epidermal cells or vascular tissues of maize. *Plant Cell* **15:** 583–596.

Simone N.L., Bonner R.F., Gillespie J.W., Emmert-Buck M.R., and Liotta L.A. 1998. Laser-capture microdissection: Opening the microscopic frontier to molecular analysis. *Trends Genet.* **14:** 272–276.

9 | Preparing DNA from Mammalian Sources

Amrik Sahota,[1] Andrew I. Brooks,[1,2] and Jay A. Tischfield[1]

[1]*Rutgers University Cell and DNA Repository, Department of Genetics, Piscataway, New Jersey 08854-8082;*
[2]*Bionomics Research and Technology Center, Environmental and Occupational Health Sciences Institute, University of Medicine and Dentistry of New Jersey–Robert Wood Johnson Medical School, Piscataway, New Jersey 08854-5635*

INTRODUCTION

The availability of high-quality DNA from a large number of individuals is a prerequisite for the success of genetic variation studies. This requirement has spurred major technological advances in DNA extraction methods. Twenty years ago, large-scale manual extractions took more than 3 days to complete. Large-scale preparations can now be completed within 3 hours using automated extraction platforms—without the need for toxic reagents and often with higher yields than in the past. Our laboratories developed an inorganic, salt precipitation method (Madisen et al. 1987; Lahiri et al. 1992) for DNA extraction from moderate to large sample volumes (3–30 ml of whole blood or 1×10^7 to 2×10^8 cultured cells). This method eliminates the hazards of phenol exposure to laboratory personnel and the environment, and all reagents are prepared in-house. We used this procedure for over 15 years, but, for convenience and increased productivity and reliability, we switched to commercial reagents about 5 years ago. Comparative studies have shown that DNA obtained using the in-house manual extraction procedure is of equal quality to that obtained using commercial reagents. The drawback of commercial reagents is that laboratory personnel know very little about the chemical composition of the reagents and kits they are using. We now use the in-house procedure as a backup, and we present it here for the benefit of laboratories that may find the purchase of commercial reagents or of automated equipment too costly.

Choice of Tissue

The amount of DNA that will eventually be required for a given project is unpredictable. Thus, we prefer to extract DNA primarily from lymphoblastoid cell lines (LCLs) and secondarily from blood. Human LCLs are established using Epstein-Barr virus (Neitze 1986; Caputo et al. 1991). DNA from these sources has been successfully assayed at multiple genotyping centers using a variety of analytical platforms (Alarcón et al. 2002; Holmans et al. 2004; Prasad et al. 2005; Suarez et al. 2006). With the recent development of two technologies, whole-genome amplification (WGA) and array-based analyses, it has been suggested that large amounts of genomic DNA may not always be necessary (Dean et al. 2002; Langmore 2002; Fathallah-Shaykh 2005; Dunphy 2006; Li et al. 2006; Quackenbush 2006; Reis-Filho et al. 2006; Tomlins et al. 2006).

There are many options for DNA extraction on a small scale (e.g., <0.5 ml of blood, dried blood spots, saliva, buccal cells, mouthwashes) (McEwen and Reilly 1994; Lum and Le Marchand 1998; Feigelson et al. 2001; Garcia-Closas et al. 2001; Heath et al. 2001; Steinberg et al. 2002). Specimens such as saliva, which can be collected using noninvasive procedures, may be viable alternatives to blood (Steinberg et al. 2002; Rylander-Rudqvist et al. 2006). Buccal swabs and mouthwashes have been used in several epidemiology studies, but the DNA yield from these sources, especially buccal swabs, is significantly lower than from saliva samples. Saliva preserved in Oragene solution (DNA Genotek, Ottawa, Ontario, Canada), as described in this chapter, is also easier to transport than mouthwash samples. Protocols for DNA extraction from buccal swabs or mouthwash samples have been documented in several publications (Lum and Le Marchand 1998; Feigelson et al. 2001; Garcia-Closas et al. 2001; Heath et al. 2001; Steinberg et al. 2002).

Choice of Anticoagulant

EDTA is the anticoagulant of choice for blood collection for DNA extractions, since it inhibits DNase activity and does not introduce volume changes. ACD tubes can be used for blood collection, but it is important to fill them to the mark to avoid causing changes in osmolarity, which can affect cell lysis. Heparin should be avoided, as it can bind to DNA during purification and can inhibit Taq polymerase used for PCR. Irrespective of the anticoagulant, the Vacutainer tube should be inverted several times to mix the blood. Blood can be shipped at ambient temperature, but if the delay between collection and extraction is more than 3 days, there will be some degradation of DNA and the yield will be lower than that from fresh blood.

Summary of Extraction

All DNA extraction procedures involve cell lysis, protein removal, and recovery of DNA. An RNA removal step may be included, depending on sample type. When DNA is isolated from whole blood, for example, nonnucleated red cells are first lysed to separate them from white cells. The white cells are then lysed using an anionic detergent, which solubilizes the cellular components. This is done in the presence of a DNA preservative that limits the activity of DNases. Cytoplasmic and nuclear proteins are removed by salt precipitation. Genomic DNA is precipitated with alcohol and dissolved in TE buffer. A portion of the DNA is diluted to a specified concentration, and the stock and diluted samples are stored in small aliquots at –70°C. This avoids repeated freezing and thawing of a large sample, which can lead to precipitate formation. Water is not used for storage because of the risk of acid hydrolysis of DNA. Any contaminants in the water may also cause DNA degradation.

DNA Quantity and Quality

For accurate DNA quantification, absorbance measurements should be taken in the range of 0.1 to 1.0. The sample dilution factor should be adjusted if the absorbance falls outside this range. Also, measurements should be taken in slightly alkaline buffer (10 mM Tris-HCl, 1 mM EDTA [pH 8.0]), since absorbance measurements and hence the $A_{260/280}$ ratio are pH dependent (Willinger et al. 1997). Lower pH can cause DNA degradation and can also result in a lower $A_{260/280}$ ratio and reduced sensitivity to protein contamination. Water should not be used for this purpose because of the wide variability in pH. In addition, the calculation factor ($1A_{260} = 50$ µg/ml DNA) is pH dependent and is not strictly valid when water is used for dilution.

DNA concentrations are routinely determined by UV spectroscopy. This method is cheap, rapid, robust, and semiautomated, but it measures total UV-absorbing material, which may include RNA in the case of DNA isolated from cell lines. (There is very little or no RNA in DNA isolated from blood.) Thus, in some cases, UV measurements may underestimate the amount of DNA. When the amount of DNA is limited, UV absorbance measurements can be made using the NanoDrop ND-1000 spectrophotometer (NanoDrop Technologies, Wilmington, Delaware). This instrument requires 1 µl of sample, and measurements can be made without dilution.

Automated procedures, such as those based on fluorescence, can also be used for determination of DNA concentration. Fluorescence-based methods such as PicoGreen have much greater sensitivity and specificity than UV-based methods, but the method of choice for a particular sample depends on the concentration range for that sample. Thus, PicoGreen would not be suitable for measuring "stock" DNA concentrations, and UV would be unsuitable for measuring "working" DNA concentrations. Alternatively, quantitative PCR (qPCR) can be used for estimating DNA concentrations where the concentration of the sample is expected to be quite low, as in forensic studies or recovery of DNA from fixed tissue (see Protocol 12).

Electrophoresis in 1% agarose gels shows that DNA isolated using the methods described in this chapter is typically >23 kb in size and is thus suitable for Southern blotting and other molecular biological procedures, as well as PCR (Lahiri et al. 1992). DNA yield depends on the size, type, age, and quality of the starting material. A sample consisting of small cells, such as transformed lymphoblasts, will have a higher cell density and hence a higher DNA content than the same-sized sample consisting of larger cells.

Protein, RNA, and Other Contaminants

Good-quality DNA has a 260/280 ratio between 1.8 and 2.0. In the event a sample has a 260/280 ratio less than 1.6 (indicative of protein contamination), it should be re-precipitated to improve its purity (Manchester 1995, 1996; Glasel 1997). An $A_{260/280}$ ratio above 2.0 is indicative of RNA contamination, but this usually is not a problem with blood samples. RNase A treatment may be especially important when analyzing pooled DNA samples (e.g., for direct estimation of allele frequencies), to ensure that the pool contains equal amounts of each DNA. Any RNA contamination of the DNA will appear as a diffuse band near the dye front in gel electrophoresis. We have found that phenol–chloroform extraction is rarely required for further DNA purification and never since we began using the Gentra automated extraction procedure. PCR amplification shows that DNA extracted from saliva is contaminated with mycoplasma DNA (A. Sahota et al., unpubl.) and contains variable amounts of bacterial DNA, which may interfere with genotyping assays, especially those involving multiplex platforms. These contaminants have not been observed in blood or cell-line DNA.

Digestion with Restriction Enzymes

As a further check on quality, we digest each DNA sample with two different restriction enzymes and determine the extent of digestion with reference to a control DNA sample (cell line K562) known to be completely digested with these enzymes. We use both a 6-base (EcoRI) and a 4-base (HaeIII) recognition sequence restriction enzyme. These digests give rise to fragment sizes spanning different molecular-weight ranges on agarose gels. Partial digestion with one or both enzymes is readily seen as higher-molecular-weight fragments outside the expected size ranges. Complete digestion with EcoRI also indicates that the input DNA is free of salt contamination, since this enzyme is highly susceptible to variations in salt concentration. Any DNA sample that does not show complete digestion is re-precipitated with high salt and isopropanol, and the digestion process is repeated.

This chapter presents a series of protocols for extraction of DNA from various sources including cell lines, tissues, blood, and saliva. The chapter also includes methods for whole-genome amplification and extraction of DNA for forensic studies. The mention of a product or company name does not imply endorsement by the authors; equally effective products may be available from other companies. The ultimate test of the quality of a DNA sample is analysis in multiple, independent laboratories. DNA prepared using the methods described here has been distributed to numerous laboratories worldwide for genotyping studies without any problems (Alarcon et al. 2002; Holmans et al. 2004; Prasad et al. 2005; Suarez et al. 2006).

ACKNOWLEDGMENTS

We thank the following colleagues for their scientific and technical contributions: Douglas A. Fugman, David A. Toke, Qi Wang, Laura Wilde, David Keller, and Hiep Tran. The work of the Rutgers University Cell and DNA Repository (RUCDR) is supported by grants and contracts from the National Institutes of Health and private organizations. RUCDR serves as the National Cell Repository for the Collaborative on the Genetics of Alcoholism project (NIAAA), Center for Collaborative Studies of Mental Disorders (NIMH), Center for Genetic Studies (NIDA), Genetic Repository (NIDDK), and the Cell and DNA Repository for several private organizations.

Protocol 1

Preparing DNA from Cell Pellets

Amrik Sahota,[1] Andrew I. Brooks,[1,2] and Jay A. Tischfield[1]

[1]Rutgers University Cell and DNA Repository, Department of Genetics, Piscataway, New Jersey 08854-8082; [2]Bionomics Research and Technology Center, Environmental and Occupational Health Sciences Institute, University of Medicine and Dentistry of New Jersey–Robert Wood Johnson Medical School, Piscataway, New Jersey 08854-5635

This protocol describes large-scale DNA extraction from cell pellets using either in-house or commercial (PUREGENE) reagents. The cell pellets are prepared from human lymphoblast cell lines (LCLs) which have been established by using Epstein-Barr virus.

MATERIALS

CAUTION: See Appendix for appropriate handling of materials marked with <!>.

Reagents

Agarose gel (1%)
Chloroform <!>
Ethanol (70%)
Ethidium bromide <!>
Human LCLs
Isopropanol <!>
NaCl (0.85% and 6 M, saturated solution)
Proteinase K (41.7 mg/ml stock solution in H_2O) <!>
PUREGENE reagents (Gentra Systems) (for commercial reagents method)
   Cell Lysis Solution (D-50K2)
   DNA Hydration Solution (D-50K4)
   Protein Precipitation Solution (D-50K3)
   Proteinase K (D-50K5)
   RNase A Solution (D-50K6)
Restriction enzymes: EcoRI and HaeIII
RNase A (DNase-free, 10 mg/ml) <!>
RPMI 1640 medium containing 15% fetal calf serum (FCS)
SDS (20%) <!>
TE buffer (1x) (10 mM Tris-HCl, 1 mM EDTA [pH 8.0])
TKM #2 buffer (10 mM Tris-HCl, 10 mM KCl, 10 mM $MgCl_2$ hexahydrate, 2 mM EDTA, 0.4 M NaCl [pH 8.0])

Equipment

Allegra GS-6 centrifuge with rotor GH-3.8 (Beckman), or a similar tabletop centrifuge
Biological waste bottle
Centrifuge tubes (50-ml conical)

Cryogenic vials (8 ml; 1-ml and 2-ml, bar-coded)
Fishhook (made from a glass Pasteur pipette)
Flasks (T-175)
Incubator preset to 37°C, 56°C
Orbital shaker
Rocking platform
Spectrophotometer
Vi-CELL Cell Viability Analyzer (Beckman Coulter)
Vortex mixer

METHOD 1

DNA Extraction Using In-House Reagents

1. Grow cultures of LCLs in T-175 flasks in RPMI 1640 medium containing 15% fetal calf serum to provide >1.5 x 10^8 cells. For quality control, randomly select 15–20% of the cultures and determine the cell count and viability using a Vi-CELL Cell Viability Analyzer.

2. Collect each culture into a 50-ml conical centrifuge tube and centrifuge at 3600 rpm for 5 minutes to collect the cells. Wash the pellets twice with 0.85% NaCl to remove culture medium and serum proteins.

3. For a pellet containing approximately 2 x 10^8 cells, add 12.8 ml of TKM #2 buffer and 400 μl of 20% SDS, followed by 50 μl of RNase A. Incubate the sample for 1 hour at 37°C.
   Adjust reagent volumes depending on the size of the cell pellet.

4. Add 100 μl of proteinase K and incubate the sample for 3 hours at 56°C. After this point, the samples can be left at room temperature overnight or over the weekend.

5. Add 4.8 ml of 6 M saturated NaCl solution (1.6 M final) to the sample. Vortex the tube for 15 seconds and then allow it to stand at room temperature for 5 minutes to denature and precipitate proteins. Centrifuge at 3000 rpm for 10 minutes and transfer the supernatant to a new 50-ml conical centrifuge tube without disturbing the protein pellet.

6. Add an equal volume of room-temperature isopropanol to the supernatant and invert the tube several times to precipitate the DNA. Using a fishhook (made from a glass Pasteur pipette), pull the DNA strands out of isopropanol and rinse them twice with 70% ethanol. Transfer the tip of the fishhook with the attached DNA into an 8-ml cryogenic vial and add 4 ml of 1x TE buffer. Then add one drop of chloroform to the vial and place it on a rocking platform until the DNA has completely dissolved (several days).

7. Determine the concentration of DNA by taking absorbance measurements at 260 nm and 280 nm in a spectrophotometer. Program the spectrophotometer to automatically calculate the DNA concentration, yield, and $A_{260/280}$ ratio, and upload these data into a DNA database.

8. Check the DNA quality as follows:

   i. Digest 1 μg of DNA with the restriction enzymes EcoRI and HaeIII (in separate reactions).

   ii. Perform electrophoresis of the digests on a 1% agarose gel.

   iii. Stain the gel with ethidium bromide.
   There should be very little or no undigested DNA. Run an aliquot of the undigested DNA on the gel as well, to check that the DNA is of high molecular weight (>23 kb).

9. Aliquot the stock DNA into two or more 2-ml bar-coded cryogenic vials and store them at –70°C. To minimize freezing/thawing of stock DNA, prepare dilutions at 100 ng/μl (or another

specified concentration) with 1x TE buffer and distribute 300 µl (30 µg of DNA) into each of six 1-ml bar-coded cryogenic vials. Store at –70°C until further analysis.

The DNA yield is 1500–2000 µg from a T-175 culture (~10 µg per 10^6 cells). The A260/280 ratio is similar to that for blood, and the DNA is readily digested with EcoRI and HaeIII.

METHOD 2

DNA Extraction Using Commercial Reagents

1. Start with 0.5×10^8 to 1.5×10^8 transformed lymphoblasts. Wash the cells in saline as described in Step 2 of Method 1, and resuspend the pellet in residual saline. Add 15 ml of Cell Lysis Solution containing RNase A (75 µl) and vortex the sample for 30 seconds to lyse the cells.

2. Incubate the sample at 37°C for 15 minutes, cool to room temperature, and then centrifuge briefly to collect all the liquid into the bottom of the tube.

3. Add 5 ml of Protein Precipitation Solution to the sample, vortex for 15 seconds, and then centrifuge at 3600 rpm for 15 minutes to collect the proteins. Pour the supernatant containing the DNA into a clean tube, add 15 ml of isopropanol, and invert the tube 20 times to precipitate the DNA.

4. Pour off the isopropanol into a waste container and drain the tube for 60 seconds. Add 15 ml of 70% ethanol and mix gently to wash the DNA. Centrifuge the samples at 4200 rpm for 15 minutes to collect the DNA. Pour off the ethanol, invert the tube for 60 seconds, and air-dry.

5. Add 4 ml of DNA Hydration Solution to hydrate the DNA, and incubate the sample in an orbital shaker at 56°C for 1 hour. Transfer the DNA into 8-ml cryogenic vials and rotate for several days.

6. Follow Steps 7 and 8 in Method 1 to determine the quantity and quality of the DNA. Store the DNA as described in Step 9.

Preparing DNA from Fixed Tissue: Extraction and Whole-Genome Amplification

Chad Brueck, Clyde Brown, Steve Michalik, Deborah Vassar-Nieto, Ernie Mueller, and Gary Davis

Sigma-Aldrich, St. Louis, Missouri 63103

Genomic DNA can be extracted from fixed tissues, although the inherent state of DNA degradation in these samples can present a significant problem in recovery. Whole-genome amplification (WGA) provides the means to amplify and effectively immortalize genomic DNA from fresh, frozen, archived, or chemically treated samples while maintaining genetic integrity. A significant advantage of this approach is its ability to amplify degraded or severely limited amounts of DNA.

One source of such nucleic acids is the potentially data-rich formalin-fixed paraffin-embedded (FFPE) tissues that are obtained during medical evaluation and are often archived for decades. These readily available clinical materials, which often have corresponding clinical and pathological data, are a notoriously difficult source from which to recover usable genomic DNA. There are

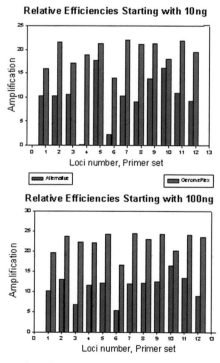

FIGURE 9-1. Results with FFPE samples. This experiment measures the relative amplification efficiencies of both PEP/DOP and MDA when amplifying damaged (FFPE) genomic DNA. The Y-axis is a measurement of 40 minus the Cycle Threshold (Ct) of a dozen gene-specific rat primers. A difference of 10 on this graph is approximately a thousand-fold loss in representation. These data demonstrate that GenomePlex WGA is better suited for DNA that has a high amount of degradation.

several hypotheses regarding DNA damage due to formalin fixation. The most common reasoning is that the DNA is progressively degraded through hydroxymethylation or irreversible cross-linkage of constituent bases. In addition, formaldehyde is known to cross-link proteins directly to DNA through peptide moieties, such as thiol, phenolic, and terminal amino groups. Another hypothesis goes further and suggests that the fixation method aggregates proteins, which subsequently bind DNA, rendering the template inaccessible to DNA polymerases and, hence, making the sequences unrecoverable by most amplification methods.

The GenomePlex WGA system is particularly well suited for these challenging degraded samples. (For a description of GenomePlex WGA strategy, see Protocol 9.) The Tissue WGA kit was formulated to amplify partially degraded genomic DNA from 1 mg of tissue. However, as little as 0.1 mg of tissue has successfully been used with this product. In this approach, FFPE tissue is lysed and genomic DNA is released in a 1-hour proteinase K incubation. During this time, the DNA is partially repaired using both chemical and enzymatic means. Then, an aliquot of lysate is used directly in the WGA reaction. The kit avoids tedious organic solvent extractions to remove excess paraffin, and produces on average 5–10 µg of WGA product. After purification, the WGA product can be analyzed in a manner similar to any other genomic or chromosomal DNA sample (see Fig. 9-1 for quantitative PCR [qPCR] analysis).

This protocol is used for extraction and subsequent amplification of the genomic contents of tissues.

MATERIALS

Reagents

Fresh, frozen, or FFPE tissue (block or slide)
GenomePlex Tissue Whole Genome Amplification Kit (Sigma), which includes:
 Amplification Master Mix (10x)
 CelLytic Y Lysis Solution
 Library Preparation Buffer
 Library Preparation Enzyme
 Library Stabilization Buffer
 Nuclease-Free H_2O
 Proteinase K Solution
 WGA Polymerase

Equipment

Balance for measuring milligram quantities of tissue
Pipettes and tips (dedicated)
Razor blade
Spectrophotometer
Thermal cycler (PE 9700 or equivalent)
Thin-Walled PCR Tubes (0.2 ml or 0.5 ml), or 96-well PCR plates

METHOD

Lysis and Fragmentation

1. Weigh out a 1-mg sample of tissue into a PCR-ready vessel. When working with FFPE tissue, remove excess paraffin with a razor blade.

   As little as 0.1 mg of tissue has been successfully used with this kit. The minimum amount of tissue that can be processed should be determined empirically, starting with 1 mg of tissue.

2. To the sample, add 24 µl of CelLytic Y Lysis Solution and 6 µl of Proteinase K Solution.

3. Incubate reaction at 60°C for 1 hour, then heat to 99°C for 4 minutes.

   The tissue may not be digested after proteinase K digestion. The GenomePlex Tissue Whole Genome Amplification Kit does not require the tissue to be completely digested. Sufficient DNA is released for WGA without completely digesting the tissue.

4. Immediately cool the sample on ice, and then centrifuge briefly to consolidate the contents.

Library Preparation

5. Combine 9 µl of nuclease-free H_2O and 1 µl of the tissue lysate from Step 4 into a fresh PCR-ready vessel.

6. Add 2 µl of 1× Library Preparation Buffer.

7. Add 1 µl of Library Stabilization Solution.

8. Mix thoroughly and place in thermal cycler at 95°C for 2 minutes.

9. Immediately cool the sample on ice, consolidate by centrifugation, and return to ice.

10. Add 1 µl of Library Preparation Enzyme, vortex thoroughly, and centrifuge briefly.

11. Place sample in a thermal cycler and incubate as follows:

    i. 16°C for 20 minutes

    ii. 24°C for 20 minutes

    iii. 37°C for 20 minutes

    iv. 75°C for 5 minutes

    v. 4°C hold

12. Remove sample from thermal cycler and centrifuge briefly. Sample may be amplified immediately or stored at −20°C for 3 days.

Amplification

13. Add the following reagents to the entire 14-µl reaction from Step 12:

    7.5 µl of 10× Amplification Master Mix

    48.5 µl of Nuclease-Free H_2O

    5 µl of WGA DNA Polymerase

14. Vortex thoroughly, centrifuge briefly, and begin thermocycling. The following profile has been optimized for a PE 9700 or equivalent thermal cycler:

Cycle number	Denaturation	Annealing/Extension
1st cycle	2 minutes at 95°C	
20 cycles	15 seconds at 94°C	4 minutes at 65°C

15. After cycling is complete, maintain the reactions at 4°C or store at −20°C until ready for purification and subsequent analysis. Sigma's PCR Cleanup Kit is recommended for WGA DNA purification. The stability of WGA DNA is equivalent to genomic DNA stored under the same conditions. Sample purification is necessary prior to DNA quantitation in order to remove unincorporated primers that will otherwise interfere with some downstream applications.

TROUBLESHOOTING

Problem: The yield is low after cycling.
Solution:

1. The WGA reaction may be inhibited by contaminants in the tissue extract. Dilute inhibitors in the tissue lysate with H_2O by a factor of 1:10 or more prior to continuing with library preparation. Alternatively, too much tissue may have been used. In this case, use less tissue. To test for inhibition, include a DNA control and/or spike 10 ng of purified genomic DNA template into the proteinase K digestion.

2. Less than 0.1–1 mg of tissue may have been weighed. Remove excess paraffin with a razor blade prior to weighing tissue. This ensures that an adequate amount of tissue is used in the subsequent proteinase K digestion.

3. The template may have been of poor quality in FFPE tissues. Formalin fixation may irreversibly damage the DNA if the fixation process is carried out for excessive periods of time, or if the FFPE samples are improperly stored. Obtain properly fixed FFPE tissue samples.

4. Extraction may be insufficient. Incubate samples at 60°C for longer than 1 hour. Samples can be incubated for 2 hours to overnight, followed by incubation at 99°C for 4 minutes.

5. Post-reaction purification may be inappropriate. Use a method that retains single- and double-stranded DNA. Sigma's PCR Cleanup Kit is recommended.

6. The tissue may not be digested after proteinase K digestion. The GenomePlex Tissue Whole Genome Amplification Kit does not require the tissue to be completely digested. Sufficient DNA is released for WGA without completely digesting the tissue.

Problem: qPCR shows significant bias in WGA representation for the gene of interest.
Solution:

1. Controls may be inappropriate. Genomic DNA can only be compared to WGA DNA once the control DNA has been sheared. Use several pooled samples that have been subjected to the fragmentation protocol, or compare against DNA subjected to hydroshearing (Thorstenson et al. 1998).

2. The DNA sample may be limited or degraded. See the solution to the "low yield" problem.

Problem: The negative (no template) control yields a product.
Solution: One or more reagents may have been contaminated with DNA from an outside source. Replace the affected components. (Although this problem may not affect the results, a clean no-template control can only be achieved by replacing the affected components.)

Protocol 3

DNA Isolation from Rat Tail or Ear

Edwin Cuppen

Hubrecht Laboratory, Utrecht, The Netherlands

This protocol describes a rapid procedure for DNA isolation from rat tail or ear punches. The simplest version of the protocol can be scaled for use in 96-well (deep-well) plates. The quality of the DNA is sufficient for any PCR-based genotyping approach.

MATERIALS

CAUTION: See Appendix for appropriate handling of materials marked with <!>.

Reagents

Ethanol (70%)
Isopropanol <!>
Phenol/chloroform (1:1) <!>(Optional, see Step 5)
ProtK lysis buffer
    100 mM Tris-HCl (pH 8.5), 200 mM NaCl, 0.2% SDS<!>, 5 mM EDTA
        Add fresh 1/100 volume of a 10 mg/ml stock of proteinase K <!> (store at −20°C).
TE (10 mM Tris [pH 7.5], 0.1 mM EDTA) (Optional, see Step 13)
Tissue sample (0.25 cm of a tail or a small ear punch)

Equipment

Incubator or water bath preset to 55°C and 80°C
Test tubes or 96-well (deep-well) plates
Vortex mixer

METHOD

1. Transfer a tissue sample immediately after collection to a tube containing 400 μl of freshly prepared ProtK lysis buffer.

2. Close tubes or seal the 96-well plate. Mix by vortexing.

3. Incubate for a minimum of 4 hours or, alternatively, overnight at 55°C. Vortex occasionally.

4. Inactivate the proteinase K by incubation for 15 minutes at 80°C. Cool the plate and centrifuge to collect the condensate.

    Steps 5 to 7 are optional (not compatible with plate format). They should be included when more pure DNA is needed.

5. Add 400 µl of phenol/chloroform (1:1).

6. Vortex for 1 minute. Centrifuge at maximum speed (tubes: 14,000g; plates: 6,000g) for 5 minutes.

7. Transfer the supernatant to a new tube or plate.

   These steps can be repeated until no more protein/SDS (white precipitate) is visible at the interface.

8. Add 300 µl of isopropanol and mix by inverting 10 times.

9. Centrifuge at maximum speed (tubes: 14,000g for 10 minutes, plates: 6,000g for 40 minutes).

10. Discard the supernatant by inverting the tube or plate. Wash with 300 µl of 70% ethanol (do not vortex).

11. Centrifuge at maximum speed (tubes: 14,000g for 5 minutes, plates: 6,000g for 15 minutes).

12. Discard the supernatant by inverting the tube or plate. Air-dry the DNA pellet.

13. Dissolve the pellet in 500 µl of MilliQ H_2O or TE.

Preparing DNA from Buccal Cells

Irena B. King

Molecular Diagnostics Program, Fred Hutchinson Cancer Research Center, Seattle, Washington 98109

DNA for epidemiological studies most frequently comes from blood collections. However, this method is considered too invasive for some participants and is expensive for large-scale studies (King et al. 2002). Collections of other specimens, including hair, fingernails, saliva, and buccal cells, have been developed. Buccal cells can be obtained using the swish method, swab, or a soft cytology brush called a cytobrush. This protocol describes the extraction of DNA from buccal cells collected with a cytobrush. It is based on the use of affinity columns that are suitable for many different biological starting materials, with the initial steps being optimized for the DNA yields in the relevant tissues. The procedure is not high-throughput. Potential drawbacks of making it high-throughput are lower recoveries of DNA and possible variability in quality of data. Such approaches may require additional quality assurance and control checks. Prior to proceeding with the protocol, become familiar with the QIAamp DNA Mini Kit handbook.

Cytobrush Collection

Typically, the number of genotypes analyzed on any given sample is limited by the amount and the quality of the starting material. Blood samples provide ample amounts of good-quality DNA, but DNA obtained from cytobrushes, especially when acquired through mailings, usually is more degraded and contains variable amounts of bacterial DNA. Although bacterial DNA does not interfere with PCR success, buccal DNA use should be validated for specific applications (Nicklas and Buel 2003). Another drawback to DNA from cytobrushes is its limited yield. Although protocols for whole-genome amplification to increase the limiting amount of DNA have been available for at least 10 years, concern persists about the fidelity of these procedures (Lee et al. 2006).

Cytobrush collections require a brief mouth rinse with ordinary tap water and gentle rub against the inside of the cheek with the cytobrush. Morning collections, prior to brushing teeth and eating, usually provide only slightly more DNA than anytime collections. To increase the total DNA yield without any detriment to the participant, 3 cytobrushes per collection are acceptable. This number is generally acceptable to participants and yields about the same amount of DNA on each cytobrush. Although it is not clear how long it takes to achieve mucosal recovery, allowing at least a week between multiple collections is advisable. The amount of total DNA recovered from one cytobrush is on average about 3 µg, with slightly higher yields obtained from participants with increased motivation or training (King et al. 2002). In a recent extraction of DNA from cytobrushes on 1002 participants in the VITAL cohort study (see Acknowledgments), the total yield ranged from 0.1 µg to 30 µg per cytobrush, as shown below.

Summary of DNA Amounts from 1002 Buccal Cytobrush Samples Received through Mail

	Concentration (ng/μl)	Total DNA yield (μg)
Minimum	0.5	0.1
1st %	1	0.2
10th %	5	0.8
25th %	10	1.5
Median	18	2.6
Mean	22	3.3
75th %	29	4.3
90th %	43	6.4
99th %	88	13.2
Maximum	201	30.2

MATERIALS

Reagents

Ethanol, absolute

Phosphate-buffered saline (PBS)

QIAamp DNA Mini Kit solutions (Qiagen) (prepare according to QIAamp DNA Mini Kit handbook)

    Buffer AE (use as supplied)

    Buffer AL (store at room temperature in the dark)

    Buffer AW1 (store at room temperature)

    Buffer AW2 (store at room temperature)

    Proteinase K solution

Salmon sperm DNA standards (ranging from 0 to 250 ng/μl in DNase-free H_2O) (Sigma)

Specimens to be analyzed

For large-scale studies, cytobrushes containing self-collected buccal cells will usually come through postal deliveries in bulk quantities that need prompt processing. Cytobrush mailings within 2–3 days of collection are acceptable. Use bar code labels for mailings to allow quick scanning of the returning specimens. Prior to scanning, refrigerate mailings, especially if scanning cannot be performed on the same day. After scanning, store batches of 25 cytobrushes in specifically designed boxes (2″ × 2″ × 6.25″) at −80°C until analysis. If these storage guidelines are followed, there are no detectable changes in DNA quantity and quality during at least 5 years of storage.

Equipment

Bar code reader (if using bar code labels) connected to a database

Boxes (2″ × 2″ × 6.25″) for storage of cytobrush samples prior to analysis

Conical polypropylene tubes (15 ml, DNase-free)

Cytobrushes (Medical Packaging Corp.)

Labels (5 per sample, white superior laser print) (Island Scientific)

Microcentrifuge tubes (1.5 ml, DNase-free)

Micropipettors and tips (100 μl and 1000 μl)

Plate reader (e.g., SpectraMax 250, Molecular Devices)

Quartz microplates (96 well) for UV reading

QIAamp DNA Mini Kit (250) (Qiagen)

    If QIAamp kits are not available, similar product such as Puregene by Gentra will produce comparable yields.

Racks for 15-ml and microcentrifuge tubes

Scissors

Tissues

Vortex mixer
Water bath, preset to 56°C

METHOD

1. Remove DNA from the cytobrush as follows:

   i. Place the cytobrush containing the sample in a 15-ml conical polypropylene centrifuge tube and cut off the plastic handle, leaving approximately 1″ from the top of the bristles. Add 400 µl of PBS to the tube, vortex, and let stand at room temperature for 15 minutes.

   ii. Add 20 µl of proteinase K solution and 400 µl of Buffer AL. Immediately vortex the mixture for 30 seconds.

   iii. Incubate the conical tube in a 56°C water bath for 30 minutes. Centrifuge the tube at 1800g for 1 minute at room temperature.

   Wipe tubes with tissue at each centrifugation step to prevent any cross-contamination during centrifugation.

   iv. Add 400 µl of 100% ethanol and vortex. Centrifuge again at 1800g for 1 minute at room temperature.

2. Remove the additional DNA from brush bristles as follows:

   i. Remove the brush from the conical tube and carefully place it inside a 1000-µl capacity pipette tip. Place the pipette tip into a second 15-ml conical tube, keeping the mucosal cell mixture in the first conical tube at room temperature.

   ii. Centrifuge the second conical tube containing the brush inside the pipette tip at 1800g for 5 minutes to remove the additional DNA from brush bristles into the conical tube through the opening of the pipette tip. Gently remove and discard the pipette tip containing the brush.

3. Carefully place 700 µl of the mixture containing the mucosal cells from the first conical tube onto a spin column assembly (provided in the kit) and microcentrifuge the spin tubes at 6800g for 1 minute at room temperature. Remove the spin column from the microcentrifuge tube and place it in a new 2-ml collection tube.

4. Combine the remaining contents of the first and second conical tubes and apply the entire volume to the prepared spin column. Microcentrifuge the column at 6800g for 1 minute again.

5. Carefully place 500 µl of Buffer AW1 on the column and centrifuge at 6800g for 1 minute.

6. Discard the wash and place the spin column (containing DNA) in a new 2-ml collection tube (provided in the kit). Carefully place 500 µl of Buffer AW1 on the column and centrifuge at 6800g for 1 minute.

7. Transfer the spin column into a new collection tube and carefully add 500 µl of Buffer AW2, being careful not to wet the rim of the tube. Centrifuge the tube containing the spin column at 20,800g for 3 minutes. Add another 500 µl of Buffer AW2 and repeat centrifugation for 1 minute.

8. Discard the collection tube containing the wash and place the spin column in a 1.5-ml microcentrifuge tube. Pipette 150 µl of Buffer AE (elution buffer) onto the column. Incubate the microcentrifuge tube at room temperature for 10 minutes before centrifuging at 6800g for 1 minute.

9. Pipette out the Elution Buffer (containing DNA) from the microcentrifuge tube and apply once more to the same spin column. (This optimizes DNA yield.) Centrifuge the spin column once more at 6800*g* for 1 minute.

10. Discard the spin column and store the eluted DNA in a labeled and tightly capped microcentrifuge tube at –80°C for long-term, or –20°C for short-term, storage.

11. Estimate DNA concentration and resulting DNA yields. The method will depend on the application needs. Options include the following:

    i. Use the fluorescent dye PicoGreen (Invitrogen). This method is the most expensive, but it is very sensitive, accurate, and uses the least amount of sample.

    ii. Compare band intensities to a standard ladder on an agarose gel. This is the least sensitive method, but it offers information on the size of DNA (DNA quality).

    iii. Estimate DNA quantity spectrophotometrically by constructing a six-point calibration curve with a salmon sperm DNA standard (0–250 ng/μl). Measure absorbance at 260 nm and at 280 nm using 10 μl of sample (diluted 1:20 with DNase-free H_2O). The OD_{260}/OD_{280} ratio R (DNA quality) is considered to be satisfactory over the range of 1.6–2.0.

    Quantifying DNA by measuring absorbance is suitable for most PCR and other downstream applications. Absorbance units should fall between 0.1 and 1.0. Lower DNA concentrations require an alternative technique, such as PicoGreen.

ACKNOWLEDGMENTS

Irena King is grateful to Dr. Emily White, the Principal Investigator of the VITAL study, for providing DNA yield distributions from extractions of 1002 cytobrushes. The VITAL (VITamins and Lifestyle) Study is a cohort study of dietary supplements and cancer risk among 75,000 men and women ages 50–74 in western Washington.

Preparing Genomic DNA from Whole Blood: Small- and Mid-Scale Extraction

Amrik Sahota,[1] Andrew I. Brooks,[1,2] Jay A. Tischfield,[1] and Irena B. King[3]

[1]Rutgers University Cell and DNA Repository, Department of Genetics, Piscataway, New Jersey 08854-8082; [2]Bionomics Research and Technology Center, Environmental and Occupational Health Sciences Institute, University of Medicine and Dentistry of New Jersey–Robert Wood Johnson Medical School, Piscataway, New Jersey 08854-5635; [3]Molecular Diagnostics Program, Fred Hutchinson Cancer Research Center, Seattle, Washington 98109

This protocol describes the extraction of genomic DNA from whole blood samples using either the QIAamp DNA Blood Mini Kits (for 200-µl samples) or the QIAamp DNA Blood Midi Kits (for 0.5-ml samples). It is based on the use of affinity columns that are suitable for many different biological starting materials, with the initial steps being optimized for the DNA yields in the relevant tissues. The procedure is not high-throughput. Potential drawbacks of making it high-throughput are lower recoveries of DNA and possible variability in data quality. Such approaches may require additional quality assurance and control checks. DNA yields for this protocol vary with the number of nucleated cells in the blood sample and the quality of the frozen sample. Expected yields for the small-scale prep are typically 4–12 µg per 200 µl of blood and 25–50 µg per 200 µl of buffy coat (data from Qiagen). Expected yields from 0.5-ml blood samples from healthy donors are between 8 µg and 35 µg. Prior to performing the DNA extraction, it is advisable to become familiar with the QIAamp DNA Blood Mini or Midi Kit handbook. DNA banking must address confidentiality and regulatory issues as specified by institutional review boards, and all blood samples should be handled in accordance with procedures specified in the Hazard Awareness Management Manual (HAMM).

MATERIALS

Reagents

Agarose gel
Ethanol, absolute
Phosphate-buffered saline (PBS, sterile)
QIAamp DNA Blood Midi Kit solutions (Qiagen) (prepare according to QIAamp DNA Blood
   Midi Kit instruction handbook)
   Buffer AE (use as supplied)
      Extra will be needed for DNA samples which are to be normalized (see Step 10iii).
   Buffer AL (store at room temperature in the dark)
   Buffer AW1 (store at room temperature)
   Buffer AW2 (store at room temperature)
   QIAGEN Protease stock solution (store at 4°C; stable for 2–3 months)
QIAamp DNA Blood Mini Kit solutions (for small-scale extraction)
   Buffer AE (use as supplied)

The following three buffers are stable for 1 year at room temperature:

Buffer AL (mix thoroughly by shaking before use)

Buffer AW1 (prepare by adding 25 ml of 96–100% ethanol)

Buffer AW2 (prepare by adding 30 ml of 96–100% ethanol)

Protease

> *Pipette 1.2 ml of the protease solvent (nuclease-free H$_2$O containing 0.04% sodium azide) into the vial containing lyophilized Qiagen Protease. Store the protease in 20-µl aliquots at –20°C to prolong the life of the reagent.*

Samples to be analyzed:

For small-scale extraction: Use 200 µl of blood or buffy coat sample.

For mid-scale extraction: Prepare frozen whole blood aliquots from freshly obtained anticoagulated blood and store at –80°C until analysis. EDTA and heparin may interfere with PCR reactions (Garcia et al. 2002). Extract batches of 16–20 samples. Two or more batches can be extracted per day, depending on the schedule.

Equipment

Bar code reader (if using bar code labels) connected to a database

Centrifuge (ambient temperature) equipped to fit 15-ml conical tubes

Conical polypropylene tubes (15 ml, 2 per sample)

Cryoboxes (9 x 9), 81 vials

Equipment for agarose gel electrophoresis

Heating block or water bath preset to 56°C (for small-scale extraction)

Labels (5 per sample, white superior laser print) (Island Scientific)

Microcentrifuge tubes (1.5-ml, DNase-free, 4 per sample)

Micropipettors and tips (100 µl, 500 µl, 1 ml)

Plate reader (e.g., SpectraMax 250, Molecular Devices)

Quartz microplates (96 well) for UV reading

QIAamp DNA Blood Midi Kit (100) (for mid-scale extraction; Qiagen)

QIAamp DNA Blood Mini Kit (for small-scale extraction; Qiagen)

Racks for 15-ml and microcentrifuge tubes

Vortex mixer

Water bath, preset to 37°C and 70°C

METHOD 1

Small-Scale DNA Extraction (A. Sahota, A.I. Brooks, and J.A. Tischfield)

1. Dissolve any precipitate in Buffer AL by incubating at 56°C. Add 200 µl of sample (blood or buffy coat) to a microcentrifuge tube containing 20 µl of Qiagen Protease (or proteinase K). If the sample volume is less than 200 µl, add enough PBS to bring the volume to 200 µl. Mix by pipetting up and down.

   > *Protease is the recommended enzyme for cell lysis, but proteinase K is preferred for tissue samples or when the lysis buffer contains high concentrations of EDTA or SDS.*

2. Add 200 µl of Buffer AL to the sample, vortex for 15 seconds, and incubate in a water bath or heating block at 56°C for 10 minutes. Pulse-centrifuge the tube after incubation to bring the sample to the bottom of the tube. Add 200 µl of 100% ethanol to the sample and vortex for 15 seconds. Pulse-centrifuge again to remove droplets from inside the lid.

3. Add the mixture (620 μl) to the QIAamp spin column without wetting the rim. Centrifuge at 8000 rpm (6000g) for 1 minute in a microcentrifuge. Discard the collection tube containing the filtrate into a bottle containing disinfectant (Lysol) and transfer the column to a new 2-ml collection tube. Add 500 μl of Buffer AW1 without wetting the rim. Centrifuge as above, discard the collection tube containing the filtrate, and transfer the column to a new 2-ml collection tube.

4. Add 500 μl of Buffer AW1 without wetting the rim. Centrifuge at 8000 rpm for 1 minute. Discard the collection tube containing the filtrate, and transfer the column to a new 2-ml collection tube. Add 500 μl of Buffer AW2 without wetting the rim. Centrifuge at 14,000 rpm (20,000g) for 3 minutes. Discard the collection tube containing the filtrate and transfer the column to a new collection tube.

5. Centrifuge at 14,000 rpm for 1 minute to remove any buffer carryover. Place the QIAamp spin column into a clean 1.5-ml microcentrifuge tube and discard the collection tube containing any remaining filtrate. Elute the DNA by adding 200 μl of Buffer AE to the column. Incubate at room temperature for 5 minutes, then centrifuge at 8000 rpm for 1 minute.

6. Quantify the DNA and analyze using PCR. Keep the DNA at 4°C while in use, or at –70°C for long-term storage.

> *Typical DNA yields are 4–12 μg per 200 μl of blood and 25–50 μg per 200 μl of buffy coat (data from Qiagen). Qiagen columns copurify DNA and RNA. This is not a problem with blood samples, which contain very little RNA. If RNA-free DNA is required, add 20 μl of an RNase A solution (20 mg/ml) to the sample prior to the addition of buffer AL.*

METHOD 2

Mid-Scale DNA Extraction (I.B. King)

1. To thaw frozen whole blood samples, incubate them for 15 minutes at 37°C, and then keep on ice during setup.

> *Anticoagulated whole blood aliquots must be thawed rapidly and then kept refrigerated or on ice.*

2. Pipette 0.5 ml of whole blood into a 15-ml conical polypropylene centrifuge tube and add 0.5 ml of cold sterile PBS. Add 100 μl of Qiagen Protease stock solution and 1.2 ml of Buffer AL to the sample. Cap and mix thoroughly by vortexing for 15 seconds. Incubate sample in a water bath at 70°C for 10 minutes.

3. Remove sample from the water bath and add 1 ml of ethanol. Mix again by vortexing for 15 seconds.

4. Carefully transfer all of the solution to the QIAamp Midi column placed in a 15-ml conical tube. Avoid spilling and do not moisten the rim of the QIAamp Midi column. Close the cap and centrifuge at 1850g (about 3000 rpm) for 3 minutes.

> *Wipe tubes with tissue at each centrifugation step to prevent any cross-contamination during centrifugation.*

5. Remove the QIAamp Midi column, discard the filtrate, and place the QIAamp Midi column back into the 15-ml centrifuge tube. Carefully, without moistening the rim, add 2 ml of Buffer AW1 to the QIAamp Midi column. Close the cap and centrifuge at 4500g (about 5000 rpm) for 1 minute.

6. Carefully, without moistening the rim, add 2 ml of Buffer AW2 to the QIAamp Midi column. Close the cap and centrifuge at 4500g (about 5000 rpm) for 15 minutes.

7. Place the QIAamp Midi column in a clean 15-ml centrifugation tube, and discard the collection tube containing the filtrate. Add 100 μl of Buffer AE, equilibrated to room temperature.

Pipette directly onto the membrane of the QIAamp Midi column and close the cap. Incubate at room temperature for 5 minutes. Centrifuge at 4500*g* (about 5000 rpm) for 5 minutes.

8. For maximum DNA yield, pipette 100 µl of fresh Buffer AE, equilibrated to room temperature, onto the membrane of the QIAamp Midi column. Incubate at room temperature for 5 minutes. Close the cap and centrifuge at 4500*g* (about 5000 rpm) for 5 minutes. Discard the QIAamp Midi column and securely cap the tube.

9. Refrigerate extracted DNA until it is quantified and aliquoted.

10. Once 90 extractions (or 6 batches) are completed, batch the extracted DNA for microplate quantification. Analyze further as follows:

    i. Perform microplate DNA quantification.

    ii. Dedicate 50–100 ng of each sample for DNA quality evaluation on an agarose gel. Score the gel visually as follows: 0 = very good, 1 = good, 2 = poor.

    iii. To prevent freeze/thaw complications, for each sample make 1 aliquot of the parent stock vial and 3 daughter aliquots of normalized vials. The original final volume of the stock vials is based on 150 µl, and daughter aliquots are arbitrarily normalized to a DNA concentration of 30 ng/µl, containing approximately 250 ng of DNA total per aliquot. To do this, add calculated volumes of buffer to suitable concentrations of DNA samples and divide into aliquots.

       When original yields are low, do not normalize; i.e., leave at original concentrations if less than 30 ng/µl.

11. Prepare the samples for freezing as follows:

    i. Use preprinted bar code labels and affix them to microcentrifuge tubes prior to filling the tubes with samples. Give each sample a unique number. Cross-check labeled vials against tracking batch forms for errors. Use unique labels for stock solutions.

    ii. Make sure all microcentrifuge vials are capped securely. Place the aliquots upright in 9 x 9 cryoboxes, starting at the upper-left corner of the box and placing one vial per box in consecutive order. Place stock and aliquot samples separately and mark clearly on the outside of the boxes. The naming convention for labels can be "Study Name, Type of Sample (whole blood DNA), Volume, and Box Number in Series." Verify vial order with the database.

    iii. Freeze at –80°C for long-term storage.

       DNA extraction, quantification, and normalization tracking forms should accompany the samples at all times, with well-defined quality control checks.

Protocol 6

Preparing DNA from Blood: Large-Scale Extraction

Amrik Sahota,[1] Andrew I. Brooks,[1,2] and Jay A. Tischfield[1]

[1]Rutgers University Cell and DNA Repository, Department of Genetics, Piscataway, New Jersey 08854-8082; [2]Bionomics Research and Technology Center, Environmental and Occupational Health Sciences Institute, University of Medicine and Dentistry of New Jersey–Robert Wood Johnson Medical School, Piscataway, New Jersey 08854-5635

This protocol describes methods for the large-scale extraction of DNA from blood using either in-house or commercial (PUREGENE) reagents. Also included is a method for DNA extraction from blood samples that are compromised or clotted.

MATERIALS

CAUTION: See Appendix for appropriate handling of materials marked with <!>.

Reagents

Agarose gel (1%)
Chloroform <!>
Ethanol (70%)
Ethidium bromide <!>
Isopropanol <!>
NaCl (0.85%)
NaCl (6 M, saturated solution)
155 mM NH_4Cl <!>/170 mM Tris (pH 7.7) (prewarmed to 37°C)
Proteinase K (41.7 mg/ml stock solution in H_2O) <!>
PUREGENE reagents (Gentra Systems) (for commercial reagent method and for compromised or clotted blood sample method)
    Cell Lysis Solution (D-50K2)
    DNA Hydration Solution (D-50K4)
    Glycogen Solution (20 mg/ml, R-5010) (for compromised or clotted blood sample method)
    Protein Precipitation Solution (D-50K3)
    RBC Lysis Solution (D-50K1)
    RNase A (D-50K6) (Optional; see Method 2, Step 3)
Restriction enzymes: EcoRI and HaeIII
RNase A <!> (DNase-free, 10 mg/ml) (Optional; see Method 1, Step 5)
SDS (20%) <!>
TE buffer (1x) (10 mM Tris-HCl, 1 mM EDTA [pH 8.0])
TKM #2 lysis buffer (10 mM Tris-HCl, 10 mM KCl, 10 mM $MgCl_2$ hexahydrate, 2 mM EDTA, 0.4 M NaCl [pH 8.0])
Whole blood collected in EDTA
    In-house reagent method: 1–3 tubes, typically 8 ml per tube
    Commercial reagent and compromised or clotted blood sample methods: 5–10 ml

Equipment

Absorbent paper (for commercial reagent method)
Allegra GS-6 centrifuge with rotor GH-3.8 (Beckman), or a similar tabletop centrifuge
Biological waste bottle containing disinfectant (e.g., Lysol)
Centrifuge tubes (50-ml conical)
Cryogenic vials (1-ml and 2-ml, bar coded)
Fishhook (made from a glass Pasteur pipette)
Heating block preset to 56°C (optional)
Ice (for compromised or clotted blood sample method)
Incubator preset to 56°C
Orbital shaker
Rocking platform
Spectrophotometer
Transfer pipette (plastic)
Vortex mixer
Water bath preset to 37°C

METHOD 1

Large-Scale Extraction Using In-House Reagents

1. Isolate white blood cells (buffy coat) from 1–3 tubes of whole blood collected in EDTA. Centrifuge the tubes at 3000 rpm for 10 minutes in an Allegra GS-6 centrifuge with rotor GH-3.8, or a similar tabletop centrifuge.

2. Using a plastic transfer pipette, transfer the buffy coat layer into a 50-ml conical centrifuge tube. Save the top layer (containing plasma and platelets) at –70°C in case it is needed for (future) biochemical studies. Pool the bottom layer (containing red cells) into one of the blood tubes and store at –20°C or –70°C.

   These samples contain sufficient numbers of immature (nucleated) red cells and white blood cells for isolation of small amounts of DNA. Use these to resolve any sample identity problems. The stored samples may be discarded once sample identity is no longer an issue.

3. Add 40 ml of 155 mM NH_4Cl/170 mM Tris (pH 7.7) (prewarmed to 37°C) to the buffy coat layer, mix slowly, and incubate for 7 minutes in a 37°C water bath.

   This step lyses the red cells in the buffy coat and releases hemoglobin. The time and temperature are critical. Red cell lysis is not complete at lower temperatures or shorter times. White blood cell lysis can occur if the sample is left to incubate longer, and this can decrease the DNA yield.

4. Centrifuge the tube at 3000 rpm for 10 minutes and discard the supernatant into a biological waste bottle. Resuspend the white cell pellet in 40 ml of 0.85% NaCl, centrifuge again, and discard the supernatant.

   This step removes any remaining hemoglobin. The pellet should be off-white at this stage; if it is not, the resulting DNA may need to be purified further after the extraction.

5. Resuspend the pellet in 6.4 ml of TKM #2 lysis buffer, followed by 200 μl of 20% SDS (0.6% final) and 50 μl of proteinase K (300 μg/ml final). Incubate the tubes at 56°C for 1–2 hours in an orbital shaker. After this point, the samples can be left at room temperature overnight or over the weekend. RNase treatment is usually not necessary for DNA isolated from whole blood. If necessary, before the addition of proteinase K, add 25 μl of DNase-free RNase (10 mg/ml) to give a final concentration of 35 μg/ml, and incubate the sample at 37°C for 30 minutes.

These reagent volumes are for a white cell pellet of approximately 1 × 10⁸ cells, which is typically obtained from about 20 ml of blood with a normal white cell count (4 × 10⁶ to 8 × 10⁶ per ml of blood). Reagent volumes may need to be increased or decreased depending on the pellet size.

6. Add 2.4 ml of 6 M saturated NaCl solution (1.6 M final) to the sample. Vortex the tube for 15 seconds and then allow it to stand at room temperature for 5 minutes to denature and precipitate proteins. Centrifuge at 3000 rpm for 10 minutes and transfer the supernatant to a new 50-ml conical centrifuge tube without disturbing the protein pellet.

7. Add an equal volume of room-temperature isopropanol to the supernatant and invert the tube several times to precipitate the DNA. Using a fishhook (made from a glass Pasteur pipette), pull the DNA strands out of isopropanol and rinse them twice with 70% ethanol. Transfer the tip of the fishhook with the attached DNA into a 2-ml bar-coded cryogenic vial and add one ml of 1× TE buffer. Then add one drop of chloroform to the vial and place it on a rocking platform until the DNA has completely dissolved (several days).

8. Determine the concentration of DNA by taking absorbance measurements at 260 nm and 280 nm in a spectrophotometer. Program the spectrophotometer to automatically calculate the DNA concentration, yield (based on original blood volume), and $A_{260/280}$ ratio, and upload these data into a DNA database.

   The DNA yield is in the range of 15–35 µg/ml whole blood (depending on white cell count) and the yield from a 20-ml blood sample is around 480 µg. The $A_{260/280}$ ratio is around 1.9.

9. Check the DNA quality as follows:

   i. Digest 1 µg of DNA with the restriction enzymes EcoRI and HaeIII (in separate reactions).

   ii. Perform electrophoresis of the digests on a 1% agarose gel.

   ii. Stain the gel with ethidium bromide.

   There should be very little or no undigested DNA. Run an aliquot of the undigested DNA on the gel as well, to check that the DNA is of high molecular weight (>23 kb).

10. Aliquot the stock DNA into two or more 2-ml bar-coded cryogenic vials and store them at –70°C. To minimize freezing/thawing of stock DNA, prepare dilutions at 100 ng/µl (or another specified concentration) with 1× TE buffer and distribute 300 µl (30 µg of DNA) into each of six 1-ml bar-coded cryogenic vials. Store at –70°C for distribution to genotyping centers, in-house PCR analysis, etc.

METHOD 2

Large-Scale DNA Extraction Using Commercial Reagents

There are many commercial reagent kits for large-scale DNA extractions. After extensive validation studies, we adopted the PUREGENE procedure for large-scale DNA extractions from blood.

1. Transfer 5–10 ml of blood collected in EDTA tubes into a 50-ml conical centrifuge tube. Add RBC Lysis Solution (total volume of blood and lysis solution should equal 40 ml). Gently invert the tube several times to mix the contents. Incubate at room temperature for 7 minutes.

2. Centrifuge the samples in an Allegra GS-6 or similar centrifuge at 3600 rpm for 5 minutes to collect the white blood cells. Pour off the supernatant into a waste beaker. Resuspend the pellet in the residual liquid and add 10 ml of Cell Lysis Solution. Mix several times with a pipette to lyse the cells. Samples can be left in Cell Lysis Solution for at least 2 years.

3. If RNase treatment is required, add 50 µl of RNase A (4 mg/ml), mix by inverting several times, and incubate the samples at 37°C for 15 minutes. Cool the samples to room temperature and add 3.33 ml of Protein Precipitation Solution. Vortex for 15 seconds to mix the Protein Precipitation Solution with the cell lysate.

4. Centrifuge samples at 3600 rpm for 5 minutes. The precipitated proteins should form a tight, dark pellet. Pour the DNA-containing supernatant into a 50-ml conical centrifuge tube and add 10 ml of isopropanol. Gently invert the tubes 20 times to precipitate the DNA. Centrifuge the samples at 3600 rpm for 5 minutes to collect the DNA. Pour the isopropanol supernatant into a waste container and invert the tube with the DNA pellet onto clean absorbent paper for 60 seconds.

5. Add 10 ml of 70% ethanol to the pellet and invert the tube several times to wash the pellet. Centrifuge the samples at 4200 rpm for 15 minutes to collect the DNA. Pour off the ethanol into a waste container, invert the tube onto clean absorbent paper for 60 seconds, and air-dry it for a few minutes.

6. Dispense 1 ml of DNA Hydration Solution to rehydrate the DNA pellet. Incubate in an orbital shaker for 1 hour at 56°C. Transfer the DNA into a 2-ml cryogenic vial and place on a rocking platform at room temperature until the DNA has completely dissolved (several days).

7. Determine DNA quantity and quality and prepare aliquots for long-term storage (see Steps 8–10 in Method 1).

    The typical yield from 10 ml of blood is 300 µg of DNA (range 150–500 µg), depending on white cell count. The DNA is of high molecular weight (>23 kb) and is readily digested with a variety of restriction enzymes, including EcoRI, HaeIII, HinDIII, and PstI. The DNA is reported by Gentra to be stable for at least 10 years at –80°C.

METHOD 3

Compromised or Clotted Blood Samples

Compromised blood samples include the following: (1) samples that spent more than 3 days in transit; (2) samples that have been stored at 4°C for more than 1 week; (3) samples that have been stored at –20°C or –70°C; and (4) samples that were not mixed properly upon collection or were collected incorrectly in a red top (clot) tube. For such samples, use the following method, which is based on Gentra Systems protocols.

1. If the compromised sample is frozen, thaw it quickly at 37°C and transfer it to a 50-ml conical centrifuge tube. Make the volume up to 40 ml with RBC Lysis Solution. Mix well and incubate at room temperature for 5 minutes, mixing at least once during the incubation. Centrifuge at 3600 rpm for 5 minutes. Pipette off most of the supernatant, leaving about 4 ml of liquid in the tube. Vortex for 30 seconds to resuspend the cells in the liquid, add 10 ml of Cell Lysis Solution, and vortex again.

2. Incubate at 37°C for at least 2 hours or at room temperature overnight. The sample should be homogeneous after the incubation. Cool the samples on ice for 5 minutes, add 4.5 ml of Protein Precipitation Solution, and vortex for 20 seconds. Centrifuge at 3600 rpm for 10 minutes. Transfer the supernatant containing DNA into a clean 50-ml conical centrifuge tube and add 13.5 ml of isopropanol and 133 µl of Glycogen Solution (20 mg/ml). Invert the tube several times to mix.

3. Centrifuge at 3600 rpm for 5 minutes to collect the DNA. Discard the isopropanol, drain the tube, and rinse the pellet with 70% ethanol. Centrifuge at 4200 rpm for 15 minutes, discard the ethanol, drain the tube, and air-dry the pellet. Rehydrate the DNA in 500 µl of DNA Hydration Solution.

4. Use the following modifications of this method for clotted samples:

 i. After adding the RBC Lysis Solution, vortex the sample at high speed and then place it on a rotator at room temperature for 5 minutes to disperse as much of the clot as possible.

 ii. Centrifuge at 3600 rpm for 5 minutes, discard the supernatant, and resuspend the pellet in 10 ml of RBC Lysis Solution. Repeat the vortexing and incubation steps.

 iii. Centrifuge the sample, discard the supernatant, and resuspend the pellet in about 400 μl of the residual liquid.

 iv. Add 10 ml of Cell Lysis Solution and 50 μl of Proteinase K Solution (20 mg/ml), vortex, then incubate at 56°C for 2 hours to overnight, until the clot has dissolved completely.

 v. Cool the sample on ice for 5 minutes, add 3.3 ml of Protein Precipitation Solution, and vortex at high speed.

 vi. Centrifuge at 3600 rpm for 10 minutes, cool the sample on ice for 5 minutes, and transfer the supernatant containing the DNA into a clean 50-ml conical centrifuge tube.

 vii. Add 10 ml of isopropanol containing 20 μl of Glycogen Solution (20 mg/ml).

 viii. Wash with 70% ethanol and rehydrate the DNA in 500 μl of DNA Hydration Solution.

DNA yield from compromised blood samples is approximately half that from fresh blood, due to the action of DNases and lysis of white cells during freezing and thawing. The yield from clotted samples is even lower.

Protocol 7

Preparing DNA from Saliva

Amrik Sahota,[1] Andrew I. Brooks,[1,2] and Jay A. Tischfield[1]

[1]*Rutgers University Cell and DNA Repository, Department of Genetics, Piscataway, New Jersey 08854-8082;* [2]*Bionomics Research and Technology Center, Environmental and Occupational Health Sciences Institute, University of Medicine and Dentistry of New Jersey–Robert Wood Johnson Medical School, Piscataway, New Jersey 08854-5635*

Saliva may sometimes be a viable alternative to blood as a source for DNA, and it has the advantage that it is collected noninvasively. This protocol describes the Oragene procedure for DNA extraction from 2-ml saliva samples. It yields high-quality DNA suitable for research and diagnostic applications.

MATERIALS

Reagents

Ethanol (95–100%)
Oragene DNA Self Collection Kit solutions (DNA Genotek)
TE buffer

Equipment

Allegra GS-6 or a similar benchtop centrifuge
Centrifuge tubes (15-ml conical)
Cryogenic vials (2-ml bar coded)
Ice
Incubator preset to 50°C, 56°C
Microcentrifuge tubes (1.5 ml)
Oragene DNA Self Collection Kit (DNA Genotek)
Orbital shaker
Rocking platform
Vortex mixer

METHOD

1. Collect 2 ml of saliva into the plastic container included in the kit. Screw the cup containing 2 ml of Oragene DNA preserving solution behind a plastic seal onto the container with the saliva. The preservative is released into the saliva as the seal is broken. Mix by inverting the container several times.

2. The saliva/Oragene lysate can be transported at ambient temperature and stored for over a year. For longer-term storage, divide the lysate into small aliquots and store in microcentrifuge tubes at −20°C.

129

3. Incubate lysate at 50°C overnight prior to DNA extraction.

The following instructions using the Oragene Purifier solution are based on the Oragene manual.

4. If the entire sample is to be extracted, transfer the lysate to a 15-ml conical centrifuge tube and add 160 µl (1/25 of the lysate volume) of Oragene Purifier solution. Mix the sample by vortexing for a few seconds, incubate on ice for 10 minutes to precipitate proteins and other impurities, and then centrifuge at 2500g for 10 minutes in an Allegra GS-6 or a similar table-top centrifuge. Transfer the supernatant to a clean 15-ml conical centrifuge tube without disturbing the pellet. Discard the pellet.

   Reduce the Purifier volume proportionally for smaller volumes of lysate.

5. Note the volume of the supernatant and add an equal volume of ethanol (95–100%). Mix the sample by inversion and keep the tube at room temperature for 10 minutes to allow DNA to precipitate.

   The DNA may appear as fibers or as a fine precipitate, depending on concentration.

6. Centrifuge the sample as in Step 4 and discard the supernatant. If necessary, carry out a second brief centrifugation to remove the remaining supernatant.

7. Resuspend the DNA in 1 ml of TE buffer and place the tube in a 56°C orbital shaker for 1 hour. Transfer the contents to a 1.5-ml microcentrifuge tube and place it on a rocking platform at room temperature until the DNA has completely dissolved (at least 1 day).

8. Centrifuge the tube at 15,000g (13,000 rpm) for 15 minutes at room temperature to remove any fine particulate material. Transfer the DNA to a 2-ml bar-coded cryogenic vial. Determine the quantity and quality of the DNA. For long-term storage, divide the DNA into small aliquots and store in microcentrifuge tubes at –70°C.

   Based on DNA Genotek data, the DNA concentration following rehydration is 2–200 µg/ml and the median yield is 110 µg (range 15–300 µg) per 2 ml of saliva. The $A_{260/280}$ ratio is around 1.6, but it increases to >1.7 after correction for absorbance at 320 nm (due to sample turbidity). (See Troubleshooting.)

   DNA from the saliva/Oragene mixture may also be extracted using PUREGENE reagents, Qiagen Mini Columns, or other methods.

TROUBLESHOOTING

Problem (Step 8): The DNA solution appears cloudy.

Solution: DNA does not absorb light at 320 nm, but if the solution appears cloudy, use measurements at this wavelength to correct for absorbance due to turbidity. Subtract the A_{320} readings from the A_{260} and A_{280} readings to give the corrected $A_{260/280}$ ratio. This often improves the $A_{260/280}$ ratio from around 1.5 to >1.7. Make this correction when calculating DNA concentration in cloudy samples also.

Protocol 8

Whole-Genome Amplification of Genomic DNA Using PCR

Chad Brueck, Clyde Brown, Steve Michalik, Deborah Vassar-Nieto, Ernie Mueller, and Gary Davis

Sigma-Aldrich, St. Louis, Missouri 63103

Most DNA analyses require microgram quantities of nucleic acid as a starting point. However, even the best DNA extraction methods are unable to produce sufficient material when starting with difficult or limited sources. Even when enough DNA is generated for the initial analytical studies, subsequent testing can rapidly deplete these modest amounts. These concerns have led many researchers to methods that overcome these obstacles.

Whole-genome amplification (WGA) is a popular solution to these problems. It provides the means to amplify and effectively immortalize the genomic DNA from fresh, frozen, archived, or chemically treated samples while maintaining genetic integrity. Currently, two WGA methods are available commercially. The first, multiple strand displacement amplification (MDA), uses a highly processive mesophilic polymerase and random priming to generate a whole genome made up of approximately 20- to 70-kb products (Dean et al. 2001). The other technique employs a combination of primer extension preamplification (PEP) (Zhang et al. 1992) and degenerate oligonucleotide-primed (DOP) (Telenius et al. 1992) PCR to produce whole-genome copies comprising short, ~440-bp products. This approach is described in detail here (schematically depicted in Fig. 9-2).

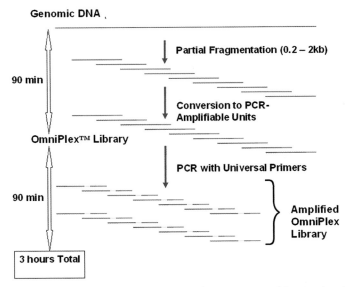

FIGURE 9-2. GenomePlex WGA scheme. GenomePlex whole-genome amplification has three main steps. First, the genomic DNA is randomly fragmented and denatured using a mild heat-based fragmentation (sequence-independently). A partially degenerate primer containing a 5′ universal priming region is then annealed to DNA, and a primer extension step is carried out to generate a DNA library representative of the entire genome. Finally, the entire library is amplified representatively with limited cycles of PCR, resulting in microgram quantities of DNA suitable for almost all downstream analyses. For tissue (i.e., FFPE) or single-cell samples, there is an initial genomic DNA extraction prior to the aforementioned steps.

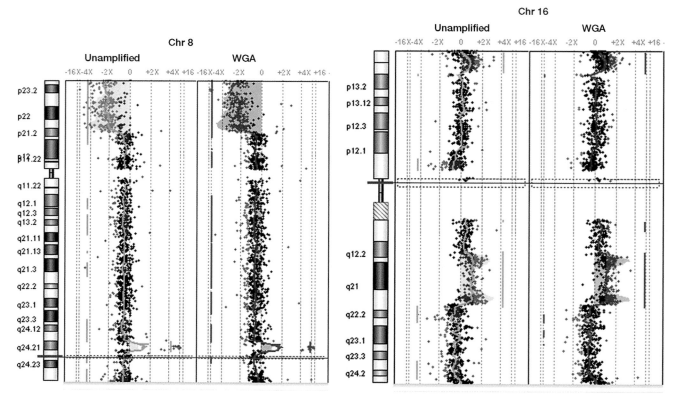

FIGURE 9-3. Comparative genome hybridization (CGH) data. Chemically treated and untreated human cells were run through the GenomePlex WGA protocol. Agilent 44K Human CGH arrays were used to compare the WGA-amplified material (*right side of each panel*) to the original unamplified genomic DNA (*left side of each panel*). Unamplified genomic DNA and WGA DNA samples were Cy3- and Cy5-labeled using BioPrime Array CGH Genomic Labeling System. On one chip, 750 ng of labeled (Cy3 and Cy5) WGA DNA was profiled; on a second chip, 2000 ng of labeled genomic DNA was profiled. These arrays were hybridized for 40 hours. After washing, the chips were analyzed on an Agilent Scanner. Chromosomes 8 and 16 are representative pieces of data obtained from this comparison. The WGA process did not produce detectable bias when compared to original unamplified DNA (Little et al. 2005).

This latter method, sold as the GenomePlex WGA line of products (Sigma-Aldrich), produces a 500- to 1000-fold unbiased amplification (Fig. 9-3). Extensive application of this platform to a range of potentially problematic samples has demonstrated the versatility of this method in amplifying DNA from a diverse array of both plant and animal tissues with fidelity. However, the method's premier value is its ability to amplify degraded or severely limited amounts of DNA (see Protocol 2 in this chapter).

This protocol is appropriate for amplifying genomic DNA that has been previously purified from various sources, including cell cultures, blood, saliva, or fresh or frozen tissues.

MATERIALS

Reagents

GenomePlex Complete Whole Genome Amplification Kit (Sigma), which includes:
    Amplification Master Mix (10x)
    Fragmentation Buffer (10x)
    Library Preparation Buffer

Library Preparation Enzyme
Library Stabilization Buffer
Nuclease-Free H$_2$O
WGA Polymerase
Purified genomic DNA from starting material of interest

> *Whole blood, cultured cells, plant tissue, and bacteria isolated via the GenElute DNA Isolation kits (Sigma) have all produced suitable templates for WGA. For specific protocols for isolating DNA from various sources, see elsewhere in this chapter for isolation from cultured cells (Protocol 1), from blood (Protocols 5, 6, 7), and from saliva (Protocol 8). See also Chapter 7 for methods for isolation of plant DNA.*

Equipment

Pipettes and tips (dedicated)
Spectrophotometer
Thermal cycler (PE 9700 or equivalent)
Thin-Walled PCR Tubes (0.2 ml or 0.5 ml), or 96-well PCR plates

METHOD

Fragmentation

1. Isolate genomic DNA and determine concentration by UV absorption (260 nm). Prepare DNA solution of 1 ng/μl.

2. Add 1 μl of 10x Fragmentation Buffer to 10 μl of DNA (1 ng/μl) in a PCR tube or plate.

3. Place the tube/plate in a thermal block or cycler at 95°C for exactly 4 minutes.
   This incubation is very time sensitive. Any deviation may alter results.

4. Immediately cool the sample on ice, then centrifuge briefly to consolidate the contents.

Library Preparation

5. Add 2 μl of 1x Library Preparation Buffer.

6. Add 1 μl of Library Stabilization Solution.

7. Mix thoroughly and place in thermal cycler at 95°C for 2 minutes.

8. Immediately cool the sample on ice, consolidate by centrifugation, and return to ice.

9. Add 1 μl of Library Preparation Enzyme, vortex thoroughly, and centrifuge briefly.

10. Place sample in a thermal cycler and incubate as follows:

    i. 16°C for 20 minutes

    ii. 24°C for 20 minutes

    iii. 37°C for 20 minutes

    iv. 75°C for 5 minutes

    v. 4°C hold

11. Remove from thermal cycler and centrifuge briefly. Reactions may be amplified immediately or stored at −20°C for 3 days.

Amplification

12. Add the following reagents to the entire 15-μl reaction from Step 11:

    7.5 μl of 10× Amplification Master Mix

    47.5 μl of Nuclease-Free H_2O

    5 μl of WGA DNA Polymerase

13. Vortex thoroughly, centrifuge briefly, and begin thermocycling. The following profile has been optimized for a PE 9700 or equivalent thermal cycler:

Cycle number	Denaturation	Annealing/Extension
1st cycle	3 minutes at 95°C	
14 cycles	15 seconds at 94°C	5 minutes at 65°C

14. After cycling is complete, maintain the reactions at 4°C or store at –20°C until ready for purification and subsequent analysis. Sigma's PCR Cleanup Kit is recommended for WGA DNA purification. The stability of WGA DNA is equivalent to genomic DNA stored under the same conditions. Sample purification is necessary prior to DNA quantitation in order to remove unincorporated primers that will otherwise interfere with some downstream applications.

TROUBLESHOOTING

Problem: There is low yield after cycling.
Solution:

1. The sample may contain PCR inhibitors or high buffer salts. Dialyze in a suitable microdialysis unit. This may dilute the inhibiting components. Quantitate the dialyzed product, as DNA may be lost in this process. Repurify or precipitate with 70% EtOH.

2. Input DNA may be severely degraded or there may be less than 10 ng of it. Amplification of insufficient DNA quantities often results in poor yield or poor representation in the final product. Use more input DNA. This renders some templates amplifiable. Successful WGA amplification has been performed with degraded samples by increasing the starting template to 25–100 ng.

3. More enzyme may be required. WGA yield suffers when limiting amounts of DNA polymerase are used. Use a minimum of 5 μl of WGA DNA Polymerase per 75 μl of reaction. This is preferable to adding cycles, which can cause amplification bias in the resulting DNA.

4. Post-reaction purification may be inappropriate. Use a method that retains single- and double-stranded DNA. Sigma's PCR Cleanup Kit is recommended.

5. The fragmentation reaction may be too long or too short. A 4-minute fragmentation time gives optimal results over a wide variety of DNA samples. Too little or no fragmentation gives low yields and poor gene representation in the resulting WGA product. A 10-minute fragmentation step will also give low yields in almost all cases, because a significant fraction of the DNA is now too small to allow efficient library production.

Problem: qPCR shows significant bias in WGA representation for the gene of interest.
Solution:

1. Controls may be inappropriate. Genomic DNA can only be compared to GenomePlex WGA once the control DNA has been sheared. Use several pooled samples that have been subjected to the fragmentation protocol, or compare against DNA subjected to hydroshearing (Thorstenson et al. 1998).

2. The DNA sample may be limited or degraded. See Step 2 above in the solution to the "low-yield" problem.

Problem: The negative (no template) control yields a product.
Solution: One or more reagents may have been contaminated with DNA from an outside source. Replace the affected components. (Although this problem may not affect the results, a clean no-template control can only be achieved by replacing the affected components.)

ACKNOWLEDGMENTS

We thank Shaukat Rangwala (MOgene, St. Louis, Missouri) for Agilent microarray service and technical troubleshooting.

Single-Cell Whole-Genome Amplification

Chad Brueck, Clyde Brown, Steve Michalik, Deborah Vassar-Nieto, Ernie Mueller, and Gary Davis

Sigma-Aldrich, St. Louis, Missouri 63103

A single human cell, although it contains only 6 pg of DNA, is the obvious level at which to study cell-specific regulation and differentiation. The unknown triggers that signal maturation, regeneration, and inheritable diseases all lie buried in a single cell that was originally part of the genetically clonal, multicellular organism. The primacy of the cell has been reinforced by several studies showing that pooled cell samples, collected in a way such that the pool was thought to be homogeneous, are often actually composed of cells with quite different phenotypes (Reyes et al. 1976).

Until recently, WGA sensitivity has been insufficient to amplify the picogram quantities of DNA in a single cell. The GenomePlex Single Cell Whole Genome Amplification Kit was created specifically to address these limitations. Sensitivity and gene content representation were enhanced by systematic optimization of the DOP and PEP methods. Although the dropout rate for the single-cell application is greater than for the standard GenomePlex application, it provides a million-fold amplification of the single-cell genome and allows genetic analysis at this fundamental level. For example, Figure 9.4 shows SNP data generated from a single U937 cell after being processed with the GenomePlex Single Cell WGA kit.

MATERIALS

Reagents

Isolated single cell
> *Isolate cells by Fluorescence-Activated Cell Sorting (FACS) or the Laser Capture Microdissection (LCM) system. Dilution is not recommended.*

GenomePlex Single Cell Whole Genome Amplification Kit (Sigma), which includes:
  Amplification Master Mix (10x)
  Library Preparation Enzyme
  Library Stabilization Buffer
  Nuclease-Free H_2O
  Proteinase K Solution
  Single Cell Lysis & Fragmentation Buffer (10x)
  Single Cell Library Preparation Buffer
  WGA Polymerase

Equipment

Pipettes and tips (dedicated)
Spectrophotometer
Thermal cycler (PE 9700 or equivalent)
Thin-walled PCR tubes (0.2 ml or 0.5 ml), or 96-well PCR plates

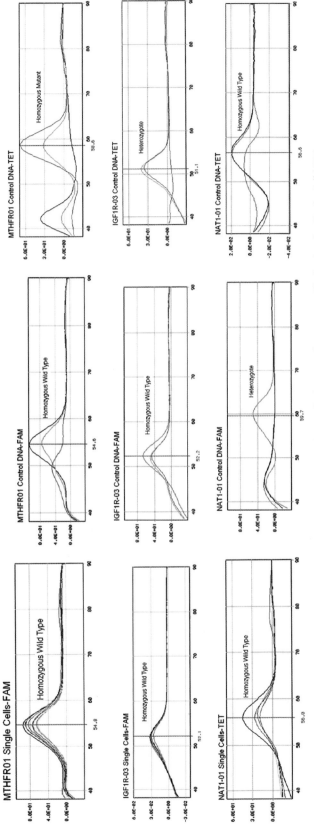

FIGURE 9-4. SNP analysis of single cells. U937 human leukemia cells were amplified using the Single Cell WGA procedure. Post-reaction products were purified by the GenElute PCR Cleanup Kit and quantified via spectrophotometry. 10 ng of resulting single-cell WGA DNA was then analyzed in a MGB Eclipse SNP assay via qPCR on the ABI 7700. The MGB Eclipse probes facilitate allelic discrimination due to their ability to allow post-PCR dissociation curves. Dissociation curves allow determination of the SNP melting point (Tm). Different alleles have substantially different Tm values. The FAM and TET fluorescent signals help distinguish homozygous wild, heterozygous, and homozygous mutant types. The DNA of the WGA-amplified single cells was tested for three SNPs: MTHFR01, IGF1R-03, and NAT1-01. Positive unamplified controls were evaluated with each group. The data indicate that the SNPs were preserved for all WGA-amplified single cells as compared to the unamplified controls (Barker et al. 2004; Gribble et al. 2004).

METHOD

Lysis and Fragmentation

1. Isolate a single cell into a PCR-ready vessel using FACS, LCM, or other method. If cells are sorted, the buffer should be of low ionic strength, such as Tris EDTA (TE) buffer, and the minimal sort volume should be used.

2. Add a sufficient volume of H_2O to the single-cell sample for a final volume of 9 µl.

3. Prepare a working Lysis & Fragmentation Buffer Solution by adding 2 µl of Proteinase K Solution to 32 µl of the 10x Single Cell Lysis & Fragmentation Buffer. Vortex thoroughly.

4. Add 1 µl of the freshly prepared working Lysis & Fragmentation Buffer Solution from Step 3 to the single cell sample. Mix thoroughly.

5. Incubate at 50°C for 1 hour, then heat to 99°C for exactly 4 minutes. Cool on ice, and centrifuge sample prior to proceeding to Library Preparation.

   This incubation is very time sensitive. Any deviation may alter results.

Library Preparation

6. Add 2 µl of 1x Single Cell Library Preparation Buffer.

7. Add 1 µl of Library Stabilization Solution.

8. Mix thoroughly and place in thermal cycler at 95°C for 2 minutes.

9. Immediately cool the sample on ice, consolidate by centrifugation, and return to ice.

10. Add 1 µl of Library Preparation Enzyme, vortex thoroughly, and centrifuge briefly.

11. Place sample in a thermal cycler and incubate as follows:

    i. 16°C for 20 minutes

    ii. 24°C for 20 minutes

    iii. 37°C for 20 minutes

    iv. 75°C for 5 minutes

    v. 4°C hold

12. Remove from thermal cycler and centrifuge briefly. Sample may be amplified immediately or stored at −20°C for 3 days.

Amplification

13. Add the following reagents to the entire 14-µl reaction from Step 12:

    7.5 µl of 10x Amplification Master Mix

    48.5 µl of Nuclease-Free H_2O

    5 µl of WGA DNA Polymerase

14. Vortex thoroughly, centrifuge briefly, and begin thermocycling. The following profile has been optimized for a PE 9700 or equivalent thermal cycler:

Cycle number	Denaturation	Annealing/Extension
1st cycle	3 minutes at 95°C	
25 cycles	30 seconds at 94°C	5 minutes at 65°C

15. After cycling is complete, maintain the reactions at 4°C or store at −20°C until ready for purification and subsequent analysis. Sigma's PCR Cleanup Kit is recommended for WGA DNA purification. The stability of WGA DNA is equivalent to genomic DNA stored under the same conditions. Sample purification is necessary prior to DNA quantitation in order to remove unincorporated primers that will otherwise interfere with some downstream applications.

TROUBLESHOOTING

Problem: The yield is low after cycling.
Solution:

1. The sample may contain salt or inhibitors. Change the sorting process so that inhibitors or salts do not carry over.

2. The template may be of poor quality. The DNA in the single cell may be degraded during the isolation process. The single cell may have been improperly stored. Perform WGA on more than one sample. This helps rule out quality issues that are random in nature, but it will not help if the sorting process damages the sample.

3. Post-reaction purification may be inappropriate. Use a method that retains single- and double-stranded DNA. Sigma's PCR Cleanup Kit is recommended.

4. A single cell may not be captured. Ensure that a single cell is present in the PCR-ready tube. Also, vortex thoroughly after adding the single cell lysis/fragmentation buffer.

5. If diluting sample to a single cell, use multiple reactions as the tube may be empty.

Problem: qPCR shows significant bias in WGA representation for the gene of interest.
Solution:

1. Controls may be inappropriate. Genomic DNA can only be compared to WGA DNA once the control DNA has been sheared. Use several pooled samples that have been subjected to the fragmentation protocol, or compare against DNA subjected to hydroshearing (Thorstenson et al. 1998).

2. The DNA sample may be limited or degraded. See the solution to the low-yield problem.

Problem: The negative (no template) control yields a product.
Solution: One or more reagents may have been contaminated with DNA from an outside source. Replace the affected components. (Although this problem may not affect the results, a clean no-template control can only be achieved by replacing the affected components.)

ACKNOWLEDGMENTS

We thank Michael Speicher (Institute Human Genetics, TU Munich) for Beta-testing the single-cell WGA protocol, Shaukat Rangwala (MOgene, St. Louis, Missouri) for Agilent microarray service and technical troubleshooting, and Barbara Pilas (UIUC, Urbana-Champaign, Illinois) and Joy Eslick (SLU, St. Louis, Missouri), both for FACS for single-cell WGA.

Whole-Genome Amplification Using Φ29 DNA Polymerase

Noël P. Burtt

Program in Medical and Population Genetics, Broad Institute of MIT and Harvard, Cambridge, Massachusetts 02139

The cornerstones of any genetic analysis study are the quality and quantity of the DNA samples. DNA is a precious limited resource, and in human disease studies, the accessibility of sample DNA is often governed by the isolation method and the human source. Additionally, forensic analysis and archaeological research are generally infeasible without intact sample DNA. Therefore, mechanisms to preserve or enhance the quantity of the DNA stock are crucial to the success of these studies. Historically, to preserve and maintain DNA stocks, costly and labor-intensive Epstein-Barr-virus-transformed cell lines were produced. The creation of cell lines can be valuable for a number of reasons, in addition to creating a renewable resource of DNA, but the cost and effort to create them, as well as the requirement of intact cells to begin with, limit the utility of this approach. Recently, several methods have been developed to amplify the entire genome in a uniform and robust manner, while limiting cost and time commitment. Two common PCR-based methods are degenerated oligonucleotide-primed PCR (DOP) (Telenius et al. 1992) and primer extension preamplification (PEP) (Zhang et al. 1992). These methods have some utility. However, the PCR step can introduce artifact and uneven genomic coverage and might result in short fragments in the resulting amplified material (Lovmar et al. 2003).

More recently, whole-genome amplification (WGA) utilizing the unique property of the enzyme, Φ29 DNA polymerase, has been reproducibly demonstrated to generate robust high-fidelity copies of the genome (e.g., Hosono et al. 2003; Luthra and Medeiros 2004; Paez et al. 2004; Pask et al. 2004; Bergen et al. 2005). WGA using Φ29 DNA polymerase allows unbiased representation of the genome via multiple-strand displacement followed by rolling-circle amplification on random primers (Dean et al. 2001). Originally proposed for use in rolling-circle amplification of plasmids, bacteriophages, and other circular genomes, this enzyme was adapted to linear genomes, harnessing its unusual strand-displacement behavior (see Fig. 9-5). Coupled with the addition of random oligonucleotide hexamer structures, 10,000× copies of the genome can produce fragment sizes far larger than previous methods (10 kb or greater). From as little as 10 ng of input DNA, 35–50 µg of product is routinely produced with commercially available kits. Good-quality input DNA is the key to success, as degraded or damaged DNA will not produce a robust product. The main strengths of this process are that cycling is not required (rather, the reaction is isothermal); the error rate of the enzyme is low, on the order of <10 errors per million bases (approximately the same as PCR); and a relatively comprehensive representation of the genome can be achieved (Paez et al. 2004).

Although this method is very robust overall, care must be taken to minimize sample cross-contamination, since any DNA in the reaction will be amplified. Additionally, quality control verification is critical prior to use of the amplified material in genetic analysis. Such measures must go beyond typical quality control measures, such as PicoGreen assays to determine DNA quantity. Commercial vendors suggest TaqMan PCR-based assays to verify that amplification has occurred and is human-DNA specific (rather than random amplification of the hexamers or contaminating DNA, which might occur in a reaction without suitable human DNA as a template). Another

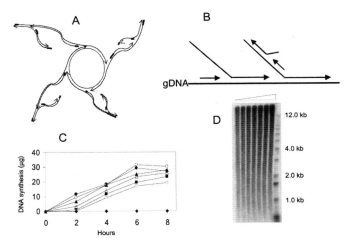

FIGURE 9-5. Multiple displacement amplification (MDA) strategy and product characterization. (*A*) Random-primed rolling-circle amplification of circular DNA templates. DNA synthesis is initiated with random oligonucleotide primers. 3′ ends are indicated by arrowheads. Thickened regions indicate primers. Secondary priming events occur on the displaced product DNA strands. (*B*) Scheme for MDA of genomic DNA. Secondary priming events are initiated from primary products. (*C*) Effect of template concentration on amplification yield. A total of 100 fg to 100 ng of human genomic DNA was amplified by MDA at 30°C. Aliquots were taken from single reactions at the times indicated to quantitate DNA synthesis. Symbols: (●) 10 ng of genomic DNA template; (○) 1 ng; (▲) 100 pg; (△) 10 pg; (■) 1 pg; (□) 100 fg; (◆) primers omitted. (*D*) Denaturing gel analysis of amplification product size. Radioactively labeled amplification products shown in *C* were electrophoresed through an alkaline agarose gel (1%), and the dried gel was exposed to a phosphor screen and imaged. Reaction products were loaded in order of increasing DNA template amount, as indicated above the gel.

valuable suggested check is the genotyping of an informative panel of genetic markers on the amplified material, as well as the native, unamplified DNA. This is obviously possible if native material is available. (We estimate that ~40 ng of DNA is needed to genotype ~50 single-nucleotide polymorphisms [SNPs] in a multiplex reaction.) The genotypes of the two products can then be verified to ensure that sample identity is maintained.

MATERIALS

CAUTION: See Appendix for appropriate handling of materials marked with <!>.

Reagents

Due to patent restrictions regarding the use of Φ29 DNA polymerase, this procedure and reagents are described in accordance with two commercially available WGA kits.

Acidic Buffer (pH 4) (Solution B; REPLI-g kit only)

Amplification Solution (REPLI-g Buffer [4×] in the REPLI-g kit; Reaction Buffer in the GenomiPhi kit)

The following are components of both kits.

37 mM Tris-HCl (pH 7.5)

50 mM KCl

10 mM MgCl$_2$

5 mM (NH$_4$)$_2$SO$_4$ <!>

1 mM dNTPs

50 μM exonuclease-resistant hexamer

1 unit/ml yeast pyrophosphatase

Φ29 DNA Polymerase

This enzyme is used in both kits. Store at –80°C. The right to use the enzyme in rolling-circle amplification is owned by GE Healthcare.

EDTA (0.5 M, pH 8)

Genomic dsDNA (100 ng)

Dilute in nuclease-free H_2O. Input DNA diluted in TE reduces the yield of the reaction.

KOH (5 M) <!>

Nuclease-free H_2O

PicoGreen

0.5× TE Buffer (200 mM Tris, 20 mM EDTA [pH 7.5])

WGA Kit (REPLI-g Kit [Qiagen] or GenomiPhi Kit [GE Healthcare])

Equipment

Ice

Macherey-Nagel NucleoFast plates (Optional; see Step 17)

Microcentrifuge tubes (for single reactions)

Special equipment (for 100 reactions in 96-well format)

96 MultiMek Robot and Stacker (e.g., Beckman Coulter Biomek FX- LP-8927A)

Use this when high-volume processing of more than 500 samples is required.

PCR plate (96 well, e.g., ABgene Thermo-Fast 96 Skirted, Low Profile-AB-0800)

Thermal cycler (e.g., Applied Biosystems Dual 96-Well GeneAmp PCR System 9700-4343176 or equivalent)

In the REPLI-g Kit handbook, the heating equipment is not specified. Alternatively, a water bath or incubator can be used.

Titer Plate Shaker (e.g., Barnstead/Lab-Line 57019-600 or equivalent)

Tubes (conical)

Vortex mixer

METHOD

The start time of the reaction depends on the equipment used to carry out the reaction. If a thermal cycler is used, program the robot to cool down to 4°C. The reaction mixture can be kept at this temperature for hours. If an incubator is used, start the reaction in the afternoon so the incubation will finish in the morning. This procedure is based on the REPLI-g kit, but we will refer to the GenomiPhi kit to indicate differences between the two protocols.

Preparation

1. Defrost Solution B at room temperature. Keep the amplification solution and the DNA polymerase on ice.

2. Add 2.5 µl of genomic DNA to a microcentrifuge tube. For 100 reactions, add 2.5 µl of genomic DNA to a 0.2-ml 96-well skirted plate.

   To optimize the yield, use genomic DNA at a concentration of 50 ng/µl. If the recommended amount (100 ng) is not available, adjust the protocol accordingly for input genomic DNA at a low concentration (>20 ng/µl).

Denaturation

This separates the genomic DNA and gives a single-strand template to facilitate the annealing of random primers.

3. Prepare the denaturing buffer (Solution A) as follows:

	1 reaction	*100 reactions*
0.5 M EDTA (pH 8)	1 µl	100 µl
5 M KOH	4 µl	400 µl
Nuclease-free H$_2$O	35 µl	3500 µl

Low Input DNA

	1 reaction	*100 reactions*
200 M KOH	5 µl	500 µl
Nuclease-free H$_2$O	35 µl	3500 µl

4. Add 2.5 µl of Solution A to the DNA sample(s) and mix thoroughly by pipetting.

   For 100 reactions, start the incubation period once the solution has been added to the last sample.

5. Incubate at room temperature for 2.5–3 minutes.

   In the GenomiPhi protocol, add 22.5 µl of sample buffer to the DNA. Heat the sample(s) to 95°C for 3 minutes and then bring the sample(s) to 4°C, by using the thermal cycler or leaving the sample(s) on ice. The REPLI-g kit does not recommend heating because of concern that heat will denature the DNA too much and cause nicking (Dean et al. 2001).

Neutralization

This is only required in the REPLI-g protocol, to stop denaturation and bring the pH of the sample(s) down to a favorable condition for the enzyme's activity.

6. Dilute Solution B 1:10 (1.8 ml of Solution B, 16.2 ml of nuclease-free H$_2$O). Prepare fresh stock for each experiment.

   Once Solution B is defrosted and at room temperature, it cannot be used again after carrying out the reaction. To ensure a homogeneous mixture, dilute the full amount of Solution B.

7. Add 5 µl of Solution B to the DNA sample(s) and mix thoroughly by pipetting.

Amplification

Combining the random primers, hexamers, and the enzyme initiates the rolling-circle amplification mechanism of MDA.

8. Prepare Master Mix on ice as follows, using the appropriate volumes from the table:

   i. In a conical tube, combine the REPLI-g Buffer (4x) and the H$_2$O. Vortex.

   ii. Add REPLI-g DNA Polymerase and vortex the solution again.

	1 reaction	*100 reactions*
REPLI-g Buffer (4x)	12.5 µl	1.25 ml
REPLI-g DNA Polymerase	0.5 µl	0.05 ml
Nuclease-free H$_2$O	27.0 µl	2.7 ml
Total	40.0 µl	4.0 ml

9. Add 40 µl of Master Mix to the DNA sample(s). Do not mix.

   The solution is viscous, and too much agitation causes bubbling.

   In the GenomiPhi protocol, add the Enzyme Mix and Reaction Buffer to the DNA sample(s).

Incubation

This is best performed with a thermal cycler, because the temperature stays constant at 30°C and the reaction across many samples is more uniform than in an incubator. The priming and amplification occur evenly across the genome and do not create bias.

10. Incubate in the thermal cycler for 16 hours at 30°C.

    In the GenomiPhi protocol, incubate for 4 hours only (due to the amount of primer and dNTPs in the Reaction Buffer as opposed to the REPLI-g Master Mix).

11. At the end of the incubation period, heat the DNA sample(s) to 65°C for 3 minutes and then bring the samples down to 4°C.

    The 3-minute spike in temperature is necessary to stop the exonuclease activity of Φ29 DNA polymerase, which would otherwise start to degrade the product.

Dilution

This makes the WGA product homogeneous and is necessary due to the initial viscosity of the reaction. The resulting working concentration will be 125 ng/μl.

12. Immediately after the reaction has finished, dilute the product 1:10 in 450 μl of 0.5x TE buffer. This is better for long-term storage than dilution in H_2O.

13. Place the DNA sample(s) on a titer plate shaker at a moderate speed overnight at room temperature.

    If the WGA product is not diluted, it is extremely difficult to resuspend. Also, quantitation of an undiluted sample (e.g., with PicoGreen) will result in highly variable concentrations.

Quantification of Yield

14. Make a 1:10 dilution from the WGA stock. If full yield occurred in the WGA reaction, this diluted product should be at a concentration of 12.5 ng/μl.

15. At this concentration range, use PicoGreen (which is automatable and consistent) to assess yield. (See Troubleshooting.)

Optional Quality Control Measures

16. Use the TaqMan copy number assay to assess DNA copy number at two sites in the genome which are susceptible to degradation (see Hosono et al. 2003).

17. Purify the WGA product. Vacuum filtration is a conservative measure that removes extra primers and hexamers. Macherey-Nagel NucleoFast plates use ultrafiltration to remove all salts and primers <10-mer long from solution.

    Purification is affordable and should be done if the DNA will be used for downstream genotyping. It will remove any extra primers or hexamers that might interfere with the chemistry of downstream reactions such as PCR.

TROUBLESHOOTING

Problem (Step 15): The concentration of the product is overestimated.

Solution: PicoGreen is the best option for quantification. Measuring the OD is not effective, because it overestimates the DNA mass and gives higher than expected concentrations. Similarly, the nanodrop method overestimates the concentration because residual hexamers and primers give added mass to the product. In addition, this procedure is not automatable.

Problem (Step 15): PicoGreen produces false positive values.

Solution: Product without DNA template will display false positive PicoGreen values. The remaining double-stranded hexamers will give PicoGreen readings equivalent to samples containing DNA template. Use an additional verification method if possible; specifically, one that indicates the presence of human amplified DNA, such as a PCR-based TaqMan assay.

DNA Extraction from Forensic Samples Using Chelex

John M. Butler

Biochemical Science Division, National Institute of Standards and Technology, Gaithersburg, Maryland 20899-8311

This protocol describes the extraction of DNA from forensic samples using Chelex, a chelating ion exchange resin suspension. It is adapted from the Armed Forces DNA Identification Laboratory (Rockville, Maryland) Standard Operating Procedure.

MATERIALS

CAUTION: See Appendix for appropriate handling of materials marked with <!>.

Reagents

Chelex solution (20%: 2 g of Chelex in 10 ml of H_2O; 5%: 0.5 g of Chelex in 10 ml of H_2O)
Dithiothreitol (DTT; 1 M) <!>
Extraction buffer (10 mM Tris [pH 8.0]; 100 mM NaCl; 50 mM EDTA [pH 8.0]; 0.5% SDS<!>)
Proteinase K (20 mg/ml) <!>
Specimen (swab or fabric cutting)

Equipment

Cryotubes (sterile, labeled)
Forceps (Optional; see Step 6)
Microcentrifuge tubes (1.7 ml)
Pipette tips
Scissors or a sterile scalpel blade
Spin-EASE Tubes
> *If Spin-EASE tubes are not available, substitute a 0.5-ml thin-wall tube with a hole pierced in the bottom (equivalent to basket portion of Spin-EASE) inside a 1.7-ml tube (equivalent to lower portion of Spin-EASE) (piggy-back spin).*

Vortex mixer
Water baths (37°C and boiling)
Weigh boat

METHOD

Recovery of Epithelial and Sperm Cells from Specimen

1. Label the lower portion of an appropriate number of Spin-EASE tubes and 1.7-ml microcentrifuge tubes with a unique identifier.

2. Add 800 µl of sterile H_2O to each labeled Spin-EASE extraction tube. Initiate a substrate control (if applicable) and a reagent blank at this time.

3. Using clean scissors or a sterile scalpel blade, dissect the swab or fabric cutting on a clean surface. Add the specimen (½ of the swab, or ~3-mm square of fabric cutting) to the tube.

4. Vortex briefly and incubate at room temperature for 30 minutes.

5. Vortex for 10 seconds to agitate the cells off the substrate. Pulse-centrifuge.

6. Transfer the substrate to the basket of the Spin-EASE tube with either forceps or a pipette tip. Place the basket containing the substrate into the lower portion of the Spin-EASE unit.

7. Centrifuge the sample in a microcentrifuge at maximum speed (10,000–15,000 rpm) for 1 minute.

8. Label the Spin-EASE basket and place it in a weigh boat in a fume hood. Allow the substrate to air-dry overnight. When it is dry, repackage it and store it as evidence.

9. Without disturbing the pellet, remove and discard all but approximately 50 µl of the supernatant. Resuspend the cellular debris pellet by stirring with a pipette tip.

    This pellet contains both epithelial and sperm cells.

Lysis of Cells and Extraction of DNA

10. Add 150 µl of sterile H_2O and 2 µl of proteinase K (20 mg/ml) to the resuspended cellular debris pellet and mix gently.

11. Incubate at 37°C for approximately 1 hour to lyse the epithelial cells, but for no more than 2 hours to minimize lysis of sperm cells.

12. Centrifuge in a microcentrifuge at 10,000–15,000 rpm for 5 minutes.

13. Without disturbing the sperm pellet, transfer 150 µl of the supernatant to a new 1.7-ml microcentrifuge tube, labeled "E" (Epithelial, or female fraction). Add 50 µl of 20% Chelex solution. Store at room temperature until Step 17.

14. Resuspend the pellet in 500 µl of extraction buffer. Vortex briefly. Centrifuge in a microcentrifuge at 10,000–15,000 rpm for 5 minutes. Remove and discard all but 50 µl of the supernatant. Repeat twice for a total of three washes.

15. After the final buffer wash, add 1 ml of sterile H_2O to the pellet. Vortex briefly and centrifuge in a microcentrifuge at 10,000 rpm for 5 minutes. Remove and discard all but 50 µl of the supernatant.

16. Add 150 µl of 5% Chelex, 2 µl of proteinase K (20 mg/ml), and 7 µl of 1 M DTT to the sperm pellet and mix gently. This is the male, or sperm, fraction of the stain. Label it with an "S."

17. Incubate at 37°C for 30–60 minutes.

18. Vortex the epithelial (E) and sperm (S) cell samples at high speed for 5–10 seconds.

19. Centrifuge the samples in a microcentrifuge at 10,000–15,000 rpm for 10–20 seconds.

20. Incubate the samples in a boiling water bath for 8 minutes.

21. Vortex the tubes for 5–10 seconds.

22. Centrifuge the samples in a microcentrifuge at 10,000–15,000 rpm for 3 minutes.

23. Quantify the samples (see Protocol 12). Then, amplify the samples.

    All nuclear criminal evidentiary extracts must be quantified before amplification.

24. Place at 2–8°C on the Chelex beads for short-term storage. To use after short-term storage on the beads, repeat Steps 20 (optional), 21, and 22. For long-term storage, transfer the supernatant from the Chelex beads to a sterile, labeled tube and freeze at least at –20°C.

Estimation of DNA Concentration in Forensic Samples Prior to Multiplex PCR Amplification

John M. Butler

Biochemical Science Division, National Institute of Standards and Technology, Gaithersburg, Maryland 20899-8311

This protocol describes the use of quantitative PCR (qPCR) for quantitation of DNA in forensics samples prior to PCR amplification. It is adapted from the Armed Forces DNA Identification Laboratory (Rockville, Maryland) Standard Operating Procedure.

MATERIALS

Reagents

Quantifiler qPCR kit (Applied Biosystems), including:
    Quantifiler Human DNA Standards
    Quantifiler Human Primer Mix
    Quantifiler PCR Reaction Mix
      See the Quantifiler Human DNA and Y Human Male DNA Quantification Kit User's Manual (Applied Biosystems 2004) for further information.
Samples (prepared as in Protocol 11)
TE buffer

Equipment

ABI Prism 7500 Sequence Detection System (SDS) (real-time qPCR)
Optical plate (96 well)
Pipettors and tips
Test tubes
Vortex mixer

METHOD

Preparation of Standards, Controls, and Samples

1. Power up the ABI Prism 7500 Sequence Detection System.

2. Prepare Quantifiler Human DNA Standards fresh daily. Label eight tubes 1 through 8, date, and initial. Add the appropriate amount of TE to each tube (see the table below) and vortex the Quantifiler Human DNA Standard to mix thoroughly. Transfer 5 μl of DNA Standard to Tube 1, pipette up and down and then vortex for 30 seconds to mix thoroughly. Change the tip, remove 5 μl from Tube 1 and add it to Tube 2 containing 10 μl of TE. Mix as described above. Repeat until 5 μl of DNA from Tube 7 has been added to Tube 8.

Make sure to pipette the DNA up and down when adding it to the tube, change the tip after addition of DNA from one tube to the next, and vortex to mix the DNA/TE solution before making the next dilution.

Standard	Concentration (ng/µl)	Dilutions	Dilution factor
Std. 1	50.000	5 µl [200 ng/µl stock] + 15 µl TE buffer	4×
Std. 2	16.700	5 µl [Std. 1] + 10 µl TE buffer	3×
Std. 3	5.560	5 µl [Std. 2] + 10 µl TE buffer	3×
Std. 4	1.850	5 µl [Std. 3] + 10 µl TE buffer	3×
Std. 5	0.620	5 µl [Std. 4] + 10 µl TE buffer	3×
Std. 6	0.210	5 µl [Std. 5] + 10 µl TE buffer	3×
Std. 7	0.068	5 µl [Std. 6] + 10 µl TE buffer	3×
Std. 8	0.023	5 µl [Std. 7] + 10 µl TE buffer	3×

3. Remove one tube of Quantifiler Human Primer Mix from the −20°C freezer, thaw, vortex to mix, and centrifuge briefly. Remove Quantifiler PCR Reaction Mix from the refrigerator. Pipette up and down to mix, but do not vortex, because this creates bubbles, which inhibit the quantitation assay.

4. Record standards and samples on a worksheet in the order in which they will be added to the plate. A maximum of 76 samples can be applied to a plate.

5. Vortex and quickly centrifuge all standards and samples that are to be quantitated.

6. Prepare the sample master mix as follows:

   Master Mix
   N = number of samples to be amplified
   N = _____

   Components
   Quantifiler Human Primer Mix (N + 4) × 10.5 µl
   Quantifiler PCR Reaction Mix (N + 4) × 12.5 µl

7. Pipette the sample master mix up and down to mix thoroughly. Pipette 23 µl of the sample master mix into the wells of the 96-well optical plate.

8. Reserve wells A2-B2 and wells G11-H11 for the 4 no-template controls (NTC) and wells A1-H1 and A12-H12 for the 16 DNA standards.

9. For the NTCs, pipette 2 µl of sterile H_2O into the designated wells.

10. Vortex the DNA standards for 15 seconds, shake down, and pipette 2 µl of the appropriate standards into the designated wells (50 ng/µl, 16.7 ng/µl, 5.56 ng/µl, 1.85 ng/µl, 0.620 ng/µl, 0.210 ng/µl, 0.068 ng/µl, 0.023 ng/µl, as in the Step 2 table).

11. For all other samples, pipette 2 µl of the samples into the designated wells.

12. Cover the plate with ABI Prism optical adhesive covers, vortex briefly, and centrifuge at 1000 rpm for 30 seconds.

13. Place the ABI compression pad on top of the optical plate and place the plate in the 7500 Detection System. (Ensure that the plate is oriented so that position A1 is at the back left of the plate.)

Detection

14. Open the Sequence Detection System software. Select File>New. In the dialog box, the Assay should be designated as Absolute Quantitation, Container should be designated as 96-Well Clear, and Template should be designated as Blank Document. Click OK to close the dialog box and open a new plate document.

15. Select "Tools>Detector Manager" and select the Quantifiler Human and internal PCR control (IPC) detectors by clicking on them while holding the "CTRL key." Select "Add to plate document" to add to plate and then select "Done" to exit "Detector Manager."

16. Apply detectors to standards as follows:

    i. Select "View>Well Inspector" to open the dialog box. This shows which detectors have been added to the plate. Make sure to choose ROX as the Passive Reference.

    ii. On the Plate tab, select the well or wells (multiple wells can be selected by holding down the "Control" key and clicking on duplicate wells) that correspond to a specific quantification standard and then go back to the "Well Inspector."

    iii. Select the "Use" boxes for the applicable detectors, IPC and Quantifiler Human.

    iv. For the IPC, select Unknown in the Task column.

    v. For the Quantifiler Human, select Standard from the drop-down list in the Task column and then select the Quantity field for the appropriate detector. Enter the quantity of DNA in the well.

    vi. Enter the Sample Name (e.g., Std. 1, Std. 2).

17. Apply detectors to unknown samples as follows:

    i. On the Plate tab, select wells that correspond to all unknown samples and then go back to the "Well Inspector."

    ii. With the well(s) selected, select View > Well Inspector and check the "Use" boxes for the applicable detectors, Quantifiler Human and IPC. Make sure to choose Rox as the Passive Reference.

    iii. With "Well Inspector" open, select on each individual well and enter Sample Name. Wells that are not in use should be designated as Not In Use.

18. When finished, click on the Instrument tab to enter the program.

    i. Delete the Stage 1 hold step. The amplification program should appear as follows:

    Hold: 95°C for 10 minutes (to activate the Taq polymerase)

    40 Cycles: 95°C for 15 seconds

    60°C for 1 minute

    ii. Enter the volume as 25 µl.

    iii. Check the box for 9600 Emulation.

    iv. Save the document as SDS Documents (*.sds). Select Start.

19. When the run is completed, verify the analysis settings by choosing Analysis Settings from Analysis. Designate the detector as All. Threshold should read 0.2, Baseline Start at 6 cycles, and Baseline end at 15 cycles. Click OK. Select Analyze from the Analysis menu.

20. Choose Export from the File Menu, and export the Results. The results will appear in a Microsoft Excel Workbook. Format the Baseline StdDev, delta Rn, and Quantity columns to four decimal places. Add a column after the Quantity column titled Concentration (ng/µl). Divide all Quantity results by a factor of 2 and record the quantity value in ng/µl on the Sample Placement and Results worksheet. Print out the sample placement and Excel worksheet and place in the appropriate folder(s).

21. Select IPC from the detector drop-down list. If the IPC has a Ct (cycle threshold) value outside the range of 26−29 cycles, print the amplification plot (ΔRn vs. cycle) for the sample and

its corresponding IPC from the SDS software and place in the appropriate case folder(s). (See Troubleshooting.)

Interpretation

22. Determine the quantity of DNA present in each sample by comparing its signal intensity to the intensity of the standard curve produced by the human DNA controls and divide that result by 2 to give the ng/µl results.

23. Under the Results tab, select the standard curve setting to display the standard curve and select Quantifiler Human from the Detector drop-down list. Make sure that the standard curve has a slope ranging from –2.9 to –3.3 and an R^2 value greater than 0.98. (See Troubleshooting.)

24. Check that the Ct values for the standards increase by approximately 1.5 cycles for every threefold reduction in DNA concentration.

25. To view the amplification plots from the standards and samples, select the Amplification Plot tab in the Results tab. Select Quantifiler Human from the Detector drop-down list. Select desired wells.

26. Make sure that the NTCs have Ct values that are greater than 38 cycles or fail to cross the threshold.

27. If a sample comes up as undetermined and the IPC crossed the threshold between 26 and 29, there is no detectable DNA present in the sample.

TROUBLESHOOTING

Problem (Step 21): The IPC has a Ct value outside the range of 26–29 cycles.

Solution: Ct values should be between 26 and 29 cycles for all standards, controls, and samples. An IPC that has a Ct value outside this range, or one that does not cross the baseline, indicates that the sample is inhibited by PCR inhibitors, too much DNA, or too much DNA and PCR inhibitors.

1. Dilute and requantitate the samples.

2. The IPC value is based on 2 µl of extract. The effects of inhibition may be enhanced with increased template volumes.

3. If the IPC never accumulates fluorescence that exceeds the unused probe (usually seen around cycle 6) and the sample has a Ct value greater than the 50 ng standard, the IPC failed due to too much unknown DNA in the reaction. The IPC is supplied at a final concentration of 5 ng, and if the unknown DNA is present at a 20-fold or greater excess (100 ng or more), the IPC can fail due to stochastic amplification of the template DNA over the IPC.

4. If the IPC accumulates fluorescence to a certain point and then hovers there without crossing the threshold, while the unknown sample accumulates fluorescence and crosses the threshold at cycle 36, then the IPC failed due to the presence of PCR inhibitors.

5. If the IPC fails, but the unknown sample does not cross the threshold until cycle 28.3 (a full 7 cycles after the sample in Step 3), IPC failure is due to a combination of PCR inhibitors and high DNA concentration. Most likely, there are PCR inhibitors in the reaction that eventually can be overcome by high DNA sample concentration. However, this leads to a false quantitation of 350 pg/µl.

Problem (Step 23): The standard curve does not comply with the stated guidelines.
Solution:

1. Eliminate no more than two outlier data points by designating those wells as not in use. To mark the standard point not in use, click on the well inspector for the desired well and check Omit Well. Reanalyze the data.

2. If the elimination of those data points does not improve the standard curve, the experiment is invalid. Repeat the experiment.

REFERENCES

Alarcón M., Cantor R.M., Liu J., Gilliam T.C., Geschwind D.H., Autism Genetic Resource Exchange Consortium. 2002. Evidence for a language quantitative trait locus on chromosome 7q in multiplex autism families. *Am. J. Hum. Genet.* **70:** 60–71.

Barker D.L., Hansen M.S., Faruqi A.F., Giannola D., Irsula O.R., Lasken R.S., Latterich M., Makarov V., Oliphant A., Pinter J.H., et al. 2004. Two methods of whole-genome amplification enable accurate genotyping across a 2320-SNP linkage panel. *Genome Res.* **14:** 901–907.

Bergen A.W., Haque K.A., Qi Y., Beerman M.B., Garcia-Closas M., Rothman N., and Chanock S.J. 2005. Comparison of yield and genotyping performance of multiple displacement amplification and OmniPlex whole genome amplified DNA generated from multiple DNA sources. *Hum. Mutat.* **26:** 262–270.

Caputo J.L., Thompson A., McClintock P., and Reid Y.A. 1991. An effective method for establishing human B lymphoblastic cell lines using Epstein-Barr Virus. *J. Tissue Culture Meth.* **13:** 39–44.

Dean F.B., Nelson J.R., Giesler T.L., and Lasken R.S. 2001. Rapid amplification of plasmid and phage dna using phi29 DNA polymerase and multiply-primed rolling circle amplification. *Genome Res.* **11:** 1095–1099.

Dean F.B., Hosono S., Fang L., Wu X., Faruqi A.F., Bray-Ward P., Sun Z., Zong Q., Du Y., Du J., et al. 2002. Comprehensive human genome amplification using multiple displacement amplification. *Proc. Natl. Acad. Sci.* **99:** 5261–5266.

Dunphy C.H. 2006. Gene expression profiling data in lymphoma and leukemia: Review of the literature and extrapolation of pertinent clinical applications. *Arch. Path. Lab. Med.* **130:** 483–520.

Fathallah-Shaykh H.M. 2005. Microarrays: Applications and pitfalls. *Arch. Neurol.* **62:** 16669–16672.

Feigelson H., Rodriguez C., Robertson A., Jacobs E., Calle E., Reid Y., and Thun M.J. 2001. Determinants of DNA yield and quality from buccal cell samples collected with mouthwash. *Cancer Epidemiol. Biomarker Preven.* **10:** 1005–1008.

Garcia M.E., Blanco J.L., Caballero J., and Gargallo-Viola D. 2002. Anticoagulants interfere with PCR used to diagnose invasive aspergillosis. *J. Clin. Microbiol.* **40:** 1567–1568.

Garcia-Closas M., Egan K., Abruzzo J., Newcomb P., Titus-Ernstoff L., Franklin T., Bender P.K., Beck J.C., Le Marchand L., Lum A., et al. 2001. Collection of genomic DNA from adults in epidemiological studies by buccal cytobrush and mouthwash. *Cancer Epidemiol. Biomarker Preven.* **10:** 687–696.

Glasel J.A. 1997. Validity of nucleic acid purities monitored by 260nm/280nm absorbance ratios. *Biotechniques* **18:** 62–63.

Gribble S., Ng B.L., Prigmore E., Burford D.C., and Carter N.P. 2004. Chromosome paints from single copies of chromosomes. *Chromosome Res.* **12:** 143–151.

Heath E.M., Morken N.W., Campbell K.A., Tkach D., Boyd E.A., and Strom D.A. 2001. Use of buccal cells collected in mouthwash as a source of DNA for clinical testing. *Arch. Pathol. Lab. Med.* **125:** 127–133.

Holmans P., Zubenko G.S., Crowe R.R., DePaulo Jr., J.R., Scheftner W.A., Weissman M.M. Zubenko W.N., Boutelle S., Murphy-Eberenz K., MacKinnon D., et al. 2004. Genomewide significant linkage to recurrent, early-onset major depressive disorder on chromosome 15q. *Am. J. Hum. Genet.* 74: 1154–1167.

Hosono S., Faruqi A.F., Dean F.B., Du Y., Sun Z., Wu X., Du J., Kingsmore S.F. Egholm M., and Lasken R.S. 2003. Unbiased whole-genome amplification directly from clinical samples. *Genome Res.* **5:** 954–964.

King I.B., Satia-Abouta J., Thornquist M.D., Bigler J., Patterson R.E., Kristal A.R., Shattuck A.L., Potter J.D., and White E. 2002. Buccal cell DNA yield, quality, and collection costs: Comparison of methods for large-scale studies. *Cancer Epidemiol. Biomark. Prev.* **11:** 1130–1133.

Lahiri D.K., Bye S., Nurnberger J.I., Jr., Hodes M.E., and Crisp M. 1992. A non-organic and non-enzymatic extraction method gives higher yields of genomic DNA from whole blood samples than do nine other methods. *J. Biochem. Biophys. Methods* **25:** 193–205.

Langmore J.P. 2002. Rubicon Genomics, Inc. *Pharmacogenomics* **3:** 557–560.

Lee C.I.P., Leong S.H., Png A.E.H., Choo K.W., Syn C., Lim D.T.H., Law H.Y., and Kon O.L. 2006. An isothermal method for whole genome amplification of fresh and degraded DNA for comparative genomic hybridization, genotyping and mutation detection. *DNA Res.* **13:** 77–88.

Li J., Harris L., Mamon H., Kulke M.H., Liu W.H., Zhu P., and Makrigiorgos G.M. 2006. Whole genome amplification of plasma-circulating DNA enables expanded screening for allelic imbalance in plasma. *J. Mol. Diagn.* **8:** 22–30.

Little S.E., Vuononvirta R., Reis-Filho J.S., Natrajan R., Iravani M., Fenwick K., Mackay A., Ashworth A., Pritchard-Jones K., and Jones C. 2005. Array CGH using whole genome amplification of fresh-frozen and formalin-fixed, paraffin-embedded tumor DNA. *Genomics* **87:** 298–306.

Lovmar L., Fredriksson M., Liljedahl U., Sigurdsson S., and Syvanen A.C. 2003. Quantitative evaluation by minisequencing and microarrays reveals accurate multiplexed SNP genotyping of whole genome amplified DNA. *Nucleic Acids Res.* **31:** e129.

Lum A. and Le Marchand L. 1998. A simple mouthwash method for obtaining genomic DNA in molecular epidemiological studies. *Cancer Epidemiol. Biomark. Prev.* **7:** 719–724.

Luthra R. and Medeiros L.J. 2004. Isothermal multiple displacement amplification: A highly reliable approach for generating unlimited high molecular weight genomic DNA from clinical specimens. *J. Mol. Diagn.* **6:** 236–242.

Madisen L., Hoar D.I., Holroyd C.D., Crisp M., and Hodes M.E. 1987. DNA banking: The effects of storage of blood and isolated DNA on the integrity of DNA. *Am J. Med. Genet.* **27:** 379–390.

Manchester K.L. 1995. Value of A260/280 ratio for measurement of purity of nucleic acids. *Biotechniques* **19:** 208–210.

———. 1996. Use of UV methods for measurement of protein and nucleic acid concentrations. *Biotechniques* **20:** 968–970.

McEwen J.E. and Reilly P.R. 1994. Stored Guthrie cards as DNA banks. *Am. J. Hum. Genet.* **55:** 196–200.

Neitze H. 1986. A routine method for the establishment of permanent growing lymphoblastoid cell lines. *Hum. Genet.* **73:** 320–326.

Nicklas J.A. and Buel E. 2003. Development of an *Alu*-based real-time PCR method for quantitation of human DNA in forensic samples. *J. Forensic Sci.* **48:** 936–944.

Paez J.G., Lin M., Beroukhim R., Lee J.C., Zhao X., Richter D.J., Gabriel S., Herman P., Sasaki H., Altshuler D., et al. 2004. Genome coverage and sequence fidelity of φ29 polymerase-based multiple strand displacement whole genome amplification. *Nucleic Acids Res.* **32:** e71.

Pask R., Rance H.E., Barratt B.J., Nutland S., Smyth D.J., Sebastian M., Twells R.C., Smith A., Lam A.C., Smink L.J., et al. 2004. Investigating the utility of combining Φ29 whole genome amplification and highly multiplexed single nucleotide polymorphism BeadArray™ genotyping. *BMC Biotechnol.* **4:** 15.

Prasad H.C., Zhu C.B., McCauley J.L., Samuvel D.J., Ramamoorthy S., Shelton R.C., Hewlett W.A., Sutcliffe J.S., and Blakely R.D. 2005. Human serotonin transporter variants display altered sensitivity to protein kinase G and p38 mitogen–activated protein kinase. *Proc. Natl. Acad. Sci.* **102:** 11545–11550.

Quackenbush J. 2006. Microarray analysis and tumor classification. *N. Engl. J. Med.* **354:** 2463–2472.

Quantifiler Human DNA and Y Human Male Quantification Kit User's Manual. 2004. Applied Biosystems, Foster City, California.

Reis-Filho J.S., Westbury C., and Pierga J.Y. 2006. The impact of expression profiling on prognostic and predictive testing in breast cancer. *J. Clin. Pathol.* **59:** 2225–2231.

Reyes F., Gourdin M.F., Lejonc J.L., Cartron J.P., Gorius J.B., and Dreyfus B. 1976. The heterogeneity of erythrocyte antigen distribution in human normal phenotypes: An immunoelectron microscopy study. *Br. J. Haematol.* **34:** 613–621.

Rylander-Rudqvist T., Hakansson N., Tybring G., and Wolk A. 2006. Quality and quantity of saliva DNA obtained from the self-administered oragene method—A pilot study on the cohort of Swedish men. *Cancer Epidemiol. Biomark. Prev.* **15:** 1742–1745.

Steinberg K., Beck J., Nickerson D., Garcia-Closas M., Gallagher M., Caggana M., Reid Y., Cosentino M., Ji J., Johnson D., et al. 2002. DNA banking for epidemiological studies: A review of current practices. *Epidemiology* **13:** 246–254.

Suarez B.K., Duan J., Sanders A.R., Hinrichs A.L., Jin C.H., Hou C., Buccola N.G., Hale N., Weilbaecher A.N., Nertney D.A., et al. 2006. Genomewide linkage scan of 409 European-ancestry and African American families with schizophrenia: Suggestive evidence of linkage at 8p23.3-p21.2 and 11p13.1-q14.1 in the combined sample. *Am. J. Hum. Genet.* **78:** 315–333.

Telenius H., Carter N.P., Bebb C.E., Nordenskjold M., Ponder B.A., and Tunnacliffe A. 1992. Degenerate oligonucleotide-primed PCR: General amplification of target DNA by a single degenerate primer. *Genomics* **13:** 718–725.

Thorstenson Y.R., Hunicke-Smith S.P., Oefner P.J., and Davis R.W. 1998. An automated hydrodynamic process for controlled, unbiased DNA shearing. *Genome Res.* **8:** 848–855.

Tomlins S.A., Mehra R., Rhodes D.R., Shah K.B., Rubin M.A., Bruening E., Makarov V., and Chinnaiyan A.M. 2006. Whole transcriptome amplification for gene expression profiling and development of molecular archives. *Neoplasia* **8:** 153–162.

Willinger W.W., Mackey M., and Chomczynski P. 1997. Effect of pH and ionic strength on the spectrophotometric assessment of nucleic acid purity. *Biotechniques* **22:** 474–481.

Zhang L., Cui X., Schmitt K., Hubert R., Navidi W. and Arnheim N. 1992. Whole genome amplification from a single cell: Implications for genetic analysis. *Proc. Natl. Acad. Sci.* **89:** 5847–5848.

10 | Intermediate-Throughput Laboratory-Scale Genotyping Solutions

Stuart J. Macdonald

Department of Ecology and Evolutionary Biology and Department of Molecular Biosciences, University of Kansas, Lawrence, Kansas 66045

INTRODUCTION

In recent years, tremendous technological advances have enabled many thousands of single-nucleotide polymorphisms (SNPs) to be genotyped simultaneously using ultrahigh-throughput genotyping platforms (for review, see Chapter 13). These technologies have been a boon for human medical geneticists, allowing previously inconceivable experiments to be carried out in an increasingly routine fashion. However, many studies do not require genotypes for thousands of SNPs, and ultrahigh-throughput genotyping platforms are inappropriate. This review is intended as a guide for those interested in carrying out genotyping projects where the total effort is on the order of a few thousand to a few hundred thousand genotypes.

The chapter begins with a survey of the available genotyping platforms, focus-

ing primarily on how each assay format discriminates between SNP alleles, and how allele-specific genotyping products are detected. For each method, the multiplexing capability is noted; i.e., how many SNPs can be genotyped in a single reaction. The template for most of the genotyping methods discussed is a PCR-amplified fragment spanning the target SNP. Multiplexing at the PCR step—by amplifying many independent fragments in the same reaction—can enable higher genotyping throughput. Designing and performing multiplex PCR procedures is neither trivial nor routine, and discussion of the techniques is outside the scope of this review. In the second section, the criteria that should be used when choosing a genotyping platform are defined, and four hypothetical genotyping projects (each requiring a different number of target SNPs and a different number of individuals) are presented. Our final discussion describes how additional, non-target DNA polymorphisms can affect genotyping accuracy. The following chapter provides a series of detailed genotyping protocols.

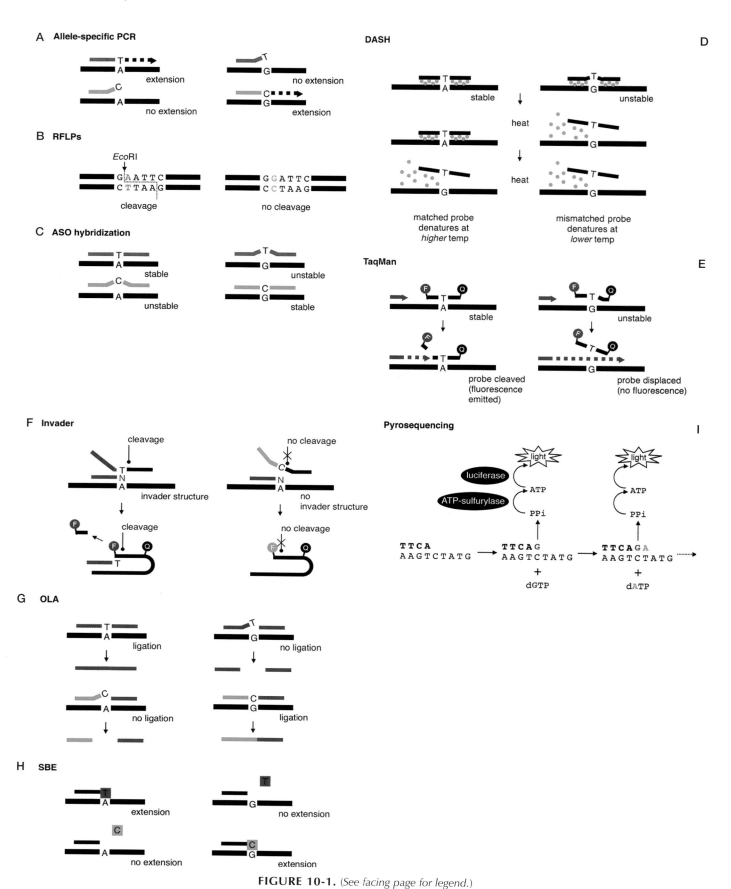

FIGURE 10-1. (See facing page for legend.)

GENOTYPING ASSAY FORMATS

Sequencing

For certain projects, it may not be necessary or worthwhile to invest time and energy exploring any SNP genotyping technology: When the number of individuals and loci to be examined is low, it may be expedient simply to sequence through the desired regions in all individuals. Sequencing provides information on all polymorphisms in the region, not only the select few SNPs initially envisioned. Furthermore, rare third (or even fourth) alleles at SNPs can be identified by sequencing. In contrast, many of the genotyping technologies discussed below assume a priori that all target SNPs are biallelic.

Allele-specific PCR

SNPs can be successfully genotyped using an allele-specific PCR approach (Newton et al. 1989; Li et al. 1990; Bottema and Sommer 1993), sometimes referred to as amplification refractory mutation system (ARMS; Newton et al. 1989). Regular PCR amplification of a stretch of DNA uses forward and reverse PCR primers that are both perfectly matched to the target sequence. With allele-specific PCR (Fig. 10-1A), two allele-specific forward oligonucleotides are developed, each terminating at the 3′-end with a nucleotide corresponding to one of the two SNP alleles. Under appropriately stringent conditions, annealing of an imperfectly matched forward PCR primer will be very unlikely, and no PCR product will form. For each sample, two reactions are carried out, each with just one of the allele-specific forward primers and the common reverse primer. PCR product presence/absence in the two reactions, as revealed by standard ethidium-bromide-stained agarose gels, will reveal the genotype of the SNP. By altering the length of one of the allele-specific forward primers, all three oligonucleotides can be used competitively in the same reaction, and the SNP genotype will be indicated by the size of the resulting PCR fragments (Li et al. 1990; Bottema and Sommer 1993). Finally, in the tetra-primer ARMS method (Ye et al. 2001), four PCR primers are used simultaneously: Two inner allele-specific primers, each ending with a 3′-discriminatory nucleotide, but each designed to prime from alternate DNA strands, are used together with a pair of perfectly matched outer primers.

FIGURE 10-1. Overview of genotyping assay formats. (*A*) Allele-specific PCR. If the allele-specific oligonucleotide (*red/green*) matches the sequence of the target (*black*), the oligonucleotide will be extended; otherwise, no PCR product will be obtained. (*B*) RFLPs. A restriction enzyme (e.g., EcoRI) is used to specifically cleave a PCR fragment if the SNP allele preserves the restriction site. (*C*) ASO hybridization. Labeled ASOs (*red/green*) are hybridized to target (*black*) DNA tethered to a solid support. ASO–template duplexes are only stable when perfectly matched. (*D*) DASH. A single allele-specific probe is hybridized to target in the presence of a double-stranded DNA-specific dye. The probe–target duplex is then heated, and a mismatched probe will denature, with a loss of fluorescence, at a lower temperature than a matched probe. (*E*) TaqMan. TaqMan FRET probes harboring a 5′ fluorophore and a 3′ quencher (Fig. 10-2) are hybridized to target and form a stable duplex if the probe perfectly matches the target. Only one of the two probes required is shown in the diagram. If the probe is matched, it is cleaved by the polymerase as a PCR primer (*blue*) is extended, releasing a fluorescent signal. If the probe is mismatched, the probe–template duplex is unstable, the probe is displaced intact, and no fluorescent signal is emitted. (*F*) Invader assay. An invader oligonucleotide (*blue*) and a pair of allele-specific oligonucleotides are allowed to hybridize to target. Only one of the two SNP alleles is shown in the diagram. When the allele-specific oligonucleotide matches template, a specific invader structure is formed and the flap (*red*) of the allele-specific oligonucleotide is released. This flap serves as an invader oligonucleotide in a secondary invasive cleavage reaction involving a FRET probe, releasing a fluorescent signal on cleavage. (*G*) OLA. A pair of allele-specific oligonucleotides (*red/green*) and a single locus-specific oligonucleotide (*blue*) are permitted to hybridize to target. When perfectly matched oligonucleotides are immediately juxtaposed to form a duplex with template, they are ligated together. Various methods are available to distinguish ligated products from unligated oligonucleotides. (*H*) SBE. A single genotyping probe is designed to anneal immediately upstream of the target SNP. DNA polymerase is used to recruit a single labeled nucleotide complementary to the SNP allele and extend the probe by a single nucleotide. (*I*) Pyrosequencing. A genotyping probe (*bold*) is designed to anneal upstream of the target SNP(s) and is extended by sequential addition of single deoxynucleotides (dATP, dCTP, dGTP, dTTP, dATP, and so on). If a complementary dNTP is incorporated, a series of enzymatic reactions takes place, resulting in the emission of light.

Each outer primer is placed a different distance from the target SNP. Following PCR, all samples will include the product amplified by the outer primers (as a control against PCR failure) and one or two additional bands, depending on the genotype of the sample.

The melting temperature-shift assay (Germer and Higuchi 1999; Wang et al. 2005) is based on allele-specific PCR. However, instead of detection by electrophoresis, allele-specific products are distinguished by melting temperature. The expected product size is kept short by placing the common primer around 20 bp downstream from the SNP. A detectable difference in melting temperature between the two short allele-specific products is engineered by incorporating GC-rich tails of unequal length (6 versus 14 nucleotides) into the allele-specific PCR primers. PCR is carried out with all three primers in the presence of the double-strand-specific DNA dye SYBR Green I. Following PCR, melting temperature curves are generated using a real-time PCR machine to measure the reduction in fluorescence as the PCR products denature. The GC-rich tails will cause the melting temperatures of the allele-specific products to differ, allowing the two products to be distinguished.

Allele-specific PCR, although cost-effective for a few SNPs, is unlikely to translate well to projects involving a large number of SNPs. First, assays cannot be multiplexed. Second, stringent conditions are required to prevent the mismatched primer from annealing with the same efficiency as the perfectly matched primer. It is unlikely that one set of PCR conditions will be routinely useful for all assays.

Restriction Fragment Length Polymorphisms (RFLPs)

A simple and direct way of discriminating SNP alleles is to use a restriction enzyme that recognizes one of the SNP alleles (Saiki et al. 1985). Following PCR amplification of a fragment of the genome containing the target SNP, the PCR product is digested with the enzyme, cutting only when the restriction site is preserved (Fig. 10-1B). The ability to genotype any given SNP with a PCR-RFLP assay is limited only by the availability of an enzyme with a suitable recognition sequence. Software is available to aid in the design of PCR-RFLP genotyping assays (SNP cutter, http://bioinfo.bsd.uchicago.edu/SNP_cutter.htm; Zhang et al. 2005). The products of digestion can be simply and cost-effectively run on standard ethidium-bromide-stained agarose electrophoresis gels, and the fragment sizes (hence, genotypes) can be scored manually. Alternatively, products can be fluorescently labeled and resolved on a capillary sequencer. The primers used in the initial PCR can be specifically labeled, or a general post-digestion fluorescent labeling method can be used (Lazzaro et al. 2002). Fluorescent PCR-RFLP assays allow a modest level of multiplexing. Multiplex PCR, using a different fluorophore for each product, and multiplex restriction digests, yields restriction fragments of varying size and fluorophore (e.g., Thomas et al. 1999).

The absence of a restriction product in a PCR-RFLP assay is consistent with both the presence of the alternate allele and technical failure of the assay. Thus, it is important to assess the rate of PCR-RFLP assay failure in known-genotype control samples and to replicate the genotyping of some or all of the samples. This constraint will prevent PCR-RFLP assays from being generally useful for large genotyping projects.

Allele-specific Hybridization

Several genotyping technologies are based on distinguishing SNP alleles via hybridization to allele-specific oligonucleotide (ASO) probes (Fig. 10-1C). Since no enzymes are involved in the allele discrimination, hybridization is one of the simplest strategies for SNP genotyping. Under high-stringency conditions, short oligonucleotide probes will only anneal to complementary DNA target sequences if they are perfectly matched—a single mismatched nucleotide in the probe is sufficient to prevent stable hybridization (Wallace et al. 1979; Conner et al. 1983). The simplest application of this approach is to PCR-amplify the target genomic region, array PCR products onto nylon membranes, and probe with each radiolabeled ASO probe (Saiki et al. 1986). The ASO probes are normally short (10–20 nucleotides) and identical except for a single central allele-specific nucleotide (for a SNP) or string of nucleotides (for an insertion/deletion polymorphism). Since a large number of samples can be arrayed, ASO can be an effective method to scan many individuals. The key technical difficulty is that the stringency conditions allowing discrimination of the match and mismatch probes must be optimized for each SNP. This potentially limits the number of SNPs that can be easily genotyped.

Dynamic allele-specific hybridization (DASH) is a SNP genotyping technology that relies on indirectly monitoring the denaturation of hybridized probe–target duplexes (Howell et al. 1999; Prince et al. 2001). Double-stranded DNA-specific dyes (e.g., ethidium bromide, SYBR Green I) can be used to generate DNA melting curves by monitoring dye fluorescence as temperature is increased (Ririe et al. 1997). Denaturation of a double-stranded product is observed as a rapid loss of fluorescence near the product's melting temperature. The DASH assay takes advantage of this observation (Fig. 10-1D). Target DNA is amplified by PCR using one primer labeled with biotin, and the biotinylated strand is subsequently bound to a streptavidin-coated surface (such as a microtiter plate well). A 15- to 21-nucleotide probe, centered on the target SNP, is developed for one of the alleles. This probe is allowed to hybridize to the single-stranded bound PCR product at low temperature in the presence of an intercalating double-strand-specific DNA dye. The amount of fluorescence emitted after hybridization is related to the amount of target–probe duplex. Following the fluorescence in a real-time PCR machine as the sample is heated reveals the melting temperature of the duplex. Because a single mismatch between the probe and target sequences results in a dramatic reduction in melting temperature, the genotype of the SNP is easily detected (Howell et al. 1999). Just as with array-based ASO methods, probe design is a crucial determinant of DASH assay success. For an overview of the criteria, see Prince et al. (2001). Because DASH only examines a single allele-specific probe in each reaction, two reactions may be required for each SNP. This lack of multiplexing implies that DASH may be limited to surveys involving only a small number of SNPs. To overcome this limitation, the DASH-2 approach (Jobs et al. 2003) allows array-based, multiplexible DASH genotyping by tethering single-stranded PCR products to the surface on an array (rather than a microtiter plate).

One of the more widely employed allele-specific hybridization methods is the 5′-nuclease, or TaqMan, assay (Fig. 10-1E) (Livak 1999; De La Vega et al. 2005). TaqMan is capable of genotyping both SNPs and small insertion/deletion polymorphisms directly from genomic DNA because PCR is an integral component of the assay. Multiplexing is difficult. A TaqMan assay consists of a pair of PCR primers that amplifies a region surrounding the target SNP, and a pair of allele-specific flu-

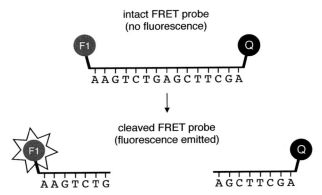

FIGURE 10-2. FRET probe. A FRET probe consists of a short stretch of nucleotides complementary to a specific DNA sequence, commonly that surrounding a specific SNP allele. The FRET probe is 5′-labeled with a fluorophore, and 3′-labeled with a quencher. The emission spectrum of the fluorescent dye must overlap with the excitation spectrum of the quencher, such that excitation energy absorbed by the fluorophore is not emitted as fluorescence, but is instead transferred to the quencher by resonance. The efficacy of the fluorescent signal quenching is strongly dependent on the distance between the fluorophore and the quencher: Provided quencher and fluorophore are in close proximity, the probe emits no fluorescence. However, if the two are separated (e.g., because the probe is cleaved), fluorescence is emitted from the fluorophore.

TABLE 10-1. Design Rules for TaqMan Assays

Primers	Probes
30–80% GC content	
no runs of > 3 consecutive G nt	
T_m of 58–60°C[a]	T_m of 65–67°C[a]
Forward/reverse primers should be as close as possible with overlapping probe	no G nt at 5′ end
5 nt at the 3′ end should include only 1–2 G or C nt	select probe from strand holding more C than G nt

Table modified, with permission, from Livak (1999 [© Elsevier]).
[a]Melting temperate (T_m) as provided by the Primer Express software (Applied Biosystems).

orescence resonance energy transfer (FRET) probes, each of which has a different 5′ fluorescent label, and a generic 3′ quencher (Fig. 10-2). In an intact probe, the quencher is physically close to the fluorophore, and the fluorescent signal is suppressed (Livak et al. 1995). Both the primers and the labeled probes are added to genomic DNA in the presence of a Taq DNA polymerase exhibiting 5′ to 3′ nuclease activity and subjected to a thermal cycling protocol. During PCR, the probes are able to hybridize to the target allele, and as the PCR primers are extended by the polymerase, the enzyme will encounter the hybridized probe. When this occurs, the probe is cleaved by the nuclease activity of the polymerase, causing the fluorophore to be dissociated from the quencher, releasing a fluorescent signal. Mismatched probes are not cleaved by the polymerase because hybridization of a mismatched probe to the target sequence is unstable, and the mismatch instead promotes displacement of the probe. Thus, there is considerably less time for cleavage to occur than with a perfectly matched probe. As PCR progresses, there will be an exponentially increasing signal from one/both of the fluorophores, depending on the genotype of the sample. This fluorescence can be detected using a real-time PCR machine. As with all allele-specific hybridization technologies, specific allele discrimination by TaqMan is strongly dependent on the design of the assay. Design guidelines are presented in Table 10-1 (reproduced from Livak 1999), and a custom assay design service is also available (http://www.appliedbiosystems.com).

Invasive Oligonucleotide Cleavage

The Invader assay relies on the cleavage, irrespective of nucleotide sequence, of a specific triplex structure that is formed by hybridization of a pair of oligonucleotides to a strand of DNA (Fig. 10-1F) (Olivier 2005). One oligonucleotide—the invader oligonucleotide—has its 3′ end at the target SNP (although the identity of the 3′-terminal nucleotide is not important) and is complementary to the sequence upstream of the SNP. The second oligonucleotide is allele-specific, harbors the nucleotide complementary to the SNP allele, and extends through the sequence downstream from the target SNP. Additionally, the allele-specific oligonucleotide has a 5′ tail (a so-called "flap") of noncomplementary nucleotides. Thermostable flap endonucleases (FENs) isolated from archaea can detect and cleave the structure formed when the two oligonucleotides hybridize simultaneously to a target strand of DNA, releasing the flap (Lyamichev et al. 1999). The invader structure does not form, and cleavage does not occur, if there is mismatch between the SNP allele and the allele-specific oligonucleotide. The allele-specific oligonucleotides can be FRET probes (Fig. 10-2), each containing a different fluorophore. Endonuclease cleavage of the invader structure releases the flap from the allele-specific oligonucleotide, the quencher can no longer suppress the fluorescent signal, and fluorescence is emitted.

Invader assay reactions are conducted at a set temperature—there is no thermal cycling. To ensure that allele-specific fluorescence reaches a detectable level, the melting temperature of the allele-specific oligonucleotides is tailored such that at the reaction temperature, the oligonucleotides continually anneal and detach from the target. In contrast, the invader oligonucleotide is designed to anneal permanently to the target. Thus, the invader structure is continually produced and continually cleaved, and the amount of fluorescence emitted increases as the reaction progresses.

A concern with this form of the Invader assay is the requirement for two expensive FRET probes for every SNP genotyped. This problem is alleviated by SISAR, the serial invasive signal amplification reaction (Hall et al. 2000). Here, the allele-specific oligonucleotides are unmodified but contain 5 flaps of different sequence. The pair of flap sequences are generic, and specific to a pair of universal FRET cassettes (Fig. 10-1F). If a flap is cleaved during the Invader assay, the released flap itself serves as an invader oligonucleotide in a secondary cleavage reaction involving a FRET cassette. The ends of the cassette and the flap invader oligonucleotide combine to form the specific invader structure, which is in turn cleaved, releasing the fluorophore from the quencher and emitting a fluorescent signal. Both forms of the Invader assay can work from PCR-amplified DNA, but SISAR procedures are sufficiently sensitive to work directly from genomic DNA (Hall et al. 2000), although a large amount is required (20–100 ng; Olivier 2005). Neither form of Invader assay is multiplexible, but development of solid-phase detection strategies may permit higher-throughput analysis. For example, labeled primary invader probes can be attached to beads, and following cleavage reactions, the beads can be analyzed by flow cytometry to detect their fluorescent signal (Rao et al. 2003). Alternatively, invader probes can be tethered to a microarray slide surface and, post-cleavage, the slide can be read using a conventional imaging system (Lu et al. 2002).

Oligonucleotide Ligation Assay (OLA)

The OLA (Landegren et al. 1988) uses a set of three oligonucleotides, in combination with a thermostable Taq DNA ligase enzyme, to discriminate SNP alleles (Fig. 10-1G). The pair of allele-specific oligonucleotides differs by a 3′-terminal discriminatory nucleotide corresponding to the SNP alleles. The common, locus-specific (5′-phosphorylated) oligonucleotide is designed to anneal immediately 3′ to the target SNP. The oligonucleotides are allowed to hybridize to target DNA in the presence of the ligase, and when a pair of perfectly matched oligonucleotides is immediately juxtaposed to form a duplex with the target DNA, they are covalently ligated together. The OLA is extremely efficient at discriminating between perfectly and imperfectly matched allele-specific oligonucleotides (Landegren et al. 1988; Luo et al. 1996; Schouten et al. 2002). Thermal cycling of the OLA reaction, between a denaturing temperature and an annealing/ligation temperature, increases the number of ligation products formed to a detectable level. By adding multiple sets of OLA oligonucleotides to a single genotyping reaction, one can achieve relatively high levels of multiplexing, genotyping tens to hundreds of SNPs or small insertion/deletion polymorphisms simultaneously. Such high-level multiplexing reactions require very little optimization.

The OLA forms the basis for a wide variety of genotyping platforms, which primarily differ in the method used to detect and discriminate allele-specific ligation products. The simplest approach is to shift the electrophoretic mobility of the ligation products by adding a small number of nucleotides to the 5′ end of one of the allele-specific oligonucleotides (Barany 1991). Ligation products (and hence genotypes) can then be distinguished by gel electrophoresis. Greater multiplexing and sample throughput can be achieved by altering the mobility of the ligation products and by fluorescently labeling the ligation reactions to run on a capillary sequencer (Grossman et al. 1994; Day et al. 1995; Eggerding 1995; Schouten et al. 2002). A potentially high-throughput OLA-based genotyping technology that uses electrophoresis to resolve differentially sized allele-specific ligation products is the SNPWave technology (van Eijk et al. 2004). It incorporates some of the principles of amplified fragment length polymorphism (AFLP) analysis (Vos et al. 1995). In SNPWave, instead of three genotyping oligonucleotides, a pair of long (80–130 nucleotides) allele-specific padlock ligation probes (Nilsson et al. 1994) are used for each SNP. These have target-specific sequences at each end separated by a variable-length linker sequence. Once the ends hybridize to the correct target allele in the sample, the padlock probe is circularized. Following multiplex OLA, all closed, circular probes are amplified with fluorescently labeled universal PCR primers, and differently sized allele-specific products can be run on a capillary sequencer. The advantages of the SNPWave method are the relatively high level of multiplexing possible (10-100 plex) and the ability to work directly from genomic DNA. However, the long padlock probes must be purified by high-pressure liquid chromatography (HPLC), which adds a significant cost for each assay developed.

A simple way to avoid electrophoresis is to detect OLA products with enzyme-linked immunosorbent assays (ELISA) (Nickerson et al. 1990; Tobe et al. 1996). Using this approach, one of the allele-specific OLA oligonucleotides is 5′-labeled with fluorescein and the other with digoxigenin, while the locus-specific oligonucleotide is biotinylated at the 3′ end. Following the OLA, ligation products are allowed to bind to a well of a streptavidin-coated microtiter plate, and alkaline phosphatase–labeled anti-fluorescein antibodies and horseradish peroxide–labeled anti-digoxigenin antibodies are added. Sequentially, an alkaline phosphatase substrate is added to detect the fluorescein reporter, and a horseradish peroxide substrate is added to detect the digoxigenin reporter. Spectrophotometric detection is performed using a plate reader, and samples can be genotyped. The OLA-ELISA approach can be a fairly cost-effective method with which to assay a large number of samples in microtiter plates. However, the assay cannot be multiplexed, and since several biochemical reactions must be carried out, it may not be particularly efficient for many labs.

Distinguishing allele-specific ligation products via size separation and fluorescent labeling may limit the level of multiplexing and narrow the range of allele-detection methods that can be used. An alternative solution is to incorporate specific nucleotide sequences, or DNA bar codes, onto the 5′ ends of the allele-specific genotyping oligonucleotides. Following OLA reactions, each ligation product then encodes a particular SNP allele via a particular bar code. Bar code tails on the oligonucleotides do not interfere with the ligation reaction, and since they are generic (i.e., the same bar code pair can be used for a different SNP in a different multiplex), they have been widely used in OLA-based genotyping technologies.

Using bar codes on the 5′ ends of allele-specific genotyping oligonucleotides permits SNP allele detection on arrays. A major advantage of array-based detection over gel- or capillary-based approaches is the relative ease of automated data extraction. Macdonald et al. (2005b) present an array-based OLA method capable of genotyping up to 16 SNPs in a single reaction. Each of the 32 allele-specific oligonucleotides contains a specific bar code just upstream of the sequence-specific region. At the 5′ end (3′ end) of the allele-specific oligonucleotides (locus-specific oligonucleotide) are tags complementary to universal PCR amplification primers. Following the OLA on the PCR-amplified template, all ligation products are amplified by PCR and printed onto nylon arrays. Hybridizing radiolabeled probes (complementary to the bar code sequences) to the arrays allows SNP genotypes to be scored after exposure of the membranes to a storage phosphor screen. This approach is particularly appropriate for examining large numbers of individuals, as it is possible to print samples from thousands of individuals onto each array. The reverse system—arraying out bar codes onto a solid substrate and probing with labeled ligation products (Gerry et al. 1999; Banér et al. 2003)—has other advantages. Bar code, or tag, arrays are universal, as bar codes can be reused in multiplexes of different sets of SNPs without any change to the array. Because the density of features on the array can be very high, the use of tag arrays along with highly multiplexed OLA reactions is an attractive approach to screen very large numbers of SNPs simultaneously. However, since each sample screened requires a separate tag array, projects may be limited to a modest number of individuals if it is difficult to push a large number of arrays. (See below for a discussion of "arrays-of-arrays," which addresses this issue.)

Finally, solid-phase detection of OLA genotyping products can be carried out using flow cytometry of microspheres, or "beads" (Iannone et al. 2000). Many microsphere types are available, distinguished by different ratios of red and orange fluorescence, and each type can be coupled to a different bar code. Following OLA, in which all allele-specific oligonucleotides harbor a different bar code, and all locus-specific oligonucleotides are fluorescein-labeled, ligation products are hybridized to the set of microspheres. Fluorescence associated with the hybridized microspheres is analyzed by flow cytometry: Red/orange fluorescence distinguishes microsphere type (i.e., the SNP allele), and green (fluorescein) fluorescence provides the amount of any given SNP allele present in a reaction. This genotyping system may be a worthwhile alternative to electrophoresis-based detection for those researchers with access to flow cytometry equipment. It may also permit more flexibility in the number of SNPs and individuals tested compared to array-based techniques.

Primer Extension Assays

A number of genotyping methods use the principle of primer extension for allelic discrimination. In single-base extension (SBE), or minisequencing, a locus-specific extension primer is allowed to anneal immediately 5′ to the target SNP. The primer is then extended one nucleotide by a DNA polymerase and incorporates the nucleotide complementary to that at the target SNP (Fig. 10-1H) (Syvänen et al. 1990). The incorporated nucleotide is a dye-terminating dideoxyribonucleoside triphosphate (ddNTP), so the primer is extended by one nucleotide only. Because SBE genotyping oligonucleotides are short, unmodified, and unlabeled, they can be purchased cheaply, and probe design is normally straightforward. However, SBE techniques will not generally be applicable to genotyping small insertion/deletion polymorphisms.

One of the most common genotyping methods of this class is template-directed primer extension detected by fluorescence polarization (FP-TDI) (Chen et al. 1999), a technique covered in detail in Chapter 11. Briefly, when a fluorophore is excited by polarized light, the emitted fluorescence is polarized also. The degree of polarization depends on several factors, including temperature of the reaction, sample viscosity, and the weight of the fluorescently labeled molecule. Following a successful SBE reaction, the fluorescent molecule is substantially heavier, and this can be detected by a change in FP. FP-TDI is carried out on PCR-amplified genomic DNA, and allele detection requires a microplate reader. The FP-TDI approach is inherently singleplex but is still amenable to screening a large number of individuals. Reactions are easy to set up, allele discrimination/detection occurs directly in the reaction plates, and SBE genotyping oligonucleotides are cheap by virtue of being short, unmodified, and unlabeled. If multiplexing is desired, the SNaPshot SBE technique (http://www.appliedbiosystems.com) allows 10-plex reactions to be carried out. Multiplex SNaPshot reactions include all four fluorescently labeled ddNTPs, and by adding different-length 5′ tails to the genotyping oligonucleotides, primer extension products can be scored on a capillary sequencer. Many genotyping methods assume SNPs are biallelic and may yield incorrect genotypes if an individual harbors a rare third allele. Because all four ddNTPs are present in the reaction mix, SNaPshot can effectively genotype SNPs with more than two alleles.

The disadvantage of the above-mentioned SBE approaches is that they permit only low levels of multiplexing. One way to increase throughput is to perform SBE on a solid substrate rather than in solution (e.g., on arrays; Pastinen et al. 1997). Briefly, following attachment of the locus-specific genotyping oligonucleotides to a microarray slide, multiplex PCR products are hybridized to the array. Upon annealing to the tethered genotyping oligonucleotides, the PCR products can act as templates for SBE reactions. Given that a microarray slide can hold thousands of different oligonucleotides, highly multiplexed SBE reactions are possible. Measurement of the incorporated ddNTPs at each of the oligonucleotide addresses using a fluorescent slide scanner allows the genotype at each SNP to be scored (Pastinen et al. 1997). An analogous form of solid-phase SBE genotyping reaction detection is offered by flow cytometric analysis of microspheres (Chen et al. 2000), similar to that described above for OLA reaction detection. The rate-limiting step for these array- and bead-based SBE methods is likely to be the level of multiplexing achievable in a single PCR.

For some projects, it will not be necessary to screen a large number of SNPs, and such highly multiplexed SBE genotyping systems may be inappropriate. Pastinen et al. (2000) adapt array-based SBE genotyping and use an "arrays-of-arrays" approach. A single microarray slide is divided into multiple independent reaction chambers, each with the same printed subarray. Each chamber is used to screen a different individual for the same small set of SNPs—the number of samples simultaneously screened is increased by sacrificing the number of SNPs tested. An arrays-of-arrays approach is used in conjunction with universal tag arrays (see above) in the SNPstream SBE-based technology (http://www.beckman.com). SNPstream is capable of performing 12-plex genotyping reactions in each well of a 384-well microtiter plate (Bell et al. 2002). Each well has a small, 16-spot tag array (4 control spots and 12 tags) printed at the bottom. Following multiplex PCR, and cleanup to remove primers and unincorporated nucleotides, SBE reactions are carried out in liquid phase in the presence of labeled ddNTPs. Each of the 12 locus-specific genotyping oligonucleotides carries a 5′ tag complementary to one of the 12 immobilized tags on the mini-array. Following hybridization of the extension reactions to the arrays, and washing of the plates to remove free, nonhybridized material, the plates are scanned to identify the fluorescent label association with each tag. The fluorescent signal reflects the genotype of the 12 target SNPs. The drawback of the SNPstream approach is that only two ddNTPs are added to each reaction; i.e., all of the SNPs simultaneously genotyped must segregate for the same pair of nucleotides. This problem can be alleviated to some degree by designing genotyping assays on the other strand of DNA, such that an A/C SNP becomes a T/G SNP.

In contrast to SBE, pyrosequencing enables the addition of multiple nucleotides to a primer (Ronaghi et al. 1996, 1998; Alderborn et al. 2000; Langaee and Ronaghi 2005). The principle is as follows (Fig. 10-1I): A sequencing primer is allowed to hybridize to a single-stranded DNA template, and is incubated with a DNA polymerase and one of the four dNTPs. The deoxynucleotide will be incorporated into the primer only if it is complementary to the next base in the template strand of DNA. If the primer is extended by the polymerase, the reaction releases pyrophosphatase (PPi), which is converted to ATP by the ATP sulfurylase enzyme. The ATP is then converted to light by the firefly luciferase enzyme, and the light level is quantified by a light-sensitive camera. Repeated cycles of deoxynucleotide addition are performed to extend the sequencing primer. Each cycle, nucleotides from the previous addition must be removed, either by washing the sample or by adding a nucleotide-degrading enzyme. In this fashion, the sequence of short fragments of DNA containing a polymorphic SNP or a short insertion/deletion polymorphism can be obtained. Although it is possible to simultaneously pyrosequence more than one template DNA fragment (Pourmand et al. 2002), pyrosequencing is generally limited to genotyping just one SNP at a time. The multiple enzymatic steps and substrates increase the complexity of the assay, as well as its cost. Furthermore, a dedicated pyrosequencing instrument must be available.

USE OF GENOTYPING PLATFORMS

Unfortunately, there is no ideal genotyping platform suitable for all projects. A researcher might choose a genotyping system on a project-by-project basis. This decision will be based on several criteria:

- How many SNPs and individuals are to be assayed?
- What is the overall cost of the proposed project? Technology development papers often give a cost per genotype *in consumables only*. It is important to calculate the *total* cost in terms of consumables, personnel, and equipment.
- Is any specialized equipment required?
- Should a commercial genotyping platform or an in-house solution be used? The latter may require substantial development and troubleshooting, especially if the research lab has limited experience with SNP genotyping.

- Is the panel of individuals to be screened fixed or variable? With a fixed DNA panel (e.g., a mapping population), all individuals will be screened simultaneously. With a variable DNA panel, individuals may be continually added; e.g., in forensic work, the same set of markers is utilized daily to genotype a new set of individuals. Some genotyping methods will be difficult to adapt to a variable DNA panel.
- Is genomic DNA a limited resource? If so, it will be important to choose a highly multiplexed method requiring very little DNA per reaction.

These concerns are discussed below, as are the platforms likely to be well-suited to four different genotyping projects covering a range of sizes. A free choice of technology for every project will not generally be available, and may even be undesirable. It may be preferable to use a genotyping technology already in use in the lab, even if it is not completely appropriate for the intended project. Furthermore, the availability of specialized equipment may limit the range of methods open to a researcher.

Small Scale: 10 SNPs–100 Individuals

A small project might be conducted as preliminary work to a much larger study, or one may be interested in a few potentially functional SNPs and their frequency in small samples from one or two populations (e.g., Wilson et al. 2001). For a small project, it is not necessary to use a genotyping method capable of multiplexing, unless genomic DNA is a very limited resource. However, an important consideration is the cost of purchasing the assay oligonucleotides/probes; i.e., the fixed cost of the assay. This cost will be spread among the individuals genotyped, and methods with high fixed assay costs (such as TaqMan) will not be cost-effective for a small sample size. Singleplex genotyping solutions such as RFLP analysis, allele-specific PCR, or the SBE-based FP-TDI are likely to be of great utility for a 10 SNP–100 individual project. Provided the requisite machinery is in place, pyrosequencing can also be used. For example, De Luca et al. (2003) genotyped 12 SNPs and small insertion/deletion events in 173 *Drosophila* inbred lines to identify polymorphic sites contributing to variation in aging. Provided a genotyping system with a low fixed cost per assay is used, the major cost for a small-scale project is likely to be labor, although this will also be relatively low.

Medium Scale: 10 SNPs–1000 Individuals

A project this size is exemplified by a follow-up experiment to a large association mapping scan, testing a small set of potentially causal SNPs for replication in a second large sample (e.g., Genissel et al. 2004; Smyth et al. 2004; Dworkin et al. 2005). With a large number of individuals, the cost of the assay oligonucleotides/probes becomes less of an issue, as the cost is pro-rated across a large sample. It is feasible to use non-multiplexible assays as the number of SNPs remains small, although this would entail individually screening 10,000 genotyping reactions, and the cost in consumables may be high. Given a sufficient thermocycler capacity and a quantitative PCR machine or fluorescent plate reader, TaqMan assays, Invader assays, or FP-TDI SBE assays can be used. These reactions are easy to set up, even without sophisticated liquid-handling systems, and the allele detection occurs directly in the genotyping reaction plates. Approaches using a capillary sequencer (e.g., SNaPshot) to resolve allele-specific products by size/fluorescent label are possible, but even in the case of a 10-plex assay, this still entails running 1000 lanes, which may not be efficient or cost-effective for many labs. An alternative approach is to use ASO-based assays on nylon membranes, by printing PCR-amplified samples and probing with radiolabeled ASO probes. For example, Zimmerman et al. (2000) used ASO to genotype 16 SNPs in 800 *Drosophila* to map quantitative trait loci (QTL) for wing shape. With this method, many samples can be printed, which improves throughput of a large number of individuals, and although hybridization conditions may need to be optimized for each probe, relatively few SNPs will be tested. Genotyping via array-based ASO is a fairly inexpensive method to collect data on 10 SNPs for 1000 individuals, but it is likely to be more labor-intensive than using an entirely liquid-phase approach (e.g., FP-TDI). One further caveat is that array-based ASO performs best when a fixed panel of individuals is screened simultaneously—genotyping a few additional individuals at a later date by the same method would be very inefficient.

Medium Scale: 100 SNPs–100 Individuals

This project scale can pose a challenge to the investigator and is commonly faced by those performing genome-wide QTL mapping or creating SNP-based linkage maps. The fixed cost of each assay must be kept low (due to the small number of individuals), but multiplexing is desirable (due to the large number of SNPs involved). Some of the multiplexible methods, such as the OLA-based SNPWave technology, stand out for 100 SNP–100 individual projects. For instance, SNPWave has been used to genotype 100 SNPs in single 100-plex reactions for 92 lines of *Arabidopsis* to map flowering-time QTL (El-Lithy et al. 2006). The high plex ensures that the genotyping is streamlined, and only a small number of lanes need to be run on a capillary sequencer. The only concern is that the genotyping probes are long, and perhaps expensive on a per-genotype basis. The SBE-based SNPstream system might be a worthwhile alternative. Here, reactions are performed and genotypes are detected directly in microtiter plates, and SNPstream is capable of 12-plex reactions. Although the technology is appropriate for this scale genotyping project, it clearly requires access to the appropriate machinery. If the researcher has experience with a particular singleplex (or low-plex) method and ready access to the appropriate equipment, it is possible to genotype 100 SNPs in a few hundred individuals without multiplexing genotyping reactions. For example, Smith et al. (2005) used SBE assays, RFLPs, and allele-specific PCR to genotype over 500 SNPs in about 100 individuals to develop a linkage map for *Ambystoma* salamanders.

Large Scale: 100 SNPs–1000 Individuals

This is the largest project size that is likely to be carried out in the average laboratory, and the only one where expenditure on novel equipment is warranted. To perform much larger experiments, many investigators may wish to explore outsourcing the genotyping to a genome center or a company, and/or use one of the ultrahigh-throughput technologies discussed in Chapter 13. It is possible to use any of the technologies discussed to collect the required set of 100,000+ genotypes, but in general, those that can be multiplexed to a high level are likely to be most efficient. It is certainly possible to use a method that detects allele-specific genotyping reaction products by size on a capillary sequencer. However, even with 100-plex reactions (e.g., via SNPWave), 1000 lanes must be run, and each must be scored for peak presence/absence, size, height, and fluorophore—a highly labor-intensive procedure with 100-plex assays. Additionally, if it is not possible in practice to multiplex to such a high level, more lanes must be run, and consumable costs will increase proportionally. Thus, it may be preferable to use a method that allows simultaneous screening of multiple samples, and more automated genotype calling, i.e., by using array-based or microtiter plate–based systems. For example, to generate genotypes for over 200 SNPs and polymorphic insertion/deletion events in a population of 2000 *Drosophila melanogaster* individuals, Macdonald et al. (2005a) used a 16-plex OLA-based genotyping technology, with array-based allele detection via radiolabeled probes. This method is inexpensive in terms of consumables and does not require highly specialized machinery but does require some labor-intensive steps. Again, this array-based method is most appropriate for a fixed panel of individuals; if extra individuals are to be continually added to the panel, array-based approaches will prove unwieldy.

THE PROBLEM OF POLYMORPHISM

SNPs are the most abundant form of sequence-level variation segregating among individuals. It is conceivable that there will be additional, but unknown, segregating polymorphisms in the vicinity of a target SNP, and perhaps in the binding region for the genotyping probes. If this is the case, the genotyping oligonucleotide will be designed to match a particular allele at the secondary SNP. Hence, in a sample harboring the alternative (disruptive) allele, the hybridized probe–target duplex will be unstable (Wallace et al. 1979), and the assay for the target SNP allele may fail (Fig. 10-3). If the disruptive secondary SNP allele is homozygous for a given individual, the assay may fail entirely, and the individual will not be assigned a genotype for the target SNP. More seriously, if the secondary SNP is a heterozygote, only one of the target SNP alleles will drop out, and a target SNP heterozygote will be scored as a homozygote (Fig. 10-3).

Secondary SNPs appear to be a problem for many, if not all, genotyping assays (Matsuzaki et al. 2004; Macdonald et al. 2005b; Koboldt et al. 2006), although the extent of the problem is somewhat platform-specific (Koboldt et al. 2006). The organism used for genotyping is also an important factor. Organisms with low nucleotide diversity (e.g., humans) will be affected far less by secondary SNPs than those with high nucleotide diversity (e.g., *Drosophila*), as it is less likely that there will be a SNP in the probe-binding region. Sequencing the target region in multiple individuals will identify the majority of common polymorphisms, and those without nearby secondary SNPs can be targeted for genotyping. It is possible to compensate for known secondary SNPs by designing degenerate genotyping oligonucleotides (Macdonald et al. 2005b), although it is not known whether this treatment is possible for other genotyping platforms. Alternatively, one might use a genotyping platform allowing some flexibility in the placement of the allele-specific probe. For instance, the assay could be designed using the alternate DNA strand (e.g., SBE methods) or by shifting the allele-specific probe slightly with respect to the target SNP (e.g., ASO, TaqMan). Nevertheless, very rare SNPs, not seen in sequencing a modest number of chromosomes, remain a potential problem (Macdonald et al. 2005b).

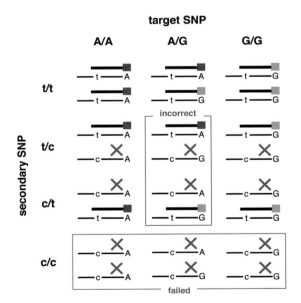

FIGURE 10-3. The problem of additional polymorphisms in genotyping. In this example, an A/G target SNP is genotyped using an SBE method, and a secondary t/c SNP segregates in the genotyping oligonucleotide-binding site. The genotyping oligonucleotide is designed to bind to the "t" allele at the secondary SNP, whereas the "c" allele completely disrupts oligonucleotide binding. It is easily seen that if both the target and secondary SNPs are heterozygous for a sample, the target SNP genotype will be incorrectly called a homozygote. In addition, the genotyping assay will fail if the secondary SNP is homozygous for the "c" allele.

REFERENCES

Alderborn A., Kristofferson A., and Hammerling U. 2000. Determination of single-nucleotide polymorphisms by real-time pyrophosphate DNA sequencing. *Genome Res.* **10:** 1249–1258.

Banér J., Isaksson A., Waldenstrom E., Jarvius J., Landegren U., and Nilsson M. 2003. Parallel gene analysis with allele-specific padlock probes and tag microarrays. *Nucleic Acids Res.* **31:** e103.

Barany F. 1991. Genetic disease detection and DNA amplification using cloned thermostable ligase. *Proc. Natl. Acad. Sci.* **88:** 189–193.

Bell P.A., Chaturvedi S., Gelfand C.A., Huang C.Y., Kochersperger M., Kopla R., Modica F., Pohl M., Varde S., Zhao R., et al. 2002. SNPstream UHT: Ultra-high throughput SNP genotyping for pharmacogenomics and drug discovery. *Biotechniques* **Suppl:** 70–77.

Bottema C.D. and Sommer S.S. 1993. PCR amplification of specific alleles: Rapid detection of known mutations and polymorphisms. *Mutat. Res.* **288:** 93–102.

Chen J., Iannone M.A., Li M.S., Taylor J.D., Rivers P., Nelsen A.J., Slentz-Kesler K.A., Roses A., and Weiner M.P. 2000. A microsphere-based assay for multiplexed single nucleotide polymorphism analysis using single base chain extension. *Genome Res.* **10:** 549–557.

Chen X., Levine L., and Kwok P.Y. 1999. Fluorescence polarization in homogeneous nucleic acid analysis. *Genome Res.* **9:** 492–498.

Conner B.J., Reyes A.A., Morin C., Itakura K., Teplitz R.L., and Wallace R.B. 1983. Detection of sickle cell β^S-globin allele by hybridization with synthetic oligonucleotides. *Proc. Natl. Acad. Sci.* **80:** 278–282.

Day D.J., Speiser P.W., White P.C., and Barany F. 1995. Detection of steroid 21-hydroxylase alleles using gene-specific PCR and a multiplexed ligation detection reaction. *Genomics* **29:** 152–162.

De La Vega F.M., Lazaruk K.D., Rhodes M.D., and Wenz M.H. 2005. Assessment of two flexible and compatible SNP genotyping platforms: TaqMan® SNP genotyping assays and the SNPlex™ genotyping system. *Mutat. Res.* **573:** 111–135.

De Luca M., Roshina N.V., Geiger-Thornsberry G.L., Lyman R.F., Pasyukova E.G., and Mackay T.F. 2003. Dopa decarboxylase (*Ddc*) affects variation in *Drosophila* longevity. *Nat. Genet.* **34:** 429–433.

Dworkin I., Palsson A., and Gibson G. 2005. Replication of an *Egfr*-wing shape association in a wild-caught cohort of *Drosophila melanogaster*. *Genetics* **169:** 2115–2125.

Eggerding F.A. 1995. A one-step coupled amplification and oligonucleotide ligation procedure for multiplex genetic typing. *PCR Methods Appl.* **4:** 337–345.

El-Lithy M.E., Bentsink L., Hanhart C.J., Ruys G.J., Rovito D., Broekhof J.L., van der Poel H.J., van Eijk M.J., Vreugdenhil D., and Koornneef M. 2006. New *Arabidopsis* recombinant inbred line populations genotyped using SNPWave and their use for mapping flowering-time quantitative trait loci. *Genetics* **172:** 1867–1876.

Genissel A., Pastinen T., Dowell A., Mackay T.F.C., and Long A.D. 2004. No evidence for an association between common nonsynonymous polymorphisms in *Delta* and bristle number variation in natural and laboratory populations of *Drosophila melanogaster*. *Genetics* **166:** 291–306.

Germer S. and Higuchi R. 1999. Single-tube genotyping without oligonucleotide probes. *Genome Res.* **9:** 72–78.

Gerry N.P., Witowski N.E., Day J., Hammer R.P., Barany G., and Barany F. 1999. Universal DNA microarray method for multiplex detection of low abundance point mutations. *J. Mol. Biol.* **292:** 251–262.

Grossman P.D., Bloch W., Brinson E., Chang C.C., Eggerding F.A., Fung S., Iovannisci D.M., Woo S., and Winn-Deen E.S. 1994. High-density multiplex detection of nucleic acid sequences: Oligonucleotide ligation assay and sequence-coded separation. *Nucleic Acids Res.* **22:** 4527–4534.

Hall J.G., Eis P.S., Law S.M., Reynaldo L.P., Prudent J.R., Marshall D.J., Allawi H.T., Mast A.L., Dahlberg J.E., Kwiatkowski R.W., et al. 2000. Sensitive detection of DNA polymorphisms by the serial invasive signal amplification reaction. *Proc. Natl. Acad. Sci.* **97:** 8272–8277.

Howell W.M., Jobs M., Gyllensten U., and Brookes A.J. 1999. Dynamic allele-specific hybridization. A new method for scoring single nucleotide polymorphisms. *Nat. Biotechnol.* **17:** 87–88.

Iannone M.A., Taylor J.D., Chen J., Li M.S., Rivers P., Slentz-Kesler K.A., and Weiner M.P. 2000. Multiplexed single nucleotide polymorphism genotyping by oligonucleotide ligation and flow cytometry. *Cytometry* **39:** 131–140.

Jobs M., Howell W.M., Stromqvist L., Mayr T., and Brookes A.J. 2003. DASH-2: flexible, low-cost, and high-throughput SNP genotyping by dynamic allele-specific hybridization on membrane arrays. *Genome Res.* **13:** 916–924.

Koboldt D.C., Miller R.D., and Kwok P.Y. 2006. Distribution of human SNPs and its effect on high-throughput genotyping. *Hum. Mutat.* **27:** 249–254.

Landegren U., Kaiser R., Sanders J., and Hood L. 1988. A ligase-mediated gene detection technique. *Science* **241:** 1077–1080.

Langaee T. and Ronaghi M. 2005. Genetic variation analyses by Pyrosequencing. *Mutat. Res.* **573:** 96–102.

Lazzaro B.P., Sceurman B.K., Carney S.L., and Clark A.G. 2002. fRFLP and fAFLP: Medium-throughput genotyping by fluorescently post-labeling restriction digestion. *Biotechniques* **33:** 539–546.

Li H., Cui X., and Arnheim N. 1990. Direct electrophoretic detection of the allelic state of single DNA molecules in human sperm by using the polymerase chain reaction. *Proc. Natl. Acad. Sci.* **87:** 4580–4584.

Livak K.J. 1999. Allelic discrimination using fluorogenic probes and the 5′ nuclease assay. *Genet. Anal. Biomol. Eng.* **14:** 143–149.

Livak K.J., Flood S.J., Marmaro J., Giusti W., and Deetz K. 1995. Oligonucleotides with fluorescent dyes at opposite ends provide a quenched probe system useful for detecting PCR product and nucleic acid hybridization. *PCR Methods Appl.* **4:** 357–362.

Lu M., Shortreed M.R., Hall J.G., Wang L., Berggren T., Stevens P.W., Kelso D.M., Lyamichev V., Neri B., and Smith L.M. 2002. A surface invasive cleavage assay for highly parallel SNP analysis. *Hum. Mutat.* **19:** 416–422.

Luo J., Bergstrom D.E., and Barany F. 1996. Improving the fidelity of *Thermus thermophilus* DNA ligase. *Nucleic Acids Res.* **24:** 3071–3078.

Lyamichev V., Mast A.L., Hall J.G., Prudent J.R., Kaiser M.W., Takova T., Kwiatkowski R.W., Sander T.J., de Arruda M., Arco D.A., et al. 1999. Polymorphism identification and quantitative detection of genomic DNA by invasive cleavage of oligonucleotide probes. *Nat. Biotechnol.* **17:** 292–296.

Macdonald S.J., Pastinen T., and Long A.D. 2005a. The effect of polymorphisms in the enhancer of split gene complex on bristle number variation in a large wild-caught cohort of *Drosophila melanogaster*. *Genetics* **171:** 1741–1756.

Macdonald S.J., Pastinen T., Genissel A., Cornforth T.W., and Long A.D. 2005b. A low-cost open-source SNP genotyping platform for association mapping applications. *Genome Biol.* **6:** R105.

Matsuzaki H., Loi H., Dong S., Tsai Y.Y., Fang J., Law J., Di X., Liu W.M., Yang G., Liu G., et al. 2004. Parallel genotyping of over 10,000 SNPs using a one-primer assay on a high-density oligonucleotide array. *Genome Res.* **14:** 414–425.

Newton C.R., Graham A., Heptinstall L.E., Powell S.J., Summers C., Kalshekerl N., Smith J.C., and Markham A.F. 1989. Analysis of any point mutation in DNA. The amplification refractory mutation system (ARMS). *Nucleic Acids Res.* **17:** 2503–2516.

Nickerson D.A., Kaiser R., Lappin S., Stewart J., Hood L., and Landegren U. 1990. Automated DNA diagnostics using an ELISA-based oligonucleotide ligation assay. *Proc. Natl. Acad. Sci.* **87:** 8923–8927.

Nilsson M., Malmgren H., Samiotaki M., Kwiatkowski M., Chowdhary B.P., and Landegren U. 1994. Padlock probes: Circularizing oligonucleotides for localized DNA detection. *Science* **265:** 2085–2088.

Olivier M. 2005. The Invader assay for SNP genotyping. *Mutat. Res.* **573:** 103–110.

Pastinen T., Kurg A., Metspalu A., Peltonen L., and Syvänen A.C. 1997. Minisequencing: a specific tool for DNA analysis and diagnostics on oligonucleotide arrays. *Genome Res.* **7:** 606–614.

Pastinen T., Raitio M., Lindroos K., Tainola P., Peltonen L., and Syvänen A.C. 2000. A system for specific, high-throughput genotyping by allele-specific primer extension on microarrays. *Genome Res.* **10:** 1031–1042.

Pourmand N., Elahi E., Davis R.W., and Ronaghi M. 2002. Multiplex Pyrosequencing. *Nucleic Acids Res.* **30:** e31.

Prince J.A., Feuk L., Howell W.M., Jobs M., Emahazion T., Blennow K., and Brookes A.J. 2001. Robust and accurate single nucleotide polymorphism genotyping by dynamic allele-specific hybridization (DASH): Design criteria and assay validation. *Genome Res.* **11:** 152–162.

Rao K.V., Stevens P.W., Hall J.G., Lyamichev V., Neri B.P., and Kelso D.M. 2003. Genotyping single nucleotide polymorphisms directly from genomic DNA by invasive cleavage reaction on microspheres. *Nucleic Acids Res.* **31:** e66.

Ririe K.M., Rasmussen R.P., and Wittwer C.T. 1997. Product differentiation by analysis of DNA melting curves during the polymerase chain reaction. *Anal. Biochem.* **15:** 154–160.

Ronaghi M., Uhlen M., and Nyren P. 1998. A sequencing method based on real-time pyrophosphate. *Science* **281:** 363–365.

Ronaghi M., Karamohamed S., Pettersson B., Uhlen M., and Nyren P. 1996. Real-time DNA sequencing using detection of pyrophosphate release. *Anal. Biochem.* **242:** 84–89.

Saiki R.K., Bugawan T.L., Horn G.T., Mullis K.B., and Erlich H.A. 1986. Analysis of enzymatically amplified β-globin and HLA-DQα DNA with allele-specific oligonucleotide probes. *Nature* **324:** 163–166.

Saiki R.K., Scharf S., Faloona F., Mullis K.B., Horn G.T., Erlich H.A., and Arnheim N. 1985. Enzymatic amplification of beta-globin genomic sequences and restriction site analysis for diagnosis of sickle cell anemia. *Science* **230:** 1350–1354.

Schouten J.P., McElgunn C.J., Waaijer R., Zwijnenburg D., Diepvens F., and Pals G. 2002. Relative quantification of 40 nucleic acid sequences by multiplex ligation-dependent probe amplification. *Nucleic Acids Res.* **30:** e57.

Smith J.J., Kump D.K., Walker J.A., Parichy D.M., and Voss S.R. 2005. A comprehensive expressed sequence tag linkage map for tiger salamander and Mexican axolotl: Enabling gene mapping and comparative genomics in *Ambystoma. Genetics* **171:** 1161–1171.

Smyth D., Cooper J.D., Collins J.E., Heward J.M., Frankly J.A., Howson J.M.M., Vella A., Nutland S., Rance H.E., Maier L., et al. 2004. Replication of an association between the lymphoid tyrosine phosphatase locus (LYP/PTPN22) with type 1 diabetes, and evidence for its role as a general autoimmunity locus. *Diabetes* **53:** 3020–3023.

Syvänen A.C., Aalto-Setala K., Harju L., Kontula K., and Söderlund H. 1990. A primer-guided nucleotide incorporation assay in the genotyping of apolipoprotein E. *Genomics* **8:** 684–692.

Thomas M.G., Bradman N., and Flinn H.M. 1999. High throughput analysis of 10 microsatellite and 11 diallelic polymorphisms on the human Y-chromosome. *Hum. Genet.* **105:** 577–581.

Tobe V.O., Taylor S.L., and Nickerson D.A. 1996. Single-well genotyping of diallelic sequence variations by a two-color ELISA-based oligonucleotide ligation assay. *Nucleic Acids Res.* **24:** 3728–3732.

van Eijk M.J.T., Broekhof J.L.N., van der Poel H.J.A., Hogers R.C.J., Schneiders H., Kamerbeek J., Verstege E., van Aart J.W., Geerlings H., Buntjer J.B., et al. 2004. SNPWave™: A flexible multiplexed SNP genotyping technology. *Nucleic Acids Res.* **32:** e47.

Vos P., Hogers R., Bleeker M., Reijans M., van de Lee T., Hornes M., Frijters A., Pot J., Peleman J., Kuiper M., and Zabeau M. 1995. AFLP: A new technique for DNA fingerprinting. *Nucleic Acids Res.* **23:** 4407–4414.

Wallace R.B., Shaffer J., Murphy R.F., Bonner J., Hirose T., and Itakura K. 1979. Hybridization of synthetic oligodeoxyribonucleotides to φχ174 DNA: The effect of single base pair mismatch. *Nucleic Acids Res.* **6:** 3543–3557.

Wang J., Chuang K., Ahluwalia M., Patel S., Umblas N., Mirel D., Higuchi R., and Germer S. 2005. High-throughput SNP genotyping by single-tube PCR with T_m-shift primers. *Biotechniques* **39:** 885–893.

Wilson J.F., Weale M.E., Smith A.C., Gratrix F., Fletcher B., Thomas M.G., Bradman N., and Goldstein D.B. 2001. Population genetic structure of variable drug response. *Nat. Genet.* **29:** 265–269.

Ye S., Dhillon S., Ke X., Collins A.R., and Day I.N. 2001. An efficient procedure for genotyping single nucleotide polymorphisms. *Nucleic Acids Res.* **29:** E88.

Zhang R., Zhu Z., Zhu H., Nguyen T., Yao F., Xia K., Liang D., and Liu C. 2005. SNP Cutter: A comprehensive tool for SNP PCR-RFLP assay design. *Nucleic Acids Res.* **33:** W489–492.

Zimmerman E., Palsson A., and Gibson G. 2000. Quantitative trait loci affecting components of wing shape in *Drosophila melanogaster. Genetics* **155:** 671–683.

WWW RESOURCE

http://bioinfo.bsd.uchicago.edu/SNP_cutter.htm SNP Cutter: SNP PCR-RFLP Assay Design, developed by Zhang et al., University of Chicago, Illinois.

Intermediate-Throughput Laboratory-Scale Genotyping Protocols

Edwin Cuppen,[1] Stuart J. Macdonald,[2] Connie Ha,[3] Pui-Yan Kwok,[4] W. Brad Barbazuk,[5] An-Ping Hsia,[6] Hsin D. Chen,[6] Yan Fu,[5] Kazuhiro Ohtsu,[6] and Patrick S. Schnable[6,7,8]

[1]Hubrecht Laboratory, Utrecht, The Netherlands; [2]Department of Ecology and Evolutionary Biology and Department of Molecular Biosciences, University of Kansas, Lawrence, Kansas 66045; [3]Cardiovascular Research Institute and Center for Human Genetics and [4]Department of Dermatology, University of California, San Francisco, California 94143-0793; [5]Donald Danforth Plant Science Center, St. Louis, Missouri 63132; [6]Department of Agronomy, [7]Department of Genetics, Development, and Cell Biology, and [8]Center for Plant Genomics, Iowa State University, Ames, Iowa 50011

INTRODUCTION

In Chapter 9, we discussed various laboratory-scale genotyping platforms as well as strategies for analysis of genotyping endeavors limited to intermediate-throughput scale. Here we provide a collection of specific genotyping protocols that describe approaches appropriate for projects that deal with analysis on the order of a few thousand to a few hundred thousand genotypes.

Protocol 1

Genotyping by Allele-specific Amplification

Edwin Cuppen

Hubrecht Laboratory, Utrecht, The Netherlands

This protocol describes the use of allele-specific amplification followed by fluorescence detection (KASPar) for genotyping. The method is compatible with any real-time PCR setup and has the advantage that no specifically labeled oligonucleotides are needed. For more information, see http://www.kbioscience.co.uk.

MATERIALS

Reagents

Genomic template DNA
$MgCl_2$ (50 mM)
Reaction mix (KASPar, Kbioscience)
Taq polymerase (Kbioscience)

Equipment

Fluorescence plate reader or real-time PCR machine (compatible with FAM, VIC, and ROX)
Klustercaller software
PCR machine
Plate in which to perform reaction

METHOD

1. Design oligonucleotides (2 allele-specific 40-mers and a common 20-mer) using the Web-based form: http://www.kbioscience.co.uk/primer-picker/.

2. Prepare the following reaction mixture (4 µl total volume):

   2 µl of genomic template DNA (10–100 ng)

   1 µl of reaction mix (KASPar)

   0.055 µl of oligonucleotide mix (12 µM each allele-specific oligonucleotide and 30 µM common oligonucleotide)

   0.013 µl of Taq polymerase

   0.032 µl of $MgCl_2$ (50 mM)

   0.9 µl of MilliQ H_2O

3. Perform PCR using the following protocol:

   94°C for 15 minutes; 20 cycles of 94°C for 10 seconds, 57°C for 5 seconds, 72°C for 10 seconds; 18 cycles of 94°C for 10 seconds, 57°C for 20 seconds, 72°C for 40 seconds.

4. Scan the plate in a fluorescence plate reader or real-time PCR machine (compatible with FAM, VIC, and ROX).

5. Analyze the data using Klustercaller software.

Protocol 2

Genotyping by Dideoxy Resequencing

Edwin Cuppen

Hubrecht Laboratory, Utrecht, The Netherlands

This protocol describes dideoxy resequencing of PCR fragments, which is a quick and flexible genotyping technique. This strategy is used to identify and confirm the presence of heterozygous single-nucleotide polymorphisms within a genomic region of interest. Sequencing data can be analyzed using a variety of packages, including Staden (Bonfield et al. 1998) or Polyphred (Stephens et al. 2006).

MATERIALS

CAUTION: See Appendix for appropriate handling of materials marked with <!>.

Reagents

Agarose gel (1%)
BigDye Dilution Buffer (2.5x, Applied Biosystems)
BigDye terminator v3.0 (Applied Biosystems)
dNTPs (10 mM each)
Ethanol (80%)
Formamide <!> (Optional, see Step 16)
Forward primer (2 mM)
Genomic template DNA
PCR buffer (10x, from supplier of Taq polymerase)
Precipitation mix

> *For 1 liter, mix 800 ml of 96% ethanol, 16 ml of Na Acetate <!> (pH 5.5), and 158 ml of MilliQ H_2O.*

Reverse primer (2 mM)
Sequencing primer (2 mM; one of the oligonucleotides used for PCR amplification can be used)
Taq polymerase (5 U/μl)

Equipment

Apparatus for agarose gel electrophoresis
Capillary sequencer (e.g., Applied Biosystems 3730 with the standard RapidSeq protocol)
Heating block preset to 80°C (Optional, see Step 15)
PCR machine
Plates or tubes in which to perform reactions
Sequence analysis software (e.g., Staden or Polyphred)
Vortex mixer

METHOD

1. Design an amplicon for PCR amplification of the genomic segment encompassing the polymorphism. Optimal size for PCR and sequencing is about 300 bp.

2. Prepare the following PCR reaction (10 µl total volume):

   5 µl of genomic template DNA (10–100 ng)

   1 µl of 10x PCR buffer

   1 µl of forward primer (2 mM)

   1 µl of reverse primer (2 mM)

   0.4 µl of dNTPs (10 mM each)

   0.4 µl of Taq polymerase (5 U/µl)

   1.2 µl of MilliQ H_2O

3. Perform PCR using the following touchdown protocol:

   94°C for 60 seconds; 12 cycles of 94°C for 20 seconds, 65°C for 20 seconds with a decrement of 0.6°C per cycle, 72°C for 30 seconds; followed by 20 cycles of 92°C for 20 seconds, 58°C for 20 seconds, and 72°C for 30 seconds; 72°C for 180 seconds.

4. Check 1–2 µl of the PCR products on a 1% agarose gel. If a single clear band is visible, proceed with sequencing without purification of the PCR product. If this is not the case, optimize PCR conditions; for example, by performing a gradient PCR.

   Although it is possible to excise fragments from an agarose gel or to purify weak products from an inefficient PCR for sequence analysis, it is not recommended to pursue these polymorphisms.

5. Dilute the remaining PCR product with 25 µl of MilliQ H_2O.

6. Prepare the following dideoxy sequencing reaction (5 µl total volume):

   1 µl of diluted PCR product

   1 µl of sequencing primer (2 mM)

   0.2 µl of BigDye terminator v3.0

   1.8 µl of 2.5x BigDye Dilution Buffer

   1 µl of MilliQ H_2O

7. Perform the following cycle sequencing protocol:

   40 cycles of 92°C for 10 seconds, 50°C for 5 seconds, and 60°C for 120 seconds.

8. Purify the products by adding 30 µl of precipitation mix.

9. Mix by vortexing for 15 seconds.

10. Centrifuge at 3,200*g* for 40 minutes (plate) or 10,000*g* for 10 minutes (tube).

11. Discard the supernatant (centrifuge plate upside down for 1 minute at 32*g*).

12. Wash the pellet by adding 25 µl of 80% ethanol.

13. Centrifuge at 3,200*g* for 5 minutes (plate) or 10,000*g* for 2 minutes (tube).

14. Discard the supernatant (centrifuge plate upside down for 1 minute at 32*g*).

15. Air-dry the pellet or heat for 10–15 minutes at 80°C (until there is no smell of alcohol; do not overdry; protect from light).

16. Dissolve the pellet in 10 µl of H_2O or formamide.

17. Analyze on a capillary sequencer (e.g., Applied Biosystems 3730).

18. Analyze the sequencing data. A variety of software packages are available (e.g., Staden or Polyphred).

Oligonucleotide Ligation Assay

Stuart J. Macdonald

Department of Ecology and Evolutionary Biology and Department of Molecular Biosciences, University of Kansas, Lawrence, Kansas 66045

This protocol describes the oligonucleotide ligation assay (OLA), which uses a set of three oligonucleotides, in combination with a thermostable Taq DNA ligase enzyme, to discriminate single-nucleotide polymorphism (SNP) alleles. Sixteen-plex OLA genotyping reactions are carried out, and allele-specific OLA products are detected on membrane arrays using radiolabeled probes. An overview of the genotyping pipeline is given in Figure 11-1.

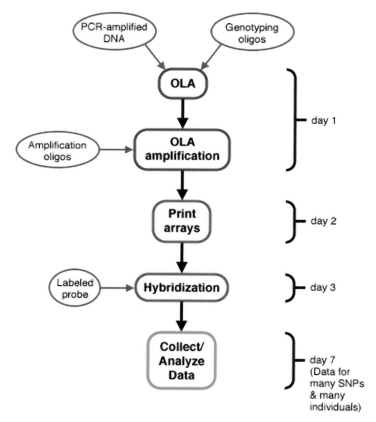

FIGURE 11-1. The genotyping pipeline.

MATERIALS

CAUTION: See Appendix for appropriate handling of materials marked with <!>.

Reagents

Adenosine 5′-triphosphate (ATP) solution (100 mM; GE Healthcare)

[γ-^{33}P]Adenosine 5′-triphosphate (250 μCi; PerkinElmer) <!>

10× Amplification buffer (500 mM KCl, 1% Triton X-100) <!>

> *A minimal amplification buffer is used because the amplification reagents are added directly to the OLA reaction, which already contains appropriate Tris-HCl and MgCl$_2$ concentrations for PCR.*

Denaturing buffer (0.5 M NaOH <!>, 1.5 M NaCl)

DNA (PCR-amplified) containing the SNPs of interest

dNTPs (25 mM)

DTT (25 mM) <!>

Herring sperm DNA, sonicated (10 μg/μl; Promega)

Hybridization buffer (prewarmed to 42°C)

0.525 M sodium phosphate (pH 7.2) <!>

7% SDS <!>

1 mM EDTA (pH 8.0)

10 mg/ml bovine serum albumin (BSA)

> *Make fresh prior to each hybridization and sterile-filter to remove any impurities that may bind to nylon membranes.*

Neutralization buffer (0.4 M Tris-HCl [pH 7.4], 2× SSC)

10× OLA buffer (500 mM Tris-HCl [pH 8.5], 500 mM KCl, 75 mM MgCl$_2$, 10 mM β-NAD)

Oligonucleotides:

Genotyping oligonucleotides (unmodified) at the lowest synthesis scale

> *See Table 11-1 for probes and bar codes.*

TABLE 11-1. Probes and Bar Codes

Probe number	Probe/bar code a	Probe/bar code b
01	ATATTCTGAGACACGCCGCG	ATACGCGATGGGATCAGACT
02	ATGCGACTCTTGACGAACGT	TTCGAGCGTCTGGCACACTT
03	GTCACTCGTGTCCAGGATGT	TATCGCGTGTCAGTGCTTGT
04	GATACCGGACCATGTTTCGC	GATGTTCGTCCATGCGACCT
05	TGATCCGCGTCGATGCTCTT	GCAGTCACGTTCTCGAATCG
06	TTTAGCCGGATCACCGTGTG	ATATGTGCAGAACCCGCGAC
07	AGAGAGACGTTGCCCAAGTC	GATGCGATACCCTGCGATCT
08	ATTTAGCGTGCAGCCGACCT	ATGCGTGGTGTCCGATCATA
09	TAAGGGTTACGAACATCGCC	TGGACTCTCATAACGGCGTC
10	GCAGCTCGTCACAGGTATTG	TACCGGATTACAGCTCGTGG
11	AGCTAATGTCGAGTCACGCT	TCTACACGAGAACGAGGCAC
12	AGCGCGACGTTGATCCAGAT	AATGAACGAGACCGCGTGAC
13	TCGGACTCGTGACGCTATTT	ATGAGAGTTCGATGACCTGT
14	ACGCACTGACGATCATTCGG	TTCGACCCGGACGACTGTAT
15	TATAGCCGTGAACCCGATGC	TAAAGCACAGTCCGTAATCT
16	ATCATGTCCCAAGCGCGGTA	AAGCCGATGTCGATCTACCT

All 20-nt probe sequences are given in the 5′ to 3′ direction. The 16-nt bar code sequences incorporated into the upstream OLA oligonucleotides are the reverse complement of the underlined portion.

M13 amplification oligonucleotides:
M13F.BRL, CCCAGTCACGACGTTGTAAAACG
M13R.BRL, AGCGGATAACAATTTCACACAGG

Stripping buffer (0.1% SDS <!>), preheated to 80°C

T4 polynucleotide kinase (10 units/µl; New England Biolabs), used with buffer provided by
manufacturer

Taq DNA ligase (40 units/µl; New England Biolabs), used with 10x OLA buffer

Taq DNA polymerase (5 units/µl; New England Biolabs), used with 10x amplification buffer

Since the amplified products are very small, almost any thermostable DNA polymerase can be used.

Washing buffer (5x SSPE, 0.1% SDS <!>), prewarmed to 40°C

Equipment

Arrayer for nylon membranes (custom built—see http://cstern.bio.uci.edu/tools/genotyping.htm)

Custom software (http://cstern.bio.uci.edu/tools/genotyping.htm) available for the free statistical
programming package R (http://www.r-project.org/)

Heating block preset to 37°C, 65°C, 80°C, 96°C

Hybridization oven preset to 42°C

Hybridization tubes

Image acquisition software package (e.g., GE Healthcare ArrayVision)

Laboratory gloves (powder-free)

Liquid-handling system (e.g., Art Robbins Instruments Hydra) (Optional, see Step 8)

Manual pin-tool (Optional, see Steps 8, 14)

PCR plates (96 well or 384 well)

PCR thermocyclers (e.g., Applied Biosystems dual-block 384-well 9700)

Phosphor imager (e.g., GE Healthcare Typhoon)

Radioactive waste container

Repeater pipette (Optional, see Step 8)

Shaking platform

Storage phosphor screens (e.g., GE Healthcare)

Ultraviolet (UV) light source (50 mJ) <!>

Water bath preset to 40°C, 80°C

METHOD

Generation of OLA Genotyping Oligonucleotides

1. Identify SNPs using sequence alignments, or extract from on-line databases.

2. Design two upstream allele-specific oligonucleotides (47 nucleotides [nt]) and a single common downstream oligonucleotide (31 nt) for each SNP as follows:

   Upstream_a, M13F + C + Barcode_a + Up_flank + Allele_a

   Upstream_b, M13F + C + Barcode_b + Up_flank + Allele_b

   Downstream, Down_flank + G + M13R.RC

   Sections of the OLA genotyping oligonucleotides are:

   M13F, GACGTTGTAAAACG

   M13R.RC, CCTGTGTGAAATTG

Barcode_a and *Barcode_b* are the 16-nt bar code sequences permitting allele discrimination during hybridization (see Table 11-1). 16 pairs are available, permitting 16-plex OLA reactions—one pair of bar codes/probes per SNP in the reaction.

Up_flank is 15 nt and specific to the region upstream of the target SNP, and *Down_flank* is 16 nt and specific to the region downstream of the target SNP. If either of these flanking regions segregates for an additional SNP, the oligonucleotide sequence at those sites can incorporate a degenerate base.

Allele_a and *Allele_b* represent the allele at the target SNP.

A "C" or "G" nucleotide adjacent to the M13 sequence ensures that multiple ligated products of different sequence are evenly amplified.

3. Resuspend unmodified genotyping oligonucleotides at the lowest synthesis scale to 100 μM.

4. Create a 1 μM 16-plex upstream OLA oligonucleotide mix by mixing 2 μl of each of the 32 upstream oligonucleotides (at 100 μM) with 136 μl of H_2O.

5. Independently 5′-phosphorylate the downstream oligonucleotides. This is required so that up- and downstream oligonucleotides can be ligated, and should not be performed en masse. (Interactions among oligonucleotides can prevent equal phosphorylation of each oligonucleotide.) For each 12.5 μl reaction use:

H_2O	8.125 μl
T4 polynucleotide kinase buffer (10×)	1.25 μl
ATP (100 mM)	0.125 μl
T4 polynucleotide kinase (10 units/μl)	1 μl
Downstream oligonucleotide (100 μM)	2 μl

Incubate these reactions for 1 hour at 37°C, followed by 20 minutes at 65°C to stop the reactions.

6. Create a 1 μM 16-plex downstream OLA oligonucleotide mix by mixing 12 μl of each of the 16 downstream oligonucleotide phosphorylation reactions.

OLA Reaction

7. Make up the following reagent mix for a sufficient number of 3-μl reactions:

H_2O	2.3 μl
OLA buffer (10×)	0.3 μl
DTT (25 mM)	0.3 μl
Taq DNA ligase (40 units/μl)	0.04 μl
Upstream oligonucleotide mix	0.03 μl
Downstream oligonucleotide mix	0.03 μl

8. Add 3 μl of the reagent mix to each well of a 384-well or 96-well PCR plate, using a repeater pipette or a liquid-handling robot. Next, use a liquid-handling robot or a manual pin-tool to spike the reactions with 0.2 μl of PCR-amplified DNA containing the SNPs of interest.

Consistency among samples is reduced with a pin-tool.

Some of the PCR samples should be no-DNA PCR blanks as controls.

9. Seal the plate(s) and centrifuge briefly. Perform ligation using the following cycling profile:

   i. Initial denaturation for 5 minutes at 95°C

   ii. 3 cycles of 30 seconds at 95°C and 25 minutes at 45°C

   iii. Storage at 4°C

OLA Amplification Reaction

10. Make up the following reagent mix for a sufficient number of 12-μl reactions:

H$_2$O	10.196 μl
Amplification buffer (10x)	1.2 μl
dNTPs (25 mM)	0.024 μl
Taq DNA polymerase (5 units/μl)	0.1 μl
M13F.BRL (50 μM)	0.24 μl
M13R.BRL (50 μM)	0.24 μl

11. Add 12 μl of reagent mix directly to each OLA ligation reaction, seal the plate(s), and centrifuge briefly. Amplify ligation products using the following cycling profile:

    i. Initial denaturation for 2 minutes at 94°C

    ii. 32 cycles of 25 seconds at 94°C, 35 seconds at 58°C, and 35 seconds at 72°C

    iii. 2 minutes at 72°C

    iv. Storage at 4°C

    > Run both control (no-DNA PCR) and positive samples: A bright band (78 bp) should be visible in the positive samples, and not in the blank controls.
    >
    > Because the concentration of M13 primer in the OLA amplification reactions is very high, blanks normally contain an accessory "primer-dimer" band that is smaller than the positive band.

Arraying

12. Dry down the OLA amplification reactions at 65°C in a thermocycler for about 1 hour.

13. Add 5 μl of sterile-filtered denaturing buffer to each well, and resuspend/denature the samples in the thermocycler using this cycling profile:

    i. 15 minutes at 65°C

    ii. 5 minutes at 95°C

14. Print samples onto nylon membranes using an arraying robot. (A manual pin-tool can be used, although spot quality suffers.) Wear powder-free laboratory gloves, and do not touch membranes with bare hands. Allow the membrane spots to dry for 10 minutes. UV crosslink the samples at 50 mJ, and gently shake the membranes in a bath of neutralization buffer for 30 minutes (to neutralize the high pH of the denaturing print buffer).

Hybridization

15. Add membrane(s) to the hybridization tube (multiple membranes can be stacked in a single tube). Add 5 ml of prewarmed (42°C) hybridization buffer and 50 μl of sonicated herring sperm DNA which has been denatured by incubation for 5 minutes at 96°C.

16. Spin the tube in a hybridization oven at 4 rpm overnight (for first use of the membranes) or for 3 hours (for all subsequent hybridizations) at 42°C.

17. Prepare radiolabeled oligonucleotide probe by end-labeling the oligonucleotide with [γ-^{33}P]ATP in the following 10-μl reaction:

H$_2$O	5 μl
T4 polynucleotide kinase buffer (10x)	1 μl

Probe oligonucleotide (10 μM)	1 μl
T4 polynucleotide kinase (10 units/μl)	1 μl
$[\gamma^{-33}P]$ATP (10 μCi/μl)	2 μl

Incubate this reaction for 40 minutes at 37°C, followed by 15 minutes at 80°C to stop the reaction.

> *It is not necessary to column- (or otherwise) purify this reaction before use.*

18. Add the radiolabeled probe reaction to the hybridization tube, and spin in a hybridization oven at 4 rpm for 4 hours at 42°C.

Washing

19. Empty the hybridization buffer/radiolabeled probe to waste. Briefly rinse the tube with a small quantity of washing buffer prewarmed to 40°C, and also discard this to waste.

20. Add 50 ml of prewarmed (40°C) washing buffer. Spin the tube at 4 rpm for 20 minutes at 40°C, and discard the buffer to waste. Perform 3–5 of these wash cycles.

> *The number of wash cycles required will depend on the number of filters hybridized per tube. Too few wash cycles will lead to an inconsistent background level of radiation across the membrane surface.*

21. Remove the membranes from the hybridization tube and rinse in a bath of prewarmed washing buffer at 40°C.

Data Collection

22. Expose the hybridized/washed membranes to a phosphor screen for 3–4 days (the actual time required will depend on the spot intensity of the hybridized membranes). Scan the screen using a phosphor imager.

23. Analyze the images with an image acquisition software package (e.g., GE Healthcare ArrayVision). Call genotypes using custom software (http://cstern.bio.uci.edu/tools/genotyping.htm) available for the free statistical programming package R (http://www.r-project.org/).

Stripping

Once membranes have been imaged, they must be stripped prior to being reprobed.

24. Move membranes from the phosphor screen cassette to a bath of neutralization buffer to ensure that they stay moist.

> *If membranes dry out while radiolabeled probe is bound, the probe can become permanently fixed to the membranes.*

25. Add the membranes to hybridization tubes with 50 ml of stripping buffer preheated to 80°C. Spin the tubes at 4 rpm for 15 minutes at 80°C.

26. Discard the buffer to waste, and either store the membranes at 4°C in neutralization buffer, or start the protocol again from Step 15.

Template-directed Dye-Incorporation Assay with Fluorescence Polarization Detection

Connie Ha[1] and Pui-Yan Kwok[1,2]

[1]Cardiovascular Research Institute and Center for Human Genetics, University of California, San Francisco, California 94143-0793;
[2]Department of Dermatology, University of California, San Francisco, California 94143-0793

The fluorescence polarization (FP) of a molecule is proportional to the molecule's rotational relaxation time (the time it takes to rotate through a 68.5° angle), which is a function of the solvent viscosity, absolute temperature, and molecular volume (Kwok 2002). If the first two properties are held constant, then FP is directly proportional to the molecular volume, which is directly proportional to the molecular weight. Thus, when excited by polarized light at the appropriate wavelength, large fluorescent molecules tumble relatively slower in space and the polarized emission is observed, whereas small fluorescent molecules tumble faster and the polarized emission is not observed (depolarized) (see Fig. 11-2).

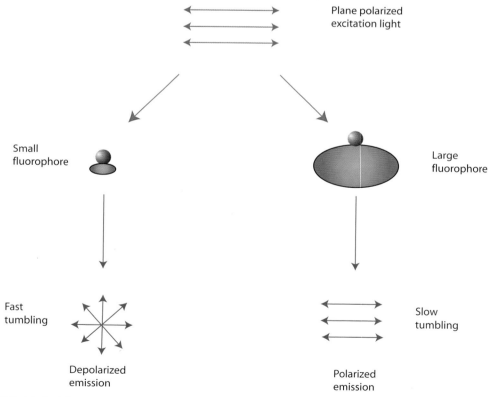

FIGURE 11-2. Schematic representation of FP. When excited by polarized light at the appropriate wavelength, large fluorescent molecules tumble relatively slower in space (polarized) whereas small fluorescent molecules tumble faster (depolarized).

This protocol describes a variation on single-base primer extension based on fluorescence polar detection. The Template-directed Dye-terminator Incorporation assay with Fluorescence Polarization detection (FP-TDI) is an economical and robust SNP genotyping method that is easy to optimize and to implement. This homogeneous assay utilizes unlabeled, unpurified PCR-grade primers, is extremely versatile, and can be used both in single-marker and moderate-throughput studies. The FP-TDI assay is a dideoxy chain-terminating DNA-sequencing protocol that ascertains the identity of the one base immediately 3′ to the unlabeled SNP primer annealed immediately upstream of the polymorphic site on the target DNA. The assay takes advantage of the specificity of DNA polymerase in extending the annealed primer by the dye terminator complementary to the polymorphic nucleotide found on the DNA target and the increase in FP when a fluorescent dye becomes part of a larger molecule as a result of the primer extension reaction (Chen et al. 1999). FP is based on the observation that when a fluorescent molecule is excited by plane-polarized light, it emits polarized fluorescent light into a fixed plane relative to the molecule itself (Perrin 1926).

The TDI assay consists of four key steps, all of which can be carried out in the same microtiter plate without further separation or purification (see Fig. 11-3). First, PCR amplification of

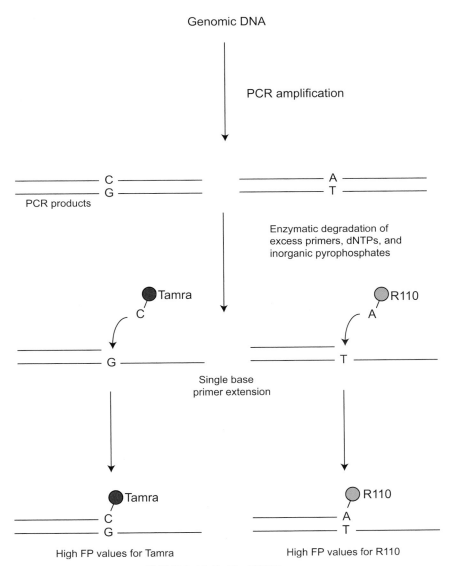

FIGURE 11-3. The FP-TDI assay.

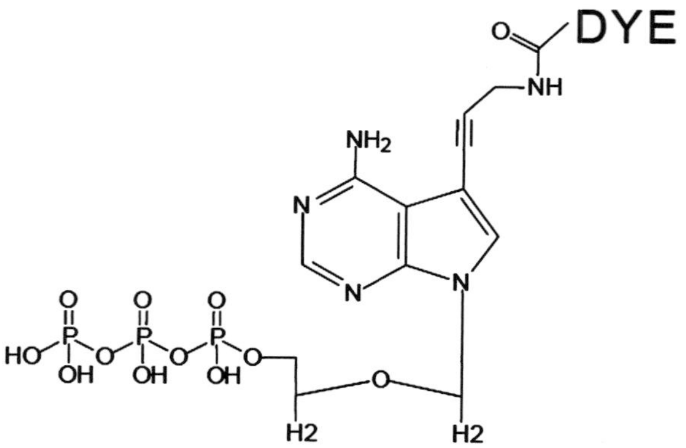

FIGURE 11-4. Structure of a dye-labeled Acyclo Terminator from PerkinElmer.

genomic DNA produces the template for the primer extension reaction. Second, a PCR cleanup enzyme cocktail with pyrophosphatase is added directly into the PCR products to remove excess PCR primers, deoxynucleoside triphosphates, and inorganic pyrophosphates. The pyrophosphatase is added to minimize the effects of misincorporation of the dye terminators (Xiao et al. 2004). Third, single-base primer extension is carried out using a SNP primer annealed to the target DNA one base upstream from the polymorphic site. Finally, the end product is scanned with an FP plate reader to determine the changes in fluorescence polarization.

The PerkinElmer AcycloPrime™ II-FP SNP Detection System, a commercial reagent kit based on the FP-TDI principle, is available at PerkinElmer Life Sciences. Instead of dideoxynucleotides, the kit contains four acyclic nucleoside triphosphates (called acycloterminators by the manufacturer), two of which are labeled with fluorescent dyes (see Fig. 11-4). It also includes Acyclopol, a mutant thermostable DNA polymerase that incorporates acycloterminators preferentially over dideoxyterminators. Six AcycloPrime kits are available to cover the six possible allele combinations (G/A, C/T, G/T, C/A, A/T, and G/C).

This genotyping assay is universal, highly specific, and very cost effective, especially in terms of assay development. In addition, FP is robust and easy to implement; at the conclusion of any of the four major sequences of steps, the reaction plate can be frozen and stored before proceeding to the next sequence. With the incorporation of pyrophosphatase in the protocol and the use of quenching as an alternative way of analyzing the fluorescence data, up to 95% of the assays work well even without optimization.

The FP-TDI method was among the genotype methods used in the International HapMap Project and produced data for over 15,000 SNPs (International HapMap Consortium 2003). A collection of on-line resources, including public databases and analytical tools, is given in Table 11-2.

TABLE 11-2. Internet Sources

Public databases for SNP sequences	
dbSNP	http://www.ncbi.nlm.nih.gov/SNP/
SNPper	http://bio.chip.org:8080/bio/
SNP consortium	http://snp.cshl.org/
PerkinElmer	http://perkinelmer.com/
Primer design freeware	
RepeatMasker	http://ftp.genome.washington.edu/RM/
Primer3	http://www.genome.wi.mit.edu/genome_software/
Data output Excel macro	
PerkinElmer	www.snpscoring.com

MATERIALS

CAUTION: See Appendix for appropriate handling of materials marked with <!>.

Reagents

AcycloPrime™-FP SNP Detection System (PerkinElmer) includes the following reagents:
  10x reaction buffer
  PCR Clean-Up reagent
  PCR Clean-Up buffer (10x)
  Acyclopol enzyme for single base extension
  AcycloTerminator mix containing equal amounts of dye-labeled R110 and Tamara
    terminators
Amplification buffer (10x)
Deoxynucleotide Set for Molecular Biology (dNTPs, each at 100 mM; Sigma-Aldrich)
$MgCl_2$
Oligonucleotides, SNP sequences, and software for primer design
  PCR primer oligonucleotides (available from Intergrated DNA Technology [IDT],
    Invitrogen, or Sigma-Aldrich)
  SNP primers are normally 18- to 25-mers complementary to either the sense or anti-sense
    PCR primers and are designed to anneal with the 3′ end immediately upstream of the
    polymorphic site on the target DNA. (SNP primers can be purchased from Intergrated
    DNA Technology.)
  SNP sequences for primer designs are available from public databases such as dbSNP
    (http://www.ncbi.nlm.nih.gov/SNP/), SNPper (http://bio.chip.org:8080/bio/), or the SNP
    consortium (TSC: http://snp.cshl.org/)
  In addition, over 38,000 designed primers for the assay are freely available at the PerkinElmer data-
  base (http://perkinelmer.com/).
  Primers are designed with freely available software such as RepeatMasker (http://ftp.genome.
    washington.edu/RM/) and Primer3 (http://www.genome.wi. mit.edu/genome_software/).
Pyrophosphatase (1 mg, Roche)
Taq DNA Polymerase (Platinum Taq) (5 units/μl with 10x amplification buffer; Invitrogen)

Equipment

Centrifuge
  Various options for centrifuges with a plate adaptor for spinning plates in between each reagent addi-
  tion can be purchased from Fisher or from VWR.
Multilabel Plate Reader (EnVision™ Multilabel Plate Reader, PerkinElmer; or Victor Multilabel
  Plate Reader, PerkinElmer)
  The plate reader generates data files from reading FP emissions after the TDI step (Steps 10 and 11).
PCR plates, black 384- or 96-well plates (LabSource or USA Scientific).
  Do not use clear PCR plates. PCR plates must be in black for the multilabel reader to detect FP emis-
  sion, and they must be compatible with the thermal cycler and the multilabel reader.
PCR sealing mats, 384-well (LabSource)
  Seals are reusable if washed in 5% bleach and rinsed thoroughly. Replace with new seals occasionally,
  depending on amount of use.
  The sealing mats can be substituted by using PCR film or foil adhesives (Thermo Fisher Scientific or
  VWR International).
PCR Thermocyclers (e.g., Auto-Lid Dual 384-Well GeneAmp PCR System 9700 or 96-Well
  GeneAmp PCR System 9700, Applied Biosystems)
  There are many different options of thermal cyclers designed for the amplification of nucleic acids,
  available from Fisher or VWR.

Pipetting platform (Evolution P³ Precision Pipetting Platform, PerkinElmer)
> *The Evolution P³ is a fast-throughput liquid-handling automation system for dispensing reagents. Depending on the design of the experiment, a regular or multichannel pipette can be substituted.*

METHOD

Primer Design

1. Obtain the sequence surrounding a targeted SNP site by submitting lists of reference SNP (rs) or submitted SNP assay (ss) numbers to dbSNP or other database by batch format. The RefSNP accession ID (rs) and submitted SNP record (ss) numbers are assigned by dbSNP for SNPs submitted to the public database. Submission of rs and ss numbers to obtain the flanking sequences of SNPs can be done simply by following the instructions on the dbSNP Web site.

2. Analyze the sequence obtained with the software RepeatMasker to avoid selecting primers within repeated regions (Vieux et al. 2002). The software is well documented and easy to use. Selecting PCR primers using the unique genome sequences will increase the success rate of PCR assay design.

3. Choose primers using the software Primer3 with an optimal set of parameters (see Table 11-3).

Amplification Reaction

For the FP-TDI assay to work well, one must generate sufficient PCR products for the primer extension reaction and minimize the amount of residual PCR primers and dNTPs such that the enzymes used in the PCR cleanup step have adequate capacity to degrade the excess primers and dNTPs. The protocol is therefore designed with these requirements in mind.

4. Dispense 2.4 ng of genomic DNA per well onto the 96- or 384-well plates; air-dry overnight. Cover plates with tissues to protect from dust.
   > *Generally, we recommend leaving two empty wells and including four sets of duplicates as controls for every set of 96 wells. Dried DNA is very stable and can be stored for over a year, depending on the quality of the original stock. Wrap dried DNA plates in plastic wrap and store in a desiccator.*

TABLE 11-3. Primer3 Parameters for Primer Design

Primer3 parameter						
Primer product size range (bp):	80–400					
Primer product opt size (bp):	250					
Primer size (bp):	Min:	20	Opt:	23	Max:	26
Primer T_m (°C):	Min:	54	Opt:	55	Max:	56
Primer GC (%):	Min:	20			Max:	50
GC Clamp:	0					
DNA concentration (mM):	40					
Primer salt concentration (mM):	50					
Primer pair wt product size LT (fraction):	0.20					
Primer pair wt product size GT (fraction):	0.50					
Primer self any:	8	Primer self end:	3			
Primer max end stability:	8					
Primer explain flag:	1					
Primer num return (primer pairs):	1					
Target start position (bp):	X					
Target length (bp):	50					

5. Prepare a PCR cocktail consisting of the following reagents:

10x amplification buffer	0.5 µl
MgCl₂ at 50 mM	0.25 µl
dNTP mix, each at 2.5 mM	0.1 µl
Platinum Taq DNA Polymerase (5 units/µl)	0.02 µl
H₂O	2.13 µl

Make all solutions on ice and add enzyme immediately before dispensing the mixture in Step 6.

6. Combine 3 µl of PCR primer mix (0.2 µM, each forward and reverse primer) with 3 µl of PCR cocktail. Dispense the total PCR mix of 6 µl into each well of dried DNA (in plates prepared in Step 4). To minimize contamination, dispense the solution toward the top of each well, then centrifuge the plate, and seal well.

> *If more than one marker is being genotyped, it is best to dispense separately the 3 µl of primer mix and 3 µl of PCR cocktail. Centrifuge the plate between each delivery to minimize contamination. Seal well to prevent evaporation of samples from the plate edges. Label each plate carefully because the same plate will be used to perform the entire protocol.*
>
> *PCR mix is relatively stable and can be saved for later use. Excess PCR mix can be stored at −20°C for 3–4 weeks without loss of activity if not subjected to repeated freeze–thaw cycles.*

7. Preheat the thermal cycler lid to 105°C. Carry out amplification according to the following program:

Number of Cycles	Denaturation	Annealing	Polymerization/Extension
1	2 minutes at 95°C		
45	10 seconds at 92°C	20 seconds at 58°C	30 seconds at 68°C
Final			10 minutes at 68°C

Store the reaction at 4°C or proceed directly to Step 8.

> *It is essential to preheat the thermal cycler lid and to keep lid heated while all programs are being run to prevent evaporation during cycling. If the thermal cycler lid is cooler than the plate temperature, condensation of vapors will contaminate the entire plate.*

PCR Cleanup with Pyrophosphatase

The PCR product must be treated to degrade the pyrophosphate produced in the reaction and the excess PCR primers and dNTPs. This is necessary because the primers, dNTPs, and pyrophosphatase will interfere with the primer extension reaction (Xiao et al. 2004). The degradation is accomplished by the addition of three enzymes: exonuclease I, shrimp alkaline phosphatase, and pyrophosphate. The exonuclease I and shrimp alkaline phosphatase are supplied as a combined reagent by the manufacturer.

8. Prepare the PCR Clean-Up Mix by combining 10x PCR Clean-Up buffer, PCR Clean-Up reagent, and pyrophosphatase enzyme in a ratio of 10.5:1.33:1.5, respectively. Deliver 2 µl of Clean-Up Mix to each PCR product from Step 7. Centrifuge the plate to mix, and seal well.

> *The PCR Clean-Up Mix may be stored at −20°C for up to 2 weeks if it is not subjected to repeated freeze–thaw cycles.*

9. Preheat the thermal cycler lid to 105°C. Incubate the reaction mixture at 37°C for 1 hour, and inactivate the enzyme by heating at 90°C for 15 minutes. Store the reaction at 4°C.

Primer Extension (TDI)

The primer extension reaction is basically DNA sequencing for just one base. The protocol is therefore that of DNA cycle sequencing.

10. Prepare a TDI cocktail for the primer extension step consisting of

Acycloprime 10x reaction buffer	2 μl
Chosen dye terminator combination	1 μl
Acyclo enzyme	0.05 μl
H_2O	4.95 μl

11. Combine 5 μl of SNP primer mix (1 μM, either the forward or reverse SNP primer) with 8 μl of TDI cocktail. Add the total mixture to each post-cleanup PCR product. Centrifuge the plate, and seal well.

    As in Step 6, if more than one marker is being genotyped, it is best to dispense the 5 μl of SNP primer mix and 8 μl of TDI cocktail separately. Centrifuge the plate between each delivery to minimize contamination. Seal the plate well to prevent evaporation of samples from the plate edges.

    Excess TDI cocktail can be stored at −20°C for 3–4 weeks if it is not subjected to repeated freeze–thaw cycles. TDI cocktails are light sensitive; wrap plates with foil before storage.

12. Preheat the thermal cycler lid to 105°C. Carry out the extension reaction according to the following program:

Number of Cycles	Denaturation	Polymerization/Extension
1	2 minutes at 95°C	
5–15	15 seconds at 95°C	30 seconds at 55°C

Store the reaction at 4°C.

Plate Reading and Data Analysis

13. Remove the plates from the thermal cycler and centrifuge them. Take off the seals and place the plates in the Envision plate reader.

14. Export the FP file from the Envision plate reader and run it against the excel macro EnVisionMacro_Excel384_4x96 (www.snpscoring.com) to generate files in excel format.

15. Call genotypes according to data clusters and FP values; a set of good data should display four distinct clusters on a scatter plot. The negative controls (wells with no DNA) have low R110 and TAMRA values and stay close to the origin, indicating that the dye terminators remain free in solution. The homozygous data points should have high FP values for one dye and low FP values for the second dye, whereas the heterozygous data points have high FP values for both R110 and TAMRA.

16. For data where the R110 values are severely quenched, reanalyzing the fluorescence data by plotting the TAMRA FP values versus the R110 intensity ratio may improve the genotype calls (see Fig. 11-5). The intensity ratio (R) of TAMRA versus R110 is defined as:

    $R = 100 * I_{total\ of\ TAMRA} / I_{total\ of\ R110}$ (Xiao et al. 2004).

17. If the scatter plot displays four clusters with incomplete separations and low FP values for both R110 and TAMRA, it may help to complete the reaction by returning the plate to the thermal cycler for more TDI cycling (as in Step 12).

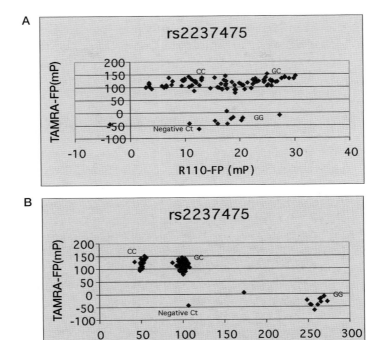

FIGURE 11-5. Quenching analysis of marker rs2237475 by plotting the TAMRA FP values versus the R110 intensity ratio. (*A*) Low R110 FP values result in poor cluster separations. (*B*) The same set of data replotted with TAMRA FP values versus the R110 intensity ratio R.

TROUBLESHOOTING

Problem (Steps 10, 11, Primer extension reaction): Genotyping by FP-TDI is a robust and accurate assay, but there are some observed issues inherent to the dye terminator properties used in the protocol. One category of assay failures is attributed to low FP signals for R110-labeled acycloterminator, resulting in poor separation of clusters.

Solution: The low R110 FP values on some extension products can be explained by the quenching of R110 after its incorporation onto the SNP primer. A systematic investigation of this phenomenon revealed that the relationship between fluorescence quenching and polarization differs for the different dyes. Fluorescence quenching happens almost exclusively in R110-labeled acycloterminators, and for some assays, the fluorescence intensities were quenched as much as 90% (Xiao et al. 2004). From our studies, the fluorescence intensities of all TAMRA-acycloterminators increase upon incorporation onto SNP primers and were quite stable, with FP values exceeding 120 mP. In contrast, the fluorescence intensities of R110-acycloterminators diminish significantly upon incorporation onto SNP primers, displaying FP values that range from 30 mP to 150 mP. Quenching effects are further exacerbated in sequences that either contain guanosine at the 3' end or several guanosines within 10 bases from the 3' end (Xiao and Kwok 2003). Assays with severe R110 quenching result in the loss of FP values and become a poor indicator of dye-terminator incorporation. The relatively slow incorporation of R110 acycloterminators by Acyclopol adds to the problem with the R110 label (Xiao et al. 2004). Acknowledging and taking the quenching phenomenon into consideration, one can easily recover fluorescence data to obtain good genotype calls. As reasoned in an earlier publication (Xiao et al. 2004), a good way to determine the true R110-acycloterminator incorporation is to plot the degree of quenching of R110. By plotting TAMRA FP values against R110 intensity ratio (quenching) rather than R110 FP, overlapping groups will separate into four clear clusters.

Problem (Steps 10, 11, Misincorporation of dye-labeled terminators): Another caveat to address in the FP-TDI protocol is the failure of assays due to misincorporation of one of the dye-labeled terminators during the primer extension reaction. Failed assays within this category are marked by the absence of one homozygous cluster accompanied by an unusually high number of heterozygotes. We noticed that when the SNP primer ends in the same base as one of the acycloterminators in the reaction, the genotype profile falls into the pattern as described.

Solution: In attempts to understand the failures and improve the assay, we discovered that the culprit behind the dye-terminator misincorporation lies in the excess inorganic pyrophosphate (PPi) generated in the PCR step. Large yields of PPi during PCR can lead to pyrophosphorolysis, the reverse reaction where DNA polymerase catalyzes the cleavage of the 3' base off the SNP primer. Primers shortened by this mechanism are then extended by dye-terminators complementary to the target sequence, giving rise to erroneous genotypes. Our observation indicates that the misincorporation induced by the pyrophosphate always happens after one of the terminators is used up and the forward extension reaction is completed (Xiao et al. 2004). Degradation of PPi prior to the primer-extension reaction is integral to the prevention

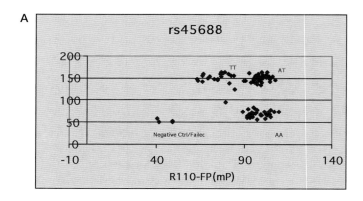

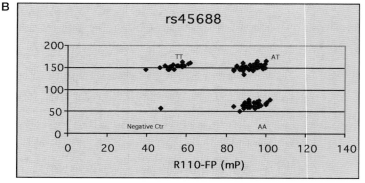

FIGURE 11-6. Genotyping data of marker rs45688 without and with pyrophosphatase added to the PCR Clean-Up incubation step. (*A*) Genotyping result without pyrophosphatase incubation at 30 TDI cycles. The homozygous TT cluster moved toward the heterozygous AT cluster and began to merge with it. (*B*) Genotyping result of the same reaction performed with incubation of pyrophosphatase. No misincorporation was observed even up to 70 cycles.

of pyrophosphorolysis and thus, to the success of the assay. In our studies, we have been able to effectively prevent misincorporation of acycloterminators by incubating the PCR product with a mixture of pyrophosphatase during the PCR cleanup step to remove the PPi generated during PCR (see Fig. 11-6). When the SNP primer ends in a base that is not one of the dye-terminators in the reaction, incorporation of an unlabeled terminator onto the cleaved primer does not interfere with the fluorescence polarization in the reaction.

ACKNOWLEDGMENTS

We thank Angie Phong for the figures and Drs. Ming Xiao and Ting-Fung Chan for their comments on the manuscript.

Protocol 5

Temperature Gradient Capillary Electrophoresis Assay

An-Ping Hsia,[1] Hsin D. Chen,[1] and Patrick S. Schnable[1,2,3]

[1]Department of Agronomy, [2]Department of Genetics, Development, and Cell Biology, [3]Center for Plant Genomics, Iowa State University, Ames, Iowa 50011

This protocol describes the use of temperature gradient capillary electrophoresis (TGCE) to test for polymorphisms between two DNA fragments (Hsia et al. 2005). Polymorphisms may be detected within an individual or between individuals, depending on the experimental design. In this assay, the alleles of interest and a "reference" allele are PCR-amplified with high-fidelity Taq polymerase and the resulting amplicons are then mixed in a 1:1 ratio. This mixture is denatured and reannealed to allow formation of homoduplexes (in the absence of polymorphisms between the allele of interest and the reference allele) and heteroduplexes (in the presence of polymorphisms). The Reveal System (SpectruMedix), which can analyze four 96-well plates at a time, is then used for electrophoresis and detection, and the data are scored and viewed with Revelation software.

MATERIALS

Reagents

AmpliTaq Gold (5 U/µl, Applied Biosystems)
Cloned Pfu DNA polymerase (2.5 U/µl, Stratagene)
DNA samples
dNTPs (2 mM) (Intermountain Scientific)
Forward primer (5 µM)
$MgCl_2$ (25 mM)
Mineral oil
PCR buffer (10x)
Reverse primer (5 µM)
TGCE reagents (SpectruMedix)
    Bottles #1 and 2: Capillary Wash (WASH-500-002)
    Bottle #3: Reveal Running Buffer without dye (BRUR-500-002)
    Bottle #4: Reveal Mutation Discovery Running Buffer (BRUR-500-001)
    Gel: Matrix, Reveal Mutation Discovery (MREV-240-001)

Equipment

The Reveal System, model RVL 9612, rev. 2.0 (SpectruMedix)
Revelation analysis software (version 2.4) to visualize and score TGCE data (SpectruMedix)
> The Reveal System and the Revelation software are now available through Transgenomic, Omaha, Nebraska.

GRAMA software (Maher et al. 2006)

METHOD

1. Prepare the following PCR reaction mixture (20-μl total volume) for amplification of target and reference alleles. Depending on the experimental design, the target and reference alleles can be within the same genome or different genomes.

20–50 ng Genomic DNA	2.5 μl
10x PCR buffer	2.0 μl
2 mM dNTPs	2.0 μl
25 mM MgCl$_2$	1.6 μl
5 μM Forward primer	2.0 μl
5 μM Reverse primer	2.0 μl
5 U/μl AmpliTaq Gold	0.09 μl
2.5 U/μl Pfu polymerase	0.02 μl
H$_2$O	7.79 μl

2. Amplify target and reference alleles using the following PCR program:

   i. 95°C for 10 minutes

   ii. 94°C for 3 minutes

   iii. 94°C for 30 seconds

   iv. 60°C for 45 seconds

   v. 72°C for 1 minute 30 seconds

   vi. Return to Step iii, 30 times

   vii. 72°C for 10 minutes

   viii. 12°C hold

3. Mix 5 μl of PCR reactions from the target and reference alleles. Denature and reanneal the samples in a thermocycler using the following program:

   i. 95°C for 2 minutes, 40 seconds

   ii. 95°C for 20 seconds (repeat 15 times, decreasing the temperature by 1°C each time)

   iii. 80°C for 1 minute (repeat 25 times, decreasing the temperature by 1°C each time)

   iv. 55°C for 18 minutes

   v. 55°C for 1 minute (repeat 10 times, decreasing the temperature by 1°C each time)

   vi. 45°C for 30 seconds (repeat 10 times, decreasing the temperature by 1°C each time)

   vii. 12°C hold

4. Add 10 μl of mineral oil to the PCR mixture to prevent evaporation of samples during electrophoresis. Load the plates and run the following TGCE instrument program:

   i. Bottle #1, Line Purge, Purge rate: 20.0 ml/minute for 15 minutes

   ii. Bottle #1, Flow rate: 22.5 ml/minute for 5 minutes

   iii. Bottle #2, Flow rate: 25.0 ml/minute for 5 minutes

   iv. Bottle #3, Line Purge, Purge rate: 10.0 ml/minute for 3 minutes

    v. Bottle #3, Flow rate: 6.0 ml/minute for 6 minutes

    vi. Gel, Line Purge, Purge rate: 12 ml/minute for 12 minutes

    vii. Gel, Gel Inject, Volume delivered/cap: 42 µl

    viii. Delivery Time: 10 minutes

    ix. Bottle #4, Gel PreRUN, Flow rate: 5.0 ml/minute, 10.0 kV, 5 minutes

    x. Bottle #4, Sample Inject, Flow rate: 5.0 ml/minute, 6.0 kV, 50 seconds

    xi. Bottle #4, Electrophoresis + Current Monitor, Flow rate: 5.0 ml/minute, 9.0 kV, 15 minutes

    xii. Bottle #4, DATA Acquisition, Flow rate: 5.0 ml/minute, 9.0 kV, 50 minutes

    Use 50 minutes and 30 minutes in this step for survey and mapping, respectively.

5. Use the Revelation Software package (ideally, supplemented with GRAMA software [Maher et al. 2006]) to score and view the TCGE data.

    See Chapter 26 for a detailed description of the software and an example of the output display and analysis.

ACKNOWLEDGMENTS

This project was supported by a competitive grant from the National Science Foundation Plant Genome Program (DBI-0321711) to P.S.S. and Hatch Act and State of Iowa funds to P.S.S.

SNP Mining from Maize 454 EST Sequences

W. Brad Barbazuk,[1] Scott Emrich,[2,3] and Patrick S. Schnable[4,5,6]

[1]Donald Danforth Plant Science Center, St. Louis, Missouri 63132; [2]Bioinformatics and Computational Biology Graduate Program, [3]Department of Electrical and Computer Engineering, [4]Department of Agronomy, [5]Department of Genetics, Development, and Cell Biology, [6]Center for Plant Genomics, Iowa State University, Ames, Iowa 50011

A massively parallel pyrosequencing technology commercialized by 454 Life Sciences Corporation was used to sequence the transcriptome of the shoot apical meristem (SAM) of B73 maize isolated using laser capture microdissection (LCM). Analysis of the data indicates that the combination of LCM and the deep sequencing possible with 454 technology enriches for SAM transcripts not present in current EST collections. RT-PCR was used to validate the expression of 27 genes whose expression had been detected in the SAM via LCM-454 technology, but that lacked orthologs in GenBank. Significantly, ~74% (20/27) of these validated SAM-expressed "orphans" were detected in the SAM but not in meristem-rich immature ears. We conclude that the coupling of LCM and 454 sequencing technologies facilitates the discovery of rare, possibly cell-type-specific transcripts.

In this protocol, 454 expressed sequence tags (ESTs) are generated by sequencing SAM cDNA from maize inbred lines on the 454 Life Sciences GS-20 sequencing system. The computational tool POLYBAYES is then used to identify single-nucleotide polymorphisms (SNPs). POLYBAYES has been used successfully to identify SNPs in many different systems, including maize (Useche et al. 2001), and is particularly recommended for identifying SNPs in 454 sequences. For a detailed discussion of this computational tool and an example of POLYBAYES-mediated SNP mining, please see Chapter 26.

MATERIALS

Reagents

Shoot apical meristem (SAM) cDNA from maize inbred lines B73 and Mo17

6–10 SAMs (which contain about 15–18,000 cells) will provide about 10 ng of RNA. The RNA samples are then subjected to two rounds of amplification, usually generating about 20–60 μg of amplified RNA (aRNA), prepared as described in Chapter 8. 20 μg of aRNA is used for cDNA synthesis to yield about 15 μg of cDNA.

Equipment

BLAST

CROSS_MATCH

454 Life Sciences GS-20 sequencing system (454 Life Sciences, http://www.454.com)

454 Life Sciences has a sequencing service center that will provide sequence from such samples as cDNA and genomic DNA. Check with them regarding requirements for cDNA quantity and quality.

POLYBAYES (http://bioinformatics.bc.edu/marthlab/polybayes.html)

METHOD

1. Generate 454 ESTs by sequencing SAM cDNA from the maize inbred lines B73 and Mo17 on the 454 Life Sciences GS-20 sequencing system.

2. Assign 454 ESTs to maize genomic anchor sequences using BLAST, by identifying the highest-scoring alignment between each 454 EST and the collection of genomic sequences (1e-8 minimum E-value).

   In place of genomic DNA, assembled ESTs can be used as an anchor. The main requirement is that the anchor sequences be of high quality, since they are driving the multiple sequence alignment (MSA). Although "best hit" criteria are used during EST to anchor assignment, poor alignments or alignments between paralogs will be caught either during formation of MSAs by CROSS_MATCH (see below) or by the internal paralog filter implemented within POLYBAYES. The genome of B73 maize is currently being sequenced, and this will provide an excellent collection of anchor sequences.

3. Run CROSS_MATCH on each anchor sequence and its associated 454 ESTs to create an anchored MSA. The following CROSS_MATCH parameters are recommended:

   -discrep_lists -tags -masklevel 5 -gap_init -1 -gap_ext -1.

   Low initiation (-gap_init) and gap extension (-gap_ext) are used to increase alignment tolerance between the short 454 ESTs and genomic anchors. Substitute higher values for gap_init and gap_ext if the anchored MSAs are unspliced (i.e., ESTs aligned to an EST anchor, or genomic sequence aligned to a genomic sequence anchor).

4. Run POLYBAYES on the MSA. Recommended POLYBAYES parameters for maize are:

   -maskAmbiguousMatches

   -nofilterParalogs

   -priorParalog 0.03

   -thresholdNative 0.75

   -screenSnps

   -considerAnchor

   -noconsiderTemplateConsensus

   -prescreenSnps

   -priorPoly 0.01

   -thresholdSnp 0.5

   It is necessary to include sequence quality files for the anchor sequence and the sequences aligned to it (member sequences). If these are unavailable or unreliable, set default quality values with:

   -anchorBaseQualityDefault

   -memberBaseQualityDefault

   Because CROSS_MATCH aligns each sequence individually to the anchor during MSA construction, and POLYBAYES assesses base quality on an individual basis, use of a stringent default rather than the base quality information provided by 454 Life Sciences is expected to increase the accuracy of polymorphism detection.

5. Perform post-processing by reading the POLYBAYES output files and deciding on appropriate rules to distinguish putative SNPs from false positives. In maize SNP mining experiments conducted by the authors, both Mo17 and B73 454 ESTs were available, and the B73 maize

MAGI assemblies were used as alignment anchors. Because Mo17 and B73 are inbreds, they should be monoallelic at every base position, with relatively rare exceptions caused by nearly identical paralogs (NIPs). Hence, putative SNPs were filtered using rules designed to substantially decrease the rate of false positives. These rules are:

i. Polymorphic sites require a minimum of 2x representation in the Mo17-454-ESTs.

ii. All Mo17 base calls at sites that are polymorphic between Mo17 454 ESTs and the B73 MAGI anchors are expected to be identical. This ensures monoallellism within the Mo17-454-ESTs.

iii. When B73-454-EST sequences also align across polymorphic sites that pass Rules 5i and 5ii, all of the B73-454-ESTs and the MAGI3.1 anchor base calls must agree. This avoids polymorphisms resulting from incorrect MAGI base calls or NIPs within B73.

ACKNOWLEDGMENTS

The development of this protocol was supported by the National Science Foundation (award numbers DBI-0321595 and CNS-0521568), ISV's Plant Science Institute, and the Donald Danforth Plant Science Center. Additional support was provided by Hatch Act and State of Iowa funds.

REFERENCES

Bonfield J.K., Rada C., and Staden R. 1998. Automated detection of point mutations using fluorescent sequence trace subtraction. *Nucleic Acids Res.* **26:** 3404–3409.

Chen X., Levine L., and Kwok P.Y. 1999. Fluorescence polarization in homogeneous nucleic acid analysis. *Genome Res.* **9:** 492–498.

Hsia A.P., Wen T.J., Chen H.D., Liu Z., Yandeau-Nelson M.D., Wei Y., Guo L., and Schnable P.S. 2005. Temperature gradient capillary electrophoresis (TGCE)—A tool for the high-throughput discovery and mapping of SNPs and IDPs. *Theor. Appl. Genet.* **111:** 218–225.

International Hapmap Consortium. 2003. The International Hapmap Project. *Nature* **426:** 789–796.

Kwok P.Y. 2002. SNP Genotyping with fluorescence polarization detection. *Hum. Mutat.* **19:** 315–323.

Maher P.M., Chou H.-H., Hahn E., Wen T.-J., and Schnable P.S. 2006. GRAMA: A genetic mapping tool for the analysis of temperature gradient capillary electrophoresis (TGCE) data.

Theor. Appl. Genet. **113:** 156–162.

Perrin F. 1926. Polarization de la lumiere de fluorescence. Vie moyenne de molecules dans l'etat excite. *J. Phys. Radium* **7:** 390–401.

Stephens M., Sloan J.S., Robertson P.D., Scheet P., and Nickerson D.A. 2006. Automating sequence-based detection and genotyping of SNPs from diploid samples. *Nat. Genet.* **38:** 375–381.

Useche F.J., Gao G., Harafey M., and Rafalski A. 2001. High-throughput identification, database storage and analysis of SNPs in EST sequences. *Genome Inform.* **12:** 194–203.

Vieux E.F., Kowk P.Y., and Miller R.D. 2002. Primer design for PCR and sequencing in high-throughput analysis of SNPs. *BioTechniques* **32:** S28–S32.

Xiao M. and Kwok P.Y. 2003. DNA analysis by fluorescence quenching detection. *Genome Res.* **13:** 932–939.

Xiao M., Phong A., Lum K., Greene R.A., Buzby P.R., and Kwok P.Y. 2004. Role of excess inorganic pyrophosphate in primer-extension genotyping assays. *Genome Res.* **14:** 1749–1755.

12 | Molecular Inversion Probes and Universal Tag Arrays: Application to Highplex Targeted SNP Genotyping

George Karlin-Neumann, Marina Sedova, Ronald Sapolsky, Jonathan Forman, Yuker Wang, Martin Moorhead, and Malek Faham

Affymetrix, South San Francisco, California 94080

INTRODUCTION

Targeted genotyping denotes the ability to genotype almost any position in a genome of interest. It is especially critical in hypothesis-driven studies or applications where one wishes to ascertain genotypes at specific genomic locations, such as candidate genes or regions from linkage studies, or at defined polymorphic loci that are known or surmised to have functional and/or phenotypic consequences. The needs of such applications can range from lower multiplex (plex) requirements as in diagnostics (tens to hundreds to several thousands of markers per sample) (Van Eerdewegh et al. 2002); to medium plex for candidate gene or linkage studies (sev-

eral thousands to ten thousand markers per sample) (Zheng et al. 2006); to high plex for studies of functional polymorphisms or follow-ups from whole-genome association studies (tens of thousands of markers per sample) (Begovich et al. 2004; Shiffman et al. 2005). At very high plex levels, targeted genotyping can be used for whole-genome association studies. These rely on the high linkage disequilibrium (LD) in the genome for the indirect assessment of a large number of SNPs by genotyping only a set of "tagging" SNPs (hundreds of thousands of markers per sample) (Risch and Merikangas 1996; de Bakker et al. 2005; International HapMap Consortium 2005).

A method suited to these needs must have the following characteristics:

- High conversion rate and minimal sequence design constraints for assay probes (i.e., probes for most target sequences can be designed and perform successfully)
- High target specificity
- High sensitivity
- High accuracy

- High level of multiplexing
- High throughput

Additional desirable features may include

- Flexible customization (including supplementation of standard panels)
- Accurate quantitation of genotyping markers

The Molecular Inversion Probes (MIP) genotyping assay exhibits all of these features: It can achieve first-pass conversion rates of up to 95% when validated HapMap SNPs are used, and it can distinguish between similar target sequences differing in as little as a single base, from DNA quantities as low as 1 μg of unamplified human genomic DNA or 50 ng of whole-genome amplified DNA. The assay routinely achieves accuracies of >99.5% as determined by trio (mother-father-child) concordance or concordance to HapMap genotypes; it can be multiplexed to <1500-plex and >50,000-plex assays per reaction which, after hybridization, are scanned on a single Affymetrix Universal Tag array in <10 minutes; and 48–96 samples can be processed by 2–3 persons in 2 days, or 192–384 samples per week. Either standard or custom panels can be created, or standard panels can be augmented with custom single-nucleotide polymorphisms (SNPs) of interest. Finally, a modified version of the assay can determine genotypes and accurately quantitate allele copy number in <100 ng of unamplified genomic DNA (see Chapter 17). A scheme of the genotyping assay, showing the sequence from annealing of probes to sample DNA through hybridization to the Universal Tag array, is depicted in Figure 12-1A.

These various features are realized in the MIP assay through the combined properties of the circularizable "padlock" probes (Nilsson et al. 1994), the enhanced specificity conferred by enzymological recognition of the SNP base, and the conversion of successful SNP recognition into a highly specific tag sequence, which is then scored by hybridization to a universal tag array (Hardenbol et al. 2005). The MIP probe achieves its initial target specificity, as well as its sensitivity and ability to multiplex at high levels, from the cooperative binding of its bipartite recognition sequence about the targeted SNP site (see Fig. 12-1B). During annealing, the probe's two terminal homology arms bind to the previously denatured genomic DNA, leaving a 1-base gap at the site of the SNP being investigated. The annealing reaction is divided into four equal aliquots, and a unique, single dNTP is added to each (dATP to the first, dCTP to the second, dGTP to the third, and dTTP to the fourth). The allele(s) present at each locus in the sample is determined by whether or not a given probe is circularized in the presence of a particular base. Specificity, and consequently accuracy, are further enhanced in this step by the combined specificities of the polymerase adding the complementary base at the gap position and the ligase covalently closing the circularized probe. Assay signal-to-noise ratio (S/N) is increased in the subsequent exonuclease digestion step, which selectively degrades excess, uncircularized probe but leaves intact the circularized probes. The padlocked probe is then inverted by cleavage at site #1 between the two universal primer-binding sites (originally positioned in the interior of the probe), now located at the termini of the inverted probe and in an orientation suitable for PCR amplification. A second, and briefer, PCR amplification uniquely labels each of the four reactions per sample so that they can be unambiguously distinguished from each other when combined onto a single array for readout of the four single-base interrogations. After hybridization to a universal tag array (which contains probe sequences complementary to the tag sequences in the MIP products), the array is washed and stained with a 4-color Qdot stain, then scanned on an Affymetrix GCS3000 4-color scanner for each of the Qdot wavelengths.

The accuracy, sensitivity, and flexibility of the MIP assay are further enhanced by the use of maximally orthogonal synthetic tags (Winzeler et al. 1999) to read out the results of the assay. The complementary tag sequences on Affymetrix Universal Tag Arrays have been selected to have very similar melting temperatures while exhibiting very low cross-hybridization and minimal secondary structure (Hardenbol et al. 2005). These Universal Tag Arrays are not purpose-specific like

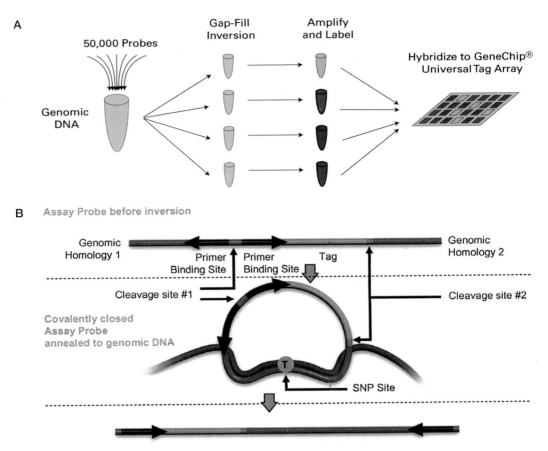

FIGURE 12-1. Schematic of highly multiplexed, 4-color MIP targeted genotyping assay. (A) Assay process from annealing up to 50,000-plex probe pool to sample DNA, through hybridization to GeneChip Universal Tag Arrays. After hybridization and washing, genotyping signals are generated by staining with Qdot conjugates and scanning with a GCS3000 4-color scanner. (B) MIP before and after inversion. (*Top*) Design of un-inverted MIP probe showing genomic homology regions which will sandwich the interrogated SNP site, universal PCR primer sites, unique Tag associated with the SNP marker, and several cleavage sites. (*Middle*) MIP probe "padlocked" onto a genomic DNA template after annealing and gapfill/ligation at a "T" SNP site. The circularized probe is subsequently inverted by cleavage at site #1. (*Bottom*) The inverted probe is ready for PCR amplification and labeling (with universal primers), and subsequent hybridization to a Universal Tag array. (Reprinted, with permission, from Karlin-Neuman et al. 2007.)

standard arrays, but can be used for an unlimited variety of panels, each having the desired biological target sequences joined with the appropriate set of tags. To further minimize unwanted cross-interactions, the genomic DNA sequence is cleaved away from its uniquely associated 21-mer tag before hybridizing the MIP products to the tag array. It is the fluorescence of this surrogate tag sequence that is then scored on a single microarray feature at each of the four wavelengths to determine the genotype at the locus. The high S/N of the assay benefits from obtaining all of the genotype information for a marker from a single feature. Although only two of the channels potentially should give signal for biallelic SNPs, fluorescence in the other two background channels is also evaluated for each probe to monitor its specificity in the assay (expressed as a signal-to-background [S/B] value)—probes having low S/B for any reason are not scored. Finally, the raw fluorescence intensities of the four chip images per sample are background-subtracted, corrected for spectral overlap among the four dyes, and normalized across the four channels. After application of the appropriate processing filters, genotypes are determined for the markers present in the assay panel. The genotypes can be called both without and with clustering; the former provides a quick, quality control (QC) summary of the sample performance and includes many useful metrics and their graphical visualization (see the Results section below); the latter employs an Expectation-Maximization (E/M) clustering algorithm for identifying marker cluster locations for homozygous and heterozygous calls and the probability that a given marker in a given sample belongs to a particular cluster (Hardenbol et al. 2005; Moorhead et al. 2006). This analysis is accomplished by the Affymetrix GeneChip® Targeted Genotyping Software (GTGS), which also includes a laboratory information management system (LIMS) for sample tracking throughout execution of the assay.

Targeted SNP Genotyping with MIP and Universal Tag Arrays

The power of this technology is illustrated here in a brief presentation of the method with representative data for a MIP 20,000-plex, nonsynonymous coding SNP (cSNP) panel read out on an Affymetrix 25K Universal Tag Array. This panel simultaneously assays the genotypes at 20,000 functional cSNPs—polymorphisms in gene-coding regions that lead to amino acid changes—distributed across >10,000 human genes with a median interval between SNPs of <70 kb. Strong evidence of selection on these SNPs is consistent with a large fraction of them having a functional consequence (Ireland et al. 2006). An early genome-wide association study using the first half of this panel (10K ns cSNP Panel 1) re-confirmed at least one of four previously known susceptibility loci for Type 1 diabetes and discovered at least one new locus, interferon-induced helicase (IFIH1) region, which was validated in six additional diverse ethnic populations (Smyth et al. 2006).

The MIP genotyping assay steps as performed for the 20K cSNP panel are briefly described below. Refer to Figure 12-1B for MIP probe transformations during the assay and to Figure 12-2 for assay workflow (corresponding steps identified in italics in Methods section below). Note that the key to successful use of the MIP assay lies in mixing reagents carefully at each step as recommended. The assay performs very robustly when instructions are followed and good molecular biology practices are observed when handling enzymes and reactions. Note that the assay is performed in two physically separate labs—the Pre-amp and the Post-amp—to prevent contamination of sample DNA with amplification products from previous reactions. Materials, notes, and equipment from the Post-amp lab should **not** be brought into the Pre-amp lab, and labcoats dedicated to each of these labs should be used conscientiously.

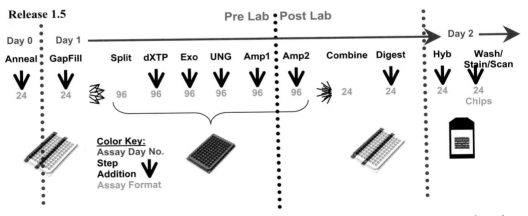

FIGURE 12-2. Workflow of MIP Rel 1.5 Assay. Assay activities starting from annealing on Day 0 through array scanning on Day 2 are pictured for a plate of 24 samples. Note that after overnight annealing, these 24 samples are split to 96 wells before dXTP addition in order to assess circularization extent of all probes with each of the 4 bases (A,C, G, and T). The 4 reactions per sample (A,C, G, and T) are then combined for hybridization to a single array.

MATERIALS

Reagents

HapMap CEPH DNA plate

These samples, obtained from the Coriell collection (Coriell Institute), consist of trios from the CEU family collection, and were used for the illustrative 20K cSNP data, described below. A full list of these samples is available at the HapMap Web site (www.hapmap.org).

Taq polymerase (Stratagene)

TITANIUM Taq polymerase (Clontech/Takara)

Equipment

The following materials and equipment are commercially available from Affymetrix to enable researchers to perform 20,000-plex cSNP genotyping in their own laboratories. Further information about other standard and custom panels, or equipment and methods for genotyping based on the MIP technology, are available at the Affymetrix Web site (http://www.affymetrix.com/products/application/targeted_genotyping.affx).

Affymetrix GeneChip Human (Panels 1 and 2) 20K cSNP

Kit contains sufficient reagents to process a total of 24 assays (including one control). This panel includes 10,000 validated (double-hit), nonsynonymous public cSNPs that code for functional changes, and an additional 10,000 independently validated at Affymetrix.

Affymetrix GeneChip Universal 25K Tag Array (6-pack or 96-pack)

Arrays have >25,000 features per array that can detect 20,000 SNPs using the Affymetrix GeneChip DNA Analysis System incorporating MIP technology.

Affymetrix GeneChip Scanner 3000 Targeted Genotyping System (GCS 3000 TG System)
Includes:

- Computer workstations and bar-code readers. For Pre-amp lab and Post-amp lab sample tracking; instrument control.
- GeneChip Hybridization Oven 640. Oven can hold up to 64 arrays with continuous rotation at hybridization temperatures between 30°C and 60°C.
- GeneChip Fluidics Station FS450 (2 per system). Each wash station will wash and stain 4 arrays at a time, unattended, in ~30 minutes.
- GeneChip Scanner 3000 7G 4C. Four-color, confocal laser scanner includes a temperature-controlled autoloader holding up to 48 arrays. Arrays are automatically scanned and tracked under control of GeneChip Operating System software (GCOS), v1.4.
- GeneChip TG Analysis Software, v1.5. Analyzes scan files and generates genotyping calls.

METHOD

An overview of the strategy and sequence of steps is given below (refer to Fig. 12-2). For complete details of the method with further commentary to optimize for different plex levels, consult the User's Guide provided with the GeneChip Targeted Genotyping System.

Pre-Amp Laboratory Sequence

1. Anneal MIP probes to sample (*Anneal*). Combine 4 µg of sample DNA with the Assay panel and treat briefly with Enzyme A. After denaturing the DNA (95°C for 5 minutes), incubate for 16–24 hours (but < 30 hours) at 58°C.

2. Circularize annealed probes on template DNA (*Gapfill, Split, dXTP*).

   i. Add Gapfill Enzyme mix to annealed template and split into four gapfill reactions (A, C, G, and T). Incubate for 10 minutes at 58°C.

   ii. Add the appropriate single dXTP to each reaction, and incubate the gapfill/ligation reactions for 10 minutes at 58°C.

3. Eliminate uncircularized probes (*Exo*). Add Exonuclease mix and incubate for 15 minutes at 37°C. Inactivate the enzymes by heating (95°C for 5 minutes).

4. Cleave and invert MIP probes (*Cleavage*). Add UNG cleavage mix and incubate for 10 minutes at 37°C.

5. Amplify inverted MIP probes with universal PCR primers (*Amp1*). Add Amp1 mix and incubate for 20 cycles using Meg 20K 1st PCR thermocycler program (runs ~1 hour).

   A QC gel may be run at this step to evaluate the quality of products before proceeding to the Amp2 labeling step.

Post-Amp Laboratory Sequence

6. Base- (or channel-) specific label the amplified MIP products (*Amp2*).

   i. Add a small volume (4 µl) of each Amp1 reaction (A, C, G, and T) to the corresponding Amp2 mix containing the appropriate allele-specific labeling primer.

   ii. Amplify for 10 cycles on a thermocycler using Meg Hypcr 10-20K 2nd PCR program (runs ~30 minutes).

7. Eliminate genomic DNA sequences from associated tags (*Digest*). Combine half of all four reactions (A, C, G, and T) per sample along with 6 µl of Digest mix and digest for 1.5 hours at 37°C. Inactivate the enzymes by heating (95°C for 5 minutes).

   A QC gel may be run at this step to evaluate the quality of products before proceeding to arrays. Note that samples for QC gel must be removed BEFORE denaturation!

8. Hybridize MIP products (alleles translated into tags in appropriate channels) to Universal Tag Array (*Hyb*). Mix the digested MIP products with hybridization cocktail, denature, and hybridize to the array for 12–16 hours at 39°C (1 sample per array).

9. Wash, stain, and scan arrays (*Wash/Stain/Scan*). (Steps i–iii automated on Fluidics station.)

   i. Optional: Manually remove the hybridization mix from array (and save, if desired).

   ii. Wash four arrays per wash station under low-, then under high-, stringency conditions.

   iii. Stain the arrays with a Qdot stain cocktail and refill with a storage buffer.

      Washing and staining of 48 arrays requires ~3 hours.

   iv. Load arrays into the scanner's autoloader and scan them.

      Scanning of all four channels requires ~6 minutes per array and produces four cel files (A,C,G, and T) per sample. Each cel file contains raw fluorescence intensity values for each feature on the array at which a single complementary tag (cTag) sequence resides.

10. Analyze the data. Import *Cel* files for each sample, along with tracked sample information, into the GTGS software where channel and chip normalizations, background subtraction, and spectral crosstalk corrections are performed.

    Pre- and post-clustering performance metrics are computed and exportable with cluster-based genotypes. Numerous data visualizations are accessible from the database (see Results section below).

RESULTS: REPRESENTATIVE DATA ANALYSIS AND VISUALIZATIONS OF 20K cSNP PANEL PERFORMANCE

In this example, 48 CEPH HapMap samples, including 5 trios and 3 repeated samples, were run with the 20K cSNP panel according to standard MIP 20K assay conditions. Raw data from scan files were imported into the GTGS Analysis software and processed with standard parameters to eliminate any probes of questionable performance. Details of the data-processing methods can be found in Moorhead et al. (2005).

Pre-Clustering Data Analysis and QC (Quality Control) Summary Views

The QC Summary analysis provides a rapid means of assessing sample performance and is presented in both tables and graphic form. After proper normalizations, background subtraction, and spectral cross-talk correction, markers passing various criteria (e.g., minimum signal, S/B, S/N, ratio of signals in assay allele channels) are scored either as homozygous for allele 1 or 2, or as heterozygous. The total number of calls, expressed as a percent of the assays in the panel, is reported as the sample "QC call rate" and can be plotted for each of the 48 CEPH samples. The plot in Figure 12-3A, showing the "Experiment Metrics Chart" view, displays the call rate for each CEPH sample. Samples with QC call rates falling below 80% are automatically failed (designated by the line at this position). The QC call rates for these 48 samples fall between ~97% and 98% with normal frequencies of homozygotes and heterozygotes (~78% and ~20%, respectively). The remaining several percent of ambiguous calls are scored as "half calls," meaning that one allele is clearly known but the other is not. Upon clustering (see below), many of these half calls can be clearly resolved. Among the additional metrics tracked in this view is the

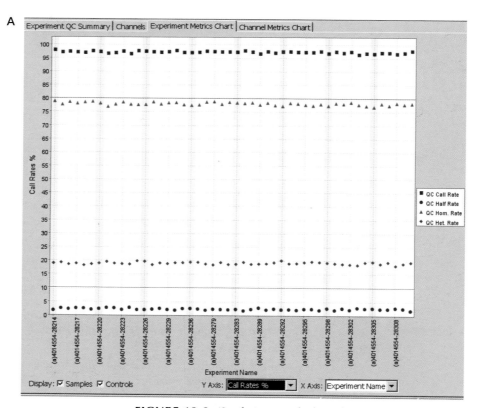

FIGURE 12-3. (*See facing page for legend.*)

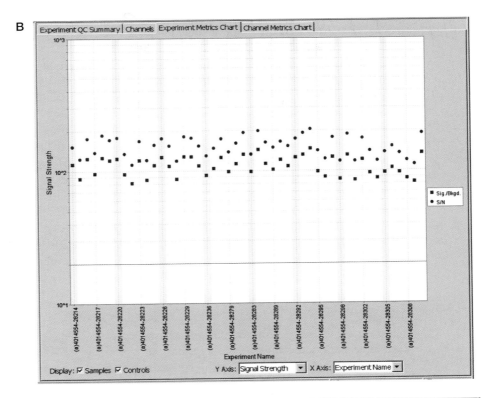

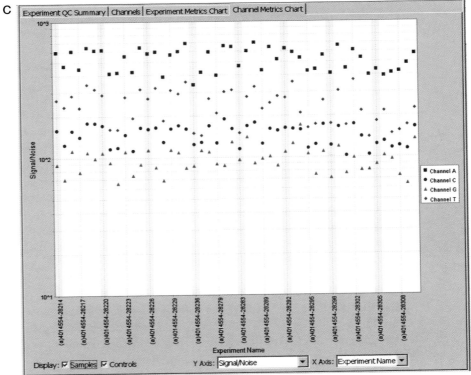

FIGURE 12-3. (*Continued.*) Quality Control (QC) Summary analysis from GTGS analysis software. (*A*) This view shows the "Experiment Metrics Chart" with QC call rate. (*B*) This view shows the "Experiment Metrics Chart" display of signal strength. (*C*) This view shows the "Channel Metrics" display of signal-to-noise (S/N) ratio. Other "Channel Metrics" views of signal, background, noise, etc. are not shown.

"Signal Strength" for each sample (Fig. 12-3B). This view reveals the array signal-to-noise ratio (S/N, blue symbols), a four-channel average based on the per-channel average corrected fluorescence signal for all markers and the channel's noise; and the signal-to-background value (S/B, red symbols), a measure of the average assay specificity based on the ratio of signals in the allele channels to those in the two background (or non-allele) channels. The minimal S/B for robust genotyping is designated by the blue line at S/B = 20; this limit is far exceeded in these samples, which are mostly near or above S/B = 100.

Similarly, numerous performance metrics for each of the four channels are tracked under the "Channel Metrics Chart" view (see Fig. 12-3C) and can be used to follow performance and diagnose any problems. Among these metrics are signal, noise, S/N, chip background, and rejected outlier rate. Figure 12-3C shows the S/N view, which is seen to be reasonably consistent and characteristic for a given channel among these samples. If a particular sample were to have failed in this run, examination of this metric as well as the others could be used to diagnose whether the problem was due to a failure of a single channel, due to some unusually high chip background or perhaps to insufficient or poor-quality input sample DNA. To facilitate such diagnosis further, additional graphical views (not shown) are available for each chip and channel that allow evaluation of raw and processed signal distributions for control and experimental samples under various transformations.

Cluster Analysis and Summary Views

Following the QC Summary analysis, the user proceeds to cluster the data according to default or modifiable parameter values. The expectation/maximization (E/M) clustering algorithm used produces genotypes that are exportable in various selectable formats. The GTGS analysis software generates a Cluster Fit Summary table as shown in Figure 12-4A for the 48 CEPH samples genotyped with the 20K cSNP panel. Note that all 48 samples (or "Experiments") were successfully genotyped (compare the "# Experiments" versus "# Genotyped Experiments") and that among the 20,127 markers attempted in the panel ("# Assays"), 99.18% were successful ("Passed Assays %") in >80% of the samples. (If a marker or assay calls successfully in <80% of samples, it is failed for the sample set.) Recall that the QC Summary call rate gave a 1–2% lower estimate in the absence of clustering. The assay precision was measured at 99.92% ("Cluster Fit Repeatability %") based on the three repeated HapMap samples among the 48 ("# Genotyped Unique Samples" = 45 versus "# Genotyped Experiments" = 48). The accuracy was measured at 99.91% ("Cluster Fit Trio Concordance %") based on informative transmitted genotypes among the five mother-father-child trios included in the 48 samples ("# Trios"). The "Completeness %" = 99.51% indicates that for the 19,962 passed assays across all 48 samples, only 0.49% (1 – Completeness) of the data was missing: i.e., almost every marker was called in almost every sample. Additional performance data and causes of marker failure are also detailed.

As with the QC Summary data, cluster genotype data can be viewed on the chip level ("Experiments" tab, not shown) to see how all markers performed in a single sample, or marker level ("Assays" tab) to see how well an individual marker clustered across all samples. Figure 12-4B shows the clusters and their locations for a representative A/C marker (rs9353689) highlighted in the "Assays" view table. In this 1-D contrast space plot, A/A homozygotes are located at or near "-1.00"; C/C homozygotes are located at or near "1.00"; and A/C heterozygotes are located at or near "0.00." The Y-axis displays the sum of the signals in the two allele channels. The solid lines running through and bracketing each cluster represent the mean cluster position ± 3 sigmas. Whether or not a marker in a given sample is considered a member of a particular cluster depends on its proximity to that cluster or any other, and on the tightness and separation of the nearest clusters from each other. In the analysis window shown, the genotypes for the trio members containing the child sample, NA10846, are marked by the dashed vertical lines. By mousing over the cluster members, it was learned that the parents are opposite homozygotes (mother = A/A, father = C/C) and the child has the expected heterozygous genotype (middle cluster). One can scroll

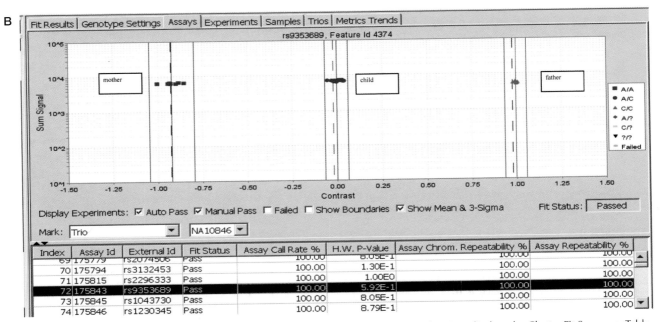

A

Property	Value
# Experiments	48
# Genotyped Experiments	48
# Unique Samples	45
# Genotyped Unique Samples	45
# Trios	5
# Assays	20,127
# Passed Assays	19,962
# Failed Assays	165
Passed Assays %	99.18
Failed by Low Call Rate %	0.82
Failed by Low H.W. P-Value %	0.00
Failed by High Het Fraction %	0.00
Failed by High Het Chi-Sq HW %	0.00
Failed by High Cluster Variance %	0.00
Failed by High Hom Sig Ratio %	0.00
Failed by High Het Sig Ratio %	0.00
Failed by Low Repeatability %	0.00
Failed by Low Trio Concordance %	0.00
Failed by Low 20%ile Call Confidenc...	0.00
# Repeated Sample Discordances	100
# Repeated Sample Concordances	118,750
Cluster Fit Repeatability %	99.92
Cluster Fit Chrom. Repeatability %	99.92
# Trio Discordances	84
# Trio Concordances	96,345
Cluster Fit Trio Concordance %	99.91
# Trio Chrom. Discordances	86
# Trio Chrom. Concordances	153,318
Cluster Fit Trio Chrom. Concordance %	99.94
Completeness %	99.51
Hom. %	79.45
Het. %	20.07
Low Call Confidence Cut %	0.00
Half %	0.19
Low S/N Cut %	0.04
Low S/BG Cut %	0.05
High Pixel CV Cut %	0.14
Rejected Outliers Cut %	0.00
Low Signal Cut %	0.01
Hom Tail Cut %	0.01
Passed Assays % x Completeness %	98.70
Polymorphic Passed Markers % x Com...	62.37

FIGURE 12-4. Cluster analysis from GTGS analysis software. (*A*) This view displays the Cluster Fit Summary Table. (*B*) This view displays the Assay Marker Cluster View. See text for explanation.

through the marker table to picture any SNP interrogated, and many additional metrics are visible for each marker in the table (e.g., assay call rate, repeatability, Hardy-Weinberg P-Value, minor allele frequency in the sample set, chromosome and chromosome position, gene locus). Other tabs disclose information on the samples used and their frequency in the sample set ("Samples"), trio performance data ("Trios"), experiment or sample performance data ("Metrics Trends"), and cluster analysis parameter values ("Genotype Settings").

SUMMARY AND CONCLUSIONS

The MIP-based Targeted Genotyping Assay is a versatile, high-throughput, multiplexed assay that can be used to simultaneously score from 1,500 to >50,000 SNP markers in sample DNA from various sources. The combination of enzymatic specificity with the intrinsic specificity of padlock probes during target recognition enables the MIP assay to achieve high accuracy, repeatability, and data completeness. Additionally, these features allow for minimal design constraints and a high conversion rate for assay probes against most target sequences desired. The design versatility of the MIP genotyping assay is further underscored by its success in genotyping even more challenging target sequences containing insertion/deletions, trialleles, and secondary SNPs for a pharmacogenomics panel (Dumaual et al. 2007). The use of synthetic hybridization tags for assay readout on Universal Tag Arrays not only contributes to assay accuracy, but also allows flexible composition of assay panels—standard or custom assay panels can be created for the same array, and either type of panel can be enlarged by merely adding additional assay probes containing unused array tags. Finally, a rich and intuitive software analysis program (GTGS Analysis Software) processes the array data to yield cluster-derived genotypes, as well as many metrics and graphic visualizations that can be used to assess performance of individual markers and samples across a study.

ACKNOWLEDGMENTS

We thank the many people from ParAllele and Affymetrix who have labored devotedly in developing the MIP technology. These contributions have come from all quarters, including the Business team for defining assay needs and capabilities; the Informatics team for probe design, algorithm development, and final software package; Manufacturing for probe panels and reagents; the Assay Development team; and Assay Services for data generation and analysis.

REFERENCES

Begovich A.B., Carlton V.E., Honigberg L.A., Schrodi S.J., Chokkalingam A.P., Alexander H.C., Ardlie K.G., Huang Q., Smith A.M., Spoerke J.M., et al. 2004. A missense single-nucleotide polymorphism in a gene encoding a protein tyrosine phosphatase (PTPN22) is associated with rheumatoid arthritis. *Am. J. Hum. Genet.* **75:** 330–337.

de Bakker P.I., Yelensky R., Pe'er I., Gabriel S.B., Daly M.J., and Altshuler D. 2005. Efficiency and power in genetic association studies. *Nat. Genet.* **37:** 1217–1223.

Dumaual C., Miao X., Daly T.M., Bruckner C., Njau R., Fu D.J., Close-Kirkwood S., Bauer N., Watanabe N., Hardenbol P., and Hockett R.D. 2007. Comprehensive assessment of metabolic enzyme and transporter genes using the Affymetrix® Targeted Genotyping System. *Pharmacogenomics* **8:**293–305.

Hardenbol P., Yu F., Belmont J., MacKenzie J., Bruckner C.,

Brundage T., Boudreau A., Chow S., Eberle J., Erbilgin A., et al. 2005. Highly multiplexed molecular inversion probe genotyping: Over 10,000 targeted SNPs genotyped in a single tube assay. *Genome Res.* **15:** 269–275.

International HapMap Consortium 2005. A haplotype map of the human genome. *Nature* **437:** 1299–1320.

Ireland J., Carlton V.E., Falkowski M., Moorhead M., Tran K., Useche F., Hardenbol P., Erbilgin A., Fitzgerald R., Willis T.D., and Faham M. 2006. Large-scale characterization of public database SNPs causing non-synonymous changes in three ethnic groups. *Hum. Genet.* **119:** 75–83.

Karlin-Neumann G., Sedova M., Falkowski M., Wang Z., Lin S., and Jain M. 2007. Application of quantum dots to multicolor micorarray experiments: Four-color genotyping. *Methods Mol. Biol.* **374:** 239–251.

Moorhead M., Hardenbol P., Siddiqui F., Falkowski M., Bruckner C., Ireland J., Jones H.B., Jain M., Willis T.D., and Faham M. 2006. Optimal genotype determination in highly multiplexed SNP data. *Eur. J. Hum. Genet.* **14:** 207–215.

Nilsson M., Malmgren H., Samiotaki M., Kwiatkowski M., Chowdhary B.P., and Landegren U. 1994. Padlock probes: Circularizing oligonucleotides for localized DNA detection. *Science* **265:** 2085–2088.

Risch N. and Merikangas K.1996.The future of genetic studies of complex human diseases. *Science* **273:** 1516–1517.

Shiffman D., Ellis S.G., Rowland C.M., Malloy M.J., Luke M.M., Iakoubova O.A., Pullinger C.R., Cassano J., Aouizerat B.E., Fenwick R.G., et al. 2005. Identification of four gene variants associated with myocardial infarction. *Am. J. Hum. Genet.* **77:** 596–605.

Smyth D.J., Cooper J.D., Bailey R., Field S., Burren O., Smink L.J., Guja C., Ionescu-Tirgoviste C., Widmer B., Dunger D.B., et al. 2006. A genome-wide association study of nonsynonomous SNPs identifies a type 1 diabetes locus in the interferon-induced helicase (*IFIH1*) region. *Nat. Genet.* **38:** 617–619.

Van Eerdewegh P., Little R.D., Dupuis J., Del Mastro R.G., Falls K., Simon J., Torrey D., Pandit S., McKenny J., Braunschweiger K., et al. 2002. Association of the ADAM33 gene with asthma and bronchial hyperresponsiveness. *Nature* **418:** 426–430.

Winzeler E.A., Shoemaker D.D., Astromoff A., Liang H., Anderson K., Andre B., Bangham R., Benito R., Boeke J.D., Bussey H., et al. 1999. Functional characterization of the *S. cerevisiae* genome by gene deletion and parallel analysis. *Science* **285:** 901–906.

Zheng S.L., Liu W., Wiklund F., Dimitrov L., Balter K., Sun J., Adami H.O., Johansson J.E., Chang B., Loza M., et al. 2006. A comprehensive association study for genes in inflammation pathway provides support for their roles in prostate cancer risk in the CAPS study. *Prostate* **66:** 1556–1564.

WWW RESOURCE

http://www.affymetrix.com/products/application/targeted_genotyping.affx Affymetrix targeted genotyping page

13 | Whole-Genome Genotyping

Stacey B. Gabriel[1] and Michael P. Weiner[2]

[1]*Broad Institute of MIT and Harvard, Cambridge, Massachusetts 02142;* [2]*RainDance Technologies, Guilford, Connecticut 06437*

INTRODUCTION

Several technologies are capable of producing data for hundreds of thousands of polymorphisms in large samples, providing the quality and cost-efficiency required to perform whole-genome association scans. Because whole-genome analysis entails a rather large financial commitment, it has been best purchased commercially as a "turn-key" solution. Until whole-genome DNA sequencing at a competitive price becomes a reality, most whole-genome studies will likely be performed using oligonucleotide arrays, as is currently the case. DNA-based arrays can be made by many methods, including chemical synthesis at fixed locations (Affymetrix, Nimblegen), spotting with piezoelectric or inkjet dispensers, or assembling beads into a random array format (Lynx, Illumina) (Walt 2002). In the sections below, we describe two of the commercial methods for whole-genome analysis; both use DNA-based arrays.

HIGH-DENSITY GENOME VARIATION ANALYSIS USING AFFYMETRIX GENECHIPS

Development of Gene Arrays

Affymetrix has supported single-nucleotide polymorphism (SNP) genotyping on GeneChip arrays since 2000, first with the 10K array, followed by the 100K array (2003), the 500K array (also called the SNP 5.0), and, most recently, the 1M SNP array (also called the SNP 6.0). The basic genotyping method has not changed throughout this product development, but the density of the array has increased due to Affymetrix's ability to decrease the feature size required for each deoxy-oligonucleotide probe on the array. The method is based on reduced representation genotyping, in which the complexity of the genome is reduced by enzymatic digestion and size selection

213

(Altshuler et al. 2000). The "reduced" portion of the genome is subject to amplification by universal PCR, and products are then labeled and hybridized to an Affymetrix array (Kennedy et al. 2003; Matsuzaki et al. 2004). Because of the restriction enzyme digestion, SNPs within the restriction fragments are genotyped. Affymetrix has developed algorithms to select enzymes for the assay that include the best distribution of SNPs possible.

Affymetrix has followed a common approach for development of content for the SNP arrays (see Fig. 13-1). First, in silico screening takes place: Sets of restriction enzyme fragments are selected which contain the optimal SNP content, in terms of density and spacing of SNPs. For both the 5.0 and 6.0 arrays, two restriction enzymes were selected for the assay, NspI and StyI, both of which reduce the complexity of the genome to about 500 Mb in total and contain about 2.5 million unique SNPs (from dbSNP). The 6.0 array was developed by experimentally screening about 2 million SNPs in a screening panel of Haplotype Map (HapMap) samples. These 2 million SNPs are distinct from the 500,000 SNPs already on the 5.0 array. In this screening stage, SNPs can be identified which do not amplify as expected, display other hybridization problems (e.g., high similarity to other regions or outliers in probe-specific hybridization), or have poor discrimination of specific genotypic classes (e.g., concordance to HapMap calls is used as a filter). Following experimental screening, 800,000 SNPs were deemed successful and polymorphic in the screening panel. Since the final SNP panel for the 6.0 includes the 500,000 SNPs on the 5.0, Affymetrix selected from the SNPs passing the screening step by determining which SNPs had the most proxies on the HapMap and, thus, along with the 5.0 SNPs, would serve as the best tag-SNPs (for discussion of tag-SNPs, see Chapter 20). In total, the 6.0 array contains about 907,000 SNPs.

The Affymetrix assay and, in turn, the product development strategy, result in a product that is not biased toward any particular population or SNP panel and contains some degree of redundancy in coverage. The strengths of this method are that the genetic power should be similar across different patient populations, the HapMap can be used to boost genetic coverage by the use of multi-marker haplotype methods, and, because the product has the highest available SNP density (1 SNP per 2.5 kb, on average), it has the greatest overall resolution to detect copy number alterations. The main disadvantage is the relative inflexibility in SNP selection created by the restriction enzyme complexity reduction. However, with any standard product, there is a trade-off in ultimate flexibility versus the low costs that come with large-scale manufacturing and a predictable market.

At the time of this writing, the 5.0 product has had the longest amount of time in the field. It has been shown that a small number of SNPs on the array fail repeatedly; about 470,000 SNPs have been consistently reliable across multiple studies. The technical performance of the product is described in terms of call rate and accuracy relative to the HapMap. According to Affymetrix,

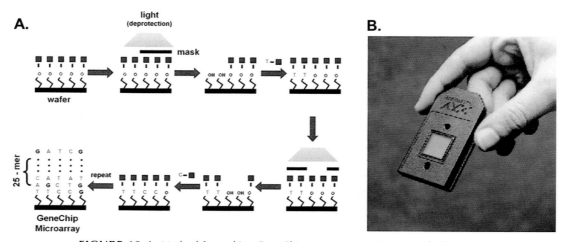

FIGURE 13-1. Method for making GeneChip MicroArrays. (Courtesy of Affymetrix.)

TABLE 13-1. Comparison of Various Genotyping Platforms

Platform	HapMap Panel					
	YRI		CEU		CHB+JPT	
	% r^2 ≥ 0.8	mean max r^2	% r^2 ≥ 0.8	mean max r^2	% r^2 ≥ 0.8	mean max r^2
Affymetrix SNP5.0	46%	0.66	68%	0.81	67%	0.80
Affymetrix SNP6.0	66%	0.80	83%	0.90	81%	0.89
Illumina HumanHap300	33%	0.56	77%	0.86	63%	0.78
Illumina HumanHap550	55%	0.73	88%	0.92	83%	0.89
Illumina HumanHap650Y	66%	0.80	89%	0.93	84%	0.90

(YRI) Yoruba; (CEU) Caucasian; (CHB) Han Chinese from Beijing, China; (JPT) Japanese from Tokyo.

the accuracy as measured by concordance with the HapMap is 99.5% and reproducibility of genotype calls is 99.9%; these figures are consistent with data from published genome-wide association studies based on the product (Diabetes Genetics Initiative 2007).

Genetic Coverage

The genetic coverage of genome-wide genotyping products is often described by testing the proportion of SNPs on the HapMap for which there is a good proxy on the genotyping array (Barrett and Cardon 2006; Pe'er et al. 2006). This is measured by reporting either the proportion of HapMap SNPs with strong correlation ($r^2 > 0.8$) or by assessing the mean correlation of SNPs on the HapMap with SNPs on the array. The latter statistic tracks well with power in an association study, because the former uses a set threshold for the correlation coefficient, and SNPs with correlation coefficients <0.8 can still be informative in association studies with large sample sizes. The Affymetrix 5.0 and 6.0 have mean max r^2 between 0.68 and 0.90, depending on the product and the HapMap population being tested (see Table 13-1) (The International HapMap Consortium 2005).

HIGH-DENSITY GENOME VARIATION ANALYSIS USING ILLUMINA BEADCHIPS

Development of Bead Arrays

In a bead-assembly approach, high-density arrays are prepared by putting beads containing oligodeoxynucleotide primers into fiber optic faceplates. The fiber cores are first partially etched with an acid that preferentially etches the fiber core (and not the optic cladding, see Fig. 13-2) to create wells. The depth of the wells can be controlled through acid concentration, time, and temperature. Each microwell is formed by the walls of the cladding and the bottom recessed end of each fiber. When the etched microwells are immersed in a solution of either latex or silica microspheres that are complementary in size to the wells, the microspheres load and assemble into them by capillary forces. For quality control, populations of microspheres are prepared in batches.

 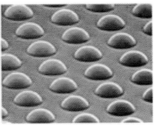

FIGURE 13-2. Fiber optic bead arrays. (*Left*) Fiber optic faceplate with hexagonal packing etched with acid selectively removes the core glass, leaving the fiber optic cladding glass relatively unaffected. (*Right*) Etched faceplate loaded with encoded beads.

Different oligonucleotide primer combinations are attached to the beads and are used as optical "bar codes" to enable the bead type to be decoded. The different bead populations are combined into a library, and the beads are then assembled into the etched wells on a fiber array. Because beads randomly self-assemble into the wells, the bar codes enable the position of each bead type to be recorded. DNA arrays containing thousands of beads can be loaded onto the ends of optical fibers.

Illumina's beads are manufactured such that unique oligonucleotide sequences are attached and are used to decode the random assembly of the beads on the arrays (see Fig. 13-3). Each bead type is represented on average 30 times; the data are averaged in each experiment. This highly redundant data collection results in high-quality data. Illumina currently manufactures multiple formats of high-density SNP genotyping arrays ("BeadChips;" for review, see Shen et al. 2005) containing from 109K (Sentrix Human-1, mostly exon-centric) to 650K SNPs.

The Infinium assay permits large-scale interrogation of variations in the human genome, allowing researchers to cost-effectively analyze a whole genome at once. The assay protocols use a single-tube sample preparation without PCR or ligation steps, significantly reducing both labor and potential sample-handling errors. An enzymatic discrimination step (described below) provides high call rates and accuracy.

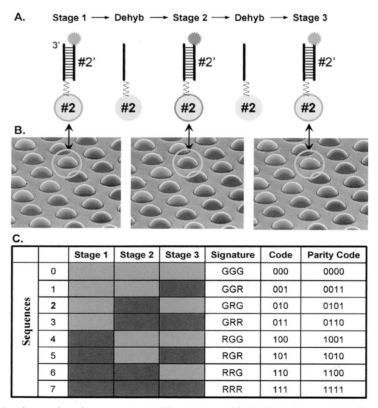

FIGURE 13-3. Bead-type decoding process. (A) The sequential hybridization process is illustrated for a single bead, of bead type 2. In stage 1, a complementary decoder hybridizes to the oligonucleotide probe that is attached to the bead. The decoder is labeled with a fluorophore (*green* in stage 1, *red* in stage 2, and *green* in stage 3). The fluorescent signal is read by imaging the entire array. The array is then dehybridized, and the process is repeated for two more stages. (B) A scanning electron micrograph of an array of beads, artificially colored to represent three sequential hybridization stages. The images, taken collectively, reveal a combinatorial code for each bead. Note that the bead circled in yellow has the color signature GRG or code 010. (C) Colors, or states, are assigned to individual decoder sequences in each stage to produce a unique combination across stages. This signature, or code, identifies each bead type. As indicated in the parity code column, an extra decoding stage (data not shown) can be performed to provide an error-checking parity bit. After three stages of decoding, all the beads are uniquely identified by their color. (Reprinted, with permission, from Gunderson et al. 2004.)

HapMap Chip

tag-SNPs are loci that can serve as proxies for many other SNPs. The use of tag-SNPs greatly improves the power of association studies. Because tag-SNP content has been employed on the Human HapMap, more statistical power and genomic coverage can be achieved using fewer SNPs and statistical tests as compared to other strategies using larger numbers of randomly chosen SNPs. The 317K BeadChip consists of more than 300,000 tag-SNP assays derived from the HapMap project with a 9-kb mean spacing between SNPs. Illumina's HumanHap-1 Genotyping BeadChip is the latest release from Illumina and enables whole-genome genotyping. Instead of reducing the complexity of the genome as Affymetrix does, Illumina amplifies the entire genome prior to the allele-discrimination step. This enables Illumina to add any SNP to the array that passes some in silico design process.

The Infinium II assay is based on whole-genome amplification followed by hybridization to a BeadChip. Deoxyoligonucleotide probes that interrogate the SNP position are attached to the beads contained in the fiber-optic faceplate. Following targeted hybridization of the amplified genomic DNA to the bead array, the arrayed locus-specific primers are extended in a two-color reaction, such that the alleles can be discriminated on the basis of which color is detected at the bead position.

The content of Illumina's products is based very strongly on tagging the HapMap. The HapMap SNPs were binned using the program LD select (see Chapter 19). tag-SNPs were selected from each bin, and priority was given to SNPs that were predicted most likely to yield a successful assay on the Illumina platform. The criteria for tagging were set so that tag-SNPs had to meet $r^2>0.8$ for bins within various distances of genes (depending on the product) or evolutionarily conserved regions, and a slightly lower threshold outside these regions. Illumina also specifically added about 8,000 missense SNPs from dbSNP and 1,500 SNPs in the major histocompatibility (MHC) region. Illumina has three products with 317,000 (as discussed above), 550,000, and 650,000 SNPs, respectively. The first two panels were selected to tag the CEU HapMap panel (Caucasian). The most dense product has added SNP content to cover the YRI HapMap panel (Yoruba). See Table 13-1 for coverage statistics.

Sample Amplification and Hybridization for BeadChips

The whole-genome amplification process requires 250–750 ng of input genomic DNA (gDNA) and creates a sufficient quantity of DNA (1000x amplification) to be used on a single BeadChip in the Infinium assay. After amplification, the product is fragmented, precipitated with 2-propanol (plus precipitating reagent), and resuspended in formamide-containing hybridization buffer. The DNA samples are denatured at 95°C for 20 minutes, loaded into Tecan flowthrough chambers, and placed in a humidified container to allow SNP loci to hybridize to the 50-mer capture probes (Steemers and Gunderson 2005; Steemers et al. 2006). Following hybridization, the BeadChip/Tecan-flow chamber assembly is placed on a temperature-controlled flowthrough rack, and all subsequent washing, extension, and staining are performed by addition of reagents to this flow chamber.

For the allele-specific primer extension (ASPE; Infinium I) assay (see Fig. 13-4), the BeadChips are first washed to remove unhybridized and nonspecifically hybridized DNA. Next, the BeadChips are blocked prior to addition of the extension mix. The extension step extends correctly matched probes hybridized to DNA on the BeadChip and incorporates biotin-labeled nucleotides. After extension, a formamide wash removes the hybridized DNA to reduce extraneous signal. The array then undergoes a multilayer staining process to signal-amplify and detect the incorporated label. Finally, the BeadChips are washed, dried, and imaged.

For the single-base extension (SBE; Infinium II) assay (see Fig. 13-4), primers are extended with a polymerase and labeled nucleotide mix, and stained with repeated application of a staining reagent. After staining is complete, the slides are washed with low-salt wash buffer, coated with XC4, and then imaged on an Illumina BeadArray Reader.

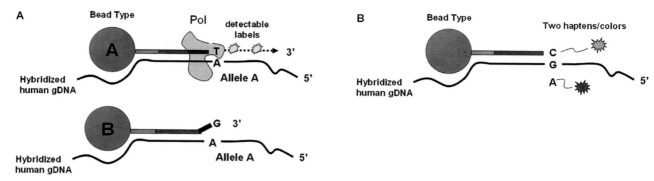

FIGURE 13-4. Infinium assay overview. (A) Infinium Assay I, allele-specific primer extension. (B) Infinium II, single-base extension reaction. See text for details.

The reader used by Illumina is a two-color (543 nm/643 nm) confocal fluorescent scanner with 0.84-μm pixel resolution. The scanner excites the fluorophors generated during signal amplification/staining of the allele-specific (one color) or single-base (two colors) extension products on the BeadChips. The image intensities are extracted using Illumina's software.

Illumina has evaluated product performance and has found a high call rate (>99%) and accuracy (99.5%). This is consistent with studies published based on Illumina products (Duerr et al. 2006; Helgadottir et al. 2007; Hunter et al. 2007).

DNA Analysis: Copy Number Variation

Genome profiling for chromosomal aberrations such as amplifications (including the loss of heterozygosity) and deletions is a crucial element of cancer biology and genetic analysis. The study of copy number variation (CNV) has become an essential component in the field of population genetics and whole-genome association studies. It is now known that numerous regions in the human genome contain large-scale CNV. These regions are thought to contribute to population diversity and possibly to influence gene expression levels.

With no changes to SNP content and assay format, both the Affymetrix SNP chip and the Infinium Genotyping BeadChips can be used to study CNV (see Chapter 17). Samples obtained from cancer patients and patients with congenital disorders can be analyzed using these chips. A weakness of using SNP arrays in copy number analysis is the probe spacing. Probes are only tiled where there is a SNP, leaving large gaps in the genome. Additionally, because SNP assays typically fail in screening when they are located at positions in the genome harboring copy number alterations, these regions may be even further underrepresented on commercial SNP arrays. However, recognizing this issue, both Affymetrix and Illumina have now supplemented the SNP content on their standard arrays by adding probes specifically designed for copy number analysis. In general, these probes are spaced evenly throughout the genome and are selected to be more isothermal in melting properties, so that a more quantitative signal can be obtained.

SUMMARY

Whole-genome analysis entails a substantial investment in equipment useful for automation. We have not discussed the laboratory information management systems (LIMS) and bioinformatics needed for data analysis. Some of this information can be found in other chapters in this manual (see, e.g., the chapters in the following Section 3), and the reader is directed to these other chapters.

We have briefly touched on whole-genome approaches as sold by two commercial entities, Affymetrix and Illumina. Alternatives are clearly on the horizon, including other chip manufacturers, microfluidics, and, eventually, whole-genome sequencing. Currently, whole-genome sequencing is too expensive to be practical for whole-genome analysis. However, methods now being developed will eventually drive the price down, thereby making it reasonable to use whole-genome sequencing as a method for chromosome variation analysis.

REFERENCES

Altshuler D., Pollara V.J., Cowles C.R., Van Etten W.J., Baldwin J., Linton L., and Lander E.S. 2000. An SNP map of the human genome generated by reduced representation shotgun sequencing. *Nature* **407:** 513–516.

Barrett J.C. and Cardon L.R. 2006. Evaluating coverage of genome-wide association studies. *Nat. Genet.* **38:** 659–662.

Diabetes Genetics Initiative of Broad Institute of Harvard and MIT, Lund University, and Novartis Institutes of BioMedical Research. 2007. Genome-wide association analysis identifies loci for type 2 diabetes and triglyceride levels. *Science* **316:** 1331–1336.

Duerr R.H., Taylor K.D., Brant S.R., Rioux J.D., Silverberg M.S., Daly M.J., Steinhart A.H., Abraham C., Regueiro M., Griffiths A., et al. 2006. A genome-wide association study identifies IL23R as an inflammatory bowel disease gene. *Science* **314:** 1461–1463.

Gunderson K.L., Kruglyak S., Graige M.S., Garcia F., Kermani B.G., Zhao C., Che D., Dickinson T., Wickham E., Bierle J., et al. 2004. Decoding randomly ordered DNA arrays. *Genome Res.* **14:** 870–877.

Helgadottir A., Thorleifsson G., Manolescu A., Gretarsdottir S., Blondal T., Jonasdottir As. Jonasdottir Ad., Sigurdsson A., Baker A., Palsson A., et al. 2007. A common variant on chromosome 9p21 affects the risk of myocardial infarction. *Science* **316:** 1491–1493.

Hunter D.J., Kraft P., Jacobs K.B., Cox D.G., Yeager M., Hankison S.E., Wacholder S., Wang Z., Welch R., Hutchinson A., et al. 2007. A genome-wide association study identifies alleles in FGFR2 associated with risk of sporadic postmenopausal breast cancer. *Nat. Genet.* (in press).

International HapMap Consortium. 2005. A haplotype map of the human genome. *Nature* **437:** 1299–1320.

Kennedy G.C., Matsuzaki H., Dong S., Liu W.M., Huang J., Liu G., Su X., Cao M., Chen W., Zhang J., et al. 2003. Large-scale genotyping of complex DNA. *Nat. Biotechnol.* **21:** 1233–1237.

Matsuzaki H., Loi H., Dong S., Tsai Y.Y., Fang J., Law J., Di X., Liu W.M., Yang G., Liu G., et al. 2004. Parallel genotyping of over 10,000 SNPs using a one-primer assay on a high-density oligonucleotide array. *Genome Res.* **14:** 414–425.

Pe'er I., de Bakker P.I., Maller J., Yelensky R., Altshuler D., and Daly M.J. 2006. Evaluating and improving power in whole-genome association studies using fixed marker sets. *Nat. Genet.* **38:** 663–667.

Shen R., Fan J.B., Campbell D., Chang W., Chen J., Doucet D., Yeakley J., Bibikova M., Wickham Garcia E., McBride C., et al. 2005. High-throughput SNP genotyping on universal bead arrays. *Mutat. Res.* **573:** 70–82.

Steemers F.J. and Gunderson K.L. 2005. Illumina, Inc. *Pharmacogenomics* **6:** 777–782.

Steemers F.J., Chang W., Lee G., Barker D.L., Shen R., and Gunderson K.L. 2006. Whole-genome genotyping with the single-base extension assay. *Nat. Methods* **3:** 31–33.

Walt D.R. 2002. Imaging optical sensor arrays. *Curr. Opin. Chem. Biol.* **6:** 689–695.

14 Comparative Genomic Hybridization to Detect Variation in the Copy Number of Large DNA Segments

Ilona N. Holcomb and Barbara J. Trask

Human Biology Division, Fred Hutchinson Cancer Research Center, Seattle, Washington 98109;
Department of Genome Sciences, University of Washington, Seattle, Washington 98105

INTRODUCTION

Genome-wide scans of the human genome reveal an extraordinary degree of variation in the copy number of large DNA segments (Iafrate et al. 2004; Sebat et al. 2004; Sharp et al. 2005). The variations observed include insertions, deletions, and duplications. A single variant can encompass multiple genes and millions of base pairs of sequence. A segment of >1 kb that is present at a variable copy number in comparison with a reference genome is called a copy number variation (CNV). Classification of CNV frequency and location, on the scale accomplished for single-nucleotide polymorphisms (SNPs), is under way and will reveal the subset of CNVs that are present in >1% of the population (defined as copy number polymorphisms [CNPs] by Feuk et

al. 2006). Ultimately, genomic catalogs of CNVs and analysis of their gene content will help us understand how this type of genomic variation contributes to normal phenotypic variation and disease susceptibility. A powerful tool that is capable of providing comprehensive analysis of large-scale variation is called array comparative genomic hybridization (array CGH).

Array CGH scans the genome at thousands of points for copy-number differences between a test and a normal reference DNA. Figure 14-1 shows an overview of array CGH. The test and reference genomes are differentially labeled with fluorescent dyes and compete for hybridization on an array of target spots, with each spot representing a given DNA sequence. The DNA in the target is either spotted or synthesized directly onto the array surface (typically a glass slide), usually in duplicate or triplicate. For each target spot, the ratio of test DNA fluorescence to reference DNA fluorescence is calculated and interpreted as a relative loss (ratio <1), gain (ratio >1), or no change (ratio ≈ 1). The strength of this whole-genome scan is that the exact genomic location in the draft sequence is known for all target spots. Consequently, we know the gene content of the locus rep-

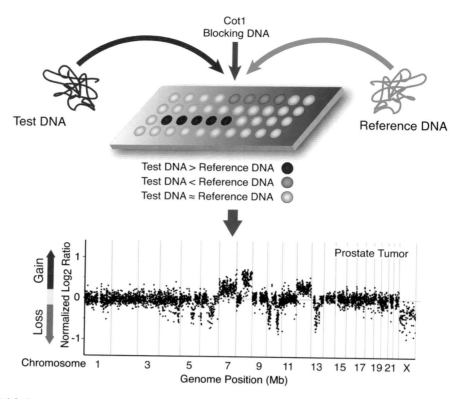

FIGURE 14-1. An overview of array CGH. The test DNA and reference DNA are differentially labeled with dNTP conjugated to fluorescent dyes. The labeled material is then cohybridized in the presence of Cot1 DNA to suppress the hybridization of repetitive elements found in either the labeled sequences or array elements. In this example showing the copy number changes for a prostate tumor, the ratio of fluorescence of test to reference DNA is calculated and log (base 2) transformed for each array element. For the plot shown, the Y-axis is the $\log_2$ ratios and X-axis is the genomic position in Mb separated by chromosome. Deviations significantly below a $\log_2$ ratio of 0 are classified as a relative reduction in the copy number of the test DNA; deviations significantly above a $\log_2$ ratio of 0 are classified as a relative increase in the copy number of the test DNA; and a $\log_2$ ratio close to 0 is classified as no change.

resented by a given target spot; i.e., the region between the previous target and the next target along the chromosome.

The concept of array CGH began with a technique called fluorescence in situ hybridization (FISH), a classic tool of cytogeneticists. In FISH, a fluorescent probe representing 5–200 kb of genomic sequence is hybridized to metaphase or interphase nuclei to ascertain the location(s) and copy number of that sequence. Conventional CGH is a whole-genome adaptation of FISH which employs the same principles that we now use for array CGH, except that the platform on which hybridization occurs in conventional CGH is metaphase chromosomes. Conventional CGH was originally developed to characterize gross copy number changes in the lesions of cancer patients (Kallioniemi et al. 1992). Although conventional CGH was a groundbreaking technology, it is time-consuming, requires the ability to accurately karyotype the human genome, and only crudely resolves the boundaries and locations of the chromosomal abnormalities.

A tremendous advance came with the advent of the first array CGH systems (Solinas-Toldo et al. 1997; Pinkel et al. 1998). These initial systems used the competitive principles of conventional CGH and an array platform composed of thousands of genomic clones. The resulting technology was high-throughput and had a resolution of approximately 1.5 Mb (versus 20 Mb for conventional CGH). Most importantly, every clone used was positioned on the draft of the human genome by a PCR-assayed sequence-tagged site (STS), an end sequence, or its full sequence. Array

CGH quickly became a staple of cancer genomics research and is increasingly used to detect abnormalities in congenital disorders. Now, array CGH is becoming an important tool in the effort to characterize the location, frequency, and gene content of CNVs.

Array CGH is an excellent tool to scan the genome for CNVs when used conscientiously. This chapter is intended to provide an understanding of the basic principles of array CGH and the different options available to the user to design their array CGH experiments. Specifically, the six subsections discuss the different array platforms available, test and reference DNA preparation, reference DNA choice, the basics of hybridization, data processing, and our current understanding of CNVs in the human genome.

ARRAY CGH PLATFORMS

Ideally, one would use the array platform with the highest resolution across the human genome (i.e., the highest number of targets per 3000 Mb) and the greatest sensitivity to copy number changes (e.g., the ability to report single-copy as well as multi-copy changes). However, these two desirable features are currently not united on the same array platform. The relationship between genomic resolution and copy number change is discussed for each of the four platforms covered in this section: bacterial artificial chromosome (BAC), oligonucleotide, SNP, and cDNA arrays. Arrays whose targets are composed of large genomic clones like BACs (100–200 kb of sequence) are very sensitive to genomic changes but typically have low genomic resolution (~1.5 Mb) (Albertson et al. 2000). A newer generation of arrays made up of oligonucleotides (20–100 bp of sequence) has a resolution down to the hundreds and even tens of kilobases, but has not been as sensitive to copy number differences as BAC arrays (Barrett et al. 2004; Carvalho et al. 2004; Sebat et al. 2004). Thus, the benefit of higher resolution is obtained possibly at the expense of sensitivity. Beyond these conceptual issues, practical considerations for platform choice include the minimum amount of test DNA required for array analysis and the relative price of each array. All of these topics for each array CGH platform are discussed in the following paragraphs.

BAC Arrays

Arrays of large genomic clones (mainly BAC clones) were the first array CGH platforms (Solinas-Toldo et al. 1997; Pinkel et al. 1998). A massive effort to sequence and map thousands of BAC clones made them optimal targets for site-specific copy number analysis (Cheung et al. 1999, 2001; Korenberg et al. 1999; Leversha et al. 1999; Kirsch and Ried 2000; Kirsch et al. 2000). Initially, thousands of BACs were sequenced-linked to the draft of the genome by BAC-end sequences or STSs, or because they had been chosen for complete sequencing as part of the Human Genome Project (Cheung et al. 2001; McPherson et al. 2001). Then, FISH mapping by the BAC resource consortium (Cheung et al. 2001) helped discriminate BACs that represented single or multiple locations (e.g., those containing recently duplicated sequences). This multitude of BACs is easily accessible from the BACPAC Resources Center (BPRC) (http://bacpac.chori.org/) and from a number of other distribution centers (http://www. ncbi.nlm.nih.gov/genome/clone/distributors. html). BAC clones contain 100–200 kb of genomic sequence, 10^4 times longer than the average sequence length of a target on an oligonucleotide array. These large genomic clones typically give more intense signals than shorter array targets, which theoretically result in better sensitivity to copy number changes. Generally, the BACs are spotted at approximately one clone per megabase across the genome. This translates to roughly 3000 array target spots.

The number and availability of well-mapped BACs have driven the production of higher-density arrays for the entire genome or specific regions. An example of the former is a newer generation of BAC arrays, called tiling arrays, which contain overlapping BACs collectively spanning the euchromatic portions of the genome (Ishkanian et al. 2004; Krzywinski et al. 2004; Li et al. 2004). Tiling arrays are an excellent tool to achieve a high-resolution analysis of a test genome, but the demands of producing and maintaining the approximately 30,000 clones that are required for a full genomic tiling path may limit their widespread use. A tiling array made up of clones from a single chromosome is easier to generate and is a very effective way to get a representative analysis of the whole genome (Woodfine et al. 2004) or to specifically interrogate at high resolution a chromosome that is implicated in a particular disease (Ammerlaan et al. 2005). Another efficient method to interrogate particular regions of the genome that are not covered well by standard BAC arrays is to use a targeted array enriched for the regions of interest. A good example of a targeted array is the segmental duplication (SD) array used by Sharp et al. to study CNVs (Sharp et al. 2005). Most array platforms are designed to exclude SDs, which are large regions (>1 kb) of high sequence identity (>90%) in more than two locations in the genome. Several lines of evidence suggest that CNVs are enriched in regions of the genome containing SDs when compared to other genomic loci (see below, Exploring the Genome for CNVs; Iafrate et al. 2004; Sebat et al. 2004; Sharp et al. 2005). Thus, SD-poor arrays may result in observations of fewer CNVs than arrays that include sequences in or near SDs.

The three benefits of using a BAC array are that they are cost-effective, compatible with considerably less starting material than other platforms, and FISH-verifiable. First, BAC arrays are relatively inexpensive, at about one-third the total setup and processing cost of most high-resolution oligonucleotide-based arrays. Second, amplified material from as little as 10 ng of DNA can produce good data on BAC arrays (Guillaud-Bataille et al. 2004; Little et al. 2006). We have found that BAC arrays are compatible with amplified material from as few as 10 cells, which is 30,000 times less starting material than is needed for most other array platforms (I. Holcomb, unpubl.). Finally, an exceptional feature of BAC arrays is that the very same BACs reporting interesting changes can be used for FISH verification. Figure 14-2 shows an example of the FISH analyses used by Iafrate et al. to confirm copy number differences observed by BAC array (Iafrate et al. 2004). In this figure, they FISH-verify loss, gain, and no change for 3 individuals for the most prevalent CNV in their study (27/55 individuals). They and other workers have also used quantitative PCR to validate their findings.

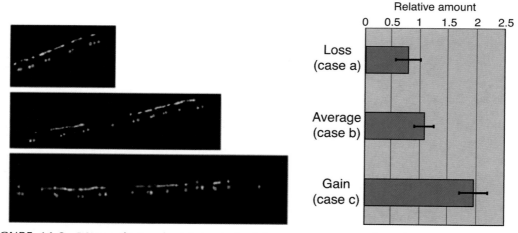

FIGURE 14-2. FISH verification by Iafrate et al. of the most common variant clone (BAC RP11-259N12) detected in their sample population. The copy number differences observed reflect differing numbers of tandem repeats in amylase genes contained within the clone. High-resolution fiber FISH was done on stretched DNA fibers using a Spectrum Green-labeled RP11-259N12 probe cohybridized with a 5′ amylase gene probe (*green*) and a 3′ amylase gene probe (*red*) to verify array CGH profiles (*not shown*) that indicated relative losses, normal ratios, or relative gains for the CNV represented by BAC RP11-259N12 in three unrelated healthy individuals. The three photos on the left show the FISH results for each case. The case with a relative loss showed 6 gene signals (a), the case with a normal ratio showed 9 gene signals (b), and the case with a relative gain showed 12 signals (c). Quantitative PCR results done on DNA from the same individuals (*shown in the panel on the right*) were consistent with the array CGH and FISH findings. (Modified, with permission, from Iafrate et al. 2004 [© Macmillan].)

Oligonucleotide Arrays

Oligonucleotide arrays are high-density arrays made up of short (20–100 bp) sequences. The exact location(s) of each sequence is known in the near-finished sequence of the human genome. These arrays are usually synthesized in situ directly onto the array surface—a procedure considerably less labor-intensive than maintaining and spotting cloned sequences. The production of an array with exceptionally high genomic resolution is possible due to the short length and ease of production of oligonucleotides. The genomic resolution of current oligonucleotide arrays is 30–50 kb, and efforts are ongoing to produce arrays with even higher resolution. In general, the oligonucleotide array targets are designed to exclude sequences that may decrease the ratio of signal to noise (e.g., SDs, short interspersed elements [SINES], long interspersed elements [LINES], and Alus). Unfortunately, the omission of SDs may decrease the observed numbers of CNVs (see above, BAC Arrays). However, the absence of the highly repetitive elements of the genome in the array targets will not only increase the signal-to-noise ratio, but may also reduce the required amount of the expensive repetitive blocker, Cot1 DNA (see below, Hybridization). To further increase signal over noise, Lucito et al. (2003) developed a method known as representational oligonucleotide microarray analysis (ROMA). ROMA is discussed in greater detail in the following chapter. In brief, ROMA involves a decrease of the complexity of test and reference DNA by Bg1II digestion and then amplification by ligation-mediated PCR. The array target sequences are designed to complement the amplified restriction fragments, thus increasing the specificity and enhancing the kinetics of the hybridization.

Two classes of oligonucleotide arrays, short (~20-mers) and long (~70-mers), are currently in use. The majority of short oligonucleotide arrays are SNP arrays. Longer oligonucleotide arrays were developed to optimize the balance between the specificity of the short sequences and the sensitivity of arrays with longer sequence targets. Long oligonucleotide arrays that are reported to be capable of detecting single-copy changes are commercially available from NimbleGen (Madison, Wisconsin) and Agilent (Palo Alto, California). Many companies that provide oligonucleotide arrays, including those that provide SNP chips, also make custom arrays that are targeted to specific regions of interest.

SNP Arrays

SNP arrays are a specialized class of short oligonucleotide arrays that were originally developed for genome-wide genotyping. More recently, SNP arrays have become suitable platforms for array CGH. Companies such as Affymetrix (Santa Clara, California) and Illumina (San Diego, California) have developed software that identifies copy gains and losses within the SNP array results. The hybridizations on SNP arrays are not true competitive experiments. To ascertain copy number differences, the hybridization intensities obtained from the test DNA are compared with average values derived from controls hybridized to separate arrays. However, the experimental design and the data obtained from SNP arrays are analogous to those of other array CGH platforms. Over time, the resolution of SNP chips has increased dramatically. The latest versions from Affymetrix (the 500K array) and from Illumina (the HumanHap500 array) have genomic resolutions of approximately 5 kb, 200 times the resolution of the average BAC array. Another advantage of using a SNP array is that this platform reports loss of heterozygosity (LOH) in addition to copy number changes. This information is generally important for cancer studies, because it allows one to distinguish between the loss of information via a deletion (the observation of both copy number loss and LOH) and segmental uniparental disomy (the observation of LOH only). Segmental uniparental disomy is the derivation of both copies of a chromosome segment from a single parent. Limiting factors for SNP arrays are that they require hundreds of nanograms of test DNA and can cost considerably more than other array platforms.

cDNA Arrays

cDNA arrays, long used for gene expression studies, have recently become amenable for array CGH use (Pollack et al. 1999). cDNA arrays are composed of clones representing whole or partial gene sequences obtained from cDNA libraries. They generally contain 10,000–30,000 array target spots. The clear benefit of these arrays is the ability to conduct CGH and expression studies on the same platform. However, one of the significant hurdles to using cDNA arrays for CGH is producing sufficient signal intensity from the array target spots to detect low-level copy number changes. A few laboratories have overcome this limitation (Pollack et al. 1999; Chen et al. 2004; Park et al. 2006), but we anticipate that the improvements in oligonucleotide array technology will obviate the need for cDNA arrays.

DNA PREPARATION

The test and reference DNA must be differentially labeled with fluorescent molecules (typically Cy3 and Cy5) for array analysis. This labeling is achieved by incorporating dye-conjugated dNTPs into sequences produced from the test or reference template by either a linear or exponential amplification. The size range of the labeled sequences must conform to the optimal size range required by the array platform being used. For most platforms, this range is from 100 bp to 1 kb, but users will need to verify the best size range for their platform and optimize their labeling method accordingly.

The quality and quantity of the test and reference DNA have profound effects on the quality of the array data. Starting with DNA that has undergone very little degradation and that is abundant (>200 ng) is the most effective way to reduce noise in the data set. DNA that meets these criteria is generally compatible with any labeling method. The simplest labeling techniques are linear amplifications that directly incorporate the dye-conjugated dNTPs: for example, random priming or nick translation. Random priming employs degenerate hexamers and a Klenow polymerase to amplify the template DNA. Efficient labeling with random priming requires that the template DNA be fully denatured. Thus, partial predigestion, especially of high-molecular-weight DNA, with a restriction enzyme that is a frequent cutter may improve the productivity of the reaction. The size range of the labeled sequences from random priming (100 bp–1 kb) is generally compatible with most array platforms. Nick translation does not use primers, but a DNase to create nick sites in the template DNA at which a polymerase initiates DNA synthesis. Due to the action of the DNase, the size range of the labeled sequences from a nick translation will decrease as the time of the reaction increases. The user may need to check the reaction products on an agarose gel at various time points until the appropriate size range is achieved.

Poor-quality DNA, such as heavily degraded DNA, is unlikely to work for array analysis. However, if the quantity of test DNA is limited (<200 ng), the sample may work with a preamplification step before labeling by one of the methods discussed above. Whole-genome PCR and multiple displacement amplification (MDA) methods can preamplify the template DNA. Two popular array-compatible PCR methods are degenerate oligonucleotide primed (DOP) (Telenius et al. 1992; Daigo et al. 2001) and ligation-mediated PCR (LMP) (Guillaud-Bataille et al. 2004). However, PCR-amplified DNA typically cannot be used with commercial oligonucleotide arrays, which usually require high-molecular-weight test DNA. Other labs prefer MDA, a relatively new, non-PCR-based method of amplification. MDA is reported to faithfully reproduce the genome as relatively long (~10-kb) sequence stretches that are compatible with both BAC and oligonucleotide arrays (Paez et al. 2004). MDA employs exonuclease-resistant random primers and DNA polymerase with high processivity (e.g., phi29) to amplify the template DNA. The user may need to test multiple techniques to assess which produces the best array data in his or her hands.

CNVs IN THE REFERENCE GENOME

With array CGH, copy number changes are observed relative to a reference DNA. Thus, CNVs in the reference may influence the frequency and detection of CNVs in the test population. Therefore, it is important to consider whether to use as a reference the genome of one individual or a pool of genomes from multiple individuals. The presence of a rare CNV in a single-genome reference is a potential problem, as it will show up in the majority of array results and lead to an overestimate of the frequency of the CNV. FISH can resolve the source of the CNV; or use of a pooled reference, by dilution of rare CNVs, can mitigate overcounting. The inability to detect CNVs shared between the test and reference genomes is a problem for both types of references. The same CNV in a test and single-genome reference will read as a ratio of 1, implying no change at the locus. The ratio value for a CNV shared by the test DNA and some individuals in the pooled reference will approach 1 (no change) as the number of reference individuals with that CNV increases. Overall, the possibility that the CNV is not detected increases as the number of individuals in the reference with the same CNV increases and as the sensitivity of the array decreases. Ideally, each genome should be tested against sufficiently many single-genome references to accurately predict population frequencies and optimize detection. Because the number of arrays needed to meet these criteria is likely to exceed the scope of most studies, an awareness of these reference effects is required.

HYBRIDIZATION

Test and reference DNA are prepared for hybridization by ridding the labeling reactions of unincorporated nucleotides and coprecipitating the labeled DNA with the repeat blocker, Cot1. Cot1 DNA is a critical component of the hybridization as it suppresses the repetitive content of the human genome (e.g., SINES, LINES, and Alus), which can dominate the hybridization and severely diminish the ability of the array target spots to respond to copy number change. Typically, 50–100 mg of Cot1 DNA is added.

A brief outline of the hybridization process is given here, but specific details are omitted, as the procedure is generally performed by an array facility. The precipitated sample is resuspended in a hybridization solution. This solution includes formamide, salt, and dextran sulfate. The formamide increases the effective hybridization temperature to reduce mismatches without the deleterious effects of actual high temperatures. The salt (e.g., saline sodium citrate [SSC]) stabilizes DNA hybrids by decreasing the effective hybridization temperature. Thus, the appropriate balance of formamide and salt is necessary to achieve proper (low mismatch) hybridization. The dextran sulfate increases the effective concentrations of the test and reference DNAs. Hybridizations are typically incubated overnight at 37°C in a humid chamber. Following hybridization, mismatched sequences are generally removed by washing the array slides in a solution of formamide and dilute salt (e.g., 2x SSC) at approximately 45°C. Lowering the salt concentration, increasing the percentage of formamide, and increasing the temperature increase the stringency of the wash. Finally, a fluorescence scanner is used to collect images of the arrays.

DATA PROCESSING

Processing array output has three steps: visual culling of bad array spots, data normalization, and copy number interpretation. First, scanned array images are inspected for flawed spots (e.g., spots that are overlapping, blank, or scratched). Commercial or publicly available software (e.g., GenePix Pro [Molecular Devices, Sunnyvale, California] and UCSF Spot and Sproc [Jain et al.

2002]) is available to support spot culling and to collate the data for further processing. The second processing step is normalization, a step used to correct for the multiple sources of systematic variations in array experiments. These variations arise from differences in labeling efficiencies, unequal amounts of total test and reference DNA, uneven hybridization across the array platform, and different intensities between the two lasers used in scanning the fluorescent output from the test and reference dyes. The end result of normalization is that the mean or median ratio is set to some standard value. In most cases, this standard value is a ratio of 1 or the log ratio of 1 (0). Commonly used normalization procedures are total intensity normalization, Lowess normalization, log centering, and ratio statistic (Quackenbush 2002). Self vs. self experiments can help identify the normalization scheme that works best for a given set of arrays.

Identifying sites of copy number change is the final step of data processing. The simplest approach is to use threshold values based on mean and standard deviation. However, several more elegant algorithms are available for copy number calling. All these methods attempt to infer the statistical significance of copy number differences and identify the boundaries of these differences. Because the array target sequences representing a particular chromosome and locus share a spatial relationship, most of these methods assess the status of each array target in the context of its neighbors. For example, the hidden Markov model (HMM) assigns spatially related target sequences to a particular state, unless the ratio value of a continuous subset of those target sequences is different enough to overcome a designated penalty value for changing states. HMM states are interpretable as losses, gains, or no change. Lai et al. (2005) provide a discussion of eleven different methods, including the HMMs, mixture models, maximum likelihood, regression, and wavelets. A list of the copy number tools assessed by Lai et al. and links to each algorithm Web site is available at http://www.chip.org/~ppark/Supplements/Bioinformatics05b.html.

EXPLORING THE GENOME FOR CNVs

From the number of recent publications, it is becoming clear that large and relatively complex forms of genetic variation are a significant part of the normal human genome. The three seminal papers in this field used competitive arrays with a panel of ethnically diverse individuals to identify a multitude of CNVs (Iafrate et al. 2004; Sebat et al. 2004; Sharp et al. 2005). A summary of some of the findings discussed here is given in Table 14-1.

The hundreds of CNVs collectively identified by Sebat et al., Sharp et al., and Iafrate et al. suggest that we have just begun to touch on the large-spectrum CNVs in the human genome. A compilation of all CNVs identified in these three studies is shown in a genome-wide map in Figure 14-3. Sebat et al. (2004) evaluated 20 individuals using the ROMA system to identify 221 array targets with copy number differences representing 76 unique CNVs; this is an average of approximately 11 CNVs per person. Impressively, Sharp et al. (2005) found 53 of the 76 CNVs identified by Sebat et al. In total, the study by Sharp et al. identified 160 CNVs across 47 individuals using a

TABLE 14-1. Summary of Genome-wide Studies of CNVs

Study	Array platform	Sample size	Reference	Total CNVs	Average # per person	Median size (kb)	Number validated
Sebat	ROMA	20	individual	221	11.1	222	11/12
Iafrate	BAC	55	pooled	255	12.4	~150[a]	18/18
Sharp	BAC	47	individual	160	3.4	~150[a]	7/9

[a]Adapted from Eichler (2006).

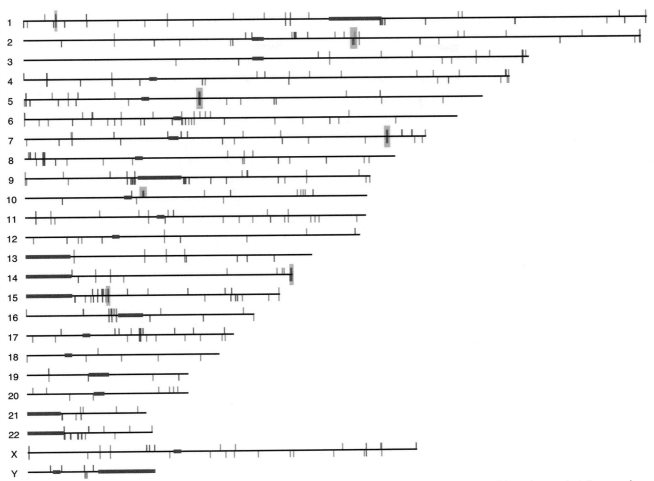

FIGURE 14-3. Genome-wide map of sites of copy number variation detected by Sebat et al., Iafrate et al., and Sharp et al. Colored bars on each chromosome indicate the location of CNVs (not to scale). The CNVs identified by Sebat et al. are in magenta, by Iafrate et al. are in green, and by Sharp et al. are in blue. Tick marks (not to scale) above the chromosome are gains, and those below are deletions. The seven sites that contain overlapping CNVs, irrespective of the type of change (i.e., loss or gain), from all three studies are highlighted by light gray boxes. Five of these seven sites contain CNVs that have concordant changes with at least one concordant CNV from each study. These five sites are on chromosomes 1, 2, 5, 7, and 10. Genome assembly gaps in pericentromeric and satellite regions are indicated by dark gray boxes. The CNV coordinates shown are from the July 2003 assembly and were obtained from http://humanparalogy.gs.washington.edu/structuralvariation/.

BAC array targeted toward SDs (3.4 CNVs per person). The large number of novel CNVs (107) identified by Sharp et al. is probably in part a result of their SD-enriched system (CNVs appear to be enriched in SD loci, see below). Surprisingly, however, Iafrate et al. (2004), who used a BAC array system enriched for non-duplicated clones, observed the greatest number of CNVs. In 55 individuals, they identified 255 CNVs (12.4 CNVs per person). Interestingly, 233 of the CNVs observed by Iafrate et al. were novel observations. Only 11 CNVs identified in Iafrate's study overlapped with the Sebat and Sharp studies, respectively. Furthermore, if the direction (i.e., loss or gain) of the variant is disregarded, only 7 sites were identified where CNVs overlapped for all three studies (see Fig. 3). These 7 sites include 5 sites for which the overlapping CNVs are changes in the same direction. The lack of concordance between the CNVs identified in the three studies may result from the diverse array platforms used or may point to valid differences in the CNV content of the individuals tested in the studies. Each study used independent methods to confirm a subset of the CNVs they found. Verification rates were high. The rate was 78% (7/9) for Sharp et al., 92% (11/12) for Sebat et al., and 100% (18/18) for Iafrate et al.

The cumulative nonoverlapping CNVs identified in these three studies represent approximately 2.7% (88.35 Mb total) of the haploid genome. Sebat et al. (2004) observed an average CNV length of 465 kb and median length of 222 kb using an array composed of 85,000 oligonucleotides, which, of the three studies, is the array with the highest genomic resolution. The majority of CNVs identified by Sharp et al. (2005) and Iafrate et al. (2004) involved a single clone on their respective BAC arrays. A single clone variant can contain all or part of the intervening sequence between it and the neighboring clones, making it more difficult to estimate the actual size of the CNV. For example, the BAC clones on the array used by the Iafrate study are approximately 1 Mb apart. Thus, the gains or losses can involve as much as 2 Mb of DNA. However, an examination of genetic variation trends gives a median CNV size of approximately 150 kb for the Iafrate and Sharp studies (Eichler 2006). All the studies found an approximately equal number of losses and gains of CNVs and that the variants are distributed throughout the genome. However, the distribution of CNVs with respect to blocks of highly homologous SDs reveals an interesting relationship.

All three studies provided support for an enrichment of CNVs in regions that contain SDs, which are considered hot spots of chromosomal rearrangement (Bailey et al. 2002). Sebat et al. (2004) found 6-fold and 12-fold higher SD sequence content within deleted and duplicated CNVs, respectively, relative to the average SD content of the entire genome. Iafrate et al. (2004) found that approximately 25% of the BACs representing CNVs in multiple individuals mapped to regions that overlapped with SDs versus the 7.3% observed for all BAC clones on their array ($p < 0.0001$). Sharp et al. (2005) identified 11.5 times more CNVs by using the SD-enriched BAC array versus a non-SD targeted BAC array. As an increasing number of studies suggest that the duplication architecture of the human genome mediates normal variation (Bailey et al. 2002; Sharp et al. 2005; de Bustos et al. 2006; Goidts et al. 2006; Redon et al. 2006), it becomes interesting to postulate a causal relationship between SDs and CNVs. However, resolving causality is tricky. What can be resolved is the nature of any preferential association of CNVs with SDs by a thorough investigation of the regions of the genome rich in SDs. These regions include pericentromeric and subtelomeric zones, regions underrepresented in the current CNV surveys.

Importantly, a subset of CNVs contains genes that could influence normal phenotypic variation and disease susceptibility. In total, Sebat et al. observed copy number variation of 70 genes, Sharp et al. found 141 genes either completely or partially overlapping their variant BACs, and Iafrate et al. identified 67 BAC clones that encompassed one or more entire genes. The CNV-related genes found by Sharp et al. (2005) include some involved in immunity and metabolism, which may significantly contribute to phenotypic variations, such as resistance to pathogens and metabolic rates. Both Iafrate et al. (2004) and Sebat et al. (2004) identified a subset of CNVs containing cancer-related genes in apparently normal individuals. Seemingly, these genes are not the direct cause of disease, but the structural variants may initiate chromosomal rearrangements or influence gene expression, which could affect tumorigenesis. Overall, the genes found within or near CNVs may have profound effects on the biology of normal or disease processes.

SUMMARY

From our review of the current literature, it appears that CNVs are a prominent part of the normal variation of the human genome. Two extremely useful resources that summarize the results of CNV research studies are found on the Web at http://humanparalogy.gs.washington.edu/structuralvariation/ and http://projects.tcag.ca/variation/. However, the process of cataloging the full spectrum of CNVs is in its infancy. We have yet to discover the exact size range of, the amount of genomic sequence subject to, and gene content of CNVs. Although by definition CNVs can encompass as little as 1 kb, some may involve as much as 2 Mb of sequence (Iafrate et al. 2004). So far, potentially 88 Mb of genomic sequence is subject to copy number variation (Eichler 2006), but this value is likely to increase as additional studies emerge. Finally, we know that CNVs can involve multiple genes (Iafrate et al. 2004; Sebat et al. 2004; Sharp et al. 2005), but we have yet to determine whether gene-containing variants or CNVs in general have an effect on normal phenotypic variation and disease susceptibility. To fully assess and curate human genetic diversity, we need to analyze CNVs on a scale comparable to that used to catalog the wealth of human SNPs, and array CGH is an essential tool in accomplishing that goal.

REFERENCES

Albertson D.G., Ylstra B., Segraves R., Collins C., Dairkee S.H., Kowbel D., Kuo W.L., Gray J.W., and Pinkel D. 2000. Quantitative mapping of amplicon structure by array CGH identifies *CYP24* as a candidate oncogene. *Nat. Genet.* **25:** 144–146.

Ammerlaan A.C., de Bustos C., Ararou A., Buckley P.G., Mantripragada K.K., Verstegen M.J., Hulsebos T.J., and Dumanski J.P. 2005. Localization of a putative low-penetrance ependymoma susceptibility locus to 22q11 using a chromosome 22 tiling-path genomic microarray. *Genes Chromosomes Cancer* **43:** 329-338.

Bailey J.A., Gu Z., Clark R.A., Reinert K., Samonte R.V., Schwartz S., Adams M.D., Myers E.W., Li P.W., and Eichler E.E. 2002. Recent segmental duplications in the human genome. *Science* **297:** 1003–1007.

Barrett M.T., Scheffer A., Ben-Dor A., Sampas N., Lipson D., Kincaid R., Tsang P., Curry B., Baird K., Meltzer P.S., et al. 2004. Comparative genomic hybridization using oligonucleotide microarrays and total genomic DNA. *Proc. Natl. Acad. Sci.* **101:** 17765–17770.

Carvalho B., Ouwerkerk E., Meijer G.A., and Ylstra B. 2004. High resolution microarray comparative genomic hybridisation analysis using spotted oligonucleotides. *J. Clin. Pathol.* **57:** 644–646.

Chen Q.R., Bilke S., Wei J.S., Whiteford C.C., Cenacchi N., Krasnoselsky A.L., Greer B.T., Son C.G., Westermann F., Berthold F., et al. 2004. cDNA array-CGH profiling identifies genomic alterations specific to stage and MYCN-amplification in neuroblastoma. *BMC Genomics* **5:** 70.

Cheung V.G., Dalrymple H.L., Narasimhan S., Watts J., Schuler G., Raap A.K., Morley M., and Bruzel A. 1999. A resource of mapped human bacterial artificial chromosome clones. *Genome Res.* **9:** 989–993.

Cheung V.G., Nowak N., Jang W., Kirsch I.R., Zhao S., Chen X.N., Furey T.S., Kim U.J., Kuo W.L., Olivier M., et al. (BAC Resource Consortium). 2001. Integration of cytogenetic landmarks into the draft sequence of the human genome. *Nature* **409:** 953–958.

Daigo Y., Chin S.F., Gorringe K.L., Bobrow L.G., Ponder B.A., Pharoah P.D., and Caldas C. 2001. Degenerate oligonucleotide primed-polymerase chain reaction-based array comparative genomic hybridization for extensive amplicon profiling of breast cancers: A new approach for the molecular analysis of paraffin-embedded cancer tissue. *Am. J. Pathol.* **58:** 1623–1631.

de Bustos C., Diaz de Stahl T., Piotrowski A., Mantripragada K.K., Buckley P.G., Darai E., Hansson C.M., Grigelionis G., Menzel U., and Dumanski J.P. 2006. Analysis of copy number variation in the normal human population within a region containing complex segmental duplications on 22q11 using high-resolution array-CGH. *Genomics* **88:** 152–162.

Devries S., Nyante S., Korkola J., Segraves R., Nakao K., Moore D., Bae H., Wilhelm M., Hwang S., and Waldman F. 2005. Array-based comparative genomic hybridization from formalin-fixed, paraffin-embedded breast tumors. *J. Mol. Diagn.* **7:** 65–71.

Eichler E.E. 2006. Widening the spectrum of human genetic variation. *Nat. Genet.* **38:** 9–11.

Feuk L., Carson A.R., and Scherer S.W. 2006. Structural variation in the human genome. *Nat. Rev. Genet.* **7:** 85–97.

Goidts V., Cooper D.N., Armengol L., Schempp W., Conroy J., Estivill X., Nowak N., Hameister H., and Kehrer-Sawatzki H. 2006. Complex patterns of copy number variation at sites of segmental duplications: An important category of structural variation in the human genome. *Hum. Genet.* **120:** 270–284.

Guillaud-Bataille M., Valent A., Soularue P., Perot C., Inda M.M., Receveur A., Smaili S., Crollius H.R., Benard J., Bernheim A., et al. 2004. Detecting single DNA copy number variations in complex genomes using one nanogram of starting DNA and BAC-array CGH. *Nucleic Acids Res.* **32:** e112.

Iafrate A.J., Feuk L., Rivera M.N., Listewnik M.L., Donahoe P.K., Qi Y., Scherer S.W., and Lee C. 2004. Detection of large-scale variation in the human genome. *Nat. Genet.* **36:** 949–951.

Ishkanian A.S., Malloff C.A., Watson S.K., DeLeeuw R.J., Chi B., Coe B.P., Snijders A., Albertson D.G., Pinkel D., Marra M.A., et al. 2004. A tiling resolution DNA microarray with complete coverage of the human genome. *Nat. Genet.* **36:** 299–303.

Jain A.N., Tokuyasu T.A., Snijders A.M., Segraves R., Albertson D.G., and Pinkel D. 2002. Fully automatic quantification of microarray image data. *Genome Res.* **12:** 325–332.

Kallioniemi A., Kallioniemi O.P., Sudar D., Rutovitz D., Gray J.W., Waldman F., and Pinkel D. 1992. Comparative genomic hybridization for molecular cytogenetic analysis of solid tumors. *Science* **258:** 818–821.

Kirsch I.R. and Ried T. 2000. Integration of cytogenetic data with genome maps and available probes: Present status and future promise. *Semin. Hematol.* **37:** 420–428.

Kirsch I.R., Green E.D., Yonescu R., Strausberg R., Carter N., Bentley D., Leversha M.A., Dunham I., Braden V.V., Hilgenfeld E., et al. 2000. A systematic, high-resolution linkage of the cytogenetic and physical maps of the human genome. *Nat. Genet.* **24:** 339–340.

Korenberg J.R., Chen X.N., Sun Z., Shi Z.Y., Ma S., Vataru E., Yimlamai D., Weissenbach J.S., Shizuya H., Simon M.I., et al. 1999. Human genome anatomy: BACs integrating the genetic and cytogenetic maps for bridging genome and biomedicine. *Genome Res.* **9:** 994–1001.

Krzywinski M., Bosdet I., Smailus D., Chiu R., Mathewson C., Wye N., Barber S., Brown-John M., Chan S., Chand S., et al. 2004. A set of BAC clones spanning the human genome. *Nucleic Acids Res.* **32:** 3651–3660.

Lai W.R., Johnson M.D., Kucherlapati R., and Park P.J. 2005. Comparative analysis of algorithms for identifying amplifications and deletions in array CGH data. *Bioinformatics* **21:** 3763–3770.

Leversha M.A., Dunham I., and Carter N.P. 1999. A molecular cytogenetic clone resource for chromosome 22. *Chromosome Res.* **7:** 571–573.

Li J., Jiang T., Mao J.H., Balmain A., Peterson L., Harris C., Rao P.H., Havlak P., Gibbs R., and Cai W.W. 2004. Genomic segmental polymorphisms in inbred mouse strains. *Nat. Genet.* **36:** 952–954.

Little S.E., Vuononvirta R., Reis-Filho J.S., Natrajan R., Iravani M., Fenwick K., Mackay A., Ashworth A., Pritchard-Jones K., and Jones C. 2006. Array CGH using whole genome amplification of fresh-frozen and formalin-fixed, paraffin-embedded tumor DNA. *Genomics* **87:** 298–306.

Lucito R., Healy J., Alexander J., Reiner A., Esposito D., Chi M., Rodgers L., Brady A., Sebat J., Troge J., et al. 2003. Representational oligonucleotide microarray analysis: A high-resolution method to detect genome copy number variation. *Genome Res.* **13:** 2291–2305.

McPherson J.D., Marra M., Hillier L., Waterston R.H., Chinwalla A., Wallis J., Sekhon M., Wylie K., Mardis E.R., Wilson R.K., et al. (International Human Genome Mapping Consortium). 2001. A physical map of the human genome. *Nature* **409:** 934–941.

Paez J.G., Lin M., Beroukhim R., Lee J.C., Zhao X., Richter D.J., Gabriel S., Herman P., Sasaki H., Altshuler D., et al. 2004. Genome coverage and sequence fidelity of phi29 polymerase-based multiple strand displacement whole genome amplification. *Nucleic Acids Res.* **32:** e71.

Park C.H., Jeong H.J., Choi Y.H., Kim S.C., Jeong H.C., Park K.H., Lee G.Y., Kim T.S., Yang S.W., Ahn S.W., et al. 2006. Systematic analysis of cDNA microarray-based CGH. *Int. J. Mol. Med.* **17:** 261–267.

Pinkel D., Segraves R., Sudar D., Clark S., Poole I., Kowbel D., Collins C., Kuo W.L., Chen C., Zhai Y., et al. 1998. High resolution analysis of DNA copy number variation using comparative genomic hybridization to microarrays. *Nat. Genet.* **20:** 207–211.

Pollack J.R., Perou C.M., Alizadeh A.A., Eisen M.B., Pergamenschikov A., Williams C.F., Jeffrey S.S., Botstein D., and Brown P.O. 1999. Genome-wide analysis of DNA copy-number changes using cDNA microarrays. *Nat. Genet.* **23:** 41–46.

Quackenbush J. 2002. Microarray data normalization and transformation. *Nat. Genet.* (suppl.) **32:** 496–501.

Redon R., Ishikawa S., Fitch K.R., Feuk L., Perry G.H., Andrews T.D., Fiegler H., Shapero M.H., Carson A.R., Chen W., et al. 2006. Global variation in copy number in the human genome. *Nature* **444:** 444–454.

Sebat J., Lakshmi B., Troge J., Alexander J., Young J., Lundin P., Månér S., Massa H., Walker M., Chi M., et al. 2004. Large-scale copy number polymorphism in the human genome. *Science* **305:** 525–528.

Sharp A.J., Locke D.P., McGrath S.D., Cheng Z., Bailey J.A., Vallente R.U., Pertz L.M., Clark R.A., Schwartz S., Segraves R., et al. 2005. Segmental duplications and copy-number variation in the human genome. *Am. J. Hum. Genet.* **77:** 78–88.

Solinas-Toldo S., Lampel S., Stilgenbauer S., Nickolenko J., Benner A., Dohner H., Cremer T., and Lichter P. 1997. Matrix-based comparative genomic hybridization: Biochips to screen for genomic imbalances. *Genes, Chromosomes & Cancer* **20:** 399–407.

Telenius H., Carter N.P., Bebb C.E., Nordenskjold M., Ponder B.A., and Tunnacliffe A. 1992. Degenerate oligonucleotide-primed PCR: General amplification of target DNA by a single degenerate primer. *Genomics* **13:** 718–725.

Woodfine K., Fiegler H., Beare D.M., Collins J.E., McCann O.T., Young B.D., Debernardi S., Mott R., Dunham I., and Carter N.P. 2004. Replication timing of the human genome. *Hum. Mol. Genet.* **13:** 191–202.

WWW RESOURCES

http://bacpac.chori.org/ BACPAC Resources Center (BPRC), Children's Hospital Oakland Research Institute, California.

http://www.ncbi.nlm.nih.gov/genome/clone/distributors.html Clone Registry, National Center for Biotechnology Information (NCBI), U.S. National Library of Medicine, National Institutes of Health, Bethesda, Maryland.

http://www.chip.org/~ppark/Supplements/Bioinformatics05b.html Lai et al. 2005. Supplementary material. *Bioinformatics*, 2005. Computational Genomics (PI: Peter J. Park).

http://humanparalogy.gs.washington.edu/structuralvariation/ Human Structural Variation Database, Department of Genome Sciences, University of Washington, Seattle.

http://projects.tcag.ca/variation/

15 Representational Oligonucleotide Microarray Analysis Detection of Genetic Variation

Rob Lucito

Cold Spring Harbor Laboratory, Cold Spring Harbor, New York 11724

INTRODUCTION

Historically, copy number variations (CNVs) were first observed by comparing karyotypes of normal and diseased individuals. Studying the karyotypes became clinically important for the diagnosis of certain syndromes, ranging from Down syndrome to cancer (Benson 1961; Stimson 1967). Generally, it is much easier to identify CNVs in cancer, since they can be so numerous, and they can vary widely in the number of copies for a specific region. With knowledge of the region, methods such as Southern blotting or fluorescence in situ hybridization (FISH) can be performed. However, if a scan of the entire genome is needed or there is no prior knowledge of a region of interest, other techniques are more appropriate. Early generation methods for CNV detection such as comparative genomic hybridization (CGH) (Thompson and Gray 1993) were useful but suffered from low resolution. With the mapping information of the genome project and the development of microarray technologies, CGH was adapted to an array platform (Pinkel et al. 1998; Pollack et al. 1999), and the completion of the human genome sequence paved the way for the development of array CGH methods with exquisite resolution (Lucito et al. 2003; Barrett et al. 2004; Selzer et al. 2005). With this increased resolution, it became apparent that CNVs in the normal population are frequent but had been undetected previously (Lucito et al. 2003; Sebat et al. 2004).

Together with Dr. Michael Wigler, we developed an array-based CGH method to detect CNVs in cancer, although we quickly learned that it was very accurate at identifying small CNVs in the normal population. This method, representational oligonucleotide microarray analysis (ROMA)(Lucito et al. 2003), was an advancement of representational difference analysis (RDA), a technique developed by Dr. Wigler and Dr. Nikolai Lisitsyn (Lisitsyn et al. 1993). Central to both methods is the practice of making a representation of the genome. A representation is reproducible sampling of the genome, prepared by cleavage of the genome with a restriction endonuclease, ligation of PCR adapters, and amplification with a PCR-adequate polymerase such as Taq polymerase. Generally, when the template is a population of mixed fragment sizes,

233

an enzyme such as Taq polymerase will preferentially amplify the smaller fragments, with resulting representations in the size range of 100–1200 bp. The lack of amplification of the larger fragments results in a decrease in complexity in the sample. In comparison to total genome hybridization, the decrease in complexity of a representation allows for an increased signal to noise, due to increased hybridization efficiency and increased labeling efficiency for fragments. The increased labeling efficiency is a result of the array being well matched to the content of the representation. In other words, the nucleotide content of the array more closely matches the representation's content than the entire genome.

Representations

The complexity of the representation can be shaped by the restriction endonuclease used for the initial digestion of the genomic DNA. For example, when a restriction endonuclease with a six-nucleotide cleavage site is used, the complexity decreases to 2–5% of the original genome. When a restriction endonuclease with a four-nucleotide cleavage site is used, the complexity decreases to 60–80%, again depending on the cleavage site of the endonuclease. The complexity of the representation can be further shaped by depletion of the representation. In this process, a second restriction endonuclease cleaves the ligation of the genomic DNA prior to PCR amplification. This allows limitless possibilities to achieve the desired complexity.

As a consequence of the required PCR amplification, representations can be easily prepared from as little as 50 ng of genomic DNA. This has made it easier to analyze samples that cannot be renewed, such as tumor specimens, where yield is often in the range of approximately 100 ng. Overamplification can occur: With extreme numbers of cycles, the measured ratios decrease. In most cases, this is not an issue. However, detection of small CNVs (covering regions containing few probes on the array) that are either hemizygous or duplicated can be problematic. We have compared representations prepared with varying number of cycles and have found no change in the level of detection of CNVs, using as many as 25 cycles. Presently, we do not recommend PCR amplification with more than 25 cycles, under the template conditions discussed above.

FACTORS AFFECTING ROMA

Several factors affect the ability to detect CNVs in a particular population, including the array itself, the representation, species, and experimental design. Representations are remarkably accurate at reflecting the copy number of the genome in relation to a reference sample. To accurately detect CNVs within a genome, all sources of system noise that can be controlled must be suppressed. The amplification of samples by PCR adds a possible source of noise that can be easily addressed. Comparable amounts of template genomic DNA should be used, and the test and reference samples should be prepared in parallel from the start of the restriction endonuclease digestion to the PCR amplification. In theory, it is possible to perform ROMA on a single sample without reference, and this is currently under investigation.

The design of the array is equally important when one is attempting to detect CNVs in normal samples. Many CNVs are found in regions that are duplicated within the genome. If these regions are removed, much less variation will be detected. Of course, the addition of regions in the genome that are found more than once must be approached carefully. As with any hybridization-based method, the more times a target is present in the genome, the less specific the hybridization will be. By analogy, repetitive regions in the genome are so numerous that they are useless for copy number detection with hybridization-based methods.

The probes on the array should perform roughly equally. If probe performance is sporadic, a CNV covered by several probes where internal probes are poor reporters could be interpreted as a cluster of single-nucleotide polymorphisms (SNPs). In this case, CNV detection and, more specifically, automated forms of detection will tend to miss such CNVs. Probes with similar performance should be used for the array, or probe performance parameters should be utilized to interpret the output data. Methods have been developed to predict probe performance, and the present ROMA array utilizes probes of similar performance.

Of equal importance is resolution on the array. If there are not enough probes for a given region, CNVs will not be detected. Many regions are fairly small (~100–200 kb) and arrays with fewer than 40,000 probes will miss many of these regions. The placement of probes could also be an issue. There are untranslated RNAs that are critical to the function of a cell and that can act in an oncogenic manner. Thus, gene-centric CGH arrays are not an optimal design. The initial ROMA arrays were roughly 82,000 probes (although much higher densities can be achieved), randomly scattered throughout the genome so that all regions (with the exception of highly repetitive regions) were equally represented on the array.

In most cases, when ROMA is used to identify genetic variation of normal samples, reference samples will be from a different individual. Since representations use restriction endonucleases to shape complexity, the method uncovers restriction fragment length polymorphisms (RFLPs). If a SNP is present in the test sample and the initial restriction endonuclease cleavage site is destroyed, in comparison to the reference sample an RFLP is created. Usually the new fragment is too large to amplify efficiently and thus is no longer in the representation. Probes recognizing such a fragment would have high-intensity measurements in the reference but very low in the test sample, producing high ratios that cause these probes to scatter from the majority of probe measurements. When these probes are numerous, it can be distracting to the human eye, but the quality of the data is not affected. In fact, since these measurements are the result of SNP detection, they can be used to advantage for informatic identification of CNVs using segmentation algorithms, which is discussed below.

The detection of SNPs is somewhat exaggerated in laboratory species such as the mouse. In contrast to the fairly heterogeneous human genome, the laboratory mouse is extremely homogeneous, except for more recent germ-cell genetic alterations that presumably occur. The mouse ROMA array was based on the first mouse genomic sequence available, which was almost entirely from C57BL6/J. Therefore, if any other species is compared to C57BL6 reference on the array, a large number of SNP events are detected. These events are clustered into patches across the genome, so closely in some regions that at first glance the clusters appear to be large CNVs. Interestingly, the genomic position of SNP clusters varies from one strain to another in comparison to C57BL6/J. Variation between different strains means that experiments must be designed carefully, since the wrong choice of reference could lead to uninterpretable results. The simplest solution is to use material from the same mouse as a reference whenever possible. A sib would be the next best choice, although there are some minor genetic differences between sibs (I. Hall, unpubl.). If a catalog is made of various strains, lineage can be delineated, using the genomic profile of a particular mouse. Comparison of germ-cell to somatic DNA allows one to identify germ-cell DNA alterations in restriction endonuclease cleavage sites over many generations. This in turn enables one to determine the rate of mutation in the animals and to identify any possible hot spots of mutation.

When human ROMA is used to analyze normal copy number variation, the reference is almost always from a different individual, unless somatic and germ-cell DNA are compared. All CNVs that are detected must be interpreted in comparison to the reference. (This is generally not an issue with SNPs.) Therefore, it is preferable to have as few reference DNAs as possible, perhaps one male and one female. Using the same reference samples makes it somewhat easier to identify the copy number for a specific CNV. In addition, whenever possible, it is useful to analyze parents to trace the inheritance of CNVs. When looking for de novo alterations, this is essential.

ARRAY PROBE DESIGN

All probes on the array are designed to be complementary to representational fragments, i.e., small restriction fragments. Probe design begins by making an in silico representation based on the available genome sequence. This is performed simply by locating and recording all instances of the restriction site of interest and size selecting to 100–1200 bp. Separately, the entire genome is annotated into N-mers of differing length, most commonly, 15- and 21-nucleotide lengths. As a precursor to probe selection, all identified representational fragments are broken down into all possible overlapping 50-mers (the length of the final probes for the array), compared to the library of genome N-mers, and annotated with N-mer frequencies representing the number of times a fragment N-mer is found in the genome library of N-mers (for more detail, see Healy et al. 2003). This is done for N-mers of length N = 21 and N = 15, using a Burrows-Wheeler transformation (Burrows and Wheeler 1994). For each fragment within the in silico digest, we quantitatively characterize every possible 50-mer. The 21-mer frequencies are used as a measure of overall uniqueness in the genome. Our most stringent criterion is that all constituent 21-mers must be unique. However, this is too stringent when analyzing normal variation. For those probes that pass the 21-mer uniqueness measures, 15-mer frequencies are minimized, since they act as a predictive measure of the likelihood of cross-hybridization.

Probes are selected in descending order of aggregate N-mer frequencies, essentially a filtration process. This limits the otherwise immense search space of candidate probes to those that are most likely to give us accurate ratio measurements and least likely to encounter cross-hybridization events or to fail the synthesis process. The probes are then tested empirically. We typically begin by designing and arraying tenfold more probes than can fit on one array. The probes are arrayed on multiple arrays with some overlap for comparison measures. Each array is used in an "enzyme-depletion" experiment where a representation is compared with an identical representation that has been further subjected to digestion with a second enzyme. Since the entire sequence of the genome is known, we have accurate predictions of which fragments should be eliminated by the second digestion step. We select those probes that have the highest signal intensities in the nondigested channel, as we have determined this to be a good measure of performance. The placements of probes on the array surface should be randomized. In this way, no hybridization artifact on the array surface will appear as a consistent genomic alteration.

POST-HYBRIDIZATION

After hybridization, the arrays are scanned on a standard scanner such as the GenePix 4000B, and intensity acquisition is performed with various software packages. The GenePix scanner has a maximal resolution of 5 μm. It is important to average pixel intensities; averaging less than 9 pixels is not advisable. With the maximal resolution, it is difficult to analyze features smaller than 15 × 15 μm. The arrays currently in use have features that are 17 × 17 μm. The Axon scanner is still applicable for analysis. For higher-resolution arrays which use smaller features, other analysis software must be used. After data acquisition, the array measurements can be normalized by several different methods. We presently use a Lowess curve fitting algorithm adapted from methods described by Dr. Terry Speed and colleagues (Yang et al. 2002), followed by an algorithm that corrects for local hybridization artifacts. The copy number data have variation in the measurements between adjacent probes, caused by several different factors, including noise in the array hybridization, labeling efficiency, washing, and the representational process. Some of this variation is removed by performing duplicate experiments. For assessing genetic variation in normals where the variation is relatively small (in number of bp) and the alterations often vary by only one copy, duplicate experiments (if not more) are strongly suggested. The hybridization can simply be repeated, but for two-color hybridizations (such as with cy3 and cy5), it is best to perform a color reversal (or dye swap), where the samples are labeled with the alternative dye as was used for the first hybridization. After normalization, the probe ratios can be averaged, or other methods can be used to remove noise. An example would be probe outliers caused by unincorporated dye. These outliers can be removed by comparing duplicate hybridizations. For a specific feature, if the ratio in one experiment is above one standard deviation, the lower ratio probe is used. This may remove some information content, but we have found that, in general, the quality of the experimental data is improved.

Even with the averaging of experiments, the data are inherently noisy. A number of different algorithms can be employed to remove this noise. The simplest is a moving average (Pollack et al. 1999), which moves in genomic order and averages adjacent probes. The number of probes being averaged can be adjusted. This improves the smoothing process but decreases the measured resolution, since averaging of probes past the breakpoint brings down the average for the probes on the end of the breakpoint. Pollack et al. improved the method by performing normal–normal comparisons which allowed the calculation of a threshold to call false discoveries. This method is quite useful for copy number detection in cancer, but it has limitations for the measurement of normal variation, since small lesions may be missed. Other methods are more involved and model the data (Hodgson et al. 2001; Autio et al. 2003; Snijders et al. 2003). We have used an algorithm developed by Olshen et al. (2004) and a second one developed by Bud Mishra for ROMA data which can be easily used for other CGH data (Daruwala et al. 2004). A third algorithm developed at Cold Spring Harbor Laboratory by Lakshmi Muthuswamy and Michael Wigler expressly for the analysis of ROMA data can also be used for analysis of other CGH platforms (Lakshmi et al. 2006). Each of these three algorithms is model-based. The method developed by Olshen and colleagues is based on binary segmentation but uses a reference distribution of the data to account for noise in the system. Mishra and colleagues use a Bayesian model that can be viewed as an optimization process minimizing a score function. Both of these methods accurately identify copy number alterations in cancer as well as normal genetic variation, but they require significant compute time to analyze an experiment as compared to the method developed by Muthuswamy and Wigler.

The method developed by Muthuswamy and Wigler is an alternative which partitions the log ratios of the intensities into segments of uniform distributions. The ratios are arranged in genome order and are divided into blocks of 100 data points with arbitrary boundaries. Then, the boundaries are iteratively moved by minimizing the variance, and further refined by using a two-distribution Komogrov-Smirnov (KS) null hypothesis test (Conover 1980). Only boundaries that give a p-value of less than 10^{-5} are accepted. Due to the nature of the KS test, no segment smaller than 3 probes in length is considered. This method is excellent for the analysis of lesions comprising 3 or more probes. A more accurate method of segmentation is being developed which utilizes a hidden Markov model. It uses the presence of SNPs to define the states of alteration between the two genomes; i.e., zero copies, one copy, three copies, or more copies. Once this is determined, the state of other alterations can be identified based on initial calculations.

After processing, the data can be visualized to identify CNVs and any SNPs present between the normal and reference samples. Since the alterations identified are in comparison to the reference, it is possible that a CNV could be present as three copies in both the test and reference sample, and the difference would cancel out. This is unlikely for rare CNVs, but much more possible for common CNVs. If one sample has a CNV that is slightly larger than that found in the other sample, the result would appear as two CNVs, since the central region would be cancelled out. These possibilities must be taken into account when analyzing data.

Large variations can be validated by other methods. Clearly, the most accurate method to validate CNV variation is fluorescent in situ hybridization (FISH). This method can assess the true copy number of an allele present in a sample, whereas the CGH data can only determine the level in comparison to the reference sample. However, FISH is somewhat slow in throughput and requires intact cells. An alternative is quantitative PCR (QPCR). This method does not require whole cells, can be performed with mid-throughput, and requires little DNA template. We have used this technique routinely to determine the relative copy number of CNVs identified in large ROMA data sets. The copy number is always calculated in comparison to a reference sample. If the reference is known to possess two copies of the region being studied, the true copy number of the test sample can accurately be determined. The most accurate determination requires a second control probe whose copy number is known for the reference as well as the test sample.

SELECTION AND ANALYSIS OF LARGE SAMPLE SETS

Although CNVs identified in a single sample can be validated, copy number detection is often used to analyze a large sample set. This may be performed on a large set of normal individuals to determine the variation in the normal population of humans (or mice). Alternatively, one can analyze a large set of individuals that has, or is suspected to have, a syndrome or disease, and the identified CNVs can be coalesced to determine whether any are associated with the particular syndrome or disease. Some history should be known about the individuals whose samples are to be analyzed. For example, to study familial susceptibility to pancreatic cancer, it would be necessary to know that the individual and at least one more member in the immediate family had pancreatic cancer. Because pancreatic cancer affects a relatively small number of individuals worldwide, two individuals within the same family would increase the probability that the cancer had a familial component (provided that environmental factors were removed). The disease or syndrome studies should be as homogeneous as possible. If a syndrome encompasses more than one true disease, it will be more difficult to determine the candidate regions for any one subsyndrome. Even in the case of familial susceptibility to pancreatic cancer, more than one gene may be involved. Careful selection of the test sample set is imperative for creating a valuable set of CNVs. Since primary cells are not available as a source of DNA, sample cells are often transformed by EBV infection so that large amounts can be grown. Such material should be approached with caution, since the transformation process and the extended period of time in culture can themselves result in genomic anomalies.

Analysis of Large Data Sets

To determine which identified CNVs are potentially associated with the disease being studied, as many normal CNVs as possible must be removed or filtered from the data set. One needs to amass a set of normals to make a catalog of CNVs and use the catalog to remove normal CNVs from the test data set. The definition of a normal CNV is somewhat arbitrary and depends on the test set; i.e., the normal set will change depending on the syndrome that is being studied. For example, if the test set is for patients who are susceptible to one or more cancers, all normals in the catalog who are from cancer-susceptible families should be removed. Thus, it is imperative that accurate family data are obtained for the individuals in the normal catalog. Selecting the catalog for comparison to a sample set with a rare cancer such as pancreatic cancer is relatively easy. Strict rules can be used to remove any "normals" who have had pancreatic cancer or have had any immediate family members who have had pancreatic cancer. However, selection is more difficult with more common cancers that have a much higher rate of sporadic occurrence or with syndromes that have low penetrance. In addition, one of the catalog members could be an affected individual who is asymptomatic. If such an individual enters the normal data set, meaningful regions could be lost during the filtering process. We address one possible way of rescuing such data below.

The size of the normal set for CNV filtering should be the largest possible, to allow the removal of common CNVs within the test set of samples. Of course, there will be rare CNVs that cannot be filtered from the population, no matter how large the filter set becomes. If the normal sample set is too small, the test data set will contain a large number of relatively common normal CNVs.

Once the selection of the normal sample set is completed, ROMA can be performed. In addition, normal data sets from other labs that are publicly available can also be used to filter common CNVs. This should only be done if accurate patient information is available.

Current ROMA arrays have 85,000 probes, but we have begun using 390,000-probe arrays. In many cases, the normal sample filter data set, as well as the test data set, can become very large. The data set, if based on all probes, would require a large data frame to represent all samples and would slow down computing time. Much of the data from the genome reports nonevents or vast regions without CNVs. By only representing the identified CNVs, the data can be drastically reduced in size without decreasing the information content. Once this transformation is performed for all samples, a master file can be produced for the normal set that can be used to filter the test samples.

For filtration, all CNVs of the normal set that are found in the test set are subtracted. Often, fragments of CNVs are left as remnants when a CNV found in the test set is larger in length (covered by more probes) than the CNV found in the normal filter set. In the worst case, the CNV in the test set would be larger on both ends, compared to the CNV found in the normal set. After the filtration, it would appear that there are two CNVs in close proximity. The entire CNV can be removed, but this must be done with caution, because valuable data could be lost. At the end of this process, the frequency that a CNV is found in the test set population can be calculated. Removal of CNVs found in the normal population is a strict constraint, especially if there is the unintentional inclusion of an asymptomatic individual. An alternative to simply removing regions that are found in the normal population is to allow some flexibility in the level of detection. If a CNV is found in the test set and in only one individual of the normal set, it will not be removed. This level of detection or flexibility can be increased up to whatever point is deemed informative for the data set. The drawback of such flexibility is that it will increase the number of rare CNVs that happen to be in both data sets as well as some common CNVs. Therefore, it is useful to analyze the filtered data by several different criteria.

Once the regions of CNVs in the test set are identified, other analyses can commence. As an example, the coordinates of the regions could be compared to results from other mapping studies using alternative technologies, such as SNP mapping and quantitative trait locus (QTL) analysis. Informatic analyses can be used to identify all genes within the regions to determine whether there are any interesting candidate genes based on gene function and the syndrome or disease being studied. It is unlikely that all CNVs will be validated, since this process is time consuming and labor intensive. However, it is important to validate a number of CNVs (possibly those of specific interest) with another method. FISH is the most accurate method for validation, but QPCR is a legitimate alternative. It is somewhat difficult and requires a large number of replicas, since the difference expected can be only one cycle. Nonetheless, it is a viable option if whole cells are not available.

REFERENCES

Autio R., Hautaniemi S., Kauraniemi P., Yli-Harja O., Astola J., Wolf M., and Kallioniemi A. 2003. CGH-Plotter: MATLAB toolbox for CGH-data analysis. *Bioinformatics* **19:** 1714–1715.

Barrett M.T., Scheffer A., Ben-Dor A., Sampas N., Lipson D., Kincaid R., Tsang P., Curry B., Baird K., Meltzer P.S., et al. 2004. Comparative genomic hybridization using oligonucleotide microarrays and total genomic DNA. *Proc. Natl. Acad. Sci.* **101:** 17765–17770.

Benson E.S. 1961. Leukemia and the Philadelphia chromosome. *Postgrad. Med.* **30:** A22–A28.

Burrows M. and Wheeler D.J. 1994. A block sorting lossless data compression algorithm. HP technical report SRC-RR-124, Hewlett-Packard, Palo Alto, California.

Conover W.J. 1980. *Practical nonparametric statistics*, 2nd edition. Wiley, New York.

Daruwala R.S., Rudra A., Ostrer H., Lucito R., Wigler M., and Mishra B. 2004. A versatile statistical analysis algorithm to detect genome copy number variation. *Proc. Natl. Acad. Sci.* **101:** 16292–16297.

Healy J., Thomas E.E., Schwartz J.T., and Wigler M. 2003. Annotating large genomes with exact word matches. *Genome Res.* **13:** 2306–2315.

Hodgson G., Hager J.H., Volik S., Hariono S., Wernick M., Moore D., Nowak N., Albertson D.G., Pinkel D., Collins C., et al. 2001. Genome scanning with array CGH delineates regional alterations in mouse islet carcinomas. *Nat. Genet.* **29:** 459–464.

Lakshmi B., Hall I.M., Egan C., Alexander J., Leotta A., Healy J., Zender L., Spector M.S., Xue W., Lowe S.W., et al. 2006. Mouse genomic representational oligonucleotide microarray analysis: Detection of copy number variations in normal and tumor specimens. *Proc. Natl. Acad. Sci.* **103:** 11234–11239.

Lisitsyn N., Lisitsyn N., and Wigler M. 1993. Cloning the differences between two complex genomes. *Science* **259:** 946–951.

Lucito R., Healy J., Alexander J., Reiner A., Esposito D., Chi M., Rodgers L., Brady A., Sebat J., Troge J., et al. 2003. Representational oligonucleotide microarray analysis: A high-resolution method to detect genome copy number variation. *Genome Res.* **13:** 2291–2305.

Olshen A.B., Venkatraman E.S., Lucito R., and Wigler M. 2004. Circular binary segmentation for the analysis of array-based DNA copy number data. *Biostatistics* **5:** 557–572.

Pinkel D., Segraves R., Sudar D., Clark S., Poole I., Kowbel D., Collins C., Kuo W.L., Chen C., Zhai Y., et al. 1998. High resolution analysis of DNA copy number variation using comparative genomic hybridization to microarrays. *Nat. Genet.* **20:** 207–211.

Pollack J.R., Perou C.M., Alizadeh A.A., Eisen M.B., Pergamenschikov A., Williams C.F., Jeffrey S.S., Botstein D., and Brown P.O. 1999. Genome-wide analysis of DNA copy-number changes using cDNA microarrays. *Nat. Genet.* **23:** 41–46.

Sebat J., Lakshmi B., Troge J., Alexander J., Young J., Lundin P., Maner S., Massa H., Walker M., Chi M., et al. 2004. Large-scale copy number polymorphism in the human genome. *Science* **305:** 525–528.

Selzer R.R., Richmond T.A., Pofahl N.J., Green R.D., Eis P.S., Nair P., Brothman A.R., and Stallings R.L. 2005. Analysis of chromosome breakpoints in neuroblastoma at sub-kilobase resolution using fine-tiling oligonucleotide array CGH. *Genes Chromosomes Cancer* **44:** 305–319.

Snijders A.M., Pinkel D., and Albertson D.G. 2003. Current status and future prospects of array-based comparative genomic hybridisation. *Brief Funct. Genomics Proteomics* **2:** 37–45.

Stimson C.W. 1967. Possible causes of mongolism (Down's syndrome) and other chromosomal aneuploidies. *Mich. Med.* **66:** 436–441.

Thompson C.T. and Gray J.W. 1993. Cytogenetic profiling using fluorescence in situ hybridization (FISH) and comparative genomic hybridization (CGH). *J. Cell. Biochem. Suppl.* **17G:** 139–143.

Yang Y.H., Dudoit S., Luu P., Lin D.M., Peng V., Ngai J., and Speed T.P. 2002. Normalization for cDNA microarray data: A robust composite method addressing single and multiple slide systematic variation. *Nucleic Acids Res.* **30:** e15.

16 | Whole-Genome Sampling Analysis to Detect Copy Number Changes in FFPE Samples

Sharoni Jacobs

Affymetrix, Santa Clara, California 95051

INTRODUCTION

WGSA and Copy Number Detection

Human populations and individuals can differ widely in both nucleotide sequence and copy number (CN) of DNA segments. Sequence differences commonly occur in the form of single-nucleotide polymorphisms (SNPs), of which millions have been identified (International HapMap Consortium 2005). Recent studies have also begun to highlight CN polymorphism frequencies (Iafrate et al. 2004; Sebat et al. 2004; Redon et al. 2006), suggesting that a comprehensive study of genetic differences between samples should combine detection of both sequence and CN variation.

The Affymetrix GeneChip® Mapping Arrays comprise a class of SNP oligonucleotide microarrays that can be used to determine both genotype and CN at specific SNP locations. The original Mapping array, known as the Mapping 10K Array and, later, the Mapping 10K 2.0 Array, interrogates over 10K SNPs on a single array. The next generation of Mapping arrays, the Mapping 100K Set, includes two arrays (the Mapping 50K Xba and the Mapping 50K Hind Arrays) that together analyze 116,204 SNPs. The third generation of these arrays, the Mapping 500K Set, which includes the Mapping 250K Nsp and the Mapping 250K Sty Arrays, examines 500,568 SNPs with a median and mean inter-SNP distance of 2.5 kb and 5.8 kb, respectively.

Genomic DNA applied to these arrays is processed using the Whole Genome Sampling Analysis (WGSA) assay (Fig. 16-1) (Kennedy et al. 2003). Importantly, this assay involves a complexity reduction step, so that instead of applying the entire genome to these arrays, a targeted subset of the genome is selectively labeled and hybridized. Reduction of DNA target assembly is accomplished as follows: During WGSA, 250 ng of genomic DNA is digested by a restriction enzyme (NspI or StyI when using the Mapping 500K Arrays), and adapters are ligated to the ends of the digested DNA. A total of 500 ng of DNA is required for application when both arrays in the

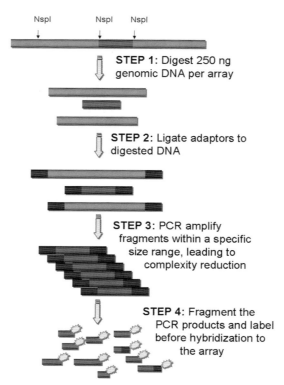

Nspl   Nspl   Nspl

STEP 1: Digest 250 ng
genomic DNA per array

STEP 2: Ligate adaptors to
digested DNA

STEP 3: PCR amplify
fragments within a specific
size range, leading to
complexity reduction

STEP 4: Fragment the
PCR products and label
before hybridization to
the array

FIGURE 16-1. The Whole Genome Sampling Analysis (WGSA) assay prepares DNA for the Mapping 500K Arrays.

Mapping 500K Set are used. The digested DNA is amplified by PCR using a universal primer set that recognizes the adapters. The PCR conditions and reagents are optimized to maximize amplification of products within a specific size range. This size range is approximately 100 bp to 1100 bp for the Mapping 10K and 500K Arrays. For the Mapping 100K Arrays, the amplicons range from about 250 bp to 2 kb.

PCR products are then fragmented with DNase I, labeled with biotin, and hybridized to the array. Hybridization efficiencies to probes on the array that recognize alternate alleles of each SNP are detected by staining the biotin-labeled DNA fragments with fluorescent-labeled streptavidin.

Differential intensities for the probes that recognize the major and minor allele are used to determine SNP genotype (Matsuzaki et al. 2004; Di et al. 2005). These probe intensities can also be used to determine genomic CN at the location of each SNP (Bignell et al. 2004; Huang 2004; Lin et al. 2004; Nannya et al. 2005). To determine CN, a test sample is compared to a reference consisting of one or many samples. The reference is assumed to represent a diploid state with CN = 2, and the ratio of probe intensities between the test and the reference is calculated in order to identify the number of copies of a DNA fragment in the test sample.

FFPE Samples for CN Detection

The WGSA assay in combination with the Mapping arrays is used widely for CN analysis of high-quality DNA samples; i.e., samples that have been collected from blood, fresh or frozen tissue, or cell lines. Formalin-fixed, paraffin-embedded (FFPE) samples, however, represent the most prevalent form of archived clinical samples (Fig. 16-2), but they provide additional challenges for molecular assays. FFPE processing is advantageous primarily to the pathologist, who desires a rapid means to fix, stain, and view the samples for a speedy and efficient diagnosis. Unfortunately,

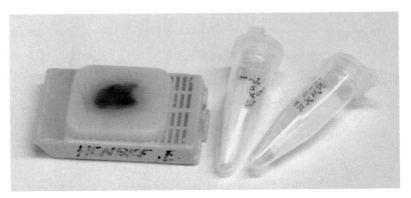

FIGURE 16-2. Example of a paraffin block containing a formalin-fixed kidney sample, and 10-μm sections from the FFPE sample contained in the eppendorf tubes.

the method usually results in the degradation of FFPE DNA and in the contamination and chemical modification of these DNA samples.

Because of these issues, FFPE DNA is not suitable for all molecular assays designed for high-quality DNA samples. A protocol recommended for processing FFPE DNA samples through WGSA and to the Mapping arrays is described below. Note that poorer quality of FFPE samples may still result in a decrease in the number of SNPs providing informative data, and therefore overall genome coverage. Guidelines on data analysis of FFPE samples are outlined in Chapter 22. The important modifications to this WGSA protocol specific to FFPE-derived samples include:

1. inclusion of a pre-WGSA PCR-based quality assessment of FFPE DNA samples

2. pooling of >3 amplification reactions to attain 90 μg of PCR product

3. quality control (QC) steps to monitor performance (expect different performance from FFPE and non-FFPE DNA samples)

DNA PREPARATION

Several high-quality commercial kits are currently on the market for extracting DNA from FFPE samples. The quality of DNA that is extracted from a given sample will vary depending on the extraction protocol, so that the same FFPE block might yield relatively poor or relatively intact DNA using alternate kits. A recommendation for extraction is the use of the Qiagen DNeasy Blood & Tissue kit (www.qiagen.com) with modifications (Wu et al. 2002). Suggested modifications to the protocol that is outlined in the Qiagen DNeasy Blood & Tissue Handbook include:

1. 95°C, 15-minute treatment prior to Proteinase K digestion, in ATL buffer

2. 3 days of Proteinase K treatment at 56°C, with daily additions of the same amount of PK used on the first day

3. NH_4OAc + EtOH cleanup of extracted DNA, as described in the "Genomic DNA Preparation" section of the GeneChip® Mapping 500K Assay Manual (www.affymetrix.com)

Another method that has demonstrated success in providing FFPE DNA of sufficient quality for the Mapping arrays is Argylla Technologies PrepParticles (beta version) (http://argylla.com/).

There are many extraction methods available that may also be used successfully. Regardless of the method, the resulting DNA sample should be assessed for quality, since not all samples will be suitable for the Mapping arrays. To pass the quality test, the DNA sample must be void of extensive contamination and degradation. Refer to the next section for more details on QC testing of FFPE DNA samples.

FFPE DNA QUALITY ASSESSMENT

FFPE DNA sample quality will differ due to a number of variables; for example, differences in the fixation protocol and the extraction protocol, and the years of storage. Some FFPE samples will fail to produce nucleic acids of sufficient quality and quantity for downstream analysis. Therefore, each FFPE DNA sample should be quality tested, and samples that fail this QC test should not be applied to the Mapping arrays. The proportion of samples that fail this QC test will vary based on the source of FFPE samples, extraction methodology, and other factors.

In WGSA, PCR is the limiting step that determines whether a given FFPE DNA sample will provide informative data on the Mapping array. Therefore, the FFPE DNA quality assessment test should include PCR. The three PCR-based QC tests in the table below are able to determine the suitability of an FFPE sample for these arrays. Any one of these three tests may be used for FFPE DNA quality assessment.

The purpose of these QC tests is to identify the largest producible amplicon size for a given FFPE sample. Importantly, these QC examples all include the amplification of a variety of fragment sizes during PCR, including larger fragment sizes up to 800 bp. Because the Mapping 500K Arrays interrogate SNPs on fragments up to 1100 bp, an FFPE sample that qualifies for array analysis should be able to provide large amplicon sizes. Although it is not necessary to amplify all fragments on the array (in silico, the SNPs on the largest fragment sizes can be excluded during analysis), it is important to amplify a significant percentage of these SNPs. Table 16-1 gives an indication of the number of SNPs present on the various fragment sizes. We do not recommend proceeding with samples that fail to produce PCR fragments greater than 300 bp.

Method	Description	Advantages and disadvantages	References
Multiplex PCR	• Simultaneous amplification of several fragments ranging in size from ~100 to 800 bp. • Each lab may design primers and reactions to amplify any region of the genome. • It is essential that the reactions are designed to amplify not only small but also large (600, 700 bp) fragments.	(+) Requires a small amount of DNA. (–) Targeted PCR amplification of specific sites may identify locus-specific quality instead of genome-wide DNA quality.	van Beers et al. (2006) This article describes multiplex PCR of fragments up to 400 bp only.
RAPD PCR	• Uses a single primer set (10-mers) to amplify a ladder of fragments. • Intact DNA will produce amplicons reaching ~2 kb. • The amplicon ladder produced from FFPE samples may reach a smaller maximum size. • This is not a kit.	(+) Requires a small amount of DNA. (+) Provides non-targeted amplification across the genome. (–) Bacterial DNA in the Taq Polymerase may be amplified to provide high background when using an enzyme preparation that is not highly purified. (–) Some variability can be seen between replicates and dilutions.	Siwoski et al. (2002), with the suggested modifications: • Visualize bands with EtBr on an agarose gel • Use a highly purified Taq polymerase, such as Qiagen HotStar HiFidelity DNA Polymerase
First steps of WGSA	• 250 ng of DNA are digested, ligated to adapters, and PCR-amplified using a universal primer set. • See Figure 16-3A for an example.	(+) Provides best correlation to Mapping array performance because it is part of the Mapping assay. (+) Products that pass QC inspection may directly continue through the WGSA assay and hybridize to the array. (–) Requires the most DNA, time, and money per sample.	GeneChip® Mapping 500K Assay Manual

TABLE 16-1. The Number of SNPs Retained on the Mapping 500K Arrays When Applying Various Fragment Size Filters

Fragment size filter	Mapping 250K Sty Array	Mapping 250K Nsp Array	Combined 500K Set	% of SNPs on 500K Set
≤ 200 bp	1,260	1,813	3,073	0.6
≤ 300 bp	15,831	13,650	29,481	5.9
≤ 400 bp	45,459	39,506	84,965	17.0
≤ 500 bp	82,085	74,387	156,472	31.3
≤ 600 bp	120,011	113,702	233,713	46.7
≤ 700 bp	155,576	153,213	308,789	61.7
≤ 800 bp	187,689	190,900	378,589	75.6
≤ 900 bp	213,302	222,317	435,619	87.0
≤ 1 kb	230,509	244,665	475,174	94.9
No filter	238,300	262,258	500,568	100.0

APPLYING FFPE DNA TO THE WGSA ASSAY

The Mapping 500K Set, or the individual Mapping 500K Arrays, should be used in preference to the other Mapping arrays when processing FFPE DNA. The Mapping 100K Arrays are not appropriate for degraded samples due to their requirement for larger amplicons during PCR. The Mapping 10K Array can be used for FFPE DNA analysis, as it does share the same distribution of PCR amplicon sizes as the Mapping 500K Arrays, but this array will provide significantly decreased resolution.

The WGSA assay is described in detail in the GeneChip® Mapping 500K Assay Manual provided by Affymetrix (www.affymetrix.com). Here, we highlight the specific steps that require special attention or that have been modified for application of FFPE DNA samples.

DNA Quantitation

For accurate quantitation of double-stranded FFPE DNA, use a method that specifically detects dsDNA, such as PicoGreen® (www.invitrogen.com). Other methods of quantitation, such as UV spectroscopy, may provide inaccurate measurements of DNA due to the higher amounts of contaminants in FFPE samples.

Digestion and Ligation

No changes are required during the digestion and ligation steps of WGSA. Begin with 250 ng of DNA, and digest with NspI for samples destined for the Mapping 250K Nsp Array, and StyI for samples directed toward the Mapping 250K Sty Array. The ligation reaction will result in 100 μl of product.

PCR

The established protocol requires three amplification reactions per DNA sample, using 10 μl of ligation product per PCR reaction. Therefore, a total of 30 μl is taken from the 100-μl ligation product. After PCR, the three reactions are pooled together, and 90 μg of PCR product is carried over to the next step (Fragmentation). Because degraded FFPE DNA may not produce larger amplicons during PCR (Fig. 16-3A), the yield of PCR products is often lower for FFPE samples, and it is necessary to pool additional PCRs to reach the suggested 90 μg. In anticipation of the need to pool additional PCRs, set up six amplification reactions per sample (using 60 μl from the 100-μl ligation product). For severely degraded samples, nine reactions may be required. Importantly, this step does not affect the input amount of DNA, which remains at 250 ng per array.

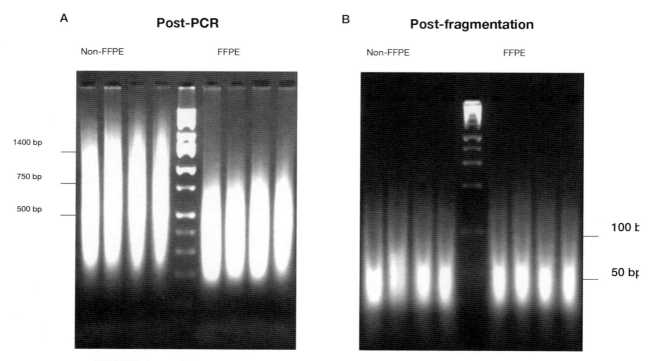

FIGURE 16-3. QC inspections of sample performance during WGSA. (A) PCR products are visualized on an agarose gel for both non-FFPE and FFPE samples. The larger fragment sizes failed to be amplified from the FFPE samples. Nonetheless, there is ample PCR product for the smaller fragments, indicating that these samples are suitable for proceeding with WGSA and hybridizing to the arrays. (B) Fragmentation products are visualized on a gel for both non-FFPE and FFPE samples. Products look essentially equivalent and are suitable for labeling and hybridization to the arrays.

Combining only three PCRs from an FFPE sample is acceptable even when this provides less than 90 μg of product, but pooling additional samples to increase the amount of DNA applied to the array generally increases the quality of the resulting data.

Purifying and Pooling Amplification Reactions

Clontech® 96-well plates are used to pool and purify PCRs for the Mapping 500K assay. Up to three reactions can be pooled together in a single well. When more than three PCRs are performed for a single DNA sample, multiple wells should be used for purification. Thus, if there are six reactions from one DNA source, there will be two wells used for PCR purification for this sample. The DNA from these wells can be combined during the DNA elution step as follows: After washing both wells three times with water and allowing the DNA to completely dry on the plate (as per instructions in the GeneChip® Mapping 500K Assay Manual), elute the first well with 45 μl of RB buffer while leaving the second well dry. Next, use the same RB buffer that now contains the DNA eluted in the first well to elute the DNA in the second well. In this way, all of the PCR products from the six reactions will be eluted into the same 45 μl of RB. If nine amplification reactions were performed for a single DNA source, use three wells on the Clontech® plate for this sample, and so on.

Fragmentation and Labeling

Fragment 90 μg of PCR product when possible. Expect decreased performance on the array if a smaller amount of DNA is used as input to fragmentation.

Data Analysis

Refer to Chapter 22 for details on how to analyze Mapping array data resulting from application of FFPE samples.

PERFORMANCE METRICS

Performance of the DNA sample can be monitored during and after the WGSA assay (see table below). It is useful to include a positive control in a batch of FFPE samples, such as the standard high-quality DNA sample provided with the Affymetrix kit, Ref103.

SUMMARY

FFPE samples represent the largest resource for archived clinical samples, but they provide special challenges for molecular assays due to DNA degradation, chemical modification, and sample contamination. Special consideration should be given when choosing an extraction method to obtain the best-quality DNA possible from a given FFPE block. All FFPE samples should be assessed for quality using a PCR-based test. Those that provide amplicons of adequate size can be applied to genome-wide analysis of genotype and CN using the Mapping 500K Arrays. Genome coverage for an FFPE sample compared to a high-quality DNA sample may be reduced in cases where PCR fails to amplify SNPs on the larger fragment sizes.

QC Inspection	Details
Image the PCR products on an agarose gel (Fig. 16-3A).	**Non-FFPE samples:** The PCR products should provide a smear from approximately 100 bp to >1 kb. **FFPE samples:** The larger PCR products will likely be missing, but there should be robust amplification of smaller PCR products. Do not proceed with samples that do not provide strong bands greater than 300 bp.
Measure the PCR yield using UV spectroscopy.	**Non-FFPE samples:** These should provide >90 μg yield from a pool of three PCR reactions. **FFPE samples:** These may have significantly reduced PCR yields. Ideally, samples should provide 90 μg when six or nine PCRs are pooled. If 90 μg is not attained with nine PCRs, expect minimal performance from this sample. Generally, higher yields of PCR products indicate larger numbers of SNPs that can be included during data analysis.
Image the fragmentation products on a gel (Fig. 16-3B).	Fragmentation products should be approximately 50 to 100 bp, for both FFPE and non-FFPE samples.
Check call rate.	**Non-FFPE samples:** The best data are found when the call rate (the percentage of SNPs assigned genotypes) is ≥ 93%[*] when using the DM algorithm and ≥ 96%[**] when using BRLMM (both algorithms are applied in the genotyping software GTYPE 4.1). **FFPE samples:** These often have reduced call rates. Nonetheless, call rates are usually above 70% for successful application. Values lower than this suggest poor data quality.

[*]With a *p*-value cutoff of 0.33.
[**]With a *p*-value cutoff of 0.33 and a BRLMM score threshold of 0.5.

There are a few modifications recommended for application of FFPE samples to the Mapping array. During WGSA, increasing the number of amplification reactions to obtain 90 µg of PCR product before fragmentation generally improves data quality. Modifications during data analysis are also necessary, and these are discussed in Chapter 22 of this manual. Whole-genome amplification of precious DNA samples can be paired with the WGSA assay when using high-quality DNA samples (Wong et al. 2004; Zhou et al. 2005); but, at this time, there are no recommendations available for performing whole-genome amplification on degraded samples before application to the Mapping assay.

REFERENCES

Bignell G.R., Huang J., Greshock J., Watt S., Butler A., West S., Grigorova M., Jones K.W., Wei W., Stratton M.R., et al. 2004. High-resolution analysis of DNA copy number using oligonucleotide microarrays. *Genome Res.* **14**: 287–295.

Di X., Matsuzaki H., Webster T.A., Hubbell E., Liu G., Dong S., Bartell D., Huang J., Chiles R., Yang G., et al. 2005. Dynamic model based algorithms for screening and genotyping over 100 K SNPs on oligonucleotide microarrays. *Bioinformatics* **21**: 1958–1963.

Huang J. 2004. Whole genome DNA copy number changes identified by high density oligonucleotide arrays. *Hum. Genomics* **1**: 287–299.

Iafrate A.J., Feuk L., Rivera M.N., Listewnik M.L., Donahoe P.K., Qi Y., Scherer S.W., and Lee C. 2004. Detection of large-scale variation in the human genome. *Nat. Genet.* **36**: 949–951.

International HapMap Consortium. 2005. A haplotype map of the human genome. *Nature* **437**: 1299–1320.

Kennedy G.C., Matsuzaki H., Dong S., Liu W.M., Huang J., Liu G., Su X., Cao M., Chen W., Zhang J., et al. 2003. Large-scale genotyping of complex DNA. *Nat. Biotechnol.* **21**: 1233–1237.

Lin M., Wei L.J., Sellers W.R., Lieberfarb M., Wong W.H., and Li C. 2004. dChipSNP: Significance curve and clustering of SNP-array-based loss-of-heterozygosity data. *Bioinformatics* **20**: 1233–1240.

Matsuzaki H., Dong S., Loi H., Di X., Liu G., Hubbell, E., Law J., Berntsen T., Chadha M., Hui H., et al. 2004. Genotyping over 100,000 SNPs on a pair of oligonucleotide arrays. *Nat. Methods* **1**: 109–111.

Nannya Y., Sanada M., Nakazaki K., Hosoya N., Wang L., Hangaishi A., Kurokawa M., Chiba S., Bailey D.K., Kennedy G.C., and Ogawa S. 2005. A robust algorithm for copy number

detection using high-density oligonucleotide single nucleotide polymorphism genotyping arrays. *Cancer Res.* **65**: 6071–6079.

Redon R., Ishikawa S., Fitch K.R., Feuk L., Perry G.H., Andrews T.D., Fiegler H., Shapero M.H., Carson A.R., Chen W., et al. 2006. Global variation in copy number in the human genome. *Nature* **444**: 444–454.

Sebat J., Lakshmi B., Troge J., Alexander J., Young J., Lundin P., Maner S., Massa H., Walker M., Chi M., et al. 2004. Large-scale copy number polymorphism in the human genome. *Science* **305**: 525–528.

Siwoski A., Ishkanian A., Garnis C., Zhang L., Rosin M., and Lam W.L. 2002. An efficient method for the assessment of DNA quality of archival microdissected specimens. *Mod. Pathol.* **15**: 889–892.

van Beers E.H., Joosse S.A., Ligtenberg M.J., Fles R., Hogervorst F.B., Verhoef S., and Nederlof P.M. 2006. A multiplex PCR predictor for aCGH success of FFPE samples. *Br. J. Cancer* **94**: 333–337.

Wong K.K., Tsang Y.T., Shen J., Cheng R.S., Chang Y.M., Man T.K., and Lau C.C. 2004. Allelic imbalance analysis by high-density single-nucleotide polymorphic allele (SNP) array with whole genome amplified DNA. *Nucleic Acids Res.* **32**: e69.

Wu L., Patten N., Yamashiro C.T., and Chui B. 2002. Extraction and amplification of DNA from formalin-fixed, paraffin-embedded tissues. *Appl. Immunohistochem. Mol. Morphol.* **10**: 269–274.

Zhou X., Temam S., Chen Z., Ye H., Mao L., and Wong D.T. 2005. Allelic imbalance analysis of oral tongue squamous cell carcinoma by high-density single nucleotide polymorphism arrays using whole-genome amplified DNA. *Hum. Genet.* **118**: 504–507.

17 Molecular Inversion Probe Targeted Genotyping: Application to Copy Number Determination

George Karlin-Neumann, Marina Sedova, Ronald Sapolsky, Steven Lin, Yuker Wang, Martin Moorhead, and Malek Faham

Affymetrix, South San Francisco, California 94080

INTRODUCTION

The ability to discover and accurately measure chromosome copy number (CN) amplifications, deletions, and loss of heterozygosity (LOH) in cancer samples has great clinical potential for diagnosis, prognosis, and treatment decisions (Slamon et al. 2001). A primary challenge in copy number determination is the ability to make quantitative as well as qualitative genotyping measurements. Cancer samples present the additional challenges of sample purity (mixtures of cancer and normal cells in the assayed tissue); low DNA sample amount; and in the case of archived formalin-fixed, paraffin-embedded (FFPE) samples, physically and chemically degraded DNA which does not amplify well and is difficult to assay with existing methods (van Beers et al. 2006). Several array-based techniques have been developed to assess DNA copy number. The ideal copy number assay would provide in a highly multiplexed fashion genotype and precise copy number estimation not only on abundant, high-quality DNA samples, but also on low amounts of degraded DNA. Chapter 12 describes the use and many advantages of molecular inversion probes (MIPs) in targeted SNP genotyping. The advantage of utilizing MIPs in the study of copy number stems from its exquisite specificity due to the requirement of the interaction between two homology sequences on the same molecule as described in Chapter 12. In addition, the small target sequence requirement (~40 bp of homology with the probe's terminal detection sequences) makes it suitable for the study of degraded samples. These features make the MIP-genotyping assay well suited to the challenge of copy number determination even in low amounts of unamplified, degraded sample DNA. The quantitative capabilities of the MIP genotyping assay are illustrated below in a modified version of the assay, not yet commercially available. This version requires <75 ng of unamplified genomic DNA and was tested with a 50,000-plex MIP genome-wide human panel on DNA derived both from normal and cancer cell lines and from FFPE samples, and it performed very well on each. To determine the copy number for each allele of a given marker in an unknown sample, each allele's signal intensity is first calibrated against copy number in a set of reference samples.

RESULTS OF CN ESTIMATES FROM MIP GENOTYPING ASSAY

To assess the accuracy and precision of MIP CN estimates, the assay was run on cell lines having various numbers of copies of the X chromosome, from 1X males to 3X, 4X, and 5X lines (cell lines were obtained from the Coriell Institute, Camden, New Jersey). The 50K-plex panel contains >900 scorable markers on the X chromosome. Plots of marker CN versus chromosome are shown in Figure 17-1 for 1X (male), 3X, 4X, and 5X cell lines. The markers on each chromosome are distinctly colored from their neighbors, and autosomes are ordered from left to right starting with Chromosome 1 on the left (red) through Chromosome 22 on the right (pink), with the X-chromosome markers on the extreme right (peach). Each marker CN value is unsmoothed but, nonetheless, shows high precision; smoothing of only two markers increases this precision still further. The CN estimates for X-chromosome markers in 1X male lines was 1.06 with a relative standard deviation (rsd) of 0.12; for the 3X cell line, CN = 3.09 with rsd = 0.10; for the 4X cell line, CN = 3.98 with rsd = 0.10; and for the 5X cell line, CN = 4.96 with rsd = 0.10. Copy number estimates are both accurate and precise within this range. It is noteworthy that the assay has been able to identify some putative regions of CN aberration in these cell lines, e.g., the apparent amplification seen in Chromosome 14 (purple) in the 3X line. Additional data with spikes of known concentrations show that the assay is linear to >50 copies (data not shown).

The ability of the assay to identify deleted regions and amplified regions within this range is further illustrated in Figure 17-2. The figure depicts the chromosome CN plot of the cancer cell line UACC812 (cell line obtained from the American Type Culture Collection [ATCC]). The chromosomes are ordered from left to right, with Chromosome 1 on the far left and X chromosome on the far right. Striking increase in copy number is evident for several markers. The levels of several of these amplifications have been confirmed also by TaqMan measurements (Wang et al. 2005).

In a third example of this application, the ability of this modified MIP assay to perform well on DNA derived from both normal and tumor FFPE samples has been investigated. In this case, DNA was purified from several 10-μm slices of all sample blocks using a standard QIAmp DNA

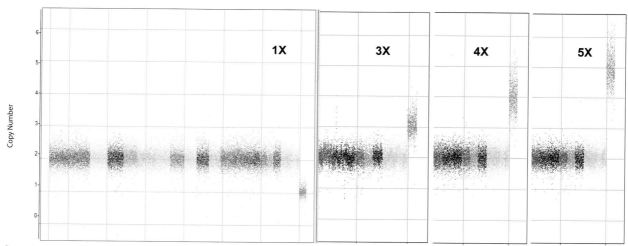

FIGURE 17-1. Accuracy and precision of CN measurements in X-chromosome titration cell lines. Four plots are shown (*left to right*): for 1X (male), and for 3X, 4X, and 5X cell lines. In each plot, the autosomes are ordered from left to right, with Chromosome 1 on the left (*red*) through Chromosome 22 on the right (*pink*), and with the X chromosome markers (*peach*) on the extreme right. All points represent the CN sum of both alleles and are derived from unsmoothed CN measurements. Details of the results are discussed in the text.

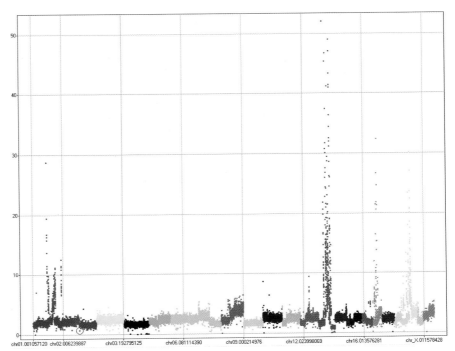

FIGURE 17-2. UACC812 cancer cell line amplifications and deletions. In this plot, also, the chromosomes in this cancer line are ordered from left to right, with a series of colored markers representing Chromosomes 1 through 22. The colored tracings depict amplification or deletion of the corresponding markers. All points represent the CN sum of both alleles and are derived from unsmoothed CN measurements.

purification procedure (Qiagen) and assayed under the same conditions as those described above for the cell line DNAs. A CN-versus-chromosome plot from a representative normal, male FFPE sample is shown in Figure 17-3. As shown in the cell line plots in Figure 17-1, each point represents a CN measurement for a single, unsmoothed marker. In Figure 17-3, the plot shows both accurate quantitation of the known 1-copy markers on the X chromosome (peach-colored, extreme right side of plot) and very good discrimination between these and the 2-copy autosomal markers (remaining series of colored markers representing Chromosomes 1 through 22). Thus, the MIP genotyping assay can provide high-quality CN measurements on DNA from both normal and FFPE samples.

SUMMARY AND CONCLUSIONS

Beyond affording highly accurate genotypes, the fundamental properties of the MIP assay system enable accurate and precise copy number measurements of highly multiplexed marker sets. Due to the assay's high specificity and large dynamic range, accurate measurement of deletions and amplifications over nearly 2 orders of magnitude can be made; the absence of cross-talk among alleles permits accurate allele-specific copy number determination without smoothing of adjacent markers. The small footprint of MIP probes on their genomic target (~40 bp) makes them well suited for copy number measurements in degraded DNA from FFPE samples. This has been demonstrated, here, at the 50,000-plex level and thus allows this technology to be applied to genome-wide assessments of deletions, amplifications, and LOH.

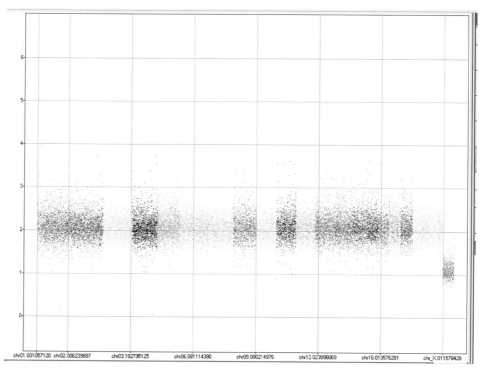

FIGURE 17-3. CN plot for normal male FFPE tissue (1X vs. 2X). DNA was purified from several 10-μm slices of sample FFPE tissue blocks using a standard QIAmp DNA purification procedure (Qiagen). The organization of chromosomes is as described in Fig. 17-1 (2X autosomes from left to right and 1X chromosome on far right), and the CN genotyping assays were performed under the same conditions as described in Fig. 17-1 for the cell line DNAs. All points represent the CN sum of both alleles and are derived from unsmoothed CN measurements. Details of the results are discussed in the text.

ACKNOWLEDGMENTS

We thank the many people from ParAllele and Affymetrix who have labored devotedly in developing the MIP technology, with special acknowledgment to the algorithm and software developers who have been instrumental in developing the copy number analysis.

REFERENCES

Slamon D.J., Leyland-Jones B., Shak S., Fuchs H., Paton V., Bajamonde A., Fleming T., Eiermann W., Wolter J., Pegram M., et al. 2001. Use of chemotherapy plus a monoclonal antibody against HER2 for metastatic breast cancer that overexpresses HER2. *N. Engl. J. Med.* **344:** 783–792.

van Beers E.H., Joosse S.A., Ligtenberg M.J., Fles R., Hogervorst F.B., Verhoef S., and Nederlof P.M. 2006. A multiplex PCR predictor for aCGH success of FFPE samples. *Br. J. Cancer* **94:** 333–337.

Wang Y., Moorhead M., Karlin-Neumann G., Falkowski M., Chen C., Siddiqui F., Davis R.W., Willis T.D., and Faham M. 2005. Allele quantification using molecular inversion probes (MIP). *Nucleic Acid Res.* **33:** e183.

18 Microsatellite Markers for Linkage and Association Studies

Jeffrey Gulcher

deCODE Genetics, Reykjavik, Iceland, and Woodridge, Illinois

INTRODUCTION

During the 1990s and the first several years of this century, microsatellites or short tandem repeats (STRs) were the workhorse genetic markers for hypothesis-independent studies in human genetics, facilitating genome-wide linkage studies and allelic imbalance studies. However, with the rise of higher-throughput and cost-effective single-nucleotide polymorphism (SNP) platforms, we are currently

in the era of the SNP for genome scans. Nevertheless, it is important to note that microsatellites remain highly informative and useful measures of genomic variation for linkage and association studies. Their continued advantage in complementing SNPs lies in their greater allelic diversity than biallelic SNPs as well as in their population history, in which single-step expansion or contraction of the tandem repeat on the background of ancestral SNP haplotypes can break up common haplotypes, leading to greater haplotype diversity within the linkage disequilibrium (LD) block of interest. In fact, microsatellites have recently starred in association studies leading to widely replicated discoveries of type 2 diabetes (TCF7L2) and prostate cancer genes (the 8q21 region) (Amundadottir et al. 2006; Grant et al. 2006). At the end of the day, it will be important to catalog all variation, including SNPs, microsatellites, copy number variations, and polymorphic inversions in human genetic studies. This chapter describes the utilities of microsatellites and experimental approaches in their use.

Use of Microsatellites in Linkage Studies

Microsatellites have long been the variant of choice in linkage studies for several reasons. First, they have great diversity of alleles, averaging five to ten alleles. This leads to a heterozygosity rate of commonly used microsatellites from 50% to 80%; in contrast, the best possible heterozygosity rate of biallelic markers such as SNPs is 50%, but in practice the heterozygosities of SNPs average 30% or lower. This increase in heterozygosity leads to much higher information content when using microsatellites than when using the same number of SNPs. Advocates of SNPs for linkage have suggested that SNPs can be much more informative if one is doing multipoint linkage and using much larger numbers of SNPs (five- to tenfold more SNPs than microsatellites). That is certainly true if

253

the SNPs can be reliably typed and quality controlled and are not at too high a density to result in LD between SNP markers. Failure to ensure that linkage markers are not in LD results in very impressive artifactual linkage peaks, since linkage programs assume that all markers are in linkage equilibrium.

Second, microsatellites were the first variant that could be measured systematically from small amounts of DNA using PCR tied to high-throughput radioactive fragment separation by poly-acrylamide electrophoresis, replacing the unwieldy Southern-blot-dependent restriction fragment length polymorphisms (RFLPs) by the early 1990s. Further scale-up of throughput by an order of magnitude came by replacing radioactive labeling with fluorescent labels of PCR fragments and automatic real-time reading of fragments separated by Applied Biosystem sequencers (Davies et al. 1994). The fluorescent labeling allowed for multiplexing markers labeled with different dyes and of different lengths in the same lane, as well as an internal molecular-weight standard within each lane on the slab gel, greatly increasing the accuracy of length assessment. Robotic mixing of DNA with PCR reagents and allele-calling software systems became necessary to keep up with this scale-up in throughput (Palsson et al. 1991). By the end of the 1990s, this high throughput was greatly expanded with the release of the capillary electrophoresis methods, which allowed auto-matic loading and prevented cross-contamination or cross-bleaching of lanes that sometimes plagued gel electrophoresis. One ABI 3730 can currently generate at least 30,000 genotypes per day. This sounds very modest compared to the current SNP chips. However, it is important to realize that microsatellite genotyping remains very cost-competitive with custom-designed SNP typing when one takes into account the increased information content per marker.

Third, microsatellites are easy to quality control. As with SNPs, microsatellites can occasionally suffer from allelic dropout (i.e., failure to measure one of the two alleles sometimes leading incor-rectly to a homozygote genotype) due to a SNP within the primer site. The rate of this occurrence is similar between microsatellites and SNPs. However, when allelic dropout occurs in a particular SNP, it is much more difficult to detect than when it occurs with a microsatellite, given the large difference between heterozygosities. About three times as many genotyping errors due to allelic dropout will be missed for SNPs as for an equal number of microsatellites. If the number of SNPs required to generate good information content (90%) for a given set of pedigree structures is 5–10 times the number of microsatellites needed to yield the same level of information content, the number of SNP markers plagued by undetected genotyping errors may be 15–30 times that from an equivalent microsatellite scan. Therefore, quality control is even more important for accurate SNP linkage scans than for microsatellite scans.

Fourth, there is a high-resolution genetic map created with microsatellites but not for SNPs. The deCODE Genetics High-Resolution Genetic Map (HRGM) published in 2002 (Kong et al. 2002) was created with 5136 curated microsatellite markers using 1257 meiotic events (average resolution, 0.6 centiMorgan [cM]). The Marshfield map, which was created using more than 8000 microsatellite markers, has much lower resolution (3 cM) because the number of meiotic events within the pedigrees was only 188. Illumina created a map using 4000 SNP markers but used fam-ilies with a similar number of meiotic events as used for the Marshfield map (Murray et al. 2004). Some have tried to interpolate genetic distance by overlaying the SNPs by physical position over deCODE's microsatellite markers. However, because physical distance is almost never propor-tional to genetic distance, significant errors may occur in estimation of genetic distance, which can lead to both false-positive and false-negative linkage signals (Gretarsdottir et al. 2002).

Microsatellite genotyping is not straightforward, and not many labs can efficiently validate and use microsatellite assays. Fortunately, there are several high-throughput laboratories that can com-petently and cost-effectively carry out microsatellite genotyping with very little error (less than 0.3–1%). Marshfield Clinic, led by the key developer of microsatellite genotyping, James Weber, was the pioneer in high-throughput microsatellite genotyping services (Weber and Broman 2001). Weber has recently moved to a private company (www.preventiongenetics.com). CIDR is another source for genotyping services, although it is focused on 10-cM framework marker scans and not custom microsatellites to follow up initial linkage signals (www.cidr.jhmi.edu). The largest microsatellite genotyping laboratory offering contract services is the author's deCODE Genetics, which provides high-density genome scans using 500, 1000, 2000, and 5000 microsatellite markers (www.decode.com/genotyping). It also offers region-specific high-density panels for selectively increasing information content within initial linkage peaks, as well as efficient custom-design microsatellite marker discovery and testing for association studies (see below).

In summary, microsatellites have been the most widely used variant for linkage studies and remain so, even in this era of high-density SNP genotyping. Methods for their use in linkage are beyond the scope of this chapter, but may be found in Gulcher and Stefansson (2006).

Use of Microsatellites in Association Studies

It has become clear that most functional variants in genes contributing to common diseases are not nonsynonymous SNPs found in coding exons, but instead, appear to affect gene expression and RNA splicing. The wealth of data from the HapMap project allows more careful selection of a subset of nonredundant SNPs that tag independent haplotypes (tag-SNPs) for each LD block to act as surrogates for functional variants. To date, Illumina is the only vendor of genome-wide high-density SNP chips that has taken advantage of the HapMap data, using it to define and select tag-SNPs for its 300K, 550K, 650K, and 1M SNP bead chips. This contrasts with the alternative approach of random SNP selection, as practiced by vendors such as Affymetrix.

tag-SNP selection is covered in detail in Chapter 19. The end result of a whole-genome scan using tag-SNPs is that LD blocks are defined, within which SNPs are highly correlated and haplotype diversity is typically low. A potential disadvantage of the limited diversity in the haplotypes that define the haplotype space within the LD block can be overcome with microsatellites. We note that many groups report using a few tag-SNPs within each LD block and find that those SNPs define at least 90% of haplotypes in the LD block. However, it is very important to realize that haplotype coverage does not equal haplotype diversity. Therefore, many studies will be underpowered to find a correlation between the given LD block in the corresponding gene and the disease of interest if the frequency of the marker allele used is much greater than the frequency of the functional variant for which it is a surrogate (Ke et al. 2004).

In practice, one is attempting to use tag-SNPs and haplotypes as surrogate markers for underlying functional variants within the same LD block. The power of that surrogate is proportional to the degree of correlation between the surrogate and the functional variant as measured by R^2 (Ke et al. 2004). Unfortunately, if the frequencies of the surrogate and functional variants are not similar, there is a substantial loss in power for case/control association studies. For example, if one used a common haplotype or tag-SNP with a population allelic frequency of 30% to detect a functional variant with a frequency of 10%, there would be a 10-fold drop in power. Thus, for a case/control association study looking at 300 patients and 300 controls, the power would be equivalent to studying 30 patients and 30 controls, essentially wasting the expensive recruitment and phenotyping effort.

Therefore, it may be prudent to define not only haplotype coverage but also haplotype diversity within each LD block in the study. It is helpful to create a measure of haplotype diversity. If our goal is to have efficient surrogates of functional variants, and if we assume that we want to be powered to find variants as low as 5–10% allelic frequency in the general population, the ideal set of haplotypes would number ten, and each would have an equal allelic frequency of 10%. Therefore, we set an arbitrary measure of haplotype diversity as the sum of the frequencies of haplotypes that are each less than or equal to 10%. For example, if there are three common haplotypes each of 20% frequency and four other haplotypes each of 10%, haplotype diversity would equal 40% and we would say the remaining 60% of haplotype space has inadequate diversity.

Within the average LD block of 60 kb, six to ten common tag-SNPs may be selected to account for much of the information of the 50–60 HapMap SNPs. This can lead to substantial savings in

TABLE 18-1. Previous Estimates of the Microsatellite Mutation Rate per Generation

Mutation rate per generation	Reference
0.000194	Huang et al. (2002).
0.00045	Whittaker et al. (2003).
0.00056	deCODE experience

genotyping costs. Assuming no historical recombinations between the markers, there are expected to be no more than N + 1 independent haplotypes tagged by N SNPs within an LD block. Thus, only six tag-SNPs within a block of strong LD would define, at most, seven haplotypes (one variant per haplotype). However, seven independent haplotypes would have an average frequency of 14%, which would not come close to satisfying the arbitrary ideal criterion of having no haplotype more frequent than 10%. One solution to this problem is to attempt to divide common tag-SNP-defined haplotypes further by adding one or two microsatellites to each LD block of interest.

Microsatellites complement SNPs as useful markers for allelic association to disease, even though they are not as stable as SNPs during population history. However, their average mutation rate is lower than originally reported by Weber (see Table 18-1) (Beckman and Weber 1992). Our estimate, based on tens of thousands of microsatellite markers first qualified as Mendelian by a plate of 30 triads, is about 5×10^{-4} or 1 mutation per 2000 generations. This is consistent with two other studies that looked at a large number of microsatellites (Huang et al. 2002; Whittaker et al. 2003). This mutation rate is not ideal for maintaining strong LD between the surrogate marker and a high-frequency underlying functional disease variant. However, this mutation rate is actually ideal for common functional variants of intermediate frequency (e.g., 2–40%). As an example, microsatellite alleles can tag HLA extended haplotypes of intermediate frequency (Vorechovsky et al. 2001). Because one microsatellite can have 5–20 variants, it may effectively split up haplotype space within a given LD block. Furthermore, microsatellites can break up common ancestral SNP haplotypes to less common microsatellite/SNP haplotypes. This provides an efficient method to greatly increase haplotype diversity in any given LD block. For a real example, Table 18-2A shows 10 independent haplotypes defined by 6 tag-SNPs within a single LD block of about 100,000 bases. About 70% of haplotype space is not adequately covered by haplotypes of

TABLE 18-2. Haplotype Diversity within a Given LD Block

		A. tag-SNPs only										B. Addition of 1 microsatellite							
RS1438	RS7627220	RS7651931	RS4894808	RS2124546	RS6787115	#Markers	#	Freq.		RS1438	RS7627220	D3S3725	RS7651931	RS4894808	RS2124546	RS6787115	#Markers	#	Freq.
3	2	1	2	4	3	6	60	0.321987		3	2	12	1	2	4	3	7	60	0.090376
1	1	3	3	2	1	6	60	0.161569		3	2	14	1	2	4	3	7	60	0.077723
1	1	3	3	4	3	6	60	0.130702		1	1	0	3	3	4	3	7	60	0.068971
3	2	1	3	4	3	6	60	0.108933		3	1	4	1	3	4	3	7	60	0.06465
3	1	1	3	4	3	6	60	0.064655		1	1	6	3	3	4	3	7	60	0.063353
3	2	1	3	2	1	6	60	0.054627		3	2	12	1	2	4	3	7	60	0.062264
3	2	1	2	2	1	6	60	0.035271		3	2	10	1	2	4	3	7	60	0.05929
3	2	1	3	4	1	6	60	0.034676		1	1	6	3	3	2	1	7	60	0.047222
3	1	3	3	2	1	6	60	0.034146		3	2	8	1	2	4	3	7	60	0.044808
1	1	1	3	4	3	6	60	0.009132		1	1	0	3	3	2	1	7	60	0.042155

(A) tag-SNPs only. (B) Addition of 1 microsatellite.

10% frequency or lower. Addition of just one microsatellite in Table 18-2B leads to diversification of the common haplotypes. Now, no SNP/microsatellite haplotype has a frequency greater than 10%. For example, the first SNP haplotype in Table 18-2A has a high frequency and is broken up into at least two haplotypes of lower frequency by the addition of the microsatellite (Table 18-2B). Note that the microsatellite can be too variable, breaking up haplotypes into too many low-frequency haplotypes that may be below the frequency of the functional variant. However, the haplotype association analysis can be done with two runs—SNPs with microsatellites, and SNPs without microsatellites with appropriate correction of *p*-values for twice the number of tests.

In practice, microsatellites have been quite useful for detecting significant associations to common diseases. For example, the widely replicated gene for type 2 diabetes, TCF7L2, was found by positional cloning through a genome-wide linkage scan of Icelandic families (Reynisdottir et al. 2003) with 1200 microsatellites. The follow-up locus-wide case/case association study of 1185 Icelandic patients and 931 controls applied 228 microsatellite markers over the two-LOD (log of the odds) drop interval (10.5 Mb) of the suggestive linkage peak on chromosome 10 (Grant et al. 2006). The average spacing of the microsatellites for the association study was 46 kb, resulting in an average of 1 or 2 microsatellites per LD block. Single-marker association using the Fisher exact test found a single microsatellite marker, DG10S478, within intron 3 of the TCF7L2 gene with significant association to the most common allele with a relative risk (RR) of 0.67 (uncorrected *p*-value of 2.1×10^{-9}). We noted that the other alleles with greater than 2% frequency showed a RR of 1.21 to 1.53 (see Table 18-3i). We merged all of the less common alleles into one common composite allele, called X, which gave a significant RR of 1.5 and a population allelic frequency of 0.276. The microsatellite association was replicated in a Danish cohort and a U.S. cohort (see Table 18-3ii). Note that the allelic frequencies and RR are quite similar across these three Caucasian populations. The overall RR is 1.56 with a corrected *p*-value of 7.8×10^{-15} and RR of 2.4 for homozygotes in a multiplicative model (Grant et al. 2006).

After we replicated this marker, we looked for HapMap SNPs in the same LD block within TCF7L2 that might show equivalent or even better risk. There were five SNPs that showed strong LD with the common microsatellite allele, and all showed significant association to type 2 diabetes, but none demonstrated as strong association or RR as the microsatellite (Grant et al. 2006). The best SNP, rs7903146, has now been replicated in about 20 populations, including Caucasians (e.g., Damcott et al. 2006; Florez et al. 2006; Groves et al. 2006) and Africans (Helgason et al. 2007). Our recent genome-wide association (GWA) scan using genotypes from a 317K Illumina bead chip array run on 1491 Icelandic T2DM patients and 4712 controls found the SNP with the best single-marker association to be the same SNP in TCF7L2, rs7903146, found through the positional cloning described above. The association had an uncorrected *p*-value of 3.7×10^{-13}, assuming a multiplicative model, which far exceeds the threshold of 1.3×10^{-7}.

TABLE 18-3. Microsatellite Association Studies

	Allele	Affected freq.	Control freq.	RR	Two-sided *P*
(i) All alleles of DG10S478					
Iceland (1185 / 931)	0	0.636	0.724	0.67	2.1×10^{-9}
	4	0.005	0.002	2.36	0.12
	8	0.093	0.078	1.21	0.090
	12	0.242	0.178	1.48	4.6×10^{-7}
	16	0.022	0.015	1.53	0.076
	20	0.001	0.003	0.39	0.17
(ii) Allele X of DG10S478					
Iceland (1185 / 931)	x	0.364	0.276	1.50 [1.31, 1.71]	2.1×10^{-9}
Denmark (228 / 539)	x	0.331	0.260	1.41 [1.11, 1.79]	0.0048
USA (361 / 530)	x	0.385	0.253	1.85 [1.51, 2.27]	3.3×10^{-9}
Combined	x	—	—	1.56 [1.41, 1.73]	4.7×10^{-18}

A second example of the utility of microsatellites for association studies is the widely replicated association of markers on chromosome 8q to prostate cancer. deCODE performed linkage studies using its genome-wide set of 1200 microsatellite markers to map prostate cancer to chromosome 8q (Amundadottir et al. 2006). Over a 10-Mb segment, 358 microsatellite assays were designed within the linkage peak at a density of 28 kb and used to genotype a case/control cohort with 869 unrelated Icelandic patients and 596 controls. Association was found to the common allele 0 of DG8S737 with a RR of 1.79 and was significant even when accounting for the number of tests. This microsatellite association was replicated in Swedish and U.S. Caucasian cohorts of prostate cancer. However, the most interesting aspect was that this microsatellite marker showed even stronger association to prostate cancer in African-Americans than in Caucasians (population-attributable risk of 16% versus 8%). This shows that microsatellites can be useful for confirmation studies across populations and even across ethnicities. The same microsatellite marker has been replicated in other populations (Freedman et al. 2006). Our recent GWA study, using the 317K SNP bead chip from Illumina, once again shows the strongest significant association to a SNP in strong LD to this microsatellite. Depending on one's point of view, these diabetes and prostate cancer studies validate one or more of the following four approaches: (1) Linkage is useful to map common disease genes, (2) GWA is useful to find common disease genes, (3) microsatellites are useful for discovery of common disease genes and their replication, and (4) the Illumina 317K bead chip produces data that appear to be unaffected by yield disparity or how alleles are called between cases and controls (which can lead to dramatic false-positive signals that drown out the true positives).

Experimental Approaches to the Use of Microsatellites in Case/Control Studies

This section gives more details on the use of microsatellites in practice for case/control association studies within candidate regions or genes. For candidate gene studies, microsatellite markers are useful complements to tag-SNPs that are selected for each LD block. A typical study selects about five tag-SNP markers in each LD block overlapping the candidate gene, using SNP selection programs such as Tagger (de Bakker et al. 2006). One or two microsatellite markers can also be added to each LD block to diversify the haplotype space.

Microsatellite markers may be found using the tandem repeat finder (TRF) program from Benson (1999). Primers are designed using adequate criteria for PCR assays, and primers that include repeats or known SNPs in the sequence are excluded. We estimate the likelihood of a successful microsatellite (i.e., a readable and polymorphic marker) by associating the characteristics of the microsatellite, including monomer length, microsatellite length, and repeat structure, with experimental success in a set of more than 25,000 microsatellite markers previously designed and tested at deCODE. We use this information to rank the putative microsatellites according to expected quality, and we have identified more than 40,000 putative microsatellites in the genome that we classify as rank 1. In our experience, more than 70% of these turn out to be usable. However, a good portion of the predicted microsatellites in the lower ranks also result in validated assays such that densities of 10–40 kb are usually achievable in candidate regions.

We have found microsatellite markers useful to ultrafine-map linkage peak intervals during the follow-up case/control association study to isolate the disease gene. Although one may choose to select only those linkage peaks that meet criteria for significant genome-wide linkage, we have found that even suggestive linkage peaks are useful for picking regions of the genome for ultrafine mapping. In fact, in the type 2 diabetes and prostate cancer studies described above which resulted in the most widely replicated genes in those diseases to date, the linkage peaks were only modest. However, it is important to point out that we apply affecteds-only allele-sharing linkage analysis without specification of the inheritance model. This greatly decreases the number of tests, since a parametric linkage study may look at more than 100 models that can lead to high rates of false-positive linkage.

We also find microsatellite markers useful to follow up on the strongest association signals in high-density GWA studies. Most of the strongest associations in GWA studies are due to very common alleles with very modest risk, such as 1.2–1.4. It is useful to follow up on such signals in the discovery population using extra markers within the same LD block (as well as other LD blocks in the same gene), including additional tag-SNPs and one or two microsatellites. This may lead to better capture of the original signal but with a higher RR, or it may lead to the discovery of other disease-associated variants in the same LD block or in other parts of the same gene. If one finds even stronger signals than the original tag-SNP on the chip, this may improve the chances of replication if it is a true disease gene.

Data from genome-wide linkage studies using microsatellites are still useful in the current GWA era, especially when one must correct the p-values by 317,000 or 550,000 SNP marker tests. One can prespecify linkage intervals that are used in the primary analysis in the GWA study. For example, if one selects the top five or six suggestive linkage peaks covering only 3% of the genome, then the correction may be for only 10,000 SNP tests within those peaks rather than for all 317,000 tests. The secondary analysis would run the rest of the genome.

For our strongest linkage peaks, the high-density SNP genotypes from the bead chip represent at the least very efficient ways to ultrafine-map the linkage intervals for the follow-up case/control association. However, it is useful to complement such high-density data with microsatellites for increased haplotype diversity, since there are only six to nine tag-SNPs per LD block on average.

Defining haplotype diversity is one method for determining adequacy of coverage of each candidate gene or interval LD block. As described above, one goal is to use enough SNP and microsatellite markers to leave no haplotype in a given LD block with a frequency greater than 10%. Inadequate haplotype diversity in an LD block may lead to the situation in which the disease-associated functional variant is on the background of a very common ancestral haplotype. If the variant is only on a portion of the common haplotype, that haplotype may not show tight enough correlation to the disease, and the gene association will be missed. In this case, the haplotype was a poor surrogate for the underlying functional variant. It is straightforward to integrate microsatellite markers into the LD patterns. The standard definitions of D′ and R^2 are extended to include microsatellite markers by determining the average of the values for all possible allele combinations of the two markers weighted by the marginal allele probabilities. Single point and haplotype associations comparing the patient group with controls are carried out as described in other chapters in this manual. Although there are no accepted criteria to define statistical significance of a haplotype or SNP association, it is clear that p-values should be corrected by the number of tests performed (including number of markers [alleles-1], phenotype runs, and statistical methods used). It is also clear that replication in other independent cohorts of the disease is necessary.

For any significant association results, one must check for population stratification. It is important to confirm the self-reported ethnicity in each group and to ensure that the genetic backgrounds of the patient group match those of the control groups. Microsatellite marker sets are especially informative in distinguishing major ethnic groups. To evaluate genetically estimated ancestry of the study cohorts, we use a panel of 75 unlinked microsatellite markers from about 2000 microsatellites genotyped in a multiethnic cohort of 35 European-Americans from Baltimore; 88 African-Americans from Pittsburgh, Baltimore, and North Carolina; 34 Chinese (Canton); and 29 Indian-Americans (Zapotec) (Helgadottir et al. 2006). Of the 2000 microsatellite markers, the selected set showed the most significant differences between the European-Americans, African-Americans, and Asians, and implicitly Indians, and also had good quality and yield. Another study by Tang et al. (2005) also used 31 of these markers.

Using genotype data from these markers on the ethnic plates (i.e., the Caucasian, West African, and Asian HapMap plates, each of 96 samples), Structure software will estimate the genetic background of test individuals (Pritchard et al. 2000; Falush et al. 2003). Structure infers the allele frequencies of K ancestral populations on the basis of multilocus genotypes from a set of individuals and a user-specified value of K, and assigns a proportion of ancestry from each of the inferred K populations to each individual. One can use the ethnicity panel in two ways. First, one can include in the analysis only those Caucasians who have at least 90% European ancestry in patients and controls. Second, we have modified the NEMO haplotype association algorithm to adjust for differences in genetic background between cases and controls so that one can include all Caucasians according to self-report (Helgadottir et al. 2006).

SUMMARY AND CONCLUSIONS

In summary, microsatellites have demonstrated utility in both linkage and association studies. The fact that they have higher mutation rates than single-base substitutions allows them to break up apparently immutable common SNP haplotypes into lower frequencies that may better match functional variants of intermediate frequency or the rare functional variants they are meant to detect.

ACKNOWLEDGMENTS

The author thanks Struan Grant for creating the haplotype diversity data in Table 18-2 while he was working at deCODE.

REFERENCES

Amundadottir L.T., Sulem P., Gudmundsson J., Helgason A., Baker A., Agnarsson B.A., Sigurdsson A., Benediktsdottir K.R., Cazier J.B., Sainz J., et al. 2006. A common variant associated with prostate cancer in European and African populations. *Nat. Genet.* **38:** 652–658.

Beckman J.S. and Weber J.L. 1992. Survey of human and rat microsatellites. *Genomics.* **12:** 627–631.

Benson G. 1999. Tandem repeats finder: A program to analyze DNA sequences. *Nucleic Acids Res.* **27:** 573–580.

Damcott C.M., Pollin T.I., Reinhart L.J., Ott S.H., Shen H., Silver K.D., Mitchell B.D., and Shuldiner A.R. 2006. Polymorphisms in the transcription factor 7-like 2 (*TCF7L2*) gene are associated with type 2 diabetes in the Amish: Replication and evidence for a role in both insulin secretion and insulin resistance. *Diabetes* **55:** 2654–2659.

Davies J.L., Kawaguchi Y., Bennett S.T., Copeman J.B., Cordell H.J., Pritchard L.E., Reed P.W., Gough S.C., Jenkins S.C., Palmer S.M., et al. 1994. A genome-wide search for human type 1 diabetes susceptibility genes. *Nature* **371:** 130–136.

de Bakker P.I., Burtt N.P., Graham R.R., Guiducci C., Yelensky R., Drake J.A., Bersaglieri T., Penney K.L., Butler J., Young S., et al. 2006. Transferability of tag SNPs in genetic association studies in multiple populations. *Nat. Genet.* **38:** 1298–1303.

Falush D., Stephens M., and Pritchard J.K. 2003. Inference of population structure using multilocus genotype data: Linked loci and correlated allele frequencies. *Genetics* **164:** 1567–1587.

Florez J.C., Jablonski K.A., Bayley N., Pollin T.I., de Bakker P.I., Shuldiner A.R., Knowler W.C., Nathan D.M., and Altshuler D. (Diabetes Prevention Program Research Group). 2006. *TCF7L2* polymorphisms and progression to diabetes in the Diabetes Prevention Program. *N. Engl. J. Med.* **355:** 241–250.

Freedman M.L., Haiman C.A., Patterson N., McDonald G.J., Tandon A., Waliszewska A., Penney K., Steen R.G., Ardlie K., John E.M., et al. 2006. Admixture mapping identifies 8q24 as a prostate cancer risk locus in African-American men. *Proc. Natl. Acad. Sci.* **103:** 14068–14073.

Grant S.F., Thorleifsson G., Reynisdottir I., Benediktsson R., Manolescu A., Sainz J., Helgason A., Stefansson H., Emilsson V., Helgadottir A., et al. 2006. Variant of transcription factor 7-like 2 (*TCF7L2*) gene confers risk of type 2 diabetes. *Nat. Genet.* **38:** 320–323.

Gretarsdottir S., Sveinbjornsdottir S., Jonsson H.H., Jakobsson F., Einarsdottir E., Agnarsson U., Shkolny D., Einarsson G., Gudjonsdottir H.M., Valdimarsson E.M., et al. 2002. Localization of a susceptibility gene for common forms of stroke to 5q12. *Am. J. Hum. Genet.* **70:** 593–603.

Groves C.J., Zeggini E., Minton J., Frayling T.M., Weedon M.N., Rayner N.W., Hitman G.A., Walker M., Wiltshire S., Hattersley A.T., and McCarthy M.I. 2006. Association analysis of 6,736 U.K. subjects provides replication and confirms *TCF7L2* as a type 2 diabetes susceptibility gene with a substantial effect on individual risk. *Diabetes* **55:** 2640–2644.

Gulcher J. and Stefansson K. 2006. Positional cloning: Complex cardiovascular traits. *Methods Mol. Med.* **128:** 137–152.

Helgadottir A., Manolescu A., Helgason A., Thorleifsson G., Thorsteinsdottir U., Gudbjartsson D.F., Gretarsdottir S., Magnusson K.P., Gudmundsson G., Hicks A., et al. 2006. A variant of the gene encoding leukotriene A4 hydrolase confers ethnicity-specific risk of myocardial infarction. *Nat. Genet.* **38:** 68–74.

Helgason A., Palsson S., Thorleifsson G., Grant S.F., Emilsson V., Gunnarsdottir S., Adeyemo A., Chen Y., Chen G., Reynisdottir I., et al. 2007. Refining the impact of *TCF7L2* gene variants on type 2 diabetes and adaptive evolution. *Nat. Genet.* **39:** 218–225.

Huang Q.Y., Xu F.H., Shen H., Deng Y.J., Liu Y.J., Liu Y.Z., Li J.L., Recker R.R., and Deng H.W. 2002. Mutation patterns at dinucleotide microsatellite loci in humans. *Am. J. Hum. Genet.* **70:** 625–634.

Ke X., Durrant C., Morris A.P., Hunt S., Bentley D.R., Deloukas P., and Cardon L.R. 2004. Efficiency and consistency of haplotype tagging of dense SNP maps in multiple samples. *Hum. Mol. Genet.* **13:** 2557–2565.

Kong A., Gudbjartsson D.F., Sainz J., Jonsdottir G.M., Gudjonsson S.A., Richardsson B., Sigurdardottir S., Barnard J., Hallbeck B., Masson G., et al. 2002. A high-resolution recombination map of the human genome. *Nat. Genet.* **31:** 241–247.

Murray S.S., Oliphant A., Shen R., McBride C., Steeke R.J., Shannon S.G., Rubano T., Kermani B.G., Fan J.B., Chee M.S., and Hansen MS. 2004. A highly informative SNP linkage panel for human genetic studies. *Nat. Methods* **1:** 113–117.

Palsson B., Palsson F., Perlin M., Gudbjartsson H., Stefansson K., and Gulcher J. 1991. Using quality measures to facilitate allele calling in high-throughput genotyping. *Genome Res.* **9:** 1002–1012.

Pritchard J.K., Stephens M., and Donnelly P. 2000. Inference of population structure using multilocus genotype data. *Genetics* **155:** 945–959.

Reynisdottir I., Thorleifsson G., Benediktsson R., Sigurdsson G., Emilsson V., Einarsdottir A.S., Hjorleifsdottir E.E., Orlygsdottir G.T., Bjornsdottir G.T., Saemundsdottir J., et al. 2003. Localization of a susceptibility gene for type 2 diabetes to chromosome 5q34-q35.2. *Am. J. Hum. Genet.* **73:** 323–335.

Tang H., Quertermous T., Rodriguez B., Kardia S.L., Zhu X., Brown A., Pankow J.S., Province M.A., Hunt S.C., Boerwinkle E., Schork N.J., and Risch N.J. 2005. Genetic structure, self-identified race/ethnicity, and confounding in case-control association studies. *Am. J. Hum. Genet.* **76:** 268–275.

Vorechovsky I., Kralovicova J., Laycock M.D., Webster A.D., Marsh S.G., Madrigal A., and Hammarstrom L. 2001. Short tandem repeat (STR) haplotypes in HLA: An integrated 50-kb STR/linkage disequilibrium/gene map between the RING3 and HLA-B genes and identification of STR haplotype diversification in the class III region. *Eur. J. Hum. Genet.* **9:** 590–598.

Weber J.L. and Broman K.W. 2001. Genotyping for human whole-genome scans: Past, present, and future. *Adv. Genet.* **42:** 77–96.

Whittaker J.C., Harbord R.M., Boxall N., Mackay I., Dawson G., and Sibly R.M. 2003. Likelihood-based estimation of microsatellite mutation rates. *Genetics* **164:** 781–787.

19 | Considerations for SNP Selection

Chris Carlson

Fred Hutchinson Cancer Research Center, Seattle, Washington 98109-1024

INTRODUCTION

The human genome is composed of roughly 3 billion base pairs (Lander et al. 2001; Venter et al. 2001). The vast majority of this sequence is identical in comparing any two copies of the genome, with approximately one sequence variant every 1000 base pairs. Single-nucleotide polymorphisms (SNPs) account for the vast majority of this variation, with single-nucleotide substitutions accounting for approximately 94% of SNPs, 1–4-bp deletions accounting for 4.5%, and 1–4-bp insertions accounting for the remainder (Bhangale et al. 2005). Other types of variation exist, such as microsatellite repeats, multiallelic variable number tandem repeats, retroposon insertions, and larger structural polymorphism (insertion/deletion and inversion). In most of this chapter, I consider SNPs, as they account for the large majority of allelic differences between individuals.

BACKGROUND AND MOTIVATION FOR TAG-SNP SELECTION

Identifying functionally important SNPs has been a crucial research topic in the human genetics community for decades. As with any scientific endeavor, the state of the art has shifted with time. The easiest functional variants to identify were rare variants with strong effects, usually associated with Mendelian diseases. Initially, these variants were identified from candidate gene analysis (e.g., sickle cell anemia), and linkage analysis allowed geneticists to screen the genome for regions harboring these variants of large effect. Many linkages were successfully narrowed to point mutations

263

by candidate gene analysis, but comprehensive lists of genes within each linkage region did not become available until the conclusion of the genome project. Even now, not all linkage regions harbor obvious candidate genes, and the obvious candidate genes do not always contain the functional variant. Nonetheless, linkage mapping and positional cloning technologies have been phenomenally successful in identifying rare variants associated with large relative risks of disease.

Linkage analysis has primarily succeeded in the arena of rare diseases. Even where it has succeeded in common disease, linkage analysis has generally done so for rare, high-penetrance phenocopies of the common disease (e.g., BRCA1). Whereas the linked variants explain much of disease risk within specific pedigrees, these rare variants explain relatively little of overall disease incidence. Thus, although a significant fraction of many disease risks appears to be heritable, relatively little of this heritability has been explained by rare variants with large effect. This suggests that the remaining risk alleles are associated with modest risks. The genetic relative risk (GRR) associated with an allele can be defined as the change in disease risk per copy of the allele carried by an individual. Theoretical analysis suggests that linkage analyses are unlikely to detect GRRs of less than fourfold (Risch and Merikangas 1996).

Association analysis differs from linkage analysis in that unrelated cases and controls are analyzed, with the object of identifying alleles with significantly different frequencies in the two groups. Whereas linkage analysis screens for regions of the genome shared by affected individuals within a pedigree, association analysis screens for alleles shared by "unrelated" individuals. Because most SNP alleles in the human genome can be traced to unique mutation events (most SNPs have not experienced recurrent mutations), this translates to screening unrelated individuals for segments of the genome that are in fact shared identically by descent from an ancient common ancestor. Whereas large segments of chromosome (megabases) are generally shared by affected individuals within a pedigree, due to the small number of recombination events, much shorter segments of chromosome (kilobases) are shared between unrelated individuals.

Because segments of chromosome flanking an ancient risk-associated SNP are short, a much larger number of markers is required to detect these segments in association studies. However, theoretical analyses suggest that even after accounting for the large number of multiple tests, association analysis is likely to be more powerful than linkage analysis in detecting common variants of modest effect (Risch and Merikangas 1996). It has been argued that common variants of modest effect are likely to be more prevalent than rare variants of modest effect (Lander et al. 2001). However, in some Mendelian diseases the same risk allele is shared by most affected pedigrees (e.g., sickle cell and cystic fibrosis), whereas risk alleles (e.g., BRCA1/2) and even risk loci (e.g., MODY) tend to be unique for other diseases. Thus, although the diseases tend to be rare, some Mendelian diseases are largely attributable to a single, "common" risk allele (cystic fibrosis, sickle cell, hemochromatosis), whereas other diseases are not (MODY, BRCA1/2, xeroderma pigmentosum). Thus, it seems likely that some modest risk alleles will be common, whereas others will be rare.

If we specify a relative risk (RR) for the risk allele homozygote and an allele frequency for a disease risk allele, it is relatively straightforward to estimate the power to detect it in a given sample size. One simply estimates the true difference in allele frequencies expected between cases and controls, and then estimates the precision with which this difference can be estimated in an available sample (Fig. 19-1). It is immediately obvious that power to detect large RR is better than power to detect small RR. However, inspecting the graph carefully shows that power to detect rare alleles (i.e., minor allele frequency [MAF] below 10% or above 90%) is weak regardless of RR. Thus, although it is perfectly reasonable to assume that rare variants with modest RR may exist, in this sample size we are underpowered to detect such variants even if they exist.

Using a simple example (borrowed from Leonid Kruglyak, pers. comm.), given a mutation rate of approximately 2×10^{-8} per base pair per generation and a genome of 3 billion base pairs, each genome carries roughly 60 new mutations per generation. Given 6 billion humans on the planet, and two copies of the genome per person, this translates to 720 billion new mutations in the current generation, or a Poisson average of 240 new mutations per base pair. Thus, every base pair in the genome where a new mutation is consistent with survival as a heterozygote probably has been

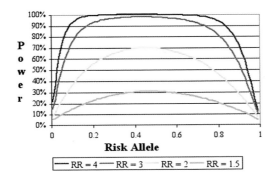

FIGURE 19-1. Relative risk (RR) detection analysis. A range of RRs in a sample of 200 cases and 200 controls is shown, using a risk inheritance model where the RR for the heterozygote is the square root of the RR specified for the risk homozygote, and not adjusting for multiple tests.

mutated in at least one person on the planet. However, if we compare any two copies of the genome, there will be roughly 3 million differences, just 120 of which represent new mutations. Most of the differences between the genomes will be considerably older, and represent allelic differences at common variants.

The preceding analyses are simply meant to provide an appreciation for what we can hope to identify in an association analysis. Association studies can be used to test the common disease/common variant (CDCV) hypothesis (Collins et al. 1997), but the possibility that rare variants with modest risk exist cannot be rejected on theoretical grounds. We simply lack the power to detect such variants in a reasonably sized study. Conveniently, the number of common SNPs in the human genome is dramatically smaller than the number of rare SNPs, so testing the CDCV hypothesis is a much easier proposition. It has been estimated that 6 to 10 million common (MAF >10%) variants exist in the human genome (Kruglyak and Nickerson 2001), and resequencing data are consistent with this estimate. Even allowing for some common variants to be population-specific, it is unlikely that more than 15 million SNPs exist which are common in at least one major geographic subpopulation.

THEORETICAL METHODS FOR TAG-SNP SELECTION

At present, most association analyses are targeted at candidate genes, regions, or pathways, although in the near future whole-genome association analyses will become substantially more frequent. It is slightly easier to grasp tag-SNP selection in the context of a limited target sequence rather than across an entire genome, so we will delve into the details for a specific target region with the understanding that the same approaches can be applied to the entire genome with only minor modifications. Within the target region, the object of an association analysis is to identify functional variants. The basic assumption behind all tag-SNP approaches is that we cannot accurately predict which SNPs are likely to have functional consequences. Although methods have been developed to predict the functional impact of some categories of functional variants a priori, such as nonsynonymous coding SNPs (ncSNPs) (Ng and Henikoff 2001, 2003), these SNPs represent a small subset of all common variants. Thus, the object of tag-SNP selection is to agnostically select a subset of SNPs that efficiently describes existing patterns of variation in a candidate region for subsequent genotyping and association analysis, without making assumptions about the potential functional impact of each SNP.

A number of algorithms for tag-SNP selection exist (Patil et al. 2001; Gabriel et al. 2002; Ke and Cardon 2003; Meng et al. 2003; Sebastiani et al. 2003; Stram et al. 2003; Weale et al. 2003; Zhang and Jin 2003; Carlson et al. 2004; Halldorsson et al. 2004a,b; Lin and Altman 2004; Schulze et al. 2004; Zhang et al. 2004, 2005; de Bakker et al. 2005; Halperin et al. 2005). Many overlap substantially in concept; therefore, we will explore only a few representative algorithms in detail. In the broadest sense, tag-SNP selection algorithms can be broken into two major categories: haplotype-based algorithms and haplotype-free algorithms. Haplotype-based algorithms require that haplotype be inferred prior to tag-SNP selection (see Box 19-1, Introduction to Haplotypes).

BOX 19-1. AN INTRODUCTION TO HAPLOTYPES

A haplotype is the pattern of alleles at multiple polymorphic sites along a single chromosome. Most applications of haplotype analysis require that the haplotypes remain stable within small pedigrees. Although an entire chromosome can technically be considered as a haplotype, generally haplotypes are only considered across tightly linked polymorphic sites spanning less than 100 kb, although exceptions do exist (e.g., human leukocyte antigen and young mutations). Haplotypes can be built using any combination of polymorphisms: variable number tandem repeats, microsatellites, insertion/deletions, or SNPs.

In comparing any pair of chromosomes within a species, the vast majority of the sequence is invariant. Most authors use a threshold between 1% and 5% when discussing polymorphism. The pattern of alleles along a single chromosome is defined as the haplotype. Given two biallelic polymorphisms, there are theoretically $2^2 = 4$ possible haplotypes. As the number of polymorphic sites included in a haplotype increases, the number of possible haplotypes increases exponentially. For n biallelic polymorphisms, there are 2^n possible haplotypes. In theory, there are four possible nucleotides at each position. In practice, there are generally only two alleles, so in order to make the data more intuitive, many authors represent haplotypes in terms of common allele and rare allele.

a   TCGRGTCATGYATCAGA

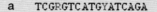

b

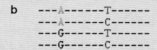
```
---A------T------
---A------C------
---G------T------
---G------C------
```

c   ATTACTCC
      GCTGCTGC
      GCAGGTGT
      GCAGCCGT

d

(a) Polymorphisms are defined as those positions where two or more variants that exceed an arbitrary frequency threshold occur in a population (shown in *red*). (b) Although haplotypes technically include the nucleotide at invariant positions, in general, only the nucleotides at polymorphic sites are reported. (c) For nearby polymorphisms, the actual number of haplotypes observed tends to be much less than the possible number of haplotypes. (d) Diagrammatic representation of the data shown in c, with blue and yellow representing common and rare alleles, respectively.

HAPLOTYPE-BASED TAG-SNP SELECTION ALGORITHMS

The first publication to describe tag-SNP selection (Johnson et al. 2001) approached the problem as one of haplotype resolution. In this approach, a limited number of common haplotypes were known to exist in a small region of the genome, and tag-SNPs were simply selected in order to discriminate efficiently between the known common haplotypes. Genotype at SNPs not selected as tag-SNPs (referred to as untyped SNPs hereafter) can efficiently be inferred from knowledge of the haplotypes carried by an individual. This approach works quite well in regions with a small number of common haplotypes, generally corresponding to regions with little or no detectable recombination (see Box 19-2, Imputation of Haplotypes; Box 19-3, Haplotypes in Nonrecombinant Regions; and Box 19-4, Founder Events). This "haplotype tagging" approach was extrapolated to larger segments of the genome by several groups (Patil et al. 2001; Gabriel et al. 2002) who recognized that the algorithm can be applied to larger regions by breaking them into a series of smaller, relatively nonrecombinant segments ("haplotype blocks"), reviewed in Cardon and Abecasis (2003).

The block paradigm is elegantly simple, but is flawed in an important way: The boundaries between recombinant regions reflect only the prevalence of recombinant chromosomes, not the frequency of recombination events. Whereas some block boundaries clearly reflect recombination

BOX 19-2. IMPUTATION OF HAPLOTYPES

Haplotype Inference

A haplotype is the pattern of alleles along a single chromosome. Thus, haplotype is equivalent to genotype in haploid organisms (e.g., bacteria) or monosomal chromosomes (e.g., the X and Y chromosomes in male mammals). Given n unlinked biallelic polymorphisms, in theory there are 2^n possible haplotypes. At the other end of the spectrum, in a nonrecombinant region, the maximum number of haplotypes is $n + 1$, and the minimum number of haplotypes is 2. For tightly linked polymorphisms (within 0.01 centimorgans), the number of extant haplotypes is usually considerably smaller than the theoretical maximum.

Humans and many other model organisms are inconveniently diploid, such that each individual carries two underlying haplotypes. If the two haplotypes carried by a diploid individual differ in the alleles carried at two or more polymorphic sites, the individual will genotype as a multiple heterozygote, and the observed genotype pattern will be consistent with multiple combinations of underlying haplotypes. For a given individual heterozygous at n polymorphisms, in theory there are 2^{n-1} possible pairs of underlying haplotypes. Resolving the haplotypes carried by an ambiguous individual can be accomplished in a number of ways. Several techniques exist for physically separating the chromosomes, including allele-specific PCR, subcloning the region into a library, and somatic cell hybrids, but these are not discussed in detail here. Alternatively, haplotypes can be inferred in silico using a variety of algorithms, including pedigree analysis, biologically naïve expectation maximization algorithms, and more complex Bayesian algorithms that incorporate the coalescent process.

In the following discussion, we use the IUPAC ambiguity codes for heterozygous genotypes at single-nucleotide substitution polymorphisms: W = Weak (A or T), S = Strong (G or C), K = Keto (G or T), M = amino (A or C), R = puRine (A or G), and Y = pYrimidine (C or T). Thus, a three-SNP genotype for an individual might be represented as YAR, indicating that the individual is heterozygous C/T at the first SNP, homozygous A at the second, and heterozygous A/G at the third. The sites need not be physically adjacent on a chromosome; intervening non-polymorphic sites will simply not be shown for convenience.

Pedigree-based Haplotype Inference

If the genotypes of parents or offspring of an ambiguous individual are available, then haplotype inference for tightly linked polymorphisms is relatively straightforward. One simply assumes that the haplotype is nonrecombinant within a pedigree, and then searches for individuals where some portion of the haplotype is unambiguous. Figure 19-2 shows a parent/child trio where the child and father are multiply heterozygous for a three-SNP genotype (YRR). Because the mother is homozygous, we infer that one of the haplotypes carried by the child is CAG and, therefore, the other haplotype must be TGA. Because the TGA haplotype in the child must have come from the father, this unambiguously resolves the father's haplotype.

hot spots (Templeton et al. 2000; Jeffreys et al. 2001; McVean et al. 2004), the vast majority of boundaries do not, with the consequence that strong linkage disequilibrium extends across many block boundaries (Wall and Pritchard 2003a,b). Although other flaws in the block model have been demonstrated, such as tag-SNP dependency on the block model (Ding et al. 2005), perhaps the most important flaw is that block definition is substantially dependent on the SNPs used to define blocks (Ke et al. 2004).

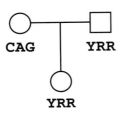

CAG   YRR

YRR

FIGURE 19-2. Pedigree-based haplotype inference. Both the father (*upper square*) and child (*lower circle*) are multiply heterozygous for a three-SNP genotype (YRR). Because the mother (*circle*) is homozygous (CAG), one of the haplotypes carried by the child must also be CAG. It can then be inferred that the other haplotype must be TGA, and that the TGA haplotype in the child must have come from the father.

BOX 19-3. HAPLOTYPES IN NONRECOMBINANT REGIONS

In a nonrecombinant region, new haplotypes are created by mutations. Because the majority of the genome is non-polymorphic at any given time, the new mutation generally occurs at a position that was not previously polymorphic. Thus, given n polymorphic sites, the maximum number of possible haplotypes in a nonrecombinant region is $n + 1$, assuming no recurrent mutations. Generally, samples of haplotypes from a population will not include all of the intermediate haplotypes, because some have drifted to extinction, so the number of haplotypes will be somewhat less than $n + 1$.

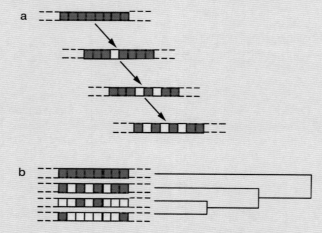

(a) Each new mutation occurs in the context of a particular existing haplotype, generating a new haplotype. (b) Although the intermediates are not observed, it is possible to reconstruct a tree or cladogram of relationships between haplotypes based on similarities between observed haplotypes.

HAPLOTYPE-FREE TAG-SNP SELECTION ALGORITHMS

As an alternative to selecting tag-SNPs on the basis of haplotype structure, a variety of alternative algorithms exist that use correlations between SNPs to select efficient sets of tag-SNPs (Weale et al. 2003; Carlson et al. 2004; Lin and Altman 2004). These methods can be described as searches for an efficient set of tag-SNPs for a region, such that genotype at untyped SNPs can be accurately inferred from genotype at a tag-SNP. Algorithms of this nature have the advantage that haplotypes need not be inferred, avoiding questions of how to define a block boundary, as well as avoiding errors from haplotype inference.

Although the block paradigm is flawed, a great diversity of useful algorithms exists for selecting tag-SNPs on the basis of haplotype information. Although these algorithms are generally less than optimally efficient (because they treat adjacent "blocks" as independent information), they do achieve economies relative to haplotype-free approaches by inferring genotype at some untyped SNPs from a combination of tag-SNPs. Thus, perhaps the most efficient approaches proposed to date are mixed algorithms, wherein tag-SNPs are selected with the object of inferring genotype at untyped SNPs (like the haplotype-free approaches), but allowing untyped SNP genotypes to be inferred from either single tag-SNPs or haplotypes of multiple tag-SNPs (de Bakker et al. 2005).

WHICH TAG-SNP SELECTION METHOD IS BEST?

Several major efforts have been made recently to compare the results from multiple tag-SNP selection algorithms. Algorithms can be compared on several metrics, including tag-SNP efficiency (the fraction of all SNPs that are selected as tag-SNPs) and tag-SNP efficacy (how accurately untyped SNP genotypes are inferred from the tag-SNPs). The best such comparison to date cleverly normalized three major algorithms on efficiency and then compared the efficacy between algorithms at a given efficiency level (Ke et al. 2005). This paper demonstrated significant differences in efficacy between algorithms, with a slight advantage to the more sophisticated haplotype-based algorithms over linkage disequilibrium (LD) algorithms (see Box 19-5), but the more important point may be that although statistically significant, the differences were quite modest. That is, at a specified tag-SNP efficiency, the efficacy of algorithms was pretty similar.

If the efficacy of various algorithms is similar at a specified efficiency level, then practical considerations such as assay design and data analysis can be more important than theoretical considerations of information content. These practical considerations can be divided into pre-genotyping and post-genotyping topics. The primary pre-genotyping topic is tag-SNP flexibility. All algorithms implemented to date are biologically naïve, treating all SNPs as equally likely to be functional and simply optimizing the number of SNPs required to capture existing common patterns of variation. This clearly has some drawbacks, as a SNP with strong a priori evidence for function may not be included in the final set of tag-SNPs. For example, ncSNPs represent a disproportionate fraction of disease-associated mutations in the Gene Mutation Database (Botstein and Risch 2003). Therefore, if a ncSNP is present in a region, it should probably be included as a tag-SNP because of higher a priori likelihood that it is functional. Thus, algorithms that allow users to specify mandatory tag-SNPs afford a clear advantage over naïve algorithms.

A second practical consideration is tag-SNP redundancy. Less efficient tag-SNP applications have increased redundancy of information within a tag-SNP set. This can actually be an advan-

---

BOX 19-4. FOUNDER EVENTS

The age of a polymorphism is related to the extent of the original haplotype associated with the derived allele. However, the age of an allele also correlates with the frequency of the allele: Young alleles tend to be rare, and common alleles tend to be ancient (Kimura and Ota 1973; Watterson 1976). Although young mutations can be associated with haplotypes extending for megabases, the extent of the original haplotype associated with common (ancient) alleles tends to be less than 20 kb.

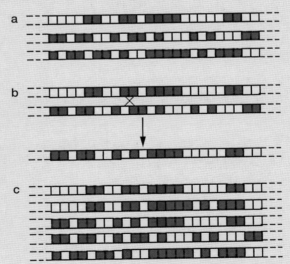

(a) In a population of existing haplotypes, a new mutation arises on a particular haplotype background (*red square*). (b) As the original haplotype is transmitted from generation to generation, recombination creates new haplotypes bearing the derived allele. (c) If the derived allele persists and expands to high frequency in the population overall, the fragment of the original haplotype shared by haplotypes bearing the derived allele is reduced each generation by recombination.

BOX 19-5. LINKAGE DISEQUILIBRIUM

Given two SNPs, A and B, each with two alleles (A1, A2 and B1, B2, respectively), there are four possible haplotypes: A1B1, A1B2, A2B1, and A2B2, which we will refer to as 11, 12, 21, and 22. Linkage disequilibrium (LD) statistics describe the deviation of the observed two SNP haplotype frequencies from expectation. The basic statistic D (Eq. 1) is simply the deviation of the observed frequency of the A1B1 haplotype (p_{11}) from that expected if genotype is independent at the two SNPs ($p_{A1}*p_{B1}$).

$$D = p_{11} - p_{A1}\,p_{B1} = p_{11}\,p_{22} - p_{12}\,p_{21} \tag{1}$$

If the frequencies of the haplotypes are exactly as predicted from allele frequencies at each SNP, $D = 0$ and the SNPs are in linkage equilibrium. For a pair of SNPs where both have a minor allele frequency (MAF) of 50%, D can range from –0.25 to 0.25, which is the largest possible range for D. However, lower MAF and/or mismatched allele frequencies will restrict the range of this statistic further, making it difficult to compare LD values between pairs of SNPs. To normalize D values between pairs of SNPs, two of the more commonly used LD statistics are D' and r^2 (Eqs. 2 and 3, respectively). D' describes the observed LD relative to the maximum possible deviation given the observed allele frequencies, whereas r^2 describes the correlation between alleles at the two SNPs.

$$D' = \frac{D}{D_{max}} \tag{2}$$

$$r^2 = \frac{D^2}{p_{A1}\,p_{A2}\,p_{B1}\,p_{B2}} \tag{3}$$

When allele frequencies are not identical, alleles are not perfectly correlated even if LD is as strong as possible. At complete LD, at least one of the four possible haplotypes is not observed, as predicted for a nonrecombinant region, where $D' = 1$, and $r^2 < 1$. When allele frequencies are identical, it is possible for genotype to be perfectly correlated, and $D' = 1$, and $r^2 = 1$, reflecting a perfect correlation of genotype at A and B. The set of alleles unique to a haplotype will be in perfect LD.

From an epidemiologic perspective, the key to tag-SNP selection is simply the correlation between a genetic risk factor (whether genotype at a functional SNP or a haplotype of functional SNPs) and the measured genetic variable (whether genotype at a tag-SNP or a haplotype composed of multiple tag-SNPs). Thus, most haplotype-free tag-SNP selection algorithms use r^2 or a more complex variant of r^2 to select tag-SNPs such that all patterns of common variation in a region are either directly assayed as a tag-SNP or strongly correlated with an assayed tag-SNP (or a haplotype of tag-SNPs).

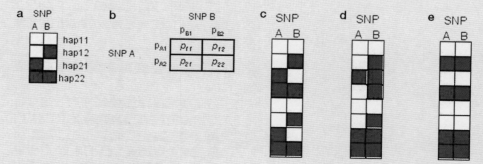

(a) Diagrammatic representation of the four possible haplotypes, encoding the alternative alleles as yellow and blue. (b) The observed frequencies of the four haplotypes can be represented as a 2 × 2 table, where p_{A1} is the observed frequency of the A1 allele, p_{B1} is the observed frequency of the B1 allele, and p_{11} is the observed frequency of the A1B1 haplotype. (c) Diagrammatic representation of linkage equilibrium. (d) Diagrammatic representation of a "complete" LD. In this example, the 21 haplotype (common [blue] A allele and rare [yellow] B allele) does not occur. (e) Diagrammatic representation of a "perfect" LD. Allele frequencies are identical, and there is a perfect correlation of genotype at A and B.

tage, because not all SNPs genotype equally effectively on any given technical platform, and a subset of 10–15% of SNPs can be expected to fail. Thus, it can be important to pay attention to how robust a set of tag-SNPs is if one of the tag-SNPs fails to genotype. For critical tag-SNPs, it can even be good practice to include a redundant tag-SNP up front, such that if one fails, the information is not lost. Weighing the value of a specific tag-SNP is tricky, but there are two situations in which it is clearly a good idea to include a redundant tag-SNP in the pool of tag-SNPs. First, redundant tag-SNPs are clearly worthwhile for SNPs with a priori inferred function (e.g., ncSNPs or SNPs in regions of strong evolutionary conservation). Second, assuming that all SNPs are equivalently likely to be functional, tag-SNPs providing information on many other SNPs are more valuable than tag-SNPs providing information on just a few SNPs. That is, in the absence of any other information, it is reasonable to assume that highly diverged haplotypes are more likely to be functionally divergent. It is therefore good practice to include a redundant tag-SNP to identify highly conserved haplotypes if they exist. By the same reasoning, when using LD-based tag-SNP approaches, it is more important to capture large bins of SNPs than small bins, because large bins will tag these highly diverged haplotypes.

The final practical consideration for tag-SNP selection is bookkeeping. Highly efficient algorithms usually involve the inference of genotype at some untyped SNPs using a haplotype from a combination of tag-SNPs. Although this is economically advantageous in terms of genotyping, it places an analytic burden on the user to track the haplotypes of tag-SNPs that need to be analyzed. Given how rapidly the cost of genotyping is decreasing, it is reasonable to ask whether fully optimizing the tag-SNP set is actually the most efficient use of resources.

ACCESS TO DATA AND TAG-SNP SELECTION APPLICATIONS

Theoretical considerations aside, there are two keys to the application of tag-SNP algorithms for most investigators: access to relevant dense genotyping data (or preferably resequencing data) and access to the tag-SNP selection software. There are two major repositories for genotype data: the HapMap (http://www.hapmap.org/cgi-perl/gbrowse/gbrowse/hapmap/) (International HapMap Consortium 2003; Altshuler et al. 2005) and dbSNP (http://www. ncbi.nlm.nih.gov/SNP/) (Sherry et al. 1999). The HapMap project has made a remarkably detailed data set available in three geographically discrete human subpopulations: Europeans, Africans, and Asians. Thus, tools built upon this resource are currently the state of the art for those regions of the genome that have not been targeted for resequencing. However, dbSNP is the central repository for genotyping information from other projects (including targeted resequencing projects, in addition to the HapMap), and should eventually supersede the HapMap as the central repository from which genotypes can be extracted for tag-SNP selection.

POPULATION SPECIFICITY

Regardless of the algorithm used, when selecting tag-SNPs for a study, it is important to keep in mind how well the population used for selection matches the samples being genotyped. Although measures of interpopulation variance indicate that common SNPs tend to be polymorphic in all major human populations, it is not at all unusual to observe large allele frequency differences between groups. Parameters such as nucleotide diversity and the extent of LD are also known to vary between populations. In broad terms, nucleotide diversity is higher in African populations than in non-African populations, which is frequently misinterpreted to indicate that the African population is older than non-African populations. Although it is clear that modern humans have inhabited Africa longer than the other continents, the higher sequence diversity within Africa simply indicates that, on an evolutionary timescale, the effective population size has been larger in Africa than outside of Africa. Similarly, LD extends over shorter distances in African populations than in non-African populations, again consistent with a larger effective population.

Because of population differences in allele frequency and LD, at the very least the samples used for tag-SNP selection should be matched to the study sample in terms of continent of origin. Within Europe, there is relatively little geographic variance in allele frequency (Rosenberg et al. 2002), so the HapMap European samples are likely appropriate for tag-SNP selection in most European populations. Relatively little difference was observed in allele frequency between the HapMap Japanese and Chinese subpopulations, so it appears that these samples are an appropriate sample for tag-SNP selection in Eastern Asian populations.

It is not always possible to match the ethnic origin of the tag-SNP selection samples to a study. For example, Mexican-American groups have significant Native American and European ancestry. If a study in such a population focuses on a few candidate genes, it may be feasible to resequence a small sample of individuals from the study for tag-SNP selection. If resequencing is not possible, then tag-SNP selection may be performed independently in each ancestral population. For example, for Mexican-Americans, it would be reasonable to use the HapMap Asian and European populations for tag-SNP selection. In such a case, it is important to examine the allele frequencies post hoc: If the observed allele frequency of a tag-SNP in the study population is not intermediate between the ancestral populations, then the region is more likely to contain population-specific variation and should be given higher priority for resequencing.

Phase 2 of the HapMap genotyped millions of SNPs in three major geographic populations: Europeans, West Africans (the Yoruban population), and East Asians (the Chinese and Japanese populations). Phase 3 of the HapMap will examine genotypes in additional populations but will genotype only a subset of SNPs from Phase 2. Thus, sparser maps of variation will be available for tag-SNP selection in these populations. Researchers working with study populations that better match the Phase 3 groups should keep this in mind. Prior to tag-SNP selection, one might consider some preliminary resequencing in candidate regions where the matched Phase 3 population shows dramatically different allele frequencies from the Phase 2 populations.

POPULATION STRUCTURE

The object of tag-SNP selection is to efficiently describe common patterns of variation within a population. However, SNPs that are common in one ancestral population may or may not be common in another. Thus, selecting tag-SNPs for significantly admixed populations can be a challenge. In general, this is best accomplished by selecting tag-SNPs independently in data from the ancestral populations, or from the best available proxies for these populations. For example, African-Americans average approximately 20% European ancestry, so it is appropriate to select tag-SNPs from both the European and African HapMap populations for this admixed group. Hispanics can be even more complex, and it may be appropriate to select tag-SNPs in African, European, and Asian data for studies of Hispanics, especially in the Caribbean. Strategies have recently been defined that help to optimize the set of tag-SNPs from multiple populations (Howie et al. 2006).

SELECTING TAG-SNPS: A USER'S MANUAL

Selecting a set of tag-SNPs requires a target region (either a single gene or the entire genome), preliminary genotype data in an appropriate tag-SNP selection population, and the tools to manipulate that information for the selection of tag-SNPs. Although some users will generate their own preliminary genotype data by resequencing a panel of representative samples from a study, it is more likely that a user will want to employ existing preliminary data to select an appropriate set of tag-SNPs. The standard repository for genotype information is dbSNP, but extracting the raw genotype data from dbSNP is not trivial. Thus, tools that allow the user to simply extract the genotypes are critical, as are tools to convert data from one format to another.

Using the Genome Variation Server

One of several tools to extract genotypes from dbSNP is the Genome Variation Server (GVS, http://gvs.gs.washington.edu/GVS/). This tool allows a user to query the genome from multiple frames of reference: using genome coordinates, a gene name, a Human Genome Organization (HUGO) gene name, or an anchor SNP (Fig. 19-3). Genome coordinates can be useful for looking at a cluster of genes, and an anchor SNP can be a useful way to find a region of interest, but in general, most queries will start with a candidate gene in mind. Once the data have been extracted, GVS also offers a built-in implementation of the LDselect haplotype-free tag-SNP selection algorithm (Carlson et al. 2004).

Let's assume that we are interested in Leptin (HUGO gene name LEP). If you click in "gene id," you are prompted to enter the gene id (Fig. 19-4). The default range retrieved is the coordinates corresponding to the longest known refseq mRNA for the gene, but you can specify additional sequence upstream (5′ of the transcription start site) or downstream (3′ of the poly(A) signal). For our example, we will specify gene name LEP, 5000 bp upstream and 2500 bp downstream, and then hit "Search."

The query returns a list of populations where genotype data have been reported for the specified region in dbSNP (Fig. 19-5). You can hotlink through the population names to learn more details of each population. For our example, we will select the PGA_YORUB-PANEL, a subset of the HapMap Yoruban samples resequenced by the SeattleSNPs PGA in this region. After selecting an appropriate population, scroll to the bottom of the screen where you can set the parameters for data visualization and/or tag-SNP selection using LDselect (Fig. 19-6) (Carlson et al. 2004).

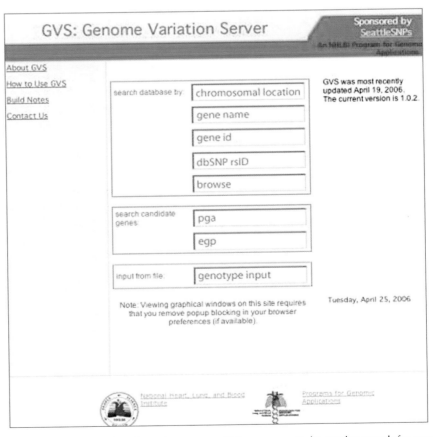

FIGURE 19-3. The Genome Variation Server home page. This program can be used to search for genes of interest using a number of different parameters.

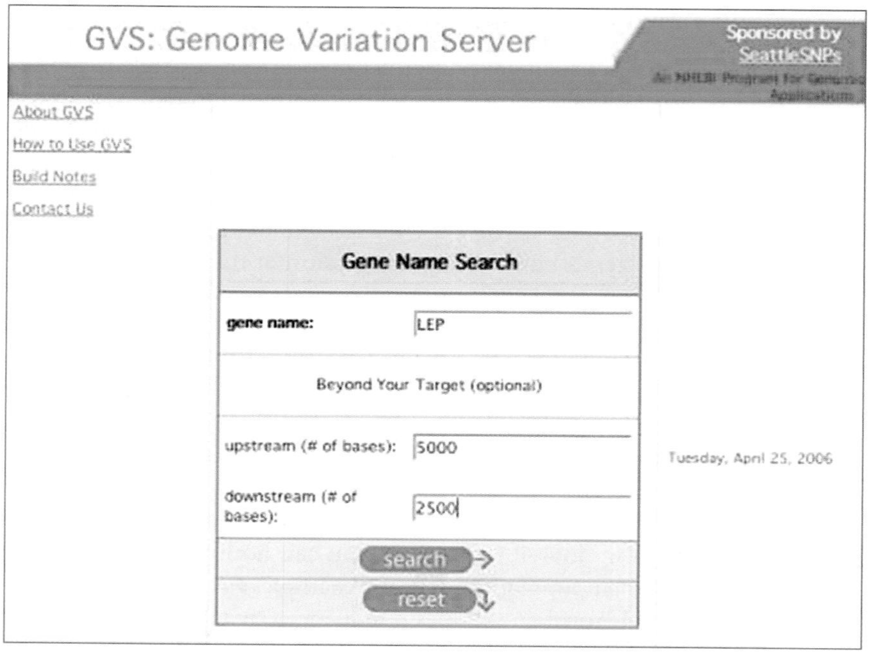

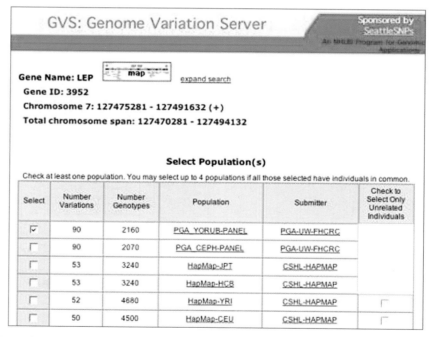

FIGURE 19-4. Example of data entry for genotype data search for the leptin gene using the GVS system. Searches can be customized to examine sequences outside the coding region.

FIGURE 19-5. The population selection screen is generated by GVS in response to the search conditions entered in the example in Fig. 19-4.

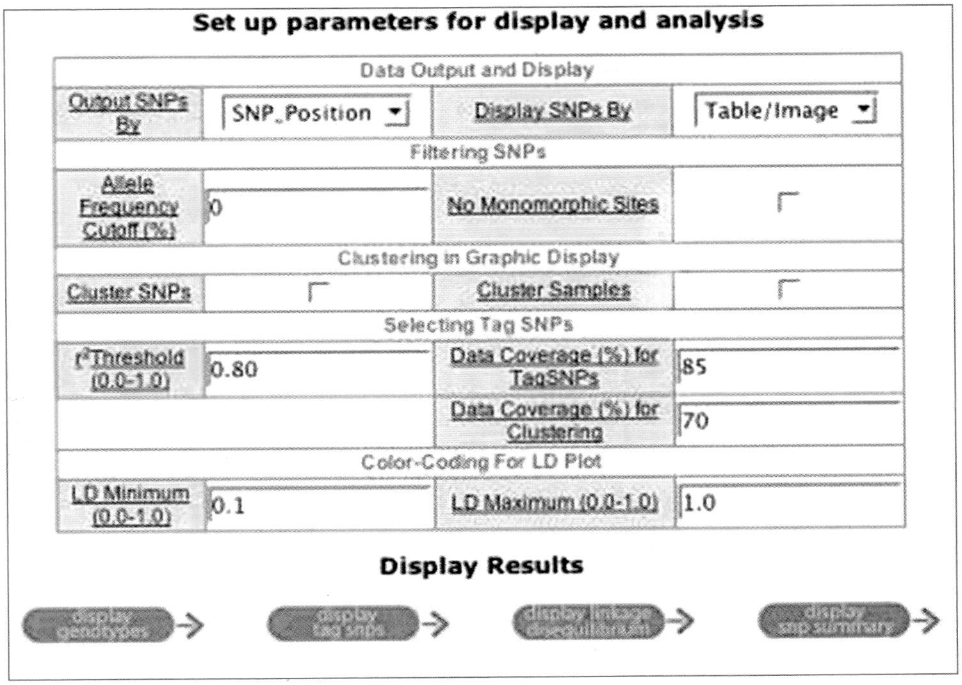

Set up parameters for display and analysis

Data Output and Display			
Output SNPs By	SNP_Position ▾	Display SNPs By	Table/Image ▾
Filtering SNPs			
Allele Frequency Cutoff (%)	0	No Monomorphic Sites	☐
Clustering in Graphic Display			
Cluster SNPs	☐	Cluster Samples	☐
Selecting Tag SNPs			
r²Threshold (0.0-1.0)	0.80	Data Coverage (%) for TagSNPs	85
		Data Coverage (%) for Clustering	70
Color-Coding For LD Plot			
LD Minimum (0.0-1.0)	0.1	LD Maximum (0.0-1.0)	1.0

Display Results

display genotypes →   display tag snps →   display linkage disequilibrium →   display snp summary →

FIGURE 19-6. Setting the parameters for tag-SNP selection. GVS provides a variety of options with which an investigator can customize search parameters and data display.

Before selecting the tag-SNPs, set "Display SNPs By:" to "Text/Image" and click on "Display Genotypes." This will bring up two pop-up windows, one showing a text format prettybase file for the displayed data, and the other showing a visual genotype (Fig. 19-7). The prettybase file is the raw input format for LDselect and multipop-TagSelect and lists the data as four tab-delimited columns: polymorphic site, individual ID, allele 1, and allele 2. Save the prettybase as a text file for later use. As an aside, if you have your own genotypes that you want to upload into GVS, you will need to convert the data into this format. A converter from prettybase to ped format (HaploView) is available at http://theta.ncifcrf.gov/gbrowse/start_db/prettybase_to_linkageformat.zip, and a converter from ped format to prettybase format is available at http://www.pharmgat.org/pharmgat.org/Documentation/help/pedtopb.

Returning to tag-SNP selection, go back to the GVS tag-SNP parameter page (Fig. 19-6). Here you can set the relevant parameters for LDselect-based tag-SNP selection within the Leptin data set. "Output SNPs By" can be either "SNP_position" in current genome build coordinates, or "RS_ID," the dbSNP refSNP ID for each polymorphism. RefSNP numbers are easier to cross-reference between studies and should always be used in reporting results from any association study, but coordinates can be easier to comprehend in terms of relative position along the gene. For the moment, leave this option set to "SNP_Position." The next parameter to set is the "Allele Frequency Cutoff (%)." The minimum allele frequency analyzed in a study should be selected on the basis of power estimates for the available samples, but in general, it is reasonable to use either 5% or 10%. Set this option to 5%. The checkbox for "No Monomorphic Sites" allows you to exclude any non-polymorphic SNP (usually a SNP that was polymorphic in another population) from display/analysis. Finally, the tag-SNP selection parameters need to be set: the r^2 threshold should also be determined on the basis of power estimates for an available sample size. The default of 0.8 is an acceptably stringent threshold for candidate gene analysis, and in general, one should not set this parameter much lower than 0.5. We will return to "Data coverage for Tag-SNPs" later. For now, simply press the "display tag-SNPs" button at the bottom of the page. Results are shown in Figure 19-8.

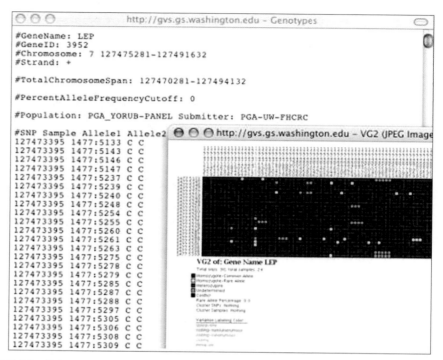

FIGURE 19-7. Visualizing genotypes in GVS. Pop-up windows in the GVS display genotypes as text (prettybase format) as well as graphically.

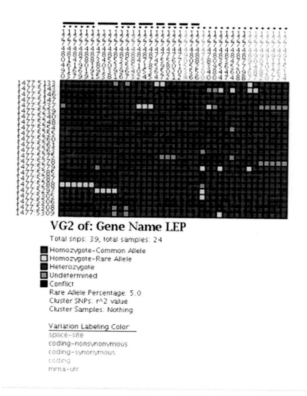

FIGURE 19-8. tag-SNP selection from Yoruban samples in the Leptin gene. The results are generated in response to the parameters specified using the selection screen shown in Fig.19-5 (see text for details). The SNPs are arranged such that SNPs in strong LD are adjacent. At the top of the picture, black bars indicate bins of SNPs, and asterisks indicate informationally equivalent tag-SNPs for each bin. SNP labels are color coded as indicated in the legend.

The tag-SNP selection results are shown in two frames: The first (Fig. 19-8) shows a new prettybase in which the SNPs have been rearranged such that SNPs in strong LD are adjacent. At the top of the picture, black bars indicate bins of SNPs, and asterisks indicate informationally equivalent tag-SNPs for each bin. Just one of the asterisked SNPs needs to be genotyped in each bin. SNP labels are color coded as indicated in the legend: For example, the third bin contains a nonsynonymous (amino-acid-changing) SNP, shown in red. Selecting the tag-SNP within each bin can then be informed by a priori function: Because the nonsynonymous SNP is a better functional candidate than the noncoding SNPs in bin 3, it makes sense to preferentially select it for genotyping.

Returning to the "coverage" options, the "Data Coverage for tag-SNPs" can be used to specify how many missing data are allowable for a tag-SNP. This can be important because SNPs with missing data are sometimes preferentially selected as tag-SNPs (e.g., the tag-SNP 127485391 in the second bin from the left). Setting this parameter up to 100 precludes the selection of this SNP as a tag-SNP and produces a bin structure with more options in the second bin. The "Data coverage for clustering" option is similar, but actually sets aside SNPs with too much missing data as their own bins.

Haploview and Tagger

Another tool for extracting genotype data and running tag-SNP selection is Haploview (http://www.broad.mit.edu/mpg/haploview/download.php; see also Chapter 21). Haploview implements much the same functionality as GVS, with several minor changes. First, Haploview is a stand-alone Java Applet that manipulates genotype data in HapMap format. Thus, in order to run tag-SNP selection in Haploview, one must first download genotype data for the appropriate region from the HapMap Web site (Fig. 19-9). For our example, we will use the same segment: Type "chr7:127475281..127491632" into the Landmark or Region box, and hit Search. You can also specify target regions using a HUGO gene name. Results are shown in Figure 19-10. Now, under "Reports & Analysis," select "Download SNP Genotype Data" and hit "Configure." Select the appropriate population (e.g., YRI for Yorubans), select "Save to Disk," and hit "Go." The file will probably be saved as "dumped_region," so you will need to figure out where the file was put on your computer.

FIGURE 19-9. The International HapMap Project Web site can be used to download genotype data.

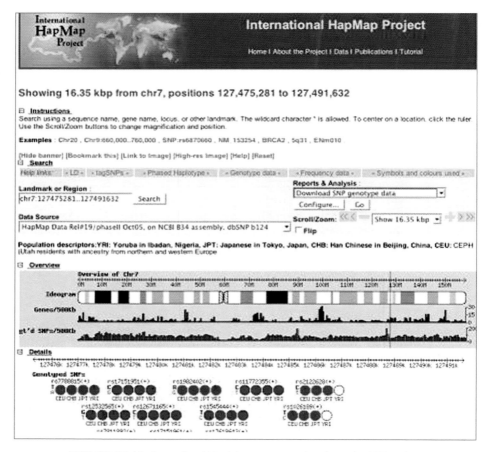

FIGURE 19-10. Example of HapMap genotype data from the LEP region.

Now open HaploView. The default screen will generally have three options: Select "Load Hapmap Data." If these options are not visible, select File-Open Genotype Data, and find your file. Once the data have opened (Fig. 19-11), select the "Tagger" tab, and you will be able to run tag-SNP selection. The "pairwise tagging only" option is essentially the LDselect algorithm (Carlson et al. 2004), and the "Force Include" and "Force Exclude" options are convenient utilities to specify SNPs that either will be genotyped (e.g., functional SNPs) or cannot be genotyped (e.g., SNPs in repetitive elements that are incompatible with most high multiplex genotyping platforms). The "aggressive tagging" options allow untyped SNPs to be inferred from haplotypes of multiple other SNPs (de Bakker et al. 2005). This can reduce the number of tag-SNPs required, but with the drawback that any tag-SNPs which fail to genotype will have a higher cost. Select "aggressive tagging: Use 2-marker haplotypes" and hit "Run Tagger." Results are shown in Figure 19-12.

Comparing GVS and Haploview

Neither Haploview nor GVS is clearly superior as a tool; both have relative strengths and weaknesses. Both implement basic pairwise r^2-based tag-SNP selection, and both also include tools for visualization of patterns of LD (which have not been discussed here). The most obvious advantage of GVS is the facility with which genotypes are served up from the dbSNP database. Although the HapMap is the best option for most of the genome, the gold standard for tag-SNP selection is resequencing data, which are available for substantial numbers of candidate genes through SeattleSNPs, the HapMap ENCODE effort, the Innate Immunity Program for Genomic Applications, and presumably, other groups in the future. The other obvious advantage of using

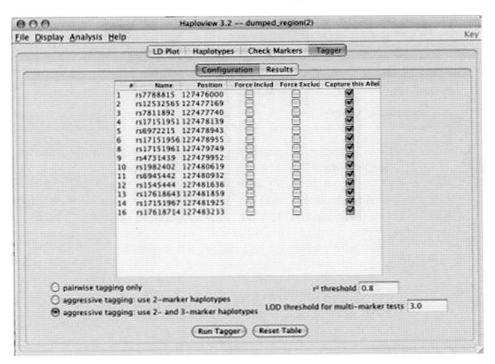

FIGURE 19-11. The HaploView program manipulates genotype data extracted from the HapMap database. The "Tagger" function analyzes genotypes for tag-SNP selection.

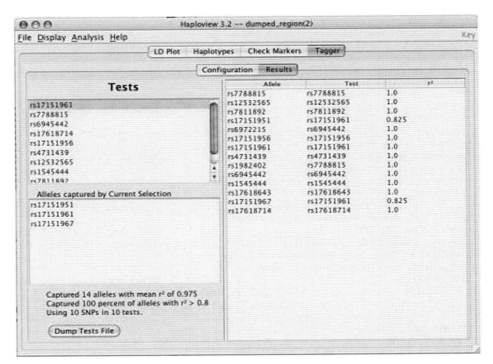

FIGURE 19-12. Tagger output from Leptin region in Yoruban HapMap samples. The Tagger output suggests a set of tag-SNPs as the "Tests" on the left side, and the right panel shows a list of the SNPs in the region (the "Allele" column) and which tag-SNPs capture each (the "Test" column).

GVS is the color coding of SNPs on the basis of a priori predicted functional class. The most obvious advantage of Haploview is the aggressive tagging feature, although the mandatory "include" and "exclude" features are also useful. Presumably, both tools will continue to evolve and incorporate new features for the foreseeable future, in a healthy competition.

THE FUTURE OF TAG-SNPS

It is now technically feasible to genotype hundreds of thousands (or even millions) of SNPs on a single sample (Matsuzaki et al. 2004; Steemers et al. 2006); however, it remains economically unfeasible to do so for a large number of samples. Thus, reducing the number of SNPs genotyped while minimizing the loss of information has been an important challenge in candidate gene analysis. It is important to keep in mind that this is fundamentally an economic problem, and that genotypes are a perfectly elastic market. In order to run an association analysis of a single gene, a pathway, or even the whole genome, the budget of a researcher is limited by the cost per sample, not cost per genotype. That is, tag-SNP selection is primarily a means to reduce the number of genotypes required for a study so that it fits within a given budget. Eventually, it seems likely that the cost of resequencing the entire genome will approach the cost of genotyping genome-wide tag-SNPs, at which point tag-SNPs will become obsolete.

REFERENCES

Altshuler D., Brooks L.D., Chakravarti A., Collins F.S., Daly M.J., and Donnelly P. 2005. A haplotype map of the human genome. *Nature* **437:** 1299–1320.

Bhangale T.R., Rieder M.J., Livingston R.J., and Nickerson D.A. 2005. Comprehensive identification and characterization of diallelic insertion-deletion polymorphisms in 330 human candidate genes. *Hum. Mol. Genet.* **14:** 59–69.

Botstein D. and Risch N. 2003. Discovering genotypes underlying human phenotypes: past successes for mendelian disease, future approaches for complex disease. *Nat. Genet.* **33 Suppl:** 228–237.

Cardon L.R and Abecasis G.R. 2003. Using haplotype blocks to map human complex trait loci. *Trends Genet.* **19:** 135–140.

Carlson C.S., Eberle M.A., Rieder M.J., Yi Q., Kruglyak L., and Nickerson D.A. 2004. Selecting a maximally informative set of single-nucleotide polymorphisms for association analyses using linkage disequilibrium. *Am. J. Hum. Genet.* **74:** 106–120.

Collins F.S., Guyer M.S., and Chakravarti A. 1997. Variations on a theme: Cataloging human DNA sequence variation. *Science* **278:** 1580–1581.

de Bakker P.I., Yelensky R., Pe'er I., Gabriel S.B., Daly M.J., and Altshuler D. 2005. Efficiency and power in genetic association studies. *Nat. Genet.* **37:** 1217–1223.

Ding K., Zhou K., Zhang J., Knight J., Zhang X., and Shen Y. 2005. The effect of haplotype-block definitions on inference of haplotype-block structure and htSNPs selection. *Mol. Biol. Evol.* **22:** 148–159.

Gabriel S.B., Schaffner S.F, Nguyen H., Moore J.M., Roy J., Blumenstiel B., Higgins J., DeFelice M., Lochner A., Faggart M., et al. 2002. The structure of haplotype blocks in the human genome. *Science* **296:** 2225–2229.

Halldorsson B.V., Istrail S., and De La Vega F.M. 2004a. Optimal selection of SNP markers for disease association studies. *Hum. Hered.* **58:** 190–202.

Halldorsson B.V., Bafna V., Lippert R., Schwartz R., De La Vega F.M., Clark A.G., and Istrail S. 2004b. Optimal haplotype block-free selection of tagging SNPs for genome-wide association studies. *Genome Res.* **14:** 1633–1640.

Halperin E., Kimmel G., and Shamir R. 2005. Tag SNP selection in genotype data for maximizing SNP prediction accuracy. *Bioinformatics* (Suppl. 1) **21:** i195–i203.

Howie B.N., Carlson C.S., Rieder M.J., and Nickerson D.A. 2006. Efficient selection of tagging single-nucleotide polymorphisms in multiple populations. *Hum. Genet.* **120:** 58–68.

International HapMap Consortium. 2003. The International HapMap Project. *Nature* **426:** 789–796.

Jeffreys A.J., Kauppi L., and Neumann R. 2001. Intensely punctate meiotic recombination in the class II region of the major histocompatibility complex. *Nat. Genet.* **29:** 217–222.

Johnson G.C., Esposito L., Barratt B.J., Smith A.N., Heward J., Di Genova G., Ueda H., Cordell H.J., Eaves I.A., Dudbridge F., et al. 2001. Haplotype tagging for the identification of common disease genes. *Nat. Genet.* **29:** 233–237.

Ke X. and Cardon L.R. 2003. Efficient selective screening of haplotype tag SNPs. *Bioinformatics* **19:** 287–288.

Ke X., Durrant C., Morris A.P., Hunt S., Bentley D.R., Deloukas P., and Cardon L.R. 2004. Efficiency and consistency of haplotype tagging of dense SNP maps in multiple samples. *Hum. Mol. Genet.* **13:** 2557–2565.

Ke X., Miretti M.M., Broxholme J., Hunt S., Beck S., Bentley D.R., Deloukas P., and Cardon L.R. 2005. A comparison of tagging methods and their tagging space. *Hum. Mol. Genet.* **14:** 2757–2767.

Kimura M. and Ota T. 1973. The age of a neutral mutant persisting in a finite population. *Genetics* **75:** 199–212.

Kruglyak L. and Nickerson D.A. 2001. Variation is the spice of life. *Nat. Genet.* **27:** 234–236.

Lander E.S., Linton L.M., Birren B., Nusbaum C., Zody M.C., Baldwin J., Devon K., Dewar K., Doyle M., FitzHugh W., et al. (International Human Genome Sequencing Consortium). 2001. Initial sequencing and analysis of the human genome. *Nature* **409:** 860–921.

Lin Z. and Altman R.B. 2004. Finding haplotype tagging SNPs by use of principal components analysis. *Am. J. Hum. Genet.* **75:** 850–861.

Matsuzaki H., Dong S., Loi H., Di X., Liu G., Hubbell E., Law J., Berntsen T., Chadha M., Hui H., et al. 2004. Genotyping over 100,000 SNPs on a pair of oligonucleotide arrays. *Nat. Methods.* **1:** 109–111.

McVean G.A., Myers S.R., Hunt S., Deloukas P., Bentley D.R., and Donnelly P. 2004. The fine-scale structure of recombination rate variation in the human genome. *Science* **304:** 581–584.

Meng Z., Zaykin D.V., Xu C.F., Wagner M., and Ehm M.G. 2003. Selection of genetic markers for association analyses, using linkage disequilibrium and haplotypes. *Am. J. Hum. Genet.* **73:** 115–130.

Ng P.C. and Henikoff S. 2001. Predicting deleterious amino acid substitutions. *Genome Res.* **11:** 863–874.

———. 2003. SIFT: Predicting amino acid changes that affect protein function. *Nucleic Acids Res.* **31:** 3812–3814.

Patil N., Berno A.J., Hinds D.A., Barrett W.A., Doshi J.M., Hacker C.R., Kautzer C.R., Lee D.H., Marjoribanks C., McDonough D.P., et al. 2001. Blocks of limited haplotype diversity revealed by high-resolution scanning of human chromosome 21. *Science* **294:** 1719–1723.

Risch N. and Merikangas K. 1996. The future of genetic studies of complex human diseases. *Science* **273:** 1516–1517.

Rosenberg N.A., Pritchard J.K., Weber J.L., Cann H.M., Kidd K.K., Zhivotovsky L.A., and Feldman M.W. 2002. Genetic structure of human populations. *Science* **298:** 2381–2385.

Sebastiani P., Lazarus R., Weiss S.T., Kunkel L.M., Kohane I.S., and Ramoni M.F. 2003. Minimal haplotype tagging. *Proc. Natl. Acad. Sci.* **100:** 9900–9905.

Schulze T.G., Zhang K., Chen Y.S., Akula N., Sun F., and McMahon F.J. 2004. Defining haplotype blocks and tag single-nucleotide polymorphisms in the human genome. *Hum. Mol. Genet.* **13:** 335–342.

Sherry S.T., Ward M., and Sirotkin K. 1999. dbSNP-database for single nucleotide polymorphisms and other classes of minor genetic variation [In Process Citation]. *Genome Res.* **9:** 677–679.

Steemers F.J., Chang W., Lee G., Barker D.L., Shen R., and Gunderson K.L. 2006. Whole-genome genotyping with the single-base extension assay. *Nat. Methods.* **3:** 31–33.

Stram D.O., Haiman C.A., Hirschhorn J.N., Altshuler D., Kolonel L.N., Henderson B.E., and Pike M.C. 2003. Choosing haplotype-tagging SNPS based on unphased genotype data using a preliminary sample of unrelated subjects with an example from the Multiethnic Cohort Study. *Hum. Hered.* **55:** 27–36.

Templeton A.R., Clark A.G., Weiss K.M., Nickerson D.A., Boerwinkle E., and Sing C.F. 2000. Recombinational and mutational hotspots within the human lipoprotein lipase gene. *Am. J. Hum. Genet.* **66:** 69–83.

Venter J.C., Adams M.D., Myers E.W., Li P.W., Mural R.J., Sutton G.G., Smith H.O., Yandell M., Evans C.A., Holt R.A., et al. 2001. The sequence of the human genome. *Science* **291:** 1304–1351.

Wall J.D. and Pritchard J.K. 2003a. Assessing the performance of the haplotype block model of linkage disequilibrium. *Am. J. Hum. Genet.* **73:** 502–515.

———. 2003b. Haplotype blocks and linkage disequilibrium in the human genome. *Nat. Rev. Genet.* **4:** 587–597.

Watterson G.A. 1976. Reversibility and the age of an allele. I. Moran's infinitely many neutral alleles model. *Theor. Popul. Biol.* **10:** 239–253.

Weale M.E., Depondt C., Macdonald S.J., Smith A., Lai P.S., Shorvon S.D., Wood N.W., and Goldstein D.B. 2003. Selection and evaluation of tagging SNPs in the neuronal-sodium-channel gene *SCN1A*: Implications for linkage-disequilibrium gene mapping. *Am. J. Hum. Genet.* **73:** 551–565.

Zhang K. and Jin L. 2003. HaploBlockFinder: haplotype block analyses. *Bioinformatics* **19:** 1300–1301.

Zhang K., Qin Z., Chen T., Liu J.S., Waterman M.S., and Sun F. 2005. HapBlock: Haplotype block partitioning and tag SNP selection software using a set of dynamic programming algorithms. *Bioinformatics* **21:** 131–134.

Zhang K., Qin Z.S., Liu J.S., Chen T., Waterman M.S., and Sun F. 2004. Haplotype block partitioning and tag SNP selection using genotype data and their applications to association studies. *Genome Res.* **14:** 908–916.

WWW RESOURCES

http://gvs.gs.washington.edu/GVS/ The Genome Variation Server is a useful tool to extract genotypes from the dbSNP using multiple frames of reference. It also offers built-in implementation of the LDselect haplotype-free tag-SNP selection algorithm.

http://theta.ncifcrf.gov/gbrowse/start_db/prettybase_to_linkageformat.zip A converter from prettybase to ped format (HaploView).

http://www.broad.mit.edu/mpg/haploview/download.php Haploview is a useful tool for extracting genotype data and running tag-SNP selection. Unlike GVS, Haploview is a stand-alone Java Applet that manipulates genotype data in HapMap format.

http://www.hapmap.org/cgi-perl/gbrowse/gbrowse/hapmap/ The HapMap project is a major repository for genotype data providing remarkably detailed data sets in three geographically discrete human subpopulations: Europeans, Africans, and Asians. Tools built upon this resource are currently the state of the art for those regions of the genome that have not been targeted for resequencing.

http://www.ncbi.nlm.nih.gov/SNP/dbSNP is the central standard repository for genotyping information, including HapMap as well as other targeted resequencing projects.

http://www.pharmgat.org/pharmgat/Documentation/help/pedtopb A converter from ped format to prettybase format.

20 | Selection and Evaluation of tag-SNPs Using Tagger and HapMap

Paul I.W. de Bakker

Program in Population and Medical Genetics, Broad Institute of MIT and Harvard, Cambridge, Massachusetts 02142

INTRODUCTION

The International HapMap Project has produced a public resource that contains more than three million single-nucleotide polymorphisms (SNPs) genotyped across the human genome in 270 DNA samples from four population samples (International HapMap Consortium 2005). By documenting not only the location of these SNPs but also the correlation structure between them, the HapMap can guide the design and analysis of genome-wide association studies. This chapter describes how to use the Tagger Web server (http://www.broad.mit.edu/mpg/tagger/; de Bakker et al. 2005) for tag-SNP selection and evaluation using HapMap data.

Tagger has also been implemented in the stand-alone Haploview program (http://www.broad.mit.edu/mpg/haploview/; Barrett et al. 2005), described in Chapter 21. Before we describe the practical aspects of tagging, we first address three key questions with regard to the utility of an incomplete resource such as HapMap and tagging approaches in general.

How Well Does HapMap Provide Information about Common Variation, Including Those Variants Not Observed in HapMap?

Using HapMap-ENCODE data with near-complete ascertainment of common variation, it was estimated that HapMap after Phase II captures 94% of common SNPs with an allele frequency $\geq 5\%$ in CEU (Utah residents with northern and western European ancestry) and CHB+JPT (Han Chinese from Beijing, China, and Japanese from Tokyo, Japan), and 81% of common SNPs in YRI (Yoruba from Ibadan, Nigeria) with a high r^2 of ≥ 0.8. Hence, most common variation is represented by SNPs listed on HapMap, albeit to a lesser degree in recent African-derived populations (International HapMap Consortium 2005). Not surprisingly, coverage of less common (i.e., rare) variants is not as complete.

283

How Is Coverage of Common Variation Diminished When Only a Subset of SNPs (Tags) Is Picked from a Reference Panel Such as HapMap?

There is an extensive correlation structure in nearby markers. Substantial efficiency gains can be obtained by exploiting the observed pair-wise correlations between SNPs. We recently demonstrated that a haplotype-based (i.e., "aggressive") tagging approach can shift this balance even further without sacrificing power (de Bakker et al. 2005). Power to detect an association also appeared to be remarkably robust relative to the (in)completeness of the reference panel.

How Well Do Tags Picked from HapMap Samples Capture Common Variation in Other Population Samples?

Many studies have evaluated the transferability of tag SNPs in various population samples (Nejentsev et al. 2004; Ahmadi et al. 2005; Mueller et al. 2005; Ramirez-Soriano et al. 2005; de Bakker et al. 2006a,b; Gonzalez-Neira et al. 2006; Huang et al. 2006; Montpetit et al. 2006; Ribas et al. 2006; Stankovich et al. 2006). Generally, these studies show that samples from the Centre d'Etude du Polymorphisme Humain collection (such as the HapMap CEU samples) capture common variation effectively in other European-derived samples. Similar results have been obtained in comparing the CHB and JPT HapMap population samples and other samples of Asian origin. For any population sample, the performance will, of course, depend on the local extent of linkage disequilibrium (LD) and on the extent of recent admixture (e.g., African-Americans). It is encouraging that HapMap samples can provide good collective coverage of common variation in many population samples across the world (Conrad et al. 2006; de Bakker et al. 2006b).

TAG-SNP SELECTION

The first step in tag-SNP selection is to define the set of alleles that are to be captured by the tags and tested for a genotype–phenotype correlation. This is achieved by specifying a specific locus (or multiple loci) in HapMap by providing physical coordinates (such as chr6:32000000–33000000 for the MHC). Currently, the Tagger Web server uses a local copy of the phased data of HapMap release 21, and coordinates must be provided relative to NCBI build 35 (UCSC build hg17), but this may be updated with future HapMap releases. This functionality will be particularly convenient if a planned study comprises multiple candidate genes.

Instead of using the local HapMap copy, it is possible to upload a reference panel as an external file in "ped" file format or as a HapMap genotype data dump. Such files can be phased automatically using the emphase program written by Nick Patterson. This program is based on the expectation-maximization algorithm and the partition-ligation approach. It also takes into account the familial relationships between samples (e.g., YRI or CEU trios). Alternatively, files with phased haplotypes can be uploaded. External genotype or haplotype files need to be accompanied by an information file (containing SNP identifiers and chromosomal positions). Note that there is a file size upload limit (see the Tagger Web site for details on file formats).

By default, Tagger uses a minimal allele frequency of 5%: All alleles (single SNPs) with ≥5% frequency are to be captured. Note that less common alleles have true allele frequencies that can fluctuate about this threshold. Inevitably, there will be some alleles observed in HapMap at <5% that may well have a population allele frequency above 5%, and vice versa. Lowering the frequency threshold will result in a larger set of alleles, which will likely require more tags. This illustrates the trade-off between genotyping cost and completeness. Note that it is also possible to specify exactly the set of alleles (single SNPs and/or haplotypes) that are to be captured, conditional on the specified frequency threshold.

The next step is to specify how tags are to be picked. Tagger picks tag-SNPs (tags) and defines allelic tests so as to capture all defined alleles at a user-defined r^2 level (minimal coefficient of determination). Setting the r^2 threshold to 1.0 results in a nonredundant set of tag-SNPs, where all alleles will have a perfect proxy tag.

Tagger combines the simplicity of a pair-wise approach (such as that of LDSelect; see Carlson et al. 2004) with the potential efficiency of multimarker (haplotype) approaches (Stram et al. 2003; Weale et al. 2003). The latter is achieved by first picking pair-wise tags, and then iteratively dropping ("peeling back") tags one by one and replacing them with a specific multimarker predictor (using any of the remaining tag SNPs). The predictor is accepted only if it can capture the alleles originally captured by the discarded tag at the required r^2. Otherwise, the provisionally dropped tag is considered indispensable and is kept. This multimarker approach essentially finds an identical set of 1 d.f. tests of association, except that it uses certain specific haplotypes as effective surrogates for single tag-SNPs, thereby requiring fewer tag-SNPs for genotyping. These multimarker predictors are explicitly recorded so that these tests can be included in association analyses (in addition to the single-marker tests).

Each investigator will have to decide whether to use the pair-wise or the multimarker method. The multimarker method can be more efficient (15–35%, depending on local LD structure and extent of the region). However, it requires greater genotyping accuracy and completeness because of the obligatory multimarker tests. Furthermore, phase uncertainty of the multimarker tests can affect the performance, although this is likely only a small effect given the generally strong LD among the SNPs within each multimarker predictor.

It is possible to force in sets of SNPs using the "include tags" function in Tagger. This is useful if more tags need to be picked in a region (in addition to SNPs that have already been genotyped) or if some SNPs are important enough to be included directly (because of functional data or a reported association). Note that Tagger will only take included SNPs into account if they exist on HapMap. (Otherwise, it is impossible to compute which alleles are captured by these included SNPs through LD.) It is also possible to specifically exclude SNPs as tags; for example, if they will likely result in genotyping failure. In addition, every SNP can be given a so-called "design score," where high-scoring SNPs will be preferentially picked. A design score threshold can be set to prevent low-scoring SNPs from being picked as tags (a "0" indicates that no SNPs will be excluded a priori).

TAG-SNP EVALUATION

The section above explains how to select tag-SNPs from scratch or how to pick additional tags in a given region. However, Tagger can also use a fixed set of SNPs to capture the observed variation in a region. To do this, Tagger requires the user to upload an "include tags" file containing SNP names (click the "evaluate" checkbox to not pick extra tags). Generally, because genotyping of selected tag-SNPs may not be successful, it is useful to reevaluate how well the working SNPs in a study capture the remaining untyped SNPs. Recently, we used this functionality to evaluate the effective coverage of some of the commercially available whole-genome genotyping products (Pe'er et al. 2006).

This following example illustrates how to evaluate the coverage of SNPs in the Affymetrix Mapping 500K GeneChip product with regard to the common variation in the *TCF7L2* gene. *TCF7L2* is located on chromosome 10, from position 114,700,200 to 114,916,057 (using NCBI build 35 coordinates). This gene is associated with type 2 diabetes risk (Grant et al. 2006), an effect that has been robustly replicated in several studies. Here we want to know how well the common SNPs in this gene are captured by pair-wise LD in the HapMap YRI population sample using the Affymetrix SNPs. We extend the conventional gene boundaries by 30 kb upstream and 30 kb downstream to include possible promoter or other regulatory elements. However, it is possible that some Affymetrix SNPs lie outside the gene and are in strong LD with SNPs inside the gene. Since such SNPs can contribute substantially to the coverage across the gene, we need to extend the region of interest even further; for example, by 200 kb up- and downstream, a reasonable distance beyond which there is typically little LD between SNPs:

1. Log on to the Tagger Web site (http://www.broad.mit.edu/mpg/tagger/) and click on the "Tagger server" link.

2. In the "chromosomal landmarks" box, enter the chromosomal positions (in this example, chr10:114,470,201–115,146,051).

3. Select YRI from the "HapMap analysis panel" pull-down menu.

4. Using the "Browse" function in the "include tag SNPs" panel, upload the 124 SNPs that lie in this genomic region from the Affymetrix database.

5. Click the "only evaluate these SNPs" box.

6. Click the "pairwise" box in the "Tagger mode" panel.

7. Submit the request to Tagger job using default values (for "minimum allele frequency," this is set at 5%).

Once the job is done, the results page will show that there are 114 SNPs as tags (that is, 10 SNPs were either monomorphic in YRI or did not pass HapMap quality control for Release 21), and 531 SNPs in the region (satisfying the ≥5% allele frequency threshold), of which 41% are captured with a pair-wise maximum r^2 ≥0.8 (with a mean maximum r^2 of 0.57). The "tests" file can be downloaded and analyzed further. This file is in tabular format and lists, for every SNP, the allelic test that captures it and the r^2 between them. For example, rs12255372 is captured with $r^2 = 0.67$, and rs7903146 with a lower r^2 of 0.37 (these two SNPs are the two most strongly associated SNPs in the original study that identified *TCF7L2* as a risk gene).

If Tagger is run again with the same parameters, but with the "aggressive (specified multimarker tests)" tagging enabled, 44% of the 531 SNPs are captured with r^2 ≥0.8. For example, SNP rs7074334 (MAF = 9%) is only poorly captured by a single tag (rs11817282) with a pair-wise r^2 of 0.26, but is effectively captured by a 2-SNP haplotype (TA) formed by rs7085980 and rs7897837 with near-perfect LD ($r^2 = 0.91$), illustrating how knowledge of the underlying haplotype structure can be exploited to gain coverage and power.

REFERENCES

Ahmadi K.R., Weale M.E., Xue Z.Y., Soranzo N., Yarnall D.P., Briley J.D., Maruyama Y., Kobayashi M., Wood N.W., Spurr N.K., et al. 2005. A single-nucleotide polymorphism tagging set for human drug metabolism and transport. *Nat. Genet.* **37:** 84–89.

Barrett J.C., Fry B., Maller J., and Daly M.J. 2005. Haploview: Analysis and visualization of LD and haplotype maps. *Bioinformatics* **21:** 263–265.

Carlson C.S., Eberle M.A., Rieder M.J., Yi Q., Kruglyak L., and Nickerson D.A. 2004. Selecting a maximally informative set of single-nucleotide polymorphisms for association analyses using linkage disequilibrium. *Am. J. Hum. Genet.* **74:** 106–120.

Conrad D.F., Jakobsson M., Coop G., Wen X., Wall J.D., Rosenberg N.A., and Pritchard J.K. 2006. A worldwide survey of haplotype variation and linkage disequilibrium in the human genome. *Nat. Genet.* **38:** 1251–1260.

de Bakker P.I.W., Graham R.R., Altshuler D., Henderson B.E., and Haiman C.A. 2006a. Transferability of tag SNPs to capture common genetic variation in DNA repair genes across multiple populations. *Pac. Symp. Biocomput.* **2006:** 478–486.

de Bakker P.I.W., Yelensky R., Pe'er I., Gabriel S.B., Daly M.J., and Altshuler D. 2005. Efficiency and power in genetic association studies. *Nat. Genet.* **37:** 1217–1223.

de Bakker P.I.W., Burtt N.P., Graham R.R., Guiducci C., Yelensky R., Drake J.A., Bersaglieri T., Penney K.L., Butler J., Young S., et al. 2006b. Transferability of tag SNPs in genetic association studies in multiple populations. *Nat. Genet.* **38:** 1298–1303.

Gonzalez-Neira A., Ke X., Lao O., Calafell F., Navarro A., Comas D., Cann H., Bumpstead S., Ghori J., Hunt S., et al. 2006. The portability of tagSNPs across populations: A worldwide survey. *Genome Res.* **16:** 323–330.

Grant S.F., Thorleifsson G., Reynisdottir I., Benediktsson R., Manolescu A., Sainz J., Helgason A., Stefansson H., Emilsson V., Helgadottir A., et al. 2006. Variant of transcription factor 7-like 2 (TCF7L2) gene confers risk of type 2 diabetes. *Nat. Genet.* **38:** 320–323.

Huang W., He Y., Wang H., Wang Y., Liu Y., Wang Y., Chu X., Wang Y., Xu L., Shen Y., et al. 2006. Linkage disequilibrium sharing and haplotype-tagged SNP portability between populations. *Proc. Natl. Acad. Sci.* **103:** 1418–1421.

International HapMap Consortium. 2005. A haplotype map of the human genome. *Nature* 437: 1299–1320.

Montpetit A., Nelis M., Laflamme P., Magi R., Ke X., Remm M., Cardon L., Hudson T.J., and Metspalu A. 2006. An evaluation of the performance of tag SNPs derived from HapMap in a Caucasian population. *PLoS Genet.* 2: e27.

Mueller J.C., Lohmussaar E., Magi R., Remm M., Bettecken T., Lichtner P., Biskup S., Illig T., Pfeufer A., Luedemann J., et al. 2005. Linkage disequilibrium patterns and tagSNP transferability among European populations. *Am. J. Hum. Genet.* 76: 387–398.

Nejentsev S., Godfrey L., Snook H., Rance H., Nutland S., Walker N.M., Lam A.C., Guja C., Ionescu-Tirgoviste C., Undlien D.E., et al. 2004. Comparative high-resolution analysis of linkage disequilibrium and tag single nucleotide polymorphisms between populations in the vitamin D receptor gene. *Hum. Mol. Genet.* 13: 1633–1639.

Pe'er I., de Bakker P.I.W., Maller J., Yelensky R., Altshuler D., and Daly M.J. 2006. Evaluating and improving power in whole-genome association studies using fixed marker sets. *Nat. Genet.* 38: 663–667.

Ramirez-Soriano A., Lao O., Soldevila M., Calafell F., Bertranpetit J., and Comas D. 2005. Haplotype tagging effi-ciency in worldwide populations in CTLA4 gene. *Genes Immun.* 6: 646–657.

Ribas G., Gonzalez-Neira A., Salas A., Milne R.L., Vega A., Carracedo B., Gonzalez E., Barroso E., Fernandez L.P., Yankilevich P., et al. 2006. Evaluating HapMap SNP data transferability in a large-scale genotyping project involving 175 cancer-associated genes. *Hum. Genet.* 118: 669–679.

Stankovich J., Cox C.J., Tan R.B., Montgomery D.S., Huxtable S.J., Rubio J.P., Ehm M.G., Johnson L., Butzkueven H., Kilpatrick T.J., et al. 2006. On the utility of data from the International HapMap Project for Australian association studies. *Hum. Genet.* 119: 220–222.

Stram D.O., Haiman C.A., Hirschhorn J.N., Altshuler D., Kolonel L.N., Henderson B.E., and Pike M.C. 2003. Choosing haplotype-tagging SNPs based on unphased genotype data using a preliminary sample of unrelated subjects with an example from the Multiethnic Cohort Study. *Hum. Hered.* 55: 27–36.

Weale M.E., Depondt C., Macdonald S.J., Smith A., Lai P.S., Shorvon S.D., Wood N.W., and Goldstein D.B. 2003. Selection and evaluation of tagging SNPs in the neuronal-sodium-channel gene SCN1A: Implications for linkage-disequilibrium gene mapping. *Am. J. Hum. Genet.* 73: 551–565.

WWW RESOURCES

http://www.broad.mit.edu/mpg/haploview/ Haploview is a useful tool for extracting genotype data and running tag-SNP selection.

http://www.broad.mit.edu/mpg/tagger/ Tagger can be used for tag-SNP selection and evaluation using HapMap data.

21 Haploview: Visualization and Analysis of SNP Genotype Data

Jeffrey C. Barrett

Wellcome Trust Centre for Human Genetics, University of Oxford, Oxford, OX3 7BN, United Kingdom

INTRODUCTION

Association studies involve accessing, parsing, generating, and analyzing large volumes of data, often carried out in many steps over many months. Large-scale surveys of genetic variation such as the International HapMap Project (Altshuler et al. 2005) and rapidly increasing volumes of SNP genotyping data have created exciting opportunities for these studies. However, they have further exacerbated the difficulty of curating and analyzing such data. Haploview is a program developed in Mark Daly's lab at the Broad Institute, designed to bundle many everyday analysis tasks into one easy-to-use package. It is an open-source program written in Java and capable of running on Windows, MacOS, and UNIX platforms. The program, its documentation, and sample files (including those used in the examples in this chapter) are available at http://www.broad.mit.edu/mpg/haploview/. Haploview has a number of features that are useful throughout association studies. Several of these are illustrated below by following a hypothetical association study from design to execution. Haploview is used to (1) analyze HapMap data and choose tag-SNPs, (2) evaluate the quality of disease genotype data, (3) test for association, and (4) evaluate a region for follow-up of a positive association.

ANALYZING HAPMAP DATA

The International HapMap Project (Altshuler et al. 2005) is a massive SNP discovery and genotyping (over 4 million SNPs genotyped in 270 samples) effort designed to illuminate genome-wide patterns of variation. Haploview is designed to integrate seamlessly with the HapMap and can load genotype files downloaded from the HapMap Web site (http://www.hapmap.org/) or directly access phased HapMap data over the Internet. The HapMap Web site can be searched by chromosomal location, or by the name of a feature such as a gene or a particular SNP.

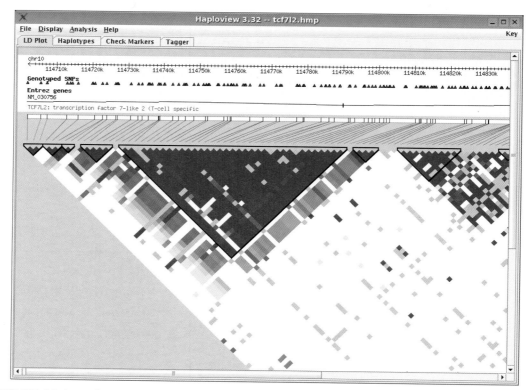

FIGURE 21-1. Haploview LD display for gene *TCF7L2* genotyped in 90 Utah residents of European descent as part of the HapMap Project. A reference track with chromosomal location, SNP positions, and gene positions runs along the top. Strong LD between markers is colored red, no LD colored white, and lack of statistical evidence colored light blue. The "block-like" pattern of LD often observed in the genome is evident in the large triangles representing regions of high LD, sharply divided by narrow areas where even adjacent markers are completely independent.

VISUALIZING LINKAGE DISEQUILIBRIUM

An increasingly clear understanding of patterns of linkage disequilibrium (LD) (Daly et al. 2001; Phillips et al. 2003; Hinds et al. 2005) has allowed a number of important advances in the design and analysis of association studies. Most of the genome falls into "blocks" of strong LD separated by sharp breakpoints formed by historical recombination hot spots (Myers et al. 2005). Figure 21-1 shows the LD pattern from HapMap data for the gene *TCF7L2* (the target gene for the example used in this chapter) after filtering out SNPs with a minor allele frequency below 5%. Users can customize color schemes, spacing, and the level of detail for the LD plot. Gene and SNP location tracks can be automatically downloaded from the Internet to provide a contextual track along the top of the display. LD blocks are calculated and shown on the display as solid black lines. The block boundaries should not be considered fixed, but rather approximations. As such, these can be customized both by block definition and by clicking and dragging along the top of the display to hand-define blocks.

CHOOSING TAGGING SNPs

Selection of tagging SNPs is perhaps the most common application of the LD pattern knowledge provided by projects like the HapMap. Since regions of strong LD are fundamentally characterized by high levels of redundancy from one SNP to the next, it is possible to capture all of the information in such a region with only a few SNPs. Haploview provides an implementation of Paul de

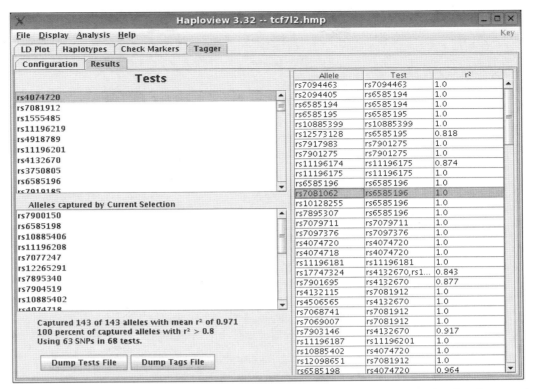

FIGURE 21-2. Haploview Tagger interface for HapMap *TCF7L2* data. The tag SNPs are listed in the upper left box, and the SNPs tagged by the currently highlighted SNP (rs4074720) are shown in the lower left box. The right side of the screen lists every SNP in the data set and its best tag. Summary information and buttons for saving output are at the bottom left.

Bakker's Tagger algorithm for choosing SNPs (de Bakker et al. 2005). At the most basic level, Haploview evaluates the LD among all the SNPs in a data set and chooses the smallest possible subset that adequately captures all the information. The process can be customized in a variety of ways, such as considering SNP "design scores," the precise threshold of LD required to consider a SNP tagged, or the inclusion of multimarker tags. More information can be found in Chapter 20, which describes the Tagger algorithm.

Running Tagger with the default settings on the minor allele frequency-filtered *TCF7L2* data set indicates that approximately 50 SNPs are required to capture all the common variation in the gene. The results panel (Fig. 21-2) shows which SNPs are captured by each tag. This data set exhibits a common pattern: A handful of the "best" (i.e., most informative) tags capture the bulk of the information, whereas the remainder of the tags capture only themselves (and are thus not very efficient) (Barrett and Cardon 2006). Since genotyping for association studies is often limited by available resources, the Tagger analysis can be re-run to choose only the best *N* markers. For example, just ten tags capture half the variation in *TCF7L2*. Haploview can export a "tags" file to list the SNPs to be genotyped and a "tests" file to list these SNPs along with any haplotype tests to be performed in our association study (see Chapter 20 for more information on haplotype tests).

ANALYZING AN ASSOCIATION STUDY

The second principal entry point to Haploview is with genotype data generated for an association study. Haploview provides a unified analysis tool to perform quality control checks on data, investigate LD and haplotype patterns, and test for association. In the *TCF7L2* study, ten tags were chosen from the HapMap, and simulated data were generated for 500 cases and 500 controls.

DATA QUALITY CONTROL

Whereas technological advances have made SNP genotyping more automated and reliable, quality control is still a key step in any analysis. Haploview calculates a number of quality control metrics, which can be used to diagnose poorly performing SNPs and DNA samples:

Genotyping call rate: Large amounts of missing data (either for a SNP or an individual sample) are an indication of poor performance or sample quality. Such markers and samples should probably be excluded from analysis to reduce the risk of artifactual errors.

Hardy-Weinberg equilibrium: Deviations from Hardy-Weinberg equilibrium can be indicative of a SNP mapping to multiple genomic locations, or an allele-calling algorithm that underperforms on one genotype class (such as failing to correctly call heterozygotes).

Mendelian inheritance errors: For family-based samples, Haploview tallies errors in transmissions from parents to offspring, another indicator of genotyping error.

Minor allele frequency: Although not a quality measure per se, it is often useful to filter SNPs on frequency for subsequent analyses.

Markers are filtered out according to some default thresholds on these tests, all of which can be adjusted manually. Markers can also be forced in or out manually, overriding the filters.

TESTING FOR ASSOCIATION

Haploview features family-based and case/control association tests for both single SNPs and haplotypes. The results show associated alleles, allele counts for cases and controls, a χ^2 test statistic, and a p value for evaluating significance. Although many other association tests (such as linear regression) are available, the basic tests offered by Haploview are a good first step in evaluating association.

By loading the saved "tests" file generated from the HapMap tagging exercise along with the association study genotypes, Haploview is automatically configured to perform this set of single SNP and haplotype tests. Haploview uses an EM algorithm (Qin et al. 2002) to generate phased haplotypes both for population frequency (the "Haplotypes" tab) and for association testing. The *TCF7L2* data set shows a strong association to SNP rs10509969, which appears worthy of close follow-up.

Because association studies involve testing numerous variants and haplotypes, it is important to consider these multiple tests when evaluating significance. Haploview generates empirical (i.e., irrespective of total number of tests) significance values by a permutation procedure. Case and control labels are permuted among individuals, and then all association tests are performed as normal. By repeating this procedure many times, Haploview builds a null distribution of test statistics that occur in the data set purely by chance. A true measure of the significance of the actual association test is the number of times one of the random data sets is more significant than our putative observed association. Using the permutation procedure on the simulated association study data shows that the detected association remains significant (Fig. 21-3) after permutation.

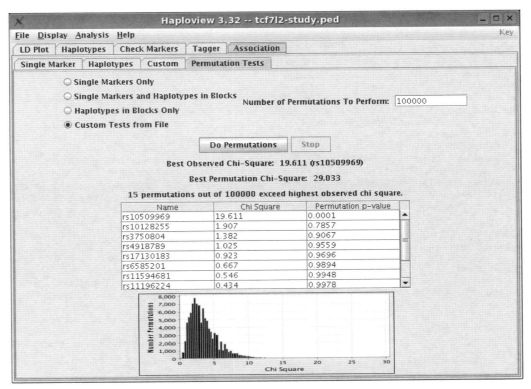

FIGURE 21-3. Haploview association permutation test interface for a simulated association study of 500 cases and 500 controls. The SNP rs10509969 is shown to be significantly associated after 100,000 permutations. The *p* value of 0.0001 is an empirical significance level, corrected for the multiple tests in the experiment.

FOLLOW-UP

Once an effect is detected, the next task is to dissect it and search for putative causal alleles. The return to and review of the original HapMap data indicates that the associated SNP is in a 12-kb block of strong LD. Options for follow-up include additional genotyping in this region, database searches for known nonsynonymous coding variation, or directed resequencing.

SUMMARY

The combination of cheap, high-throughput genotyping and public catalogs of common variation such as the HapMap is yielding large amounts of data and promising opportunities to discover genes involved in complex human diseases. The challenge of converting this raw resource into biological understanding requires tools that provide an easy interface to the most important analysis tasks. Haploview is a useful tool for a wide range of these challenges, such as choosing tag-SNPs, evaluating LD, and testing for association. A comprehensive study will use many software tools to glean as much information as possible from the data, but Haploview provides a good starting point for many of the most common questions.

ACKNOWLEDGMENTS

Haploview is developed and maintained by Jeffrey Barrett, Julian Maller, David Bender, and Mark Daly. Thanks to B. Herrera and M. Lincoln for comments on the manuscript.

REFERENCES

Altshuler D., Brooks L.D., Chakravarti A., Collins F.S., Daly M.J., and Donnelly P. 2005. A haplotype map of the human genome. *Nature* **437:** 1299–1320.

Barrett J.C. and Cardon L.R. 2006. Evaluating coverage of genome-wide association studies. *Nat. Genet.* **38:** 659–662.

Daly M.J., Rioux J.D., Schaffner S.F., Hudson T.J., and Lander E.S. 2001. High-resolution haplotype structure in the human genome. *Nat. Genet.* **29:** 229–232.

de Bakker P.I.W., Yelensky R., Pe'er I., Gabriel S.B., Daly M.J., and Altshuler D. 2005. Efficiency and power in genetic association studies. *Nat. Genet.* **37:** 1217–1223.

Hinds D.A., Stuve L.L., Nilsen G.B., Halperin E., Eskin E., Ballinger D.G., Frazer K.A., and Cox D.R. 2005. Whole-genome patterns of common DNA variation in three human populations. *Science* **307:** 1072–1079.

Myers S., Bottolo L., Freeman C., McVean G., and Donnelly P. 2005. A fine-scale map of recombination rates and hotspots across the human genome. *Science* **310:** 321–324.

Phillips M.S., Lawrence R., Sachidanandam R., Morris A.P., Balding D.J., Donaldson M.A., Studebaker J.F., Ankener W.M., Alfisi S.V., Kuo F.S., et al. 2003. Chromosome-wide distribution of haplotype blocks and the role of recombination hot spots. *Nat. Genet.* **33:** 382–387.

Qin Z.S., Niu T., and Liu J.S. 2002. Partition-ligation-expectation-maximization algorithm for haplotype inference with single-nucleotide polymorphisms. *Am. J. Hum. Genet.* **71:** 1242–1247.

WWW RESOURCES

http://www.broad.mit.edu/mpg/haploview/ Haploview program home page, Broad Institute of MIT and Harvard. Haploview bundles many everyday analysis tasks into one easy-to-use package. It is an open-source program written in Java and capable of running on Windows, MacOS, and UNIX platforms.

http://www.hapmap.org/ International HapMap Project home page. The HapMap project is a major repository for genotype data providing remarkably detailed data sets in three geographically discrete human subpopulations: Europeans, Africans, and Asians. Tools built upon this resource are currently the state of the art for those regions of the genome that have not been targeted for resequencing.

22 Considerations for Copy Number Analysis of FFPE Samples

Sharoni Jacobs

Affymetrix, Santa Clara, California 95051

INTRODUCTION

WGSA, the Mapping Arrays, and Copy Number Detection

The Whole Genome Sampling Analysis (WGSA) assay can simultaneously genotype thousands of single-nucleotide polymorphisms (SNPs) using the Affymetrix GeneChip® Mapping Arrays (Kennedy et al. 2003). Using the same Mapping array data, quantitative analysis of copy number (CN) can also be performed for each SNP (Bignell et al. 2004; Huang 2004; Lin et al. 2004; Nannya et al. 2005). Thus, in one experiment, the Mapping 500K Array set, comprising the Mapping 250K Nsp Array and the Mapping 250K Sty Array, can provide genotype and CN data for more than 500,000 SNPs.

The Mapping Arrays consist of short (25-mer) oligonucleotide probes that specifically anneal to alternate alleles at a given SNP. During WGSA, DNA is digested with a restriction enzyme appropriate for

the targeted microarray (either NspI or StyI for the Mapping 500K Set). Digested ends are ligated to short oligonucleotides, which are the templates for primer annealing in the following PCR step. During PCR, a subset of the genome is selectively amplified based on the size of the digested DNA fragments—specifically, only fragments of ~100–1100 bp are amplified. Following this, PCR amplicons are fragmented by DNase I digestion, labeled with biotin, and hybridized to the microarrays. The arrays are then stained with fluorescent-labeled streptavidin, and fluorescent probe intensities are measured. These intensity values are compared to ascertain the genotype for each SNP represented on the array.

CN at a given SNP location can also be determined using SNP probe intensity values (Huang 2004). When the SNP is assigned a heterozygous genotype, separate intensity measurement of labeled DNA hybridized to probes specific to the different alleles enables identification of allele-specific CN (LaFramboise et al. 2005; Huang et al. 2006). A benefit of allele-specific CN detection is that one can detect regions of a chromosome that have been duplicated, but where the homologous region has been deleted (thus no overall CN change has occurred) (as seen in the p-terminal in Fig. 22-1). When DNA does exhibit an overall CN change, allele-specific CN can add valuable information by distinguishing different types of chromosomal changes, such as a gain of both alleles (not shown), a gain of a single allele (as seen in the q-terminal in Fig. 22-1), and a gain of a single allele accompanied by a loss of the alternate allele (as seen in the region denoted as "3+0" in Fig. 22-1).

Quantitative Analysis of DNA Content Using WGSA

CN detection provides a quantitative assessment of DNA copies. It is, therefore, essential that the process of WGSA does not skew the relative representation of DNA segments. The steps of digestion, ligation, and fragmentation occur without bias to particular DNA regions. Additionally, biotin labeling is performed as an end-labeling reaction using Terminal DeoxyTransferase (TDT), which preserves the proportional representation of DNA fragments.

In addition, PCR, which presents the highest potential for changing the relative quantities of DNA fragments, is terminated while still within the linear amplification range. Observation of data from the Mapping 500K Arrays indicates that there is minimal effect on relative quantities of DNA fragments when applying WGSA to high-quality DNA fragments, and most CN analysis tools are able to ignore even this minimal effect (Huang 2004; Lin et al. 2004).

FFPE and Copy Number Detection

The use of formalin-fixed, paraffin-embedded (FFPE) samples as a source of DNA requires special considerations for sample processing and data analysis. These samples are often degraded, and the extent of degradation can vary greatly between samples; thus, some FFPE sources yield extensively degraded DNA (providing strands no greater than 100 bp), whereas other FFPE samples provide relatively intact DNA. In addition to the variable state of degradation, the DNA may be chemically modified. Finally, FFPE DNA samples often include remnant contaminants that can inhibit downstream molecular reactions. Guidelines for application of FFPE samples to the Mapping arrays are described in Chapter 16.

The step in WGSA most affected by FFPE DNA quality is PCR. As described in the Introduction, high-quality DNA samples are amplified into fragments up to 1100 bp during

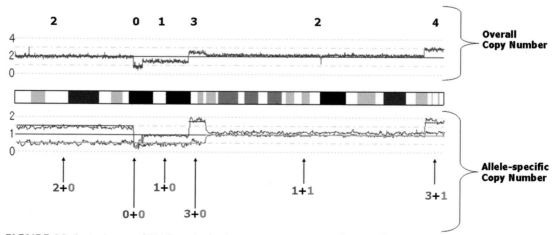

FIGURE 22-1. An image of CN for a single chromosome in CNAG. The overall CN prediction is presented above in blue as a moving average of 10 SNPs. Below, allele-specific CN is displayed as a red and a green line for the more abundant and less abundant allele, respectively. Black numbers represent the overall CN for each distinct region, and red and green numbers represent the allele-specific CN for these regions.

PCR. Degraded FFPE samples, however, often do not provide templates of this size. Therefore, the larger amplicons and associated SNPs may be absent after PCR. Additionally, the presence of contaminants and DNA modifications can interfere with the reaction, decreasing amplification efficiency and introducing fragment size biases in which smaller fragments are better amplified than larger fragments.

This fragment size bias can create a situation where, by default and independent of the true presence of chromosomal aberrations, small fragments can indicate CN gains and large fragments can indicate deletions (Fig. 22-2, top row). It is possible to correct this fragment size bias in silico, as well as to accommodate a complete failure of large fragment amplification, by excluding them from analysis. In this chapter, we describe the steps needed to obtain reliable CN predictions from degraded and contaminated FFPE samples.

SOFTWARE REQUIREMENTS

Numerous software packages are available for Affymetrix Mapping Array data CN analysis. These include the Affymetrix Copy Number Tool (CNAT; Huang 2004), available for free at the Affymetrix Web site www.affymetrix.com; academic freeware such as dChipSNP (Lin et al. 2004),

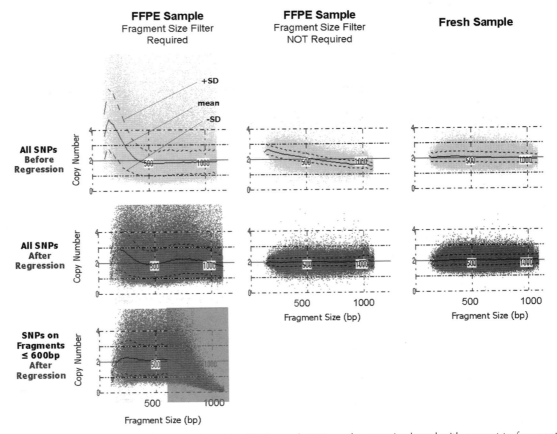

FIGURE 22-2. CN versus fragment size plots. CN for each SNP on the array is plotted with respect to fragment size. In the top row, raw CN is displayed: The solid line denotes the mean CN; the two dotted lines mark the S.D. The mean CN fluctuates according to fragment size for FFPE samples, whereas fluctuation is minimal in the fresh sample. In the middle row, quadratic regression is applied to correct for fragment size bias. This results in a consistent mean CN regardless of fragment size for one of the FFPE samples (*middle column*), indicating that no fragment size filter is required. Regression is insufficient to remove the bias for the other FFPE sample (*left column*). For this sample, exclusion of SNPs on fragment sizes >600 bp in combination with regression removes most of the fragment size bias from CN predictions for the SNPs maintained in the analysis (*bottom row*).

CNAG (Nannya et al. 2005), and GEMCA (Ishikawa et al. 2005; Komura et al. 2006); and commercial tools such as Partek® Genomics Suite, Sapio Science's Exemplar for Copy Number, and Stratagene's ArrayAssist® Copy Number. Any of these can be used for CN detection when using high-quality DNA. At the time of writing, only CNAT 4.0 and CNAG 2.0 are suitable for the analysis of FFPE samples; therefore, this chapter focuses on their usage (earlier versions of CNAT should not be applied to FFPE sample analysis). In the future, other tools may be modified so that they can also be used for FFPE analysis.

For proper analysis of FFPE samples, two characteristics in a software tool are required (Jacobs et al. 2007). The first is the ability to compensate against fragment size bias. This can be done by quadratic regression, which is automatically implemented in both CNAT and CNAG. The second is the ability to filter out SNPs based on fragment size. Filtering can be accomplished by manual input of a fragment size threshold. Currently, most applications allow users to assign a fragment size threshold, but only CNAT and CNAG correct for fragment size bias.

REFERENCE CONSIDERATIONS

Choice of Reference Type, Quantity, and Gender

To determine quantitatively the number of copies of DNA present in a sample, a reference with a known ploidy is required. The source of these samples is typically from "normal," nondiseased individuals (e.g., HapMap samples); non-tumor tissue from patients with cancer; or unaffected parents of children with a congenital disease. The reference can be a single DNA sample (as in the case of tumor–normal paired analysis), or a combination of multiple DNA samples. When using CNAT for unpaired analyses, a minimum of 25 references is recommended to obtain the highest signal-to-noise ratio. For CNAG, when manually selecting a pool of references, a minimum of 5 references is suggested.

The software tools assume that reference samples are diploid. Thus, if the X chromosome is to be analyzed, a reference set should consist of a single gender, preferably female. If male samples are used as reference, it should be noted that the assumption of two copies is incorrect for the X chromosome, and the interpretation of CN for this chromosome must be adjusted.

When no references are available (or for reference pool supplementation), data for 48 HapMap samples that were run on the Mapping 500K Set are available for download from the Affymetrix Web site (www.affymetrix.com).

Batch Effects

For CN data with the highest signal-to-noise ratio, references should be generated in the same laboratory, by the same user, and in the same batch as the case samples. Batch effects due to differences in hybridization times, wash cycles, and other factors may introduce a small amount of noise to the CN predictions, but the effect is relatively minimal (as compared to expression data, which are more adversely affected by differences between batches). Therefore, when references from the same batch are unavailable, alternate references may be used.

Paired Versus Unpaired Analyses

It is possible to perform paired analyses in which a test DNA is compared to a matching reference sample. The identification of somatic changes may be done by comparison of DNA collected from both tumor and non-tumor sources from a single patient. For germ-line mutations, a child's DNA may be compared to the parents' DNA to detect de novo chromosomal aberrations. When non-tumor DNA from the patient or parental DNA is unavailable, unpaired analyses can be performed using unmatched reference samples. Typically, in this case, a pool of multiple references is used. Both paired and unpaired analyses can be performed in CNAT and CNAG.

Automatic Selection of References in CNAG

When performing an unmatched analysis, CNAG can automatically select the references that best match a test DNA sample. When using this option, CNAG computes the standard deviation (S.D.) associated with CN predictions for a test DNA sample when compared to each reference sample that has been extracted into the CNAG folder. Additionally, CNAG will compare the test DNA sample to multiple combinations of references until the reference or set of references that provides a CN prediction with the smallest SD is determined. To see which reference samples were chosen during automatic selection, click on the "Info" icon at the top of the CNAG window once the CN data are presented.

FFPE and Non-FFPE References

When analyzing FFPE samples for CN, the references can be non-FFPE or FFPE samples, but for reasons described in the next section, using non-FFPE references is preferred.

COPY NUMBER ANALYSIS OF FFPE SAMPLES

Compensation against Fragment Size Bias

A fragment size bias is typically seen in CN predictions of FFPE samples; small fragments predict higher CNs and larger fragments predict lower CNs. This type of bias can be seen in a CN-by-fragment-size graph (Fig. 22-2, top row). Quadratic regression, automatically implemented by CNAT and CNAG, reduces the effect fragment size exerts on the mean CN (Fig. 22-2, middle row). For degraded samples that do not provide templates for larger fragment sizes, this compensation is insufficient at completely removing the influence of fragment size on CN predictions. In such cases, an additional step of excluding larger fragment sizes from analysis is also required (Fig. 22-2, bottom row).

Application of Fragment Size Filter

Exclusion of larger fragment sizes can be performed in both CNAT and CNAG. When using CNAG, choose "SetFragmentLengthRange" in the Parameters menu item, and enter the range of fragment sizes that should be included in the analysis. In CNAT, click on the "Advanced Analysis Options" icon within the CNAT4 Batch Analysis window; click the checkbox next to the statement "Restrict Analysis to SNPs on Fragment Sizes Ranging:"; and fill in the minimum and maximum fragment sizes desired for this analysis.

In CNAG, fragment size selection can be adjusted after the samples have been analyzed. In CNAT, the fragment size filter is selected before analysis of the sample or set of samples. Changing the fragment size filter requires reanalysis of the sample. In CNAT, multiple samples analyzed in a single batch will all share the same fragment size filter.

Selection of Fragment Size Filter

Currently, guidelines for optimal fragment size filter selection can only be performed in CNAG. To choose the fragment size filter for a given FFPE sample, follow the work flow outlined in Figure 22-3, using the CN versus fragment size plots. These can be viewed by selecting "Fragment Length Plot" in the Parameters menu when CN data are displayed for a given sample. Since the SNPs are grouped together during the normalization step, those SNPs that were not amplified during PCR can influence CN predictions of the smaller fragment SNPs that were amplified. Therefore, these large fragment SNPs must be excluded from analysis. To determine the upper limit for SNP inclusion, sequentially exclude more and more SNPs (i.e., include first SNPs ≤1000 bp, then SNPs ≤900 bp, ≤800 bp, etc.). Each exclusion will change the CN-versus-fragment-size plots so that once the proper exclusion is applied, SNPs on fragment sizes retained in the analysis should provide a consistent mean CN across the X axis (Fig. 22-3B). This indicates that fragment size no longer exerts an effect on CN analysis and that reliable CN can be predicted for the remaining SNPs. Once the proper filter has been chosen, exit the CN-versus-fragment-size plots by again selecting "Fragment Length Plot" in the Parameters menu.

Samples requiring exclusions of a large percentage of SNPs may provide data with a higher amount of noise, increasing the necessity for some data smoothing. Samples that require exclusions of SNPs below 500 bp are unlikely to provide any satisfactory CN data. Note that the fragment size threshold is not directly predicted by the QC test described in Chapter 16.

Because of the variability in quality of FFPE DNA, different samples will require different fragment size filters; thus, poorer-quality samples will require exclusion of more SNPs. When a large number of FFPE samples are analyzed in batch, a single fragment size filter may be implemented. Note that increased noise for poorer samples and decreased resolution for the samples that could provide good data for the larger fragment SNPs may occur when applying a general filter across samples.

FFPE Samples as References

A problem in determining the fragment size cutoff may arise when FFPE samples are used as references in CN prediction. This occurs because large fragment SNPs may be equally absent after PCR in both the test FFPE sample and the reference FFPE sample. Note that CN is predicted by using a ratio of intensities from these two samples, with an assumption that the reference sample represents a diploid state, regardless of true CN. With equivalently decreased representation of SNPs on large fragments in both test and reference DNA, the software will predict that the test FFPE sample is diploid as well. The CN-versus-fragment-size plot will then no longer be useful for determining

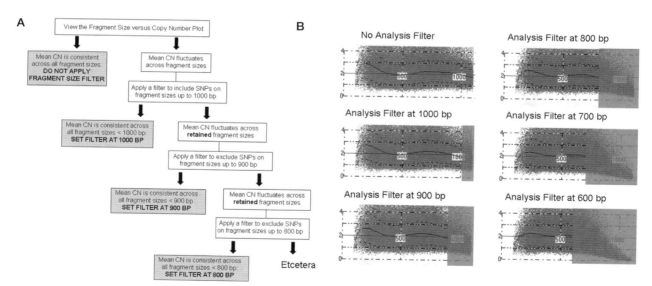

FIGURE 22-3. Identification of the optimal fragment size filter for an FFPE sample. (*A*) Outlined here are the steps to perform in CNAG to identify the optimal fragment size filter for a given FFPE sample. (*B*) Sequentially more stringent fragment size filters are applied to an FFPE sample while viewing the CN versus fragment size plots. Although the mean CN fluctuates as a function of fragment size before filtration and when moderate filters are applied, the mean CN stabilizes close to a value of 2 once the filter is lowered to 600–700 bp (for SNPs retained in the analysis). For this sample, 600 bp would be chosen as the fragment size cutoff.

noninformative fragment sizes because the large fragments will present a consistent prediction of CN = 2 with a very small S.D. Therefore, although FFPE references can be used when analyzing FFPE samples, the ability to determine the necessary fragment size cutoffs will be compromised and there may be a false assumption of reliability of data from larger fragment SNPs (Fig. 22-4).

CONCLUSION/SUMMARY

DNA degradation and contamination present special challenges when identifying CN changes in FFPE DNA samples using Mapping arrays, but these can be largely overcome by in silico correction. These in silico steps include compensation against fragment size bias and exclusion of larger fragment size SNPs. Correction of fragment size bias is automatically implemented in both CNAT 4.0 and CNAG (freely available copy number analysis software tools). Cutoff selection for fragment size exclusion is possible using CNAG, and this exclusion can then be applied in either of these software programs. Although genomic coverage may be reduced for FFPE samples due to any fragment size exclusions, reliable CN predictions are possible for the SNPs retained in analysis. These modifications thereby enable a whole-genome, high-resolution approach to CN detection of FFPE samples.

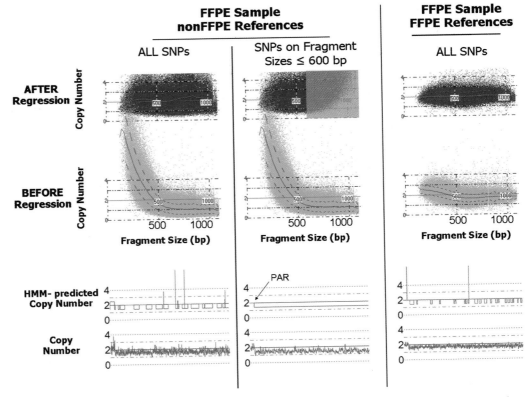

FIGURE 22-4. Comparison of FFPE and non-FFPE references in CN detection of FFPE samples. A single FFPE sample was compared to non-FFPE references (*left* and *middle columns*). CN versus fragment size plots indicate that a fragment bias remains after regression when all SNPs are included in analysis (*left*) but is removed from the retained SNPs when a filter is set at 600 bp (*middle*). This male sample has noisy CN predictions on the X chromosome when all SNPs in analysis are included, as seen by both Hidden Markov Method (HMM) predictions and CN predictions. Excluding the SNPs on fragment sizes >600 bp results in a prediction of CN = 1 except in the pseudo-autosomal region (PAR), which is diploid. When the same sample is compared to FFPE references (*right column*), the CN-versus-fragment-size plot suggests that there is no fragment size bias and indicates that all SNPs can be included for CN detection, but HMM plots and CN predictions are nonetheless noisy.

REFERENCES

Bignell G.R., Huang J., Greshock J., Watt S., Butler A., West S., Grigorova M., Jones K.W., Wei W., et al. 2004. High-resolution analysis of DNA copy number using oligonucleotide microarrays. *Genome Res.* **14:** 287–295.

Huang J. 2004. Whole genome DNA copy number changes identified by high density oligonucleotide arrays. *Hum. Genomics* **1:** 287–299.

Huang J., Wei W., Chen J., Zhang J., Liu G., Di X., Mei R., Ishikawa S., Aburatani H., Jones K.W., and Shapero M.H. 2006. CARAT: A novel method for allelic detection of DNA copy number changes using high density oligonucleotide arrays. *BMC Bioinformatics* **7:** 83.

Ishikawa S., Komura D., Tsuji S., Nishimura K., Yamamoto S., Panda B., Huang J., Fukayama M., Jones K.W., and Aburatani H. 2005. Allelic dosage analysis with genotyping microarrays. *Biochem. Biophys. Res. Commun.* **333:** 1309–1314.

Jacobs S., Thompson E.R., Nannya Y., Yamamoto G., Pillai R., Ogawa S., Bailey D.K., and Campbell I.G. 2007. Genome-wide, high resolution detection of copy number, loss of heterozygosity, and genotypes from formalin-fixed, paraffin-embedded tumor tissue using micro arrays. *Cancer Res.* **67:** 2544–2551.

Kennedy G.C., Matsuzaki H., Dong S., Liu W.M., Huang J., Liu G., Su X., Cao M., Chen W., Zhang J., et al. 2003. Large-scale genotyping of complex DNA. *Nat. Biotechnol.* **21:** 1233–1237.

Komura D., Shen F., Ishikawa S., Fitch K.R., Chen W., Zhang J., Liu G., Ihara S., Nakamura H., Hurles M.E., et al. 2006. Genome-wide detection of human copy number variations using high-density DNA oligonucleotide arrays. *Genome Res.* **16:** 1575–1584.

LaFramboise T., Weir B.A., Zhao X., Beroukhim R., Li C., Harrington D., Sellers W.R., and Meyerson M. 2005. Allele-specific amplification in cancer revealed by SNP array analysis. *PLoS Comput. Biol.* **1:** e65.

Lin M., Wei L.J., Sellers W.R., Lieberfarb M., Wong W.H., and Li C. 2004. dChipSNP: Significance curve and clustering of SNP-array-based loss-of-heterozygosity data. *Bioinformatics* **20:** 1233–1240.

Nannya Y., Sanada M., Nakazaki K., Hosoya N., Wang L., Hangaishi A., Kurokawa M., Chiba S., Bailey D.K., Kennedy G.C., and Ogawa S. 2005. A robust algorithm for copy number detection using high-density oligonucleotide single nucleotide polymorphism genotyping arrays. *Cancer Res.* **65:** 6071–6079.

23 | Assessing Significance in Genetic Association Studies

Mark J. Daly

Center for Human Genetic Research, Massachusetts General Hospital, Boston, Massachusetts 02114; Broad Institute of Harvard and MIT, Cambridge, Massachusetts 02142

INTRODUCTION

A critical step in each genetic association study is the estimation of statistical significance. It is on the basis of these estimates, frequently in the form of the traditional *p*-value, that most studies attempt to draw a qualitative conclusion regarding the association, or lack of association, of the tested genetic variants. This chapter introduces basic concepts and fundamental considerations relevant to designing appropriate significance tests and discusses what these tests can and cannot tell us about genetic association.

Basic Statistical Concepts and Methods

Significance testing or hypothesis testing is among the most pervasive of mathematical concepts in science today. Generally speaking, when an experiment is conducted, one or more hypotheses are tested, and we desire to compute, on the basis of the observed data, the likelihood that the hypothesis is true. When multiple competing models are proposed, most often with variable numbers of parameters, we customarily engage in likelihood comparisons. The probability of each model, given the observed data, is computed and compared in light of the complexity of each one in order to determine which model fits the observations best. Most commonly in genetic association studies, a general question is posited: "Is gene X associated with the disease or phenotype of interest?"

Although this captures the spirit and goal of most association studies, it is imprecise in several important aspects. First, many studies focus not on one gene only, but on the entire genome (i.e., a genome-wide association study) and address diseases for which it is not known whether a small or large number of genes is relevant. Next, whereas genes are the functional entity of interest in most studies, they themselves are not directly tested in association studies. Genetic variation only rarely has an obvious functional impact on a gene product (e.g., recognizable changes in protein coding or invariate splice junction bases). The vast majority of polymorphisms, even those proximal to a region encoding a gene of interest, have little or no annotation with respect to what

303

influence they may have on that gene. This introduces complications—both in terms of how to structure the hypothesis to address multiple variants and their combinations, and how to interpret results from multiple hypothesis tests rather than simply one. Third, even if only a single genetic variant were targeted, the question as posed does not propose a specific model for how the gene or variant is associated to phenotype.

Fortunately, there is a commonly accepted surrogate for the third of these problems. Let us propose for a moment that we collect genotype data for a single-nucleotide polymorphism (SNP) resulting in a coding change (R100W) in a gene under evaluation. We want to answer the question as to whether this substitution is associated with a disease by comparing the proportion of Rs and Ws found in 1000 cases of the disease with that found in 1000 healthy control individuals. Although we have not yet specified a model for the form of this association, we do have a "model" for the alternative null hypothesis of no association—specifically, that the proportions of R and W among cases and controls do not differ. Therefore, we can invert our question into the following, "Can we reject the model that the proportion of Rs and Ws is the same in these two groups?" which can be readily answered. In this most basic of significance tests, rather than affirmatively calculating association, we are calculating the probability of the data assuming no effect.

	W	R		
Cases	246	1754	2000	12.3%
Controls	200	1800	2000	10.0%
	446	3554	4000	

χ^2 = SUM over all cells $(O–E)^2/E$ = **5.34**, where O is the observed value and E is the expected value of each cell

degrees of freedom ≥ (nrows–1)*(ncols–1) = **1**

p (χ^2 ≥ 5.34, d.f. = 1) = **0.02**

For this, we might frequently employ the chi-square goodness-of-fit test, developed by Karl Pearson. Here, observations are compared with expectations to generate a deviate of a well-known distribution, referred to as a chi-square (χ^2), which has a canonical probability associated with it. Greater detail on basic statistical methods is given in numerous textbooks (see, e.g., Freedman et al. 1998; Venables and Ripley 2002). An excellent presentation of the historical development of statistical methods can be found in *The Lady Tasting Tea* by Salsburg (2001). Shortly after the chi-square test was introduced, R.A. Fisher introduced many of the basic concepts of significance testing, including the invaluable concept of a p-value, a probability that can be used to reject a proposed hypothesis when this probability is sufficiently low. Arbitrarily, the tradition in statistics has been to use a p-value of 0.05 to reject a hypothesis; i.e., 1 time in 20, one would expect to falsely reject a hypothesis. In the example above, the observed p-value of 0.02 indicates that 1 time in 50, we would expect data to show an equivalent or greater difference between cases and controls simply by chance. The 2 × 2 chi-square test is among the simplest and most frequently used tools for association testing. Fisher's exact test (suitable for tables with one or more small cell counts) and the Cochran-Mantel-Haenszel test (extending the 2 × 2 chi-square test to scenarios involving multiple strata, such as distinct population groups) (Agresti 1990) are among the related tests that can be employed and interpreted similarly.

Although the calculation of p-values in this scenario is simple, the qualitative interpretation of them is quite complex. In part, this is because our goal is to establish connections between gene variants and disease phenotypes as a stepping-stone to developing biological insights and medical treatments and preventives. The goal is not simply to wander the genome, rejecting the null hypothesis at an arbitrary statistical threshold. Researchers seeking to confirm reported findings, not to mention anyone who wishes to explore the medical ramifications of them, need to know what hypothesis is being proposed when one rejects the model of no association. In strictly mathematical terms, Neyman and Pearson pointed out that one can only accept or reject a hypothesis in comparison with an alternative hypothesis—not in a vacuum. Several considerations emerge from this thinking and influence many of the points discussed below.

The other most common formulation of the earlier association question—as a formal hypothesis test—can be readily performed using logistic regression techniques. Here we propose a model with a single parameter (often an additive term representing the excess log-risk increase due to each copy of the allele being tested for association) and compare the likelihood of the model to explain outcome (e.g., case/control status) with and without this term. (Standard numerical techniques are used to find the most likely value of this parameter.) Now we have reformulated the question as a specific hypothesis test, albeit one where the effect is not specified in advance but is discovered, resulting in a likelihood difference that can be evaluated as a chi-square:

2 ln (likelihood [model with fit parameters] / likelihood [null model with 0 parameters]) = chi-square (N degrees of freedom, where N = excess # of parameters)

At the very least, this gives us a more formal starting point and a specific hypothesis that can be tested parsimoniously in subsequent studies. The above regression is more intuitive and widely used for quantitative phenotypes, where a linear parameter is used to describe the phenotypic difference attributable to the allele being tested, with the subsequent likelihood and significance considerations the same. Often, both an additive and dominance parameter are fit to accommodate both genetic models, albeit at a statistical cost via an increase in the degrees of freedom.

Special Considerations in Genetic Association Studies

This does not address, however, several related and fundamental considerations about significance evaluation in genetic association studies. First, when we query a single SNP or other genetic variant, we consider it very unlikely a priori that this variant (or one in close linkage disequilibrium [LD] with it) will be truly associated with disease. After all, there are roughly 20,000 genes and millions of common genetic variants in the genome, and we expect very few of them to be causally linked with any individual disease. Thus, if we truly want to compare the hypotheses of association and no association, we may reasonably wish to consider how likely these alternatives are before we collect data. Moreover, rather than a single-hypothesis test, many genetic variants are tested in each gene under evaluation. Given that individual researchers or collaborative groups are studying many genes simultaneously, and furthermore, that technology has now emerged to enable genome-wide association studies of hundreds of thousands of SNPs simultaneously, the statistical threshold for accepting an alternate hypothesis as true may need to be quite prohibitively low. An introduction to many of the practical and statistical considerations in genome-wide association can be found in Hirschhorn and Daly (2005) and Wang et al. (2005).

These considerations bring us to the oft-discussed "multiple testing" problem, which has generated considerable methodologic attention in recent years. In a nutshell, the best of many tests performed under the null hypothesis is likely to display a very strong *p*-value. (Not surprisingly, the best *p*-value observed by chance in *N* independent trials is expected to be roughly 1/*N*.) Thus, interpreting the significance of any observation (e.g., the odds of achieving a more extreme result in the experiment simply by chance under the null hypothesis) requires some consideration. Although multiple testing is not so keenly attended to in all fields of biology, geneticists performing genome scans (first for linkage, now for association) have been conditioned to be extremely attentive to the problem. This may be due to the hypothesis-free nature of genome scanning (in contrast with clinical trial outcomes or environmental causality in epidemiological studies).

Permutation Testing

One simple approach is the invocation of a Bonferroni correction; given that the best result in an experiment is expected to be $1/N$, a traditional significance threshold of 0.05 might then be set at $0.05/N$ in order to define study-wide significance. To achieve the same outcome, one might multiply extreme p-values by N. (In fact, $0.05/N$ is an approximation of the exact probability α that no hits are observed in 95% of similar experiments by chance: $0.05 = 1 - (1 - \alpha)^N$.) Such a correction factor is likely to be over-conservative in the case of genetic association studies, since the assumption of independence among the N trials is violated. SNPs may frequently be correlated (in LD) with other tested SNPs, and this has the effect of increasing the lowest p-value expected by chance study-wide. A more accurate threshold for study-wide significance is frequently obtained by permutation testing—where one might generate data sets matched in all properties (including the maintenance of LD relationships among markers and sporadic data error and missing data) by randomly swapping the phenotype values (e.g., case and control labels) among study participants and repeating the precise statistical analysis on the permuted data that was performed on the observed data. Analyzing thousands of permuted studies in this way provides accurate thresholds for extreme values unlikely to be observed by chance study-wide.

This category of test (refinements of Bonferroni are available) is concerned with controlling the chance that any null hypothesis in the experiment is falsely rejected. This is sensible, as a false "discovery" that a certain gene underlies a disease can quickly precipitate costly and time-consuming research in pursuit of a specious functional hypothesis of disease etiology. However, thresholds for declaring significance, although they are useful guidelines for interpretation, are prohibitively low when we consider the case of a genome-wide association study. Testing 500,000 SNPs might recommend use of a significance threshold of 10^{-7} ($= 0.05/500,000$). Such a threshold greatly limits the power to provide compelling evidence of association in a single study. For example, a study comparing 1000 cases to 1000 controls would have nearly a 95% chance of detecting a 10% allele with an associated 1.5× relative risk at $p < 0.01$; however, only 12% of the time would that same p-value be less than 10^{-7}. Alternatively, in this setting, the observed result will often be in an intermediate significance range such as $0.00001 < p < 0.001$. Although this is substantial evidence in normal circumstances, after 500,000 tests (the overwhelming majority of which truly represent the null, or no association, hypothesis), there will be tens to hundreds of similar results purely by chance. This has two ramifications. First, any individual result in this probability range cannot be distinguished from many chance events that do not correspond to true association, and second, the presence of one result (or even a handful of results) in this range may not have a discernable impact on the overall distribution of p-values.

False Discovery Rate Analysis

This brings us to an alternative and now frequently used approach to interpreting data sets with many independent (or partially so) data points, referred to as False Discovery Rate (FDR) analysis. The original FDR test was introduced by Benjamini and Hochberg in 1995, but several variations now exist. Generally, here one does not attempt a Bonferroni or permutation style analysis of each test individually but evaluates the distribution of p-values with the hypothesis that there may be multiple, modestly positive signals present and that a significant excess of tests establishes that some of them must represent true positives. In contrast with the significance testing described at the outset (where we were evaluating the probability of significance given no association), here we are assessing the probability of a true association given a significance level. Informally, let us imagine that in the earlier 500,000 SNP study, no SNP individually met our criteria for significance by exceeding 10^{-7}—yet 10 SNPs fell between 10^{-6} and 10^{-7}. According to a null distribution of p-values, less than 1 test should fall below 10^{-6} by chance, and thus one could conclude that the majority of the top 10 actually constitute true positives, even though none can individually have overwhelming certainty.

In the evaluation of such results, the importance of permutation testing cannot be overstated. LD dependencies among markers as well as inevitable study imperfections in data quality and population structure, and distant relatedness among "unrelated" samples, can alter the analytic expectations and variances substantially. In genome-wide studies in particular, one should always determine whether an analysis is largely consistent with a null distribution, rather than presuming it is. (The majority of the distribution should be consistent with a null distribution, since there are likely to be few true positive associations in a genome-wide study.)

The FDR and related tests attempt to estimate and control the proportion of false rejections, instead of eliminating them. In the above example, an FDR analysis might conclude that if one accepts the top 10 results as true, only 5–10% of those assumptions are likely incorrect. Although one may not wish to pursue any of those results without further confirmation of association in an independent study, the study clearly provides compelling evidence of multiple genetic factors that will undoubtedly soon be resolved by follow-up studies. More specifically, one can set a specific "acceptable" false discovery rate, and the FDR procedure will find the largest set of results that meets that criterion. Storey and Tibshirani (2003) have extended these concepts and have developed the now widely used q-value, which is defined for each result in, for example, a genome-wide data set, indicating the average proportion of false positives expected when that particular result is used as the threshold value. In the above example, if the tenth-ranking result had a q-value of 0.1, we could now formally propose that 90% of the top 10 (plus or minus a healthy dose of sampling variance) might actually be associated with the phenotype being examined. Application of such methodology has been widespread and effective in microarray analysis, where one is often comparing two settings in which hundreds or thousands of genes may actually vary. It remains to be seen whether these approaches are helpful in genome-wide association, where the actual number of true associations for some phenotypes may be relatively modest and thus not induce considerable deviation from the null distribution. It seems likely, however, that these approaches will prove useful in interpreting the most significant tails of genome-wide distributions.

Other Approaches

Other approaches have recently been proposed which supplement those already available to tackle the challenges of association and multiple testing. Altering the prior probability of each individual test (based on prior data, power, evidence of functionality, LD patterns) has been proposed by several authors to refine genome-wide studies in light of considerable emerging data (Wacholder 2005; Pe'er et al. 2006). Treating the set of tests (whether SNP- or haplotype-based) at a gene as a single unit in association and epistasis testing has been proposed as a way of managing the number of correlated tests (which, of course, vary from study to study) by collapsing them into a direct test of the true hypothesis under examination in a candidate gene study. Several authors have pointed out that Hotelling's T^2 statistic is a more powerful approach to evaluating a group of markers (Xiong et al. 2002), particularly in the common setting in which multiple independent alleles in a gene are associated with a phenotype. Such a state is not readily considered by individual marker or haplotype-based testing. Finally, Patterson suggests a logical extension to a Bayesian "whole-genome" statistic, in which an entire experiment could be queried as to whether there is evidence of association anywhere in the genome (Patterson et al. 2004). Depending on the nature and scope of the association study, which hypotheses are being tested, and what prior information can be usefully brought to bear, any of the above approaches could be very useful.

SUMMARY

Rather than provide blanket recommendations about methodology and simple thresholds to apply, this chapter has given novice researchers background understanding and references with which to design and evaluate significance testing in a specific experimental setting. Grounding in the fundamental concepts (in far more detail than space allows here) and a healthy dose of common sense will serve the new geneticist well. Although software and methods abound for taking in data and returning *p*-values, it is in understanding how to pose the question and how to interpret the results that valid and biologically relevant findings can be reached.

ACKNOWLEDGMENTS

The author acknowledges frequent thoughtful discussions with (as well as comments from) Shaun Purcell and Nick Patterson, and comments on this manuscript from Clay Stephens.

REFERENCES

Agresti A. 1990. *Categorical data analysis.* John Wiley and Sons, New York, pp. 100–102.

Benjamini Y. and Hochberg Y. 1995. Controlling the false discovery rate: A practical and powerful approach to multiple testing. *J. R. Stat. Soc. B* **57:** 289–300.

Freedman D., Pisani R., and Purves R. 1998. *Statistics*, 3rd edition. W.W. Norton, New York.

Hirschhorn J.N. and Daly M.J. 2005. Genome-wide association studies for common diseases and complex traits. *Nat. Rev. Genet.* **6:** 95–108.

Patterson N., Hattangadi N., Lane B., Lohmueller K.E., Hafler D.A., Oksenberg J.R., Hauser S.L., Smith M.W., O'Brien S.J., Altshuler D., et al. 2004. Methods for high-density admixture mapping of disease genes. *Am. J. Hum. Genet.* **74:** 979–1000.

Pe'er I., de Bakker P.I.W., Maller J., Yelensky R., Altshuler D., and Daly M.J. 2006. Evaluating and improving power in whole-genome association studies using fixed marker sets. *Nat. Genet.* **38:** 663–667.

Salsburg D. 2001. *The lady tasting tea: How statistics revolutionized science in the twentieth century.* Henry Holt and Co., New York.

Storey J.D. and Tibshirani R. 2003. Statistical significance for genomewide studies. *Proc. Natl. Acad. Sci.* **100:** 9440–9445.

Venables W.N. and Ripley B.D. 2002. *Modern applied statistics with S*, 4th edition. Springer, New York.

Wacholder S. 2005. Publication environment and broad investigation of the genome. *Cancer Epidemiol. Biomarkers Prev.* **14:** 1361.

Wang W.Y., Barratt B.J., Clayton D.G., and Todd J.A. 2005. Genome-wide association studies: Theoretical and practical concerns. *Nat. Rev. Genet.* **6:** 109–118.

Xiong M., Zhao J., and Boerwinkle E. 2002. Generalized T^2 test for genome association studies. *Am. J. Hum. Genet.* **70:** 1257–1268.

24 Assessing Human Variation Data for Signatures of Natural Selection

Mike Bamshad[1] and J. Claiborne Stephens[2]

[1]*Departments of Pediatrics and Genome Sciences, Division of Genetics and Developmental Medicine, University of Washington School of Medicine, Seattle, Washington 98195;* [2]*Motif BioSciences Inc., New York, New York 10017*

INTRODUCTION

Humans differ from one another in many ways, such as their physical appearance, behavior, and susceptibility to disease. Some of this phenotypic variability results from chance or differences in environmental exposures, but part of it is due to genetic differences among individuals. These genetic variants together constitute the fraction of the human genome exposed to natural selection. Natural selection for beneficial genetic variants—those variants that make it more likely for their carriers to reproduce and thus become more numerous over time—is called positive or adaptive selection, whereas moderately to severely deleterious gene variants tend to be eliminated by a process called purifying selection.

Positive natural selection, as first described by Darwin and Wallace, is thought to have played a particularly important role in shaping human variability over the past approximately 100,000 years, during which anatomically modern humans emigrated from Africa and adapted to widely varied environments (Darwin and Wallace 1858; Klein 1999). Extant as well as newly arising functional variants that were advantageous in these environments were potential targets of positive selection. An important corollary is that these targets of positive selection serve as divining rods of sorts that point toward functional variants amid the background of genetic variants with little, if any, phenotypic effects (i.e., neutral variants). Accordingly, finding variants that have been subject to selection can provide insights about which genes and noncoding regulatory elements influence human phenotypic variability, including differences in individual health. Therefore, understanding how evolutionary forces and demographic processes have shaped the geographic distribution of genetic variation among humans provides both a perspective for interpreting patterns of genetic variation and the basis for experimental strategies to identify medically relevant variation.

One of the most important sources of information about the effects of natural selection on human genes is found in the genes themselves. Population genetics models predict that different types of natural selection leave different "signatures" in the pattern of variation of a gene. Therefore, the effects of natural selection potentially can be inferred from patterns of variation, a strategy that has become more popular recently because of the wealth of new variation data available from humans and many other species. The validity of the inferences of whether positive or balancing selection has acted on a gene, based on the statistical analysis of patterns of genetic variation, has been confirmed in model organisms (e.g., most notably *Drosophila*) and domestic crops/animals (e.g., corn) exposed to "artificial" selection (for review, see Doebley et al. 2006). Such experiments obviously cannot be performed in humans, but there is no a priori reason to expect that inferences based on analyses of human diversity data using these same statistics would be less valid.

In this chapter, we highlight some of the different types of natural selection, their effects on patterns of DNA variation, and some of the statistical tests that are commonly used to detect such effects. We also explain some of the relative strengths and weaknesses of different strategies that can be used to detect signatures of natural selection at individual loci. These strategies are illustrated by their application to empirical data from gene variants that are often associated with differences in disease susceptibility. We briefly outline some of the methods proposed to scan the genome for evidence of selection. Finally, we discuss some of the problems associated with identifying signatures of selection and with making inferences about the nature of the selective process. A number of important philosophical issues (e.g., the units of selection), theoretical concepts, and empirical studies in other species are beyond the scope of this review. For further information, the interested reader is referred to a variety of excellent resources (see, e.g., Sober 1993; Li 2006).

METHODS FOR DETECTING SELECTION

Mutation is ultimately the source of genetic variation, but the fate of new mutations is determined by natural selection and population history. Selection can eliminate new mutations, maintain them at intermediate frequencies, or sweep them to fixation (a variant is fixed when its frequency reaches 100%) within a population. When natural selection acts on a genetic variant, it alters the pattern of variation in the surrounding DNA sequence so as to leave a molecular signature. These signatures can be identified in several ways, and many statistical tests have been proposed to detect signatures of selection. The applications of many of these tests have been summarized in several excellent reviews (Kreitman 2000; Schlotterer 2002; Vallender and Lahn 2004; Nielsen 2005; Biswas and Akey 2006; Harris and Meyer 2006; Sabeti et al. 2006).

Statistical tests for selection are typically based on comparison to expectations under the null distribution of a probabilistic model (i.e., parametric approaches), the empirical distribution of variation across the genome, or combinations thereof. Power is determined by carrying out simulations under a restricted range of demographic models and parameters (Simonsen et al. 1995; Fu 1997; Wall 1999; Prezworski 2002; Sabeti et al. 2002). This is important, because the effects of population demographic history such as subdivided populations, periods of reduced population size (i.e., bottlenecks), and population growth can result in patterns of DNA sequence variation that resemble those caused by natural selection (Ptak and Przeworski 2002). For example, the proportion of variants with a low frequency (i.e., rare alleles) is predicted to increase in an expanding population because new mutations are lost at a lower rate. Indeed, such an excess of rare variants is seen in humans and has been interpreted as evidence of a rapid expansion in human population size. However, positive selection can produce a similar excess of low-frequency variants (Braverman et al. 1995). To this end, an understanding of human population history is crucial for identifying regions of the genome that have been subject to selection. Moreover, for all test statistics, it is prudent to consider factors such as how the variation data were ascertained and the influence of local genomic factors such as recombination rates.

Comparison of Polymorphisms across Species

Much of what is known about the impact of adaptive evolution on the human genome comes from comparisons of patterns of variation between humans and other species. Specifically, positive selection over a prolonged period of time can increase the rate of fixation of beneficial function-altering variants. This increase can be detected by comparing the rate of nonsynonymous (i.e., amino acid-altering) to synonymous (i.e., silent) changes across lineages (Bush 2001; Yang 2002; Wong et al. 2004). The current wealth of DNA sequence data is making this approach increasingly popular, and a particular emphasis has been placed on comparing loci between humans and chimpanzees for evidence of positive selection (Bustamante et al. 2005; Nielsen 2005). Identification of such loci might provide further insights about the nature of the changes, such as the origin of speech, a key adaptation in the evolution of modern humans (Enard et al. 2002).

Statistical tests used to detect this signature include the K_A/K_s test and the McDonald-Kreitman test (Li et al. 1985; Nei and Gojobori 1986; McDonald and Kreitman 1991). These tests are generally conservative because the substitution rates are summarized across all the amino acid sites tested, and sites which are likely under different functional constraints. Similar tests can be applied to separate domains of a protein that are functionally important, individual amino acid residues, and noncoding regulatory sequences (Suzuki and Gojobori 1999).

Comparison of Polymorphisms within Species

An excess of functional changes identifies a region of the genome on which natural selection has acted on a species as a whole. Such signatures provide important information about how a species evolved to distinguish itself from other species but reveal little about whether natural selection has played a role in shaping phenotypic differences among individuals within a species (e.g., differences in appearance, disease susceptibility). To investigate whether natural selection influences inter-individual differences, polymorphism data from different populations within a species can be compared to look for signatures of selection (Bamshad and Wooding 2003). The development of statistical tests to detect such signatures is a very active area of investigation, and over the past several decades many different tests have been proposed. These tests can be broadly categorized into tests that look for (1) a reduction or excess in genetic diversity; (2) an excess or lack of differentiation among populations; and (3) extended haplotypes.

For two different models of selection, signatures are manifest as a reduction in variation at linked sites. The first model, background selection, removes deleterious mutations and eliminates variation at linked sites (Charlesworth et al. 1993). The strength of this effect varies with the recombination rate, magnitude of selection, and mutation rate (Hudson and Kaplan 1995). The second model, genetic hitchhiking, predicts that when a mutation increases in frequency in a population as a result of positive selection, linked neutral variation is dragged along with it (Maynard-Smith and Haigh 1974; Fay and Wu 2000). As a consequence, variation not linked to the adaptive mutation is eliminated from nearby regions, resulting in a so-called selective sweep. Diversity is eventually restored, but this occurs slowly because new mutations are rare. Therefore, positive selection causes an overall reduction in genetic diversity with an excess of rare variants compared to other regions of the genome. Statistical tests commonly used to detect this signature include Tajima's D (Tajima 1989), Fu and Li's D* (Fu and Li 1993), and the Hudson-Kreitman-Aguade (HKA) test (Hudson et al. 1987).

For example, the genomic region surrounding *CYP3A4* and *CYP3A5* (which encode enzymes that metabolize a large fraction of common pharmaceuticals) exhibits an overall reduction in diversity and an excess of rare alleles compared to other human loci (Thompson et al. 2004). The specific target of selection is unclear, but a functional *CYP3A5* allele that influences salt/water homeostasis and risk for hypertension has an unusual geographic distribution, whereby its frequency is correlated with physical distance from the equator. A variant in the unlinked *AGT* gene also associated with hypertension demonstrates a similar relationship. This result suggests that a shared selective pressure which correlates with latitude influenced variation in both genes. Other regions of the genome that exhibit low levels of diversity (suggesting that they have been targets of selection) include those which contain *MC1R* (melanocortin-1 receptor; Makova et al. 2001), *LCT* (lactase phlorizin hydrolase; Bersaglieri et al. 2004), and *KEL* (Kel antigen; Akey et al. 2002).

In a selective sweep, many of the new mutations that arise and are linked to the advantageous allele will reach high frequency due to hitchhiking but will not reach fixation. These new alleles are younger and thus predicted to be "derived" compared to the older, ancestral alleles. Therefore, another signature of positive selection is the presence of a region containing an excess of high-frequency derived alleles compared to expectations (Watterson and Guess 1977). Inference of whether an allele is ancestral or derived is based on comparison to an outgroup. Chimpanzees and gorillas, the two species with which humans most recently shared a common ancestor, are the most commonly used outgroups. A test that is often used to identify regions with an excess of high-frequency derived alleles is Fay and Wu's H (Fay and Wu 2000).

For instance, for four SNPs in *CYP1A2*, which encodes another drug-metabolizing enzyme, the minor SNP in humans was the fixed allele in chimpanzee and gorilla (Wooding et al. 2002). Thus, the common SNPs in humans, each of which had a frequency greater than 90%, were inferred to be the derived state. Compared to the expected fraction of variants at each frequency estimated under the neutral model, *CYP1A2* exhibited an excess of both low- and high-frequency derived alleles. This indicates that *CYP1A2* might have been influenced by both positive selection and recent population growth, although the relative strengths of each are unclear.

The size of the genomic region affected by a sweep depends on the strength of selection and the local rate of recombination. Sweeps caused by strong selective advantages or in regions with a low recombination rate can affect large tracts of the genome. This facilitates detection of a selection signature but makes it more difficult to identify the causal variant. As noted previously, population size expansion can also cause an excess of rare alleles, making it a challenge to distinguish between the effects of positive selection and demographic history.

When a mutation arises, it does so on an existing background haplotype characterized by complete linkage disequilibrium (LD) between the new mutation and linked polymorphisms. Over time, new mutations and recombination reduce the size of this haplotype such that, on average, older and typically common mutations will be found on smaller haplotypes (i.e., there is only short-range LD between the mutation and linked polymorphisms). Younger, low-frequency mutations can be associated with either small or large haplotypes. However, selective sweeps can produce an allele that both has a high frequency and occurs in a region of extended LD.

Developing statistical tests to detect such signatures has been an active area of investigation (Sabeti et al. 2002; Toomajian et al. 2003; Kim and Nielsen 2004; Hanchared et al. 2006; Voight et al. 2006; Wang et al. 2006). One of the first and most popular such tests available is the long-range haplotype (LRH) test (Sabeti et al. 2002). This test can narrow the genomic region on which positive selection has acted to a relatively short distance. One limitation of this test is that it can only detect selection signatures that are relatively recent. Nevertheless, it still has widespread applicability, because local adaptive evolution is thought to have been particularly important over the past approximately 10,000 years, as humans changed from hunter-gatherers to an agriculture-based subsistence.

For example, most mammals, including humans, lose the capacity to metabolize lactose after weaning from breast milk. Accordingly, lactase nonpersistence appears to be the ancestral state in humans and is typical in most parts of the world (Swallow 2003). However, lactase persistence is common in many populations of Northern Europe (e.g., >90% in Swedes and Danes) and is present in a handful of populations in sub-Saharan Africa (e.g., ~90% in the Tutsi), where dairy products became a staple food source. Accordingly, it has long been suspected that *LCT* had been the target of recent positive selection coincident with the domestication of cattle (Hollox et al. 2001; Bersaglieri et al. 2004; Myles et al. 2005). In European populations, differences in *LCT* expression in adults are caused by a C/T SNP at −13910 located about 14 kb upstream (Enattah et al. 2002), whereas three novel SNPs (G/C-14010, T/G-13915, and C/G-13907) located upstream of *LCT* in sub-Saharan African populations appear to increase transcription of *LCT* in vitro (Tishkoff et al. 2006). These variants are associated with regions of extended LD in Europeans and Africans, indicating that recent positive selection has rapidly increased the frequency of several different *LCT* variants independently in populations from at least two parts of the world. In each case, the selective force appears to have been a local adaptive response to the higher fitness afforded by the ability to consume dairy products.

Local adaptive evolution may result in the high frequency of a beneficial allele in one population but not in another. Large differences in allele frequencies between populations are therefore another potential signature of positive selection. For example, the *FY*O* allele of the chemokine receptor, Duffy, which imparts resistance to *Plasmodium vivax*, has been driven to near-fixation in sub-Saharan Africans but is rare in non-Africans (Hamblin et al. 2002). Interestingly, the *FY*O* allele appears to have arisen independently and been subject to selection in sub-Saharan Africans and New Guinea highlanders. Such cases of convergent evolution are perhaps the strongest evidence of the effects of natural selection. The genomic regions surrounding *T2R16*, a bitter-taste receptor (Soranzo et al. 2005), and *SLC24A*, a gene that encodes a putative cation exchanger which plays a role in human pigmentation (Lamason et al. 2005), are highly differentiated between African and non-African populations. Such differences can only arise between populations that are at least partially isolated reproductively. Thus, in humans, they are relevant only to selective events occurring since anatomically modern humans emigrated out of Africa 50,000–100,000 years ago.

Sometimes, selection tends to maintain two or more alleles at a locus in one or more populations. This is known as balancing selection, because the frequencies of alleles are maintained in a balance as a result of some form of rare allele advantage. Balancing selection can lead to substantially less differentiation between populations than expected and can lead to an excess of alleles at intermediate frequencies, because the variations at linked loci also accumulate as a result of genetic hitchhiking (Lewontin and Hubby 1966; Kaplan et al. 1988). In many plants and animals, balancing selection appears to play a role in maintaining diversity at loci that coordinate recognition between self and non-self (Richman and Kohn 1999). In humans, this has been best studied at loci involved with host–pathogen responses, including *HLA* class I and II genes (Hughes and Yeager 1998), *TAS2R38* (Wooding et al. 2004), *G6PD* (Verrelli et al. 2002), and the *cis*-regulatory region of *CCR5* (Bamshad et al. 2002).

GENOME SCANS

The approaches that are used to detect selection at individual loci are adaptable to scanning the entire human genome for signatures of selection (Lewontin and Krakauer 1973). This has been made feasible by the avalanche of polymorphism data that have become available over the past few years from the International Haplotype Map (HapMap) Project (International HapMap Consortium 2005), Perlegen Sciences (Hinds et al. 2005), Applera Project (Bustamante et al. 2005), and the Genaissance Resequencing Project (Stephens et al. 2001). Nearly a dozen scans for signatures of selection have been completed to date (Cargill et al. 1999; Sunyaev et al. 2000; Akey et al. 2002; Payseur et al. 2002; Bamshad and Wooding 2003; Bustamante et al. 2005; Carlson et al. 2005; Weir et al. 2005; Bubb et al. 2006; Voight et al. 2006; Wang et al. 2006). They vary in strategy by, for example, using different markers, different statistical tests, and/or screening anonymous regions of the genome versus coding regions (McVean and Spencer 2006). Nevertheless, comparison across the most comprehensive of these scans enables cautious generalization about the overall impact of selection on the genome and provides information about heretofore unknown functional variation.

Most of the completed genome scans have searched for signatures of positive selection within humans. Several scans looking for selection signatures based on differences in allele frequencies among populations have been completed. For example, an initial analysis of the HapMap data revealed 926 SNPs in 27 genes for which differentiation among populations was more extreme than for the *FY* (i.e., *Duffy*) locus (International HapMap Consortium 2005). More than 30 of these SNPs were predicted to cause nonsynonymous amino acid substitutions, including rs1426654 in *SLC24A5*, which (as previously noted) partly explains variation in skin pigmentation between Africans and non-Africans (Lamason et al. 2005). Weir et al. (2005) estimated population differentiation in sliding windows of about 5.0 megabases across the genome, using both the HapMap and Perlegen data, and found about 300 regions containing putative targets of selection. Several other scans have been performed using statistics similar to the long-range haplotype test which are based on the long-range decay of LD. Wang et al. (2006) identified 25,386 SNPs clustered around 1799 genes in the Perlegen data for which long-range LD decay was significantly different from the genome-wide average. Voight et al. (2006) developed a similar statistic and found about 300 regions containing 455 genes that were predicted to have been targets of positive selection.

Together, these genome scans identified about 2300 genes that were putative targets of positive selection within humans, the majority of which had not been previously evaluated as candidates. Indeed, fewer than half of previously reported positive selection candidate genes were identified in these scans. Only a modest number of genes were reported to have a signature of selection in more than one analysis, and selection signatures were frequently confined to a single population rather than shared among populations. Some of these targets are likely to be false positives, but direct comparison among studies is difficult because of differences in experimental design. Interestingly, many of the genomic regions that appear to have been targets of selection contain few, if any, genes. This observation underscores the growing evidence that adaptive evolution on noncoding sequence has played a bigger role than previously appreciated in shaping patterns of human variation (Ponting and Lunter 2006).

CONCLUSIONS

Our interpretation of how selection has influenced the human genome is relatively simple, when, in fact, the effects of selection are likely to be much more complex. Selection intensity can fluctuate over time, and different selective effects may overlap or interfere with one another. Moreover, the expectations of population genetic models are dependent on assumptions about demographic parameters for which estimates remain fairly ambiguous. Therefore, continued progress toward identifying genes subject to selection will depend on understanding more about the demographic structure of human populations.

For most signatures that are identified, confirmation and validation of the result poses special challenges. For example, replication of a result is difficult because individuals sampled from the same populations have overlapping ancestors, and different measures of genetic variation are not independent of one another. The most rigorous evidence that a selection signature is genuine is to demonstrate that a candidate target has a functional consequence on the phenotype on which natural selection is thought to have acted. In such cases, function can be assessed via population-based association studies, in vitro molecular and biochemical studies, and testing in model organisms. This is also, perhaps, the most difficult approach.

The increasing enthusiasm for characterizing signatures of natural selection is wagered, in part, on how well these signatures will be able to predict the location of gene variants of biomedical importance with little, if any, a priori knowledge of their functional significance. To this end, the study of population variation will continue to be of great interest and relevance to researchers and clinicians.

REFERENCES

Akey J.M., Zhang G., Zhang K., Jin L., and Shriver M.D. 2002. Interrogating a high-density SNP map for signatures of natural selection. *Genome Res.* 12: 1805–1814.

Bamshad M.J. and Wooding S.W. 2003. Signatures of natural selection in the human genome. *Nat. Rev. Genet.* 4: 99–111.

Bamshad M.J., Mummidi S., Gonzalez E., Ahuja S.S., Dunn D.M., Watkins W.S., Wooding S., Stone A.C., Jorde L.B., Weiss R.B., and Ahuja S.K. 2002. A strong signature of balancing selection in the 5' cis-regulatory region of *CCR5*. *Proc. Natl. Acad. Sci.* 99: 10539–10544.

Bersaglieri T., Sabeti P.C., Patterson N., Vanderploeg T., Schaffner S.F., Drake J.A., Rhodes M., Reich D.E., and Hirschhorn J.N. 2004. Genetic signatures of strong recent positive selection at the lactase gene. *Am. J. Hum. Genet.* 74: 1111–1120.

Biswas S. and Akey J.M. 2006. Genomic insights into positive selection. *Trends Genet.* 22: 437–446.

Braverman J.M., Hudson R.R., Kaplan N.L., Langley C.H., and Stephen W. 1995. The hitchhiking effect on the site frequency spectrum of DNA polymorphisms. *Genetics* 140: 783–796.

Bubb K.L., Bovee D., Buckley D., Haugen E., Kibukawa M., Paddock M., Palmieri A., Subramanian S., Zhou Y., Kaul R., et al. 2006. Scan of human genome reveals no new loci under ancient balancing selection. *Genetics* 173: 2165–2177.

Bush R.M. 2001. Predicting adaptive evolution. *Nat. Rev. Genet.* 2: 387–392.

Bustamante C.D., Fledel-Alon A., Williamson S., Nielsen R., Hubisz M.T., Glanowski S., Tanenbaum D.M., White T.J., Sninsky J.J., Hernandez R.D., Civello D., et al. 2005. Natural selection on protein-coding genes in the human genome. *Nature* 437: 1153–1157.

Cargill M., Altshuler D., Ireland J., Sklar P., Ardlie K., Patil N., Shaw N., Lane C.R., Lim E.P., Kalyanaraman N., et al. 1999. Characterization of single-nucleotide polymorphisms in coding regions of human genes. *Nat. Genet.* 22: 231–238.

Carlson C.S., Thomas D.J., Eberle M.A., Swanson J.E., Livingston R.J., Rieder M.J., and Nickerson D.A. 2005. Genomic regions exhibiting positive selection identified from dense genotype data. *Genome Res.* 15: 1553–1565.

Charlesworth B., Morgan M.T., and Charlesworth D. 1993. The effect of deleterious mutations on neutral molecular variation. *Genetics* 134: 1289–1303.

Darwin C. and Wallace A.R. 1858. On the tendency of species to form varieties; and on the perpetuation of varieties and species by natural means of selection. *J. Proc. Linn. Society Lon. Zool.* 3: 46–50.

Doebley F., Gaut B.S., and Smith B.D. 2006. The molecular genetics of crop domestication. *Cell* 127: 1309–1321.

Enard W., Przeworski M., Fisher S.E., Lai C.S., Wiebe V., Kitano T., Monaco A.P., and Paabo S. 2002. Molecular evolution of *FOXP2*, a gene involved in speech and language. *Nature* 418: 869–872.

Enattah N.S., Sahi T., Savilahti E., Terwilliger J.D., Peltonen L., and Jarvela I. 2002. Identification of the variant associated with adult-type hypolactasia. *Nat. Genet.* 30: 233–237.

Fay J.C. and Wu C.I. 2000. Hitchhiking under positive Darwinian selection. *Genetics* 155: 1405–1413.

Fu Y.X. 1997. Statistical tests of neutrality of mutations against population growth, hitchhiking and background selection. *Genetics* 147: 915–925.

Fu Y.X. and Li W.H. 1993. Statistical tests of neutrality of mutations. *Genetics* 133: 693–709.

Hamblin M.T., Thompson E.E., and Di Rienzo A. 2002. Complex signatures of natural selection at the Duffy blood group locus. *Am. J. Hum. Genet.* 70: 369–383.

Hanchard N.A., Rockett K.A., Spencer C., Coop G., Pinder M., Jallow M., Kimber M., McVean G., Mott R., and Kwiatkowski D.P. 2006. Screening for recently selected alleles by analysis of human haplotype similarity. *Am. J. Hum. Genet.* 78: 153–159.

Harris E.E. and Meyer D. 2006. The molecular signature of selection underlying human adaptations. *Yearb. Phys. Anthropol.* 49: 89–130.

Hinds D.A., Stuve L.L., Nilsen G.B., Halperin E., Eskin E., Ballinger D.G., Frazer, K.A., and Cox D.R. 2005. Whole-genome patterns of common DNA variation in three human populations. *Science* 307: 1072–1079.

Hollox E.J., Poulter M., Zvarik M., Ferak V., Krause A., Jenkins T., Saha N., Kozlov A.I., and Swallow D.M. 2001. Lactase haplotype diversity in the Old World. *Am. J. Hum. Genet.* 68: 160–172.

Hudson R.R. and Kaplan N.L. 1995. Deleterious background selection with recombination. *Genetics* 141: 1605–1617.

Hudson R.R., Kreitman M., and Aguade M. 1987. A test of neutral molecular evolution based on nucleotide data. *Genetics* 116: 153–159.

Hughes A.L. and Yeager M. 1998. Natural selection at major histocompatibility complex loci of vertebrates. *Annu. Rev. Genet.* 32: 415–435.

International HapMap Consortium. 2005. A haplotype map of the human genome. *Nature* 437: 1299–1320.

Kaplan N.L., Darden T., and Hudson R.A. 1998. The coalescent process in models with selection. *Genetics* 120: 819–829.

Kim Y. and Stephan W. 2002. Detecting a local signature of genetic hitchhiking along a recombining chromosome. *Genetics* 160: 765–777.

Klein R.G. 1999. *The human career: Human biological and cultural origins*, 2nd edition. University of Chicago Press, Chicago, Illinois.

Kreitman M. 2000. Methods to detect selection in populations with applications to the human. *Annu. Rev. Genomics Hum. Genet.* 1: 539–559.

Lamason R.L., Mohideen M.A., Mest J.R., Wong A.C., Norton H.L., Aros M.C., Jurynec M.J., Mao X., Humphreville V.R., Humbert J.E., et al. 2005. SCL24A5, a putative cation exchanger, affects pigmentation in zebrafish and humans. *Science* 310: 1782–1786.

Lewontin R.C. and Hubby J.L. 1966. A molecular approach to the study of genetic heterozygosity in natural populations. II. Amount of variation and degree of heterozygosity in natural populations of *Drosophila pseudoobscura*. *Genetics* 54: 595–609.

Lewontin R.C. and Krakauer J. 1973. Distribution of gene frequency as a test of the theory of the selective neutrality of polymorphisms. *Genetics* 74: 175–195.

Li W.H., Wu C.I., and Luo C.C. 1985. A new method for estimating synonymous and nonsynonymous rates of nucleotide substitution considering the relative likelihood of nucleotide and codon changes. *Mol. Biol. Evol.* 2: 150–174.

Li W. 2006. *Molecular evolution*. Sinauer, Sunderland, Massachusetts.

Makova K.D., Ramsay M., Jenkins T., and Li W.H. 2001. Human DNA sequence variation in a 6.6-kb region containing the melanocortin 1 receptor promoter. *Genetics* 158: 1253–1268.

Maynard-Smith J. and Haigh J. 1974. The hitch-hiking effect of a favorable gene. *Genet. Res.* 23: 23–35.

McDonald J.H. and Kreitman M. 1991. Adaptive protein evolution at the *Adh* locus in *Drosophila*. *Nature* 351: 652–654.

McVean G. and Spencer C.C.A. 2006. Scanning the human genome for signals of selection. *Curr. Opin. Genet. Dev.* 16: 624–629.

Myles S., Bouzekri N., Haverfield E., Cherkaoui, M. Dugoujon J.M., and Ward, R. 2005. Genetic evidence in support of a shared Eurasian-North African dairying origin. *Hum. Genet.* **117:** 34–42.

Nei M. and Gojobori T. 1986. Simple methods for estimating the numbers of synonymous and nonsynonymous substitutions. *Mol. Biol. Evol.* **3:** 418–426.

Nielsen R. 2001. Statistical tests of selective neutrality in the age of genomics. *Heredity* **86:** 641–647.

———. 2005. Molecular signatures of natural selection. *Annu. Rev. Genet.* **39:** 197–218.

Payseur B.A., Cutter A.D., and Nachman M.W. 2002. Searching for evidence of positive selection in the human genome using patterns of microsatellite variability. *Mol. Biol. Evol.* **19:** 1143–1153.

Ponting C.P. and Lunter G. 2006. Signatures of adaptive evolution within human non-coding sequence. *Hum. Mol. Genet.* **15:** R170–R175.

Przeworski M. 2002. The signature of positive selection at randomly chosen loci. *Genetics* **160:** 1179–1189.

Ptak S.E. and Przeworski M. 2002. Evidence for population growth in humans is confounded by fine-scale population structure. *Trends Genet.* **18:** 559–563.

Richman A.D. and Kohn J.R. 1999. Self-incompatibility alleles from *Physalis*: Implications for historical inference from balanced genetic polymorphisms. *Proc. Natl. Acad. Sci.* **96:** 168–172.

Sabeti P.C., Schaffner S.F., Fry B., Lohmueller J., Varilly P., Shamovsky O., Palma A., Mikkelsen T.S., Altshuler D., and Lander E.S. 2006. Positive natural selection in the human lineage. *Science* **312:** 1614–1620.

Sabeti P.C., Reich D.E., Higgins J.M., Levine H.Z., Richter D.J., Schaffner S.F., Gabriel S.B., Platko J.V., Patterson N.J., McDonald G.J., et al. 2002. Detecting recent positive selection in the human genome from haplotype structure. *Nature* **419:** 832–837.

Schlotterer C. 2002. Towards a molecular characterization of adaptation in local populations. *Curr. Opin. Genet. Dev.* **12:** 683–687.

Simonsen K.L., Churchill G.A., and Aquadro C.F. 1995. Properties of statistical tests of neutrality for DNA polymorphism data. *Genetics* **141:** 413–429.

Sober E. 1993. *The nature of selection: Evolutionary theory in philosophical focus.* University of Chicago Press, Chicago, Illinois.

Soranzo N., Bufe B., Sabeti P.C., Wilson J.F., Weale M.E., Marguerie R., Meyerhof W., and Goldstein D.B. 2005. Positive selection on a high-sensitivity allele of the human bitter-taste receptor TAS2R16. *Curr. Biol.* **15:** 1257–1265.

Stephens J.C., Schneider A.J., Tanguay D.A., Choi J., Acharya T., Stanley S.E., Jiang R., Messer C.J., Chew A., Han J., et al. 2001. Haplotype variation and linkage disequilibrium in 313 human genes. *Science* **293:** 489–493.

Sunyaev S.R., Lathe W.C., III, Ramensky V.E., and Bork P. 2000. SNP frequencies in human genes an excess of rare alleles and differing modes of selection. *Trends Genet.* **16:** 335–337.

Suzuki Y. and Gojobori T. 1999. A method for detecting positive selection at single amino acid sites. *Mol. Biol. Evol.* **16:** 1315–1328.

Swallow D.M. 2003. Genetics of lactase persistence and lactose intolerance. *Ann. Rev. Genet.* **37:** 197–229.

Tajima F. 1989. Statistical method for testing the neutral mutation hypothesis by DNA polymorphism. *Genetics* **123:** 585–595.

Tishkoff S.A., Reed F.A., Ranciaro A., Voight B.F., Babbitt C.C., Silverman J.S., Powell K., Mortensen H.M., Hirbo J.B., Osman M., et al. 2006. Convergent adaptation of human lactase persistence in Africa and Europe. *Nat. Genet.* **39:** 31–40.

Thompson E.E., Kuttab-Boulos H., Witonsky D., Yang L., Roe B.A., and DiRienzo A. 2004. *CYP3A* variation and the evolution of salt-sensitivity variants. *Am. J. Hum. Genet.* **75:** 1059–1069.

Toomajian C. and Kreitman M. 2002. Sequence variation and haplotype structure at the human *HFE* locus. *Genetics* **161:** 1609–1623.

Toomajian C., Ajioka R.S., Jorde L.B., Kushner J.P., and Kreitman M. 2003. A method for detecting recent selection in the human genome from allele age estimates. *Genetics* **161:** 1609–1623.

Vallender E.J. and Lahn B.T. 2004. Positive selection on the human genome. *Hum. Mol. Genet.* **13:** 245–254.

Verrelli B.C., McDonald J.H., Argyropoulos G., Destro-Bisol G., Froment A., Drousiotou A., Lefranc G., Helal A.N., Loiselet J., and Tishkoff S.A. 2002. Evidence for balancing selection from nucleotide sequence analyses of human *G6PD*. *Am. J. Hum. Genet.* **71:** 1112–1128.

Voight B.F., Kudaravalli S., Wen X., and Pritchard J.K. 2006. A map of recent positive selection in the human genome. *PLos Biol.* **4:** 446–458.

Wall J.D. 1999. Recombination and the power of statistical tests of neutrality. *Genet. Res.* **74:** 65–79.

Wang E.T., Kodama G., Baldi P., and Moyzis R.K. 2006. Global landscape of recent inferred Darwinian selection for *Homo sapiens*. *Proc. Natl. Acad. Sci.* **103:** 135–140.

Watterson G.A. and Guess H.A. 1977. Is the most frequent allele the oldest? *Theor. Popul. Biol.* **11:** 141–160.

Weir B.S., Cardon L.R., Anderson A.D., Nielsen D.M., and Hill W.G. 2005. Measures of human population structure show heterogeneity among genomic regions. *Genome Res.* **15:** 1468–1476.

Wooding S.P., Watkins W.S., Bamshad M.J., Dunn D.M., Weiss R.B., and Jorde L.B. 2002. DNA sequence variation in a 3.7-kb non-coding sequence 5′ of the *CYP1A2* gene: Implications for human population history and natural selection. *Am. J. Hum. Genet.* **71:** 528–542.

Wooding S., Kim U.K., Bamshad M.J., Larsen J., Jorde L.B., and Drayna D. 2004. Natural selection and molecular evolution in PTC, a bitter taste receptor gene. *Am. J. Hum. Genet.* **74:** 637–646.

Wong W.S., Yang Z., Goldman N., and Nielsen R. 2004. Accuracy and power of statistical methods for detecting adaptive evolution in protein coding sequences and for identifying positively selected sites. *Genetics* **168:** 1041–1051.

Yang Z. 2002. Inference of selection from multiple species alignments. *Curr. Opin. Genet. Dev.* **12:** 688–694.

25 | *Arabidopsis*

Yan Li[1] and Justin O. Borevitz[1,2]

[1]*Department of Ecology and Evolution and* [2]*Committee on Genetics, University of Chicago, Chicago, Illinois 60637*

INTRODUCTION

For the past 20 years, *Arabidopsis thaliana* has been the model plant to study genetics, development, and physiology. The geographically wide distribution and the available population genomic tools now make *A. thaliana* the organism of choice for studies of ecology and evolution as well. A complete genome sequence and saturating molecular markers allow ecotypes and accessions to be clearly distinguished. Resources and tools include collections of thousands of wild accessions, several sets of recombinant inbred lines (RILs), a gene expression atlas of genes (Schmid et al. 2005b) and genomic regions (Ecker 2004; Stolc et al. 2005), a nearly complete set of knockout lines (Alonso et al. 2003), and a 250K single-nucleotide polymorphism (SNP) tiling array (our current work). Although

Arabidopsis is a mostly self-fertilizing (selfing) species, enough outcrossing has occurred to allow fine mapping via linkage disequilibrium (LD). In *A. thaliana*, LD decays within approximately 25–50 kilobases (Nordborg et al. 2005), which is comparable to humans. However, *A. thaliana* has a higher SNP density. Population structure in *A. thaliana* is substantial on the local scale but also exists at the regional and worldwide levels. Two recent studies which utilize a core set of accessions, spanning these geographic levels, required a good statistical model to control the population structure in association mapping. Utilization of these tools over the next few years will uncover the genetic and genomic basis for the natural variation in *A. thaliana* and thus enhance our understanding of the ecological function of the genetic variants likely to be under natural selection.

GENETIC AND PHYSICAL MAPS OF *A. THALIANA*

The first genetic map of *A. thaliana* was released in 1983 (Koornneef et al. 1983), helping *Arabidopsis* to gain momentum as a model plant in the early 1980s (Meinke et al. 1998). Later, several genetic maps were generated from RILs (Lister and Dean 1993; Liu et al. 1996; Alonso-Blanco et al. 1998; Wilson et al. 2001; Loudet et al. 2002; Weinig et al. 2003; Clerkx et al. 2004; Symonds et al. 2005; Werner et al. 2005). These are very useful for mapping quantitative trait loci (QTL) and for fine mapping.

A. thaliana is the first plant species to have its whole genome sequenced. The genome of the reference accessions, Columbia (Col), was reported to contain 25,498 genes in the 115.4 megabase (Mb) sequenced regions of the 125-Mb genome (Arabidopsis Genome Initiative 2000), and was reannotated by the *Arabidopsis* Information Resource (TAIR) (Wortman et al. 2003), which now has been updated with 31,407 genes (TAIR6, November 11, 2005). Another ecotype, Landsberg erecta (Ler), was sequenced by Cereon Genomics at approximately 2x coverage (http://www.arabidopsis.org/browse/Cereon/index.jsp) (Jander et al. 2002). 1,267 amplified fragment-length polymorphism (AFLP) markers have been located on the *Arabidopsis* genome sequence (Peters et al. 2001). Recently, 876 short fragments (later 1,238 released, http://walnut.usc.edu/apache2-default/2010/index-old.html) from 96 *A. thaliana* accessions were sequenced as well (Nordborg et al. 2005). This sequence information from different accessions certainly accelerates the study of the genetic and genomic variation in *A. thaliana*. For example, many more polymorphic markers were identified which will reduce the time toward fine mapping and gene cloning (Borevitz and Chory 2004), and the densely spaced markers enable biologists to obtain insight into the genome-wide haplotype and population structure of *A. thaliana* (Nordborg et al. 2005; Schmid et al. 2006).

Polymorphic Markers for the Genetic Map

In addition to the conventional molecular markers such as restriction fragment length polymorphisms (RFLPs), AFLPs, and codominant cleaved amplified polymorphic sequences (CAPS), thousands of SNPs and insertion/deletions (indels) are available as well (Jander et al. 2002). Since the Col and Ler sequences became publicly available, a total of 56,670 polymorphisms (including 37,344 SNPs, 18,579 indels, and 747 large indels) are now predicted between Col and Ler by Cereon Genomics, and these collections are available through the Web site http://www.arabidopsis.org/browse/Cereon/index.jsp. The SNPs and indels have several advantages compared to the microsatellite markers (Torjek et al. 2003): They are highly abundant, mostly codominant, and generally phenotypically neutral (Berger et al. 2001). Another 8,051 predicted SNPs identified in 12 *A. thaliana* ecotypes are also publicly available through the Max-Planck *Arabidopsis* SNP Consortium (MASC) SNP database (Schmid et al. 2003). Recently, over 17,000 SNPs and indels have been identified by resequencing 876 short fragments in 96 accessions (Nordborg et al. 2005). These polymorphisms have been submitted to the *Arabidopsis* Information Resource (TAIR) (http://www.arabidopsis.org/news/monthly/TAIR_News_ Sept04.jsp) and can be viewed through http://walnut.usc.edu/2010 or queried at http://msqt.weigelworld.org/ (N. Warthmann, unpubl.).

The Affymetrix GeneChip microarray provides a high-throughput platform both for discovering polymorphic markers and for genotyping them. High-density oligonucleotide arrays contain millions of 25-mer features. Each feature could be a potential marker when the arrays are hybridized with labeled total genomic DNA. About 4,000 single-feature polymorphisms (SFPs) were identified between Col and Ler at a 5% error rate, and from five *Arabidopsis* accessions, 4,713 SFPs were identified as polymorphic in multiple accessions when compared to Col (Borevitz et al. 2003), using arrays with only 100,000 features. The Borevitz and Nordborg labs have recently generated a new array that incorporates 250,000 nonsingleton SNPs (R. Clark and D. Weigel, in prep.) and more than 1.5 million tiling array probes (that can detect novel SFPs and indels). In addition, the gene and tiling probes generated from sequences of several plant pathogens (including *Pst, Psm, Psy, Psx*, Agrobacteria, Xanthomonas, *H. parasitica*, and many viruses) were also included on this array.

GEOGRAPHIC AND POPULATION STRUCTURE OF *A. THALIANA*

In early studies, the genetic diversity in *A. thaliana* was examined using limited numbers of molecular markers including allozymes (Abbott and Gomes 1989), RFLPs (Bergelson et al. 1998), AFLPs (Miyashita et al. 1999; Sharbel et al. 2000), and microsatellites (Kuittinen et al. 1997). With the development of high-density molecular markers across the *Arabidopsis* genome, two research groups recently surveyed the genetic variation among global populations (Nordborg et al. 2005; Schmid et al. 2006). Both studies suggested there is significant global population structure in *A. thaliana*, as indicated by differences in genetic diversity between geographic regions (Nordborg et al. 2005; Schmid et al. 2006). However, many polymorphisms are shared between populations (Nordborg et al. 2005). There is considerable genetic variation within and among local populations as well (Nordborg et al. 2005; Bakker et al. 2006), refuting the original idea that *A. thaliana* existed as clonal patches. Some individual populations contain much of the variation present species-wide (Nordborg et al. 2005). The U.S. midwest population was noticed to be a heterogeneous collection, as indicated by extensive haplotype sharing with accessions from other geographic regions (Nordborg et al. 2005). In another study, the Central and Eastern European accessions were found as admixed populations (Schmid et al. 2006).

As for the pattern of molecular polymorphism, the allele frequency distribution fails to fit standard neutral models due to an excess of rare alleles (Nordborg et al. 2005; Schmid et al. 2006). The levels of molecular polymorphism average about seven nucleotides per kb (nucleotide diversity is about 0.007) (Nordborg et al. 2005; Schmid et al. 2005a), but vary between different genomic regions, which is generally positively correlated with segmental duplications and negatively correlated with gene density (Nordborg et al. 2005). The level of polymorphism also varies between populations. For example, Central Asia has a low level of polymorphism relative to the Iberian Peninsula and Central Europe (Schmid et al. 2006). Therefore, sampling strategies and the molecular markers used should be considered to have an effect on the observed genetic variation patterns.

TOOLS TO DISCOVER THE GENETIC BASIS OF NATURAL VARIATION

The functions of less than 10% of the genes in *Arabidopsis* have been successfully identified (Ostergaard and Yanofsky 2004), mainly by gene knockout through laboratory mutagenesis. Although this approach will continue to be important for gene discovery, natural variation is experiencing a rebirth in its popularity as a method for gene discovery and functional genomics. Natural variation overcomes the limitations of laboratory mutagenesis, such as no obvious phenotypic variation, unstable phenotypes, narrow genetic background, and limited range of phenotypic variation (Tonsor et al. 2005). It also has several advantages in discovering new genes or new alleles, describing the nature of gene function in the context of broad genetic range and environments, and understanding the genetic architecture of complex traits.

Several genes responsible for naturally occurring variation have been identified: the genes involved in disease resistance, including *RTM1, RPS2, RPM1* (Mindrinos et al. 1994; Stahl et al. 1999; Chisholm et al. 2000); genes involved in flowering, such as *FRI* (Johanson et al. 2000; Gazzani et al. 2003), *FLC* (Gazzani et al. 2003; Michaels et al. 2003), *EDI* (El-Assal et al. 2001), *FLM* (Werner et al. 2005); genes involved in hypocotyl length/light response, *PHYA* (Maloof et al. 2001) and *PHYD* (Aukerman et al. 1997); and several others (Koornneef et al. 2004). In most cases, the natural variation among accessions is multigenic (Koornneef et al. 2004). Traditionally, QTL analysis is the first step to dissect the genetic basis of this natural variation. RILs have proved to be an effective approach in QTL analysis where the phenotypes can be measured using multiple replicates which increase the power to detect QTL (Koornneef et al. 2004). Many traits have been analyzed using RILs in *Arabidopsis*, including resistance/tolerance to biotic factors, developmental traits, physiological traits, chemical contents, and enzyme activity (Koornneef et al. 2004). The most commonly used Ler x Col RIL population has been used for many traits including expression QTL (eQTL) mapping (DeCook et al. 2006). The Ler x Cvi RIL population has been analyzed for more than 40 phenotypic traits (Koornneef et al. 2004) and now for more than 2,000 metabolites (Keurentjes et al. 2006). Mapping multiple traits in the same population helps compare QTL map positions. Currently, there are nine different RIL populations that are publicly available (http://www.arabidopsis. org/abrc/catalog/recombinant_inbred_set_1.html), but many more are in development (http:// naturalvariation.org/genetic.html, http://www.inra.fr/internet/Produits/vast/RILs.htm).

Bulk segregant analysis is a quick method to map genes or QTLs with large effects. By array hybridization with DNA from pools of lines with extreme phenotypes, the mutations with large effects, such as big deletions, can be located quickly (Borevitz et al. 2003; Hazen et al. 2005a,b; Werner et al. 2005). Recently, the feasibility of using LD mapping to identify/map loci responsible for natural variation has been investigated in *Arabidopsis* (Nordborg et al. 2002, 2005; Schmid et al. 2003, 2005a, 2006; Olsen et al. 2004; Aranzana et al. 2005; Rosenberg and Nordborg 2006; Zhao et al. 2007). LD mapping would be very useful in selfing species such as *Arabidopsis*, since the genotyped core set of mapping populations can be used repeatedly by many researchers to study different traits. The recent studies showed that LD mapping will be a powerful approach to detect the genetic basis for natural variation once the population structure can be controlled and a good statistical model is established. We are currently developing several large core sets (>384) of accessions that will be genotyped with the 250K SNP array as a community platform for high-resolution LD mapping.

QTLs can be confirmed and fine mapped in near isogenic lines (NILs) or heterogeneous inbred families (HIFs) (Borevitz and Chory 2004). The density of polymorphic markers and the recombination events are critical factors for fine mapping. In some cases, the QTLs can be mapped directly to the gene (Fridman et al. 2000; Kroymann et al. 2001). Usually, the selection of candidate genes can begin once QTLs have been localized to a relatively narrow region of 3 centiMorgans (cM) or less (Borevitz and Chory 2004). The candidate genes can be chosen based on polymorphism information, such as changes/deletions in genes or gene expression profiles. Once candidate genes are identified, functional studies (loss/gain of function) can be conducted by using the mutants (http://signal.salk.edu) (Sessions et al. 2002; Alonso et al. 2003) and creating transgenic lines (Jander et al. 2002).

HIGH-THROUGHPUT GENOTYPING, PHENOTYPING, AND RELATED PROTOCOLS

With the development of molecular markers and collections of wild accessions in *Arabidopsis*, high-throughput methods are needed for both genotyping and phenotyping. Cost is another factor to consider if a small laboratory wants to study natural variation. In our laboratory, we have been using methods in high-throughput format for DNA extraction, genotyping, and phenotyping, although the phenotyping is still the limiting step so far. The protocols and related resources are listed below.

DNA Extraction

The genomic DNA purification kit from PUREGENE (Gentra System) is the choice for high-throughput methods with good DNA quality. We have used this for SNP genotyping with good results. We modified the protocol to the 96-well plate format (see Chapter 7).

FIGURE 25-1. Natural variation in flowering time within a local field (MNF-CHE in the U.S. Midwest population). (DTF) Days to flower, which is the days from transferring seeds to soil until the appearance of the first floral buds. The labels were created using bar code technology.

SNP Genotyping and Array Genotyping

We have been using 149 SNP markers at intermediate allele frequencies for mapping in F_2 crosses of various genetic backgrounds (www.naturalvariation.org). The SNP genotyping was done by Sequenom in iPLEX pools of approximately 40 SNPs each. Over 5,000 accessions (including over 800 accessions from the *Arabidopsis* Biological Resource Center, Columbus, Ohio) have been tested using these SNPs, and the genotypes are available at our Web site www.naturalvariation.org. The methods for array genotyping have been described previously (Borevitz et al. 2003; Werner et al. 2005).

Phenotyping

We use bar code technology to record phenotypes (Fig. 25-1). The picture of each accession is taken when the plant flowers (the appearance of the first floral buds), and these photos are available at www.naturalvariation.org. The bar-code-reading software (Softek Software, www.barcode.com) is used to rename the photos with the bar code name (a modified version is available at www.naturalvariation. org). In 2007, we and other workers will begin time-lapse photography to profile variation in developmental programs. Phenotypic space can be quantified with morphometric analysis.

RESOURCES

The databases and Web sites for wild collections, sequence, polymorphisms, and mutants in *Arabidopsis* are listed in Table 25-1. The publicly available RIL sets are listed in Table 25-2.

TABLE 25-1. Resources to Study Natural Variation in *Arabidopsis*

Database or Web site	Functions
http://www.arabidopsis.org/	sequences, polymorphisms, seeds
http://msqt.weigelworld.org/;	polymorphic SNPs in the Nordborg 96 accessions
http://walnut.usc.edu/2010;	
http://www.arabidopsis.org/	
http://www2.mpiz-koeln.mpg.de/masc/	polymorphic SNPs in 12 ecotypes
http://www.arabidopsis.org/browse/Cereon/index.jsp	polymorphisms between Col/Ler
http://www.naturalvariation.org	wild collections and their pictures/genotypes, SNPs, array genotyping methods
http://signal.salk.edu	mutants

SUMMARY

A. thaliana has served as a model plant system for more than 20 years, and the development of the large number of polymorphic markers, high-throughput methods, and wild collections will enable it to continue its contributions in genetics, development, physiology, ecology, and evolution. Association mapping will be utilized widely in *Arabidopsis* to identify candidate genes for many traits, especially the ecologically relevant traits and those related to yield/disease resistance. The methods established in *Arabidopsis* can be used in other plant systems, especially the selfing crops.

ACKNOWLEDGMENTS

We thank Dr. Joy Bergelson and Dr. Magnus Nordborg for their collaboration, Dr. Randy Scholl and Luz Rivero for providing 850 stock center lines, and Dr. Eric Holub and Dr. Dianne Byers for providing wild *Arabidopsis* accessions in association mapping projects. This work is supported by National Institutes of Health grant R01 GM073822.

TABLE 26-2. RILs That Are Publicly Available through TAIR
(http://www.arabidopsis.org/abrc/catalog/recombinant_inbred_set_1.html)

Parental lines (female × male)	Donor(s)	Traits/Uses[a]
Col-3 × Nd-1 and Col-5(gl1) × Nd-1	Eric Holub Jim Beynon Ian Crute	Disease resistance
Van-0 × Col-0(WT-2)	Justin Borevitz Evadne Smith Julin Maloof Detlef Weigel Joanne Chory	Flowering time
Ler-0 / No-0[b]	Jorge Casal Javier Botto Alan Lloyd	Hypocotyl growth
Bay-0 × Sha	Olivier Loudet Sylvain Chaillou	Nitrogen use efficiency, light responses, root growth, carbohydrate content, cell wall composition etc.
Ler-0 × Col-4	Clare Lister Caroline Dean	Developing genetic and physical maps
Ler-0 × Sha	Maarten Koornneef	Flowering time
Cvi-1 / Ler-2[b]	Maarten Koornneef	Insect resistance, freezing tolerance, flowering time, plant and seed size, seed dormancy, etc.
Ler × Ws-1	Pablo Scolnik	Developing genetic map
Col(gl1) × Kas-1	Shauna Somerville	Powdery mildew resistance, flowering time

[a] Information is obtained from TAIR Web site.
[b] The direction of the cross is unknown.

REFERENCES

Abbott R.J. and Gomes M.F. 1989. Population genetic-structure and outcrossing rate of *Arabidopsis thaliana* (L.) Heynh. *Heredity* **62**: 411–418.

Alonso J.M., Stepanova A.N., Leisse T.J., Kim C.J., Chen H.M., Shinn P., Stevenson D.K., Zimmerman J., Barajas P., Cheuk R., et al. 2003. Genome-wide insertional mutagenesis of *Arabidopsis thaliana*. *Science* **301**: 653–657.

Alonso-Blanco C., Peeters A.J.M., Koornneef M., Lister C., Dean C., van den Bosch N., Pot J., and Kuiper M.T.R. 1998. Development of an AFLP based linkage map of Ler, Col and Cvi *Arabidopsis thaliana* ecotypes and construction of a Ler/Cvi recombinant inbred line population. *Plant J.* **14**: 259–271.

Arabidopsis Genome Initiative 2000. Analysis of the genome sequence of the flowering plant *Arabidopsis thaliana*. *Nature* **408**: 796–815.

Aranzana M.J., Kim S., Zhao K.Y., Bakker E., Horton M., Jakob K., Lister C., Molitor J., Shindo C., Tang C.L., et al. 2005. Genome-wide association mapping in *Arabidopsis* identifies previously known flowering time and pathogen resistance genes. *Plos Genet.* **1**: 531–539.

Aukerman M.J., Hirschfeld M., Wester L., Weaver M., Clack T., Amasino R.M., and Sharrock R.A. 1997. A deletion in the *PHYD* gene of the *Arabidopsis* Wassilewskija ecotype defines a role for phytochrome D in red/far-red light sensing. *Plant Cell* **9**: 1317–1326.

Bakker E.G., Stahl E.A., Toomajian C., Nordborg M., Kreitman M., and Bergelson J. 2006. Distribution of genetic variation within and among local populations of *Arabidopsis thaliana* over its species range. *Mol. Ecol.* **15**: 1405–1418.

Bergelson J., Stahl E., Dudek S., and Kreitman M. 1998. Genetic variation within and among populations of *Arabidopsis thaliana*. *Genetics* **148**: 1311–1323.

Berger J., Suzuki T., Senti K.A., Stubbs J., Schaffner G., and Dickson B.J. 2001. Genetic mapping with SNP markers in *Drosophila*. *Nat. Genet.* **29**: 475–481.

Borevitz J.O. and Chory J. 2004. Genomics tools for QTL analysis and gene discovery. *Curr. Opin. Plant Biol.* **7**: 132–136.

Borevitz J.O., Liang D., Plouffe D., Chang H.S., Zhu T., Weigel D., Berry C.C., Winzeler, E., and Chory, J. 2003. Large-scale identification of single-feature polymorphisms in complex genomes. *Genome Res.* **13**: 513–523.

Chisholm S.T., Mahajan, S.K., Whitham, S.A., Yamamoto, M.L., and Carrington, J.C. 2000. Cloning of the *Arabidopsis RTM1* gene, which controls restriction of long-distance movement of tobacco etch virus. *Proc. Natl. Acad. Sci.* **97**: 489–494.

Clerkx E.J.M., El-Lithy M.E., Vierling E., Ruys G.J., Blankestijn-De Vries H., Groot S.P.C., Vreugdenhil D., and Koornneef M. 2004. Analysis of natural allelic variation of *Arabidopsis* seed germination and seed longevity traits between the accessions Landsberg *erecta* and Shakdara, using a new recombinant inbred line population. *Plant Physiol* **135**: 432–443.

DeCook R., Lall S., Nettleton D., and Howell S.H. 2006. Genetic regulation of gene expression during shoot development in *Arabidopsis*. *Genetics* **172**: 1155–1164.

Ecker J.R. 2004. Genome-wide discovery of transcription units and functional elements in arabidopsis. *Mol. Biol. Cell* **15**: 122A.

El-Assal S.E.D., Alonso-Blanco C., Peeters A.J.M., Raz V., and Koornneef M. 2001. A QTL for flowering time in *Arabidopsis* reveals a novel allele of *CRY2*. *Nat. Genet.* **29**: 435–440.

Fridman E., Pleban T., and Zamir D. 2000. A recombination hotspot delimits a wild-species quantitative trait locus for tomato sugar content to 484 bp within an invertase gene. *Proc. Natl. Acad. Sci.* **97**: 4718–4723.

Gazzani S., Gendall A.R., Lister C., and Dean C. 2003. Analysis of the molecular basis of flowering time variation in *Arabidopsis* accessions. *Plant Physiol.* **132**: 1107–1114.

Hazen S.P., Schultz T.F., Pruneda-Paz J.L., Borevitz J.O., Ecker J.R., and Kay S.A. 2005a. *LUX ARRHYTHMO* encodes a Myb domain protein essential for circadian rhythms. *Proc. Natl. Acad. Sci.* **102**: 10387–10392.

Hazen S.P., Borevitz J.O., Harmon F.G., Pruneda-Paz J.L., Schultz T.F., Yanovsky M.J., Liljegren S.J., Ecker J.R., and Kay S.A. 2005b. Rapid array mapping of circadian clock and developmental mutations in *Arabidopsis*. *Plant Physiol.* **138**: 990–997.

Jander G., Norris S.R., Rounsley S.D., Bush D.F., Levin I.M., and Last R.L. 2002. Arabidopsis map-based cloning in the post-genome era. *Plant Physiol.* **129**: 440–450.

Johanson U., West, J., Lister, C., Michaels, S., Amasino, R., and Dean, C. 2000. Molecular analysis of *FRIGIDA*, a major determinant of natural variation in *Arabidopsis* flowering time. *Science* **290**: 344–347.

Keurentjes J.J.B., Fu J.Y., de Vos C.H.R., Lommen A., Hall R.D., Bino R.J., van der Plas L.H.W., Jansen R.C., Vreugdenhil D., and Koornneef M. 2006. The genetics of plant metabolism. *Nat. Genet.* **38**: 842–849.

Koornneef M., Alonso-Blanco C., and Vreugdenhil D. 2004. Naturally occurring genetic variation in *Arabidopsis thaliana*. *Annu. Rev. Plant Biol.* **55**: 141–172.

Koornneef M., Vaneden J., Hanhart C.J., Stam P., Braaksma F.J., and Feenstra W.J. 1983. Linkage map of *Arabidopsis thaliana*. *J. Hered.* **74**: 265–272.

Kroymann J., Textor S., Tokuhisa J.G., Falk K.L., Bartram S., Gershenzon J., and Mitchell-Olds T. 2001. A gene controlling variation in arabidopsis glucosinolate composition is part of the methionine chain elongation pathway. *Plant Physiol.* **127**: 1077–1088.

Kuittinen H., Mattila A., and Savolainen O. 1997. Genetic variation at marker loci and in quantitative traits in natural populations of *Arabidopsis thaliana*. *Heredity* **79**: 144–152.

Lister C. and Dean C. 1993. Recombinant inbred lines for mapping RFLP and phenotypic markers in *Arabidopsis thaliana*. *Plant J.* **4**: 745–750.

Liu Y.G., Mitsukawa N., Lister C., Dean C., and Whittier R.F. 1996. Isolation and mapping of a new set of 129 RFLP markers in *Arabidopsis thaliana* using recombinant inbred lines. *Plant J.* **10**: 733–736.

Loudet O., Chaillou S., Camilleri C., Bouchez D., and Daniel-Vedele F. 2002. Bay-0 × Shahdara recombinant inbred line population: A powerful tool for the genetic dissection of complex traits in *Arabidopsis*. *Theor. Appl. Genet.* **104**: 1173–1184.

Maloof J.N., Borevitz J.O., Dabi T., Lutes J., Nehring R.B., Redfern J.L., Trainer G.T., Wilson J.M., Asami T., Berry C.C., Weigel D., and Chory J. 2001. Natural variation in light sensitivity of *Arabidopsis*. *Nat. Genet.* **29**: 441–446.

Meinke D.W., Cherry J.M., Dean C., Rounsley S.D., and Koornneef M. 1998. *Arabidopsis thaliana*: A model plant for genome analysis. *Science* **282**: 662, 679–682.

Michaels S.D., He Y.H., Scortecci K.C., and Amasino R.M. 2003. Attenuation of FLOWERING LOCUS C activity as a mechanism for the evolution of summer-annual flowering behavior in *Arabidopsis*. *Proc. Natl. Acad. Sci.* **100**: 10102–10107.

Mindrinos M., Katagiri F., Yu G.L., and Ausubel F.M. 1994. The *A. thaliana* disease resistance gene *RPS2* encodes a protein containing a nucleotide-binding site and leucine-rich repeats. *Cell* **78**: 1089–1099.

Miyashita N.T., Kawabe A., and Innan H. 1999. DNA variation in the wild plant *Arabidopsis thaliana* revealed by amplified fragment length polymorphism analysis. *Genetics* **152:** 1723–1731.

Nordborg M., Borevitz J.O., Bergelson J., Berry C.C., Chory J., Hagenblad J., Kreitman M., Maloof J.N., Noyes T., Oefner P.J., Stahl E.A., and Weigel D. 2002. The extent of linkage disequilibrium in *Arabidopsis thaliana*. *Nat. Genet.* **30:** 190–193.

Nordborg M., Hu T.T., Ishino Y., Jhaveri J., Toomajian C., Zheng H.G., Bakker E., Calabrese P., Gladstone J., Goyal R., et al. 2005. The pattern of polymorphism in *Arabidopsis thaliana*. *Plos Biol.* **3:** 1289–1299.

Olsen K.M., Halldorsdottir S.S., Stinchcombe J.R., Weinig C., Schmitt J., and Purugganan M.D. 2004. Linkage disequilibrium mapping of *Arabidopsis CRY2* flowering time alleles. *Genetics* **167:** 1361–1369.

Ostergaard L. and Yanofsky M.F. 2004. Establishing gene function by mutagenesis in *Arabidopsis thaliana*. *Plant J.* **39:** 682–696.

Peters J.L., Constandt H., Neyt P., Cnops G., Zethof J., Zabeau M., and Gerats T. 2001. A physical amplified fragment-length polymorphism map of *Arabidopsis*. *Plant Physiol.* **127:** 1579–1589.

Rosenberg N.A. and Nordborg M. 2006. A general population-genetic model for the production by population structure of spurious genotype-phenotype associations in discrete, admixed or spatially distributed populations. *Genetics* **173:** 1665–1678.

Schmid K.J., Ramos-Onsins S., Ringys-Beckstein H., Weisshaar B., and Mitchell-Olds T. 2005a. A multilocus sequence survey in *Arabidopsis thaliana* reveals a genome-wide departure from a neutral model of DNA sequence polymorphism. *Genetics* **169:** 1601–1615.

Schmid K.J., Torjek O., Meyer R., Schmuths H., Hoffmann M.H., and Altmann T. 2006. Evidence for a large-scale population structure of *Arabidopsis thaliana* from genome-wide single nucleotide polymorphism markers. *Theor. Appl. Genet.* **112:** 1104–1114.

Schmid K.J., Sorensen T.R., Stracke R., Torjek O., Altmann T., Mitchell-Olds T., and Weisshaar B. 2003. Large-scale identification and analysis of genome-wide single-nucleotide polymorphisms for mapping in *Arabidopsis thaliana*. *Genome Res.* **13:** 1250–1257.

Schmid M., Davison T.S., Henz S.R., Pape U.J., Demar M., Vingron M., Scholkopf B., Weigel D., and Lohmann J.U. 2005b. A gene expression map of *Arabidopsis thaliana* development. *Nat. Genet.* **37:** 501–506.

Sessions A., Burke E., Presting G., Aux G., McElver J., Patton D., Dietrich B., Ho P., Bacwaden J., and Ko C. 2002. A high-throughput *Arabidopsis* reverse genetics system. *Plant Cell* **14:** 2985–2994.

Sharbel T.F., Haubold B., and Mitchell-Olds T. 2000. Genetic isolation by distance in *Arabidopsis thaliana*: Biogeography and postglacial colonization of Europe. *Mol. Ecol.* **9:** 2109–2118.

Stahl E.A., Dwyer G., Mauricio R., Kreitman M., and Bergelson J. 1999. Dynamics of disease resistance polymorphism at the *Rpm1* locus of *Arabidopsis*. *Nature* **400:** 667–671.

Stolc V., Samanta M.P., Tongprasit W., Sethi H., Liang S.D., Nelson D.C., Hegeman A., Nelson C., Rancour D., Bednarek S., et al. 2005. Identification of transcribed sequences in *Arabidopsis thaliana* by using high-resolution genome tiling arrays. *Proc. Natl. Acad. Sci.* **102:** 4453–4458.

Symonds V.V., Godoy A.V., Alconada T., Botto J.F., Juenger T.E., Casal J.J., and Lloyd A.M. 2005. Mapping quantitative trait loci in multiple populations of *Arabidopsis thaliana* identifies natural allelic variation for trichome density. *Genetics* **169:** 1649–1658.

Tonsor S.J., Alonso-Blanco C., and Koornneef M. 2005. Gene function beyond the single trait: Natural variation, gene effects, and evolutionary ecology in *Arabidopsis thaliana*. *Plant Cell Environ.* **28:** 2–20.

Torjek O., Berger D., Meyer R.C., Mussig C., Schmid K.J., Sorensen T.R., Weisshaar B., Mitchell-Olds T., and Altmann T. 2003. Establishment of a high-efficiency SNP-based framework marker set for *Arabidopsis*. *Plant J.* **36:** 122–140.

Weinig C., Stinchcombe J.R., and Schmitt J. 2003. QTL architecture of resistance and tolerance traits in *Arabidopsis thaliana* in natural environments. *Mol. Ecol.* **12:** 1153–1163.

Werner J.D., Borevitz J.O., Warthmann N., Trainer G.T., Ecker J.R., Chory J., and Weigel D. 2005. Quantitative trait locus mapping and DNA array hybridization identify an *FLM* deletion as a cause for natural flowering-time variation. *Proc. Natl. Acad. Sci.* **102:** 2460–2465.

Wilson I.W., Schiff C.L., Hughes D.E., and Somerville S.C. 2001. Quantitative trait loci analysis of powdery mildew disease resistance in the *Arabidopsis thaliana* accession Kashmir-1. *Genetics* **158:** 1301–1309.

Wortman J.R., Haas B.J., Hannick L.I., Smith R.K., Maiti R., Ronning C.M., Chan A.P., Yu C.H., Ayele M., Whitelaw C.A., White O.R., and Town C.D. 2003. Annotation of the Arabidopsis genome. *Plant Physiol.* **132:** 461–468.

Zhao K., Aranzana M.J., Kim S., Lister C., Shindo C., Tang C., Toomajian C., Zheng H., Dean C., Marjoram P., and Nordborg M. 2007. An *Arabidopsis* example of association mapping in structured samples. *Plos Genet.* **3:** e4.

WWW RESOURCES

http://www.arabidopsis.org/abrc/catalog/recombinant_inbred_set_1.html Sets of Recombinant Inbred Lines, Arabidopsis Biological Resource Center (ABRC)

http://www.arabidopsis.org/news/monthly/TAIR_News_Sept04.jsp The Arabidopsis Information Resource (TAIR) news release

http://www.inra.fr/internet/Produits/vast/RILs.htm V.A.S.T. (Variation and Abiotic Stress Tolerance) Lab tries to list available and upcoming Arabidopsis RILs (or NILs), which are useful for quantitative genetics

http://walnut.usc.edu/apache2-default/2010/index-old.html A genomic survey of polymorphism and linkage disequilibrium, funded by the NSF 2010 Project. J. Bergelson, M. Kreitman, Department of Ecology & Evolution, University of Chicago; M. Nordborg, Program in Molecular & Computational Biology, University of Southern California

26 | Maize

W. Brad Barbazuk,[1] An-Ping Hsia,[2] Hsin D. Chen,[2] Yan Fu,[1]
Kazuhiro Ohtsu,[2] and Patrick S. Schnable[2,3,4]

[1]Donald Danforth Plant Science Center, St. Louis, Missouri 63132; [2]Department of Agronomy,
[3]Department of Genetics, Development, and Cell Biology, [4]Center for Plant Genomics,
Iowa State University, Ames, Iowa 50011

INTRODUCTION

Maize is both a globally important crop and a classic system for studying genome structure and function. The nutritional, agricultural, and industrial uses of maize ensure that its continued genetic improvement remains a goal of ongoing research programs. The recent push to secure renewable bioenergy sources provides another motivation for a thorough understanding of the biology and genetics of this important crop. Archaeological and molecular data suggest that maize was domesticated 6,000 and 9,000 years ago from *Zea mays* ssp. Parviglumis (Piperno and Flannery 2001; Matsuoka et al. 2002). Maize differs from the related grass teosinte at the morphological level in many respects; the genes that control these traits were targets for selection by the Native Americans who domesticated maize, and plant breeders have further molded this species to meet human needs. As such, maize is also an excellent model for studies of domestication and artificial selection.

The maize genome is large. It is estimated that up to 50,000 genes are distributed among its ten chromosomes and contribute to an estimated haploid genome size of approximately 2.5 billion nucleotides. Maize genes appear to be distributed throughout the genome in small clusters separated by large stretches of repetitive DNA (Martienssen et al. 2004; Barbazuk et al. 2005; Rabinowicz and Bennetzen 2006). There are a growing number of maize genome data sets and genomic tools, such as genetic (Cone et al. 2002; Lee et al. 2002; Sharopova et al. 2002; Fu et al. 2006) and physical (Coe et al. 2002; Cone et al. 2002) maps, and sequence sets including more than 1 million expressed sequence tags (ESTs) and more than 1 million Genome Survey Sequence (GSS) sequences. A publicly funded project to determine the whole genome sequence of the B73 maize inbred line is also under way (Rabinowicz and Bennetzen 2006).

Maize is genetically very diverse; single-nucleotide polymorphism (SNP) and indel polymorphism (IDP) frequencies between inbred lines and land races average one variation per 124 bases for coding regions (Ching et al. 2002) and per 28 bases for all associated regions (Tenaillon et al. 2001). This exceptional genetic diversity underlies the phenotypic diversity that permits maize to be adapted to a wide range of environments, and contributes to heterosis. Heterosis, or "hybrid vigor," is the occurrence of superior characteristics due to serendipitous interactions among alleles and genes. Understanding the role of genetic diversity and the mechanisms underlying heterosis is critical to the success of future plant breeding efforts. This diversity easily permits the identification of genetic polymorphisms—including IDPs and SNPs, which can be converted into genetic markers that can be inexpensively assayed in a high-throughput manner (Gut 2001; Kwok 2001). Due to their abundance, it is possible to use IDP- and SNP-based markers to generate dense genetic maps (Rafalski 2002). Such maps enable studies of genome organization and function. They address fundamental questions related to evolution and meiotic recombination and can be applied to marker-assisted selection programs. IDPs and SNPs can also be used for genome-wide linkage disequilibrium and association studies that associate genes with specific functions or traits. Furthermore, transcript-associated SNPs can be used to develop allele-specific assays for the examination of *cis*-regulatory variation within maize (Cowles et al. 2002; Bray et al. 2003; Guo et al. 2004; Pastinen et al. 2004; Stupar and Springer 2006).

In this chapter, we discuss the history and construction of maize genetic maps and the current status of the maize genetic map. We also present a strategy for the rapid construction of maize IDP markers to increase the density of markers on genetic maps. IDP markers occur in high frequency and can be detected using high-throughput technologies. Importantly, the actual polymorphisms underlying an IDP marker need not be identified prior to mapping. Also included in this chapter are brief discussions of protocols which appear elsewhere in this volume: DNA extraction from freeze-dried plant tissue (Chapter 7); capturing discrete cell populations from plant tissue sections, and isolation and amplification of RNA from these cells (Chapter 8); and testing for polymorphisms using temperature gradient capillary electrophoresis (TGCE) and mining of SNP data (both in Chapter 10). We also introduce the new massively parallel DNA sequencing platform developed by 454 Life Sciences (Margulies et al. 2005) and explain how this can be used to rapidly identify maize transcript-associated SNPs.

MOLECULAR GENETIC MAPS OF MAIZE

Genetic maps enable the mapping of quantitative trait loci (QTLs) or genes without prior knowledge beyond the phenotypic effects of the underlying genetic variation. The first published maize genetic map (Emerson et al. 1935) contained just 62 loci, each identified by a morphological mutant. Multiple genetic maps have been generated since. The first molecular map of maize was constructed using restriction fragment length polymorphism (RFLP) molecular markers and an F_2 population (Helentjaris et al. 1986), whereas Burr et al. (1988) mapped RFLPs against a set of recombinant inbred lines (RILs). Davis et al. (1999) constructed a genetic map produced by placement of core bin markers, ESTs, RFLPs, and sequence-tagged sites (STSs) onto a mapping population consisting of 54 immortalized F_2s from Tx303 × CO159. This map, known as the UMC98 map, has been a useful and central resource to many initiatives in the plant sciences. However, it has limited resolution, and the dependence of RFLP markers on blot hybridization prevents its application to high-throughput genetic mapping.

Lee et al. (2002) developed the intermated B73 × Mo17 (IBM) population by randomly intermating an F_2 population derived from the single cross of the inbreds B73 and Mo17 for several generations prior to extraction of intermated recombinant inbred lines (IRILs). The resolution in the resulting mapping population was greatly enhanced because additional opportunities for recombination were provided during the generations of intermating. Using this IBM population and the available maize DNA sequence resources, over 2300 PCR-based markers, i.e., more than 1000 microsatellite markers (Sharopova et al. 2002) and more than 1300 IDP markers (Fu et al. 2006), have been developed and mapped. The Web site http://magi.plantgenomics.iastate.edu

provides access to the ISU (Iowa State University)-IDP map, IDP primer sequences, sequences from which IDP primers were designed, optimized marker-specific PCR conditions, and polymorphism data for all IDP markers. Common framework markers integrate data from several other maps (SSR T218 x GT119, SSR Tx303 x Co159, UMC98, BNL 96 frame, BNL 2002) (Cone et al. 2002), adding a further 5400 off-frame markers (http://www.maizemap.org). Seeds for the IBM lines are available from the Maize Genetics Cooperation Stock Center (Urbana, Illinois), and DNA from a subset of 94 IBM lines is available to the public in 96-well microtiter plate format (http://www.maizemap.org/dna_kits.htm). Web interfaces to recent and widely used maize molecular genetic maps are available at the Maize Genetics and Genomics Database (MaizeGDB; http://www.maizegdb.org). Table 26-1 summarizes Web sites and software packages referred to in this chapter and in the maize protocols in this volume.

Mapping a gene requires identification of polymorphisms among alleles of that gene and the detection of the polymorphisms among the members of a mapping population. Various techniques can be used to detect the many types of DNA polymorphisms (Kristensen et al. 2001). The choice of technique depends on whether the sequences associated with the polymorphism are known in advance. SNPs or small IDPs that have been previously identified via sequence analyses can be mapped using platforms such as those based on pyrosequencing, PCR, microarray, and mass spectrometry technologies (e.g., that marketed by Sequenom). As an alternative, heteroduplex analyses can detect SNPs or small IDPs that exist between the parents of a mapping population even when the sequences underlying the polymorphism are unknown. IDPs occur at high frequencies and can be captured by PCR primers that amplify 3′-untranslated regions (UTRs) or across introns. Candidate regions for primer design are identified by the following strategy, which can be applied to any plant or animal system.

Primer Design Strategy for Capturing IDPs

1. If EST and genomic sequence data are available, align these to identify exons. Partial assemblies of the maize genome are available at http://magi.plantgenomics.iastate.edu/ and other sites. GeneSeqer (Brendel et al. 2004) and GMAP (Wu and Watanabe 2005) are splice-aware tools for aligning ESTs to genomic DNA and are freely available.

TABLE 26-1. Software and Web Resources

Software packages	
POLYBAYES	http://bioinformatics.bc.edu/marthlab/polybayes.html
GeneSeqer	http://deepc2.psi.iastate.edu/cgi-bin/gs.cgi
GMAP	http://www.gene.com/share/gmap
Web sites	
http://www.maizemap.org/	Maize Mapping Project Web site
http://www.maizegdb.org	Maize Genome Database—A Maize Genetics and Genomics Database
http://www.tigr.org/tdb/e2k1/osa1	Rice Genome Annotation Database at The Institute for Genomic Research (TIGR)
http://compbio.dfci.harvard.edu/tgi	Gene Index Portal—EST assemblies from several plants, animals, and microbes
http://www.plantgdb.org	The Plant Genome Database—Resources for plant comparative genomics
http://www.454.com	454 Life Sciences Web site—Includes links to technology platforms and the Sequencing Service center
http://magi.plantgenomics.iastate.edu/	Assembled maize genomic sequence assemblies, and other maize sequence resources
http://www.sequenom.com	Sequenom home page with links to technology platforms
www.agcol.arizona.edu	Arizona Genomics Computational Laboratory—Links to maize sequence data and SyMAP, a synteny mapping and analysis program
http://www.panzea.org	Panzea Web site: Molecular and functional diversity in maize

2. When only EST sequences are available, estimate gene structures by aligning the ESTs to genomic sequences from other related species (Wei et al. 2005), such as the rice (*Oryza sativa* spp. Nipponbarre) assembly available at the Institute for Genome Research (TIGR; http://www.tigr.org/tdb/e2k1/osa1/). GeneSeqer (Brendel et al. 2004) and GMAP (Wu and Watanabe 2005) can perform cross-species splice-aware alignments.

3. When only genomic sequence is available, estimate gene structure by aligning the genomic sequence to ESTs from other related species (e.g, rice, sorghum, wheat, barley). The TIGR gene indices (Quackenbush et al. 2001) (at http://compbio.dfci.harvard.edu/tgi/) and the Plant Genome Database (Dong et al. 2005) (at http://www.plantgdb.org) provide collections of ESTs and EST assemblies for several plant species.

Analysis of Polymorphism

DNA can be extracted from freeze-dried plant tissue with a modified version of DNA isolation (Dietrich et al. 2002) using hexadecyltrimethylammonium bromide (Rogers and Blendich 1985) and 96-well plates. TGCE can be used to analyze the DNA for polymorphism as it provides a high-throughput, multiplexed platform for heteroduplex analysis. TGCE has the capacity to run twelve 96-well assays per day and acquire up to 2304 data points if two assays are multiplexed (Hsia et al. 2005). This sensitive assay detects even a single SNP in amplicons of over 800 bp and 1-bp IDPs in amplicons of approximately 500 bp (Hsia et al. 2005). It reliably detects SNPs, simple sequence repeats (SSRs), and IDPs (Hsia et al. 2005). The data are scored and viewed with the Revelation software (version 2.4). A new software package called genetic recombinant analysis mapping assistant (GRAMA) has been created to facilitate analysis of TGCE data (Maher et al. 2006). GRAMA presents mapping data in an intuitive fashion and reduces manual processing. It also checks data calling with an alternative algorithm and validates data output from Revelation. It is available for download at http://www.complex.iastate.edu/download/GRAMA/index.html. The electropherograms produced by the Revelation software package display one peak in the absence of polymorphism and multiple peaks otherwise. For example, in Figure 26-1, PCR amplicons of MAGI_65972 (polymorphic between B73 and Mo17) from 12 maize IBM (B73 × Mo17) IRILs (Lee et al. 2002) were mixed with amplicons from inbred line B73. The inset in the upper left corner of Figure 26-1 shows the expected peak pattern for monomorphic (only the B73 allele present) and dimorphic (both B73 and Mo17 alleles present) PCR mixtures. Accordingly, the genotype of each recombinant inbred line (RIL) can be determined based on its electrophoretic pattern (see genotype scores next to well position labels in Fig. 26-1).

High-Throughput Sequencing of Maize Transcriptomes with 454 Technology for SNP Discovery

SNP-based markers have been mined by comparing genomic sequences (Yamasaki et al. 2005) or ESTs (Batley et al. 2003) derived from two or more different lines of maize. Recent advances in high-throughput sequencing technology provide a rapid and cost-effective means to generate sequence data from diverse maize lines for SNP discovery. 454 Life Sciences has developed a scalable, highly parallel DNA sequencing system that is 100 times faster than standard sequencing methods and is currently capable of sequencing over 200,000 fragments per 4-hour run (Margulies et al. 2005). Although 454 sequence reads are short, this does not pose a problem when resequencing genomic DNA or aligning 454 ESTs to a genomic template. 454 has been used to sequence transcripts (Bainbridge et al. 2006; Emrich et al. 2006; Gowda et al. 2006). It is well-suited for this because transcriptomes are substantially smaller than the genomes from which they are derived, they contain less repetitive DNA, and 454 technology does not require clone-based library construction. The use of laser-capture microdissection (LCM; for review, see Schnable et al. 2004) to isolate transcripts that accumulate in specific cell types for 454 sequencing can further reduce the size of a target transcriptome and can capture rare transcripts. If required, amplification can be used

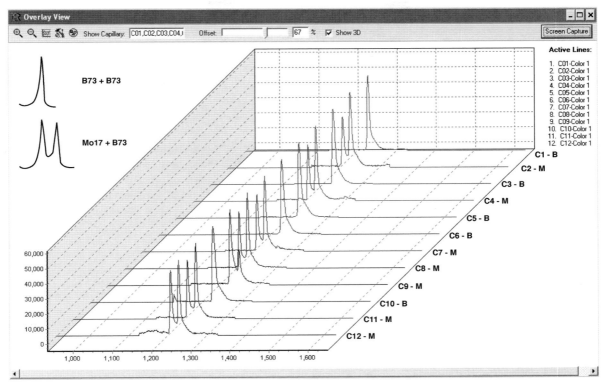

FIGURE 26-1. Genotyping MAGI_65972 via TGCE with primers 31bMAGI_65972L5 and 31bMAGI_65972R4. Inset in the upper left corner shows the expected peak pattern for B73 and Mo17 alleles. The top and bottom graphs are data obtained via amplification products from the inbred lines B73 (coded here as "B") and Mo17 ("M"), respectively, mixed with amplification products obtained from B73 ("B"). Note evidence of heteroduplex molecules (multiple peaks) in the bottom graph (Mo17 + B73 amplicons), but not in the top (only B73 amplicons). Well positions C1–C12 contained amplicons from RIs M0034, M0039, M0043, M0045, M0048, M0052, M0055, M0058, M0061, M0063, M0067, and M0075 mixed with the amplicons from the inbred line B73. Electropherograms that exhibit evidence of heteroduplexes (multiple peaks) are derived from RIs that carry the Mo17 allele; those with single peaks carry the B73 allele. The genotype scores (B vs. M) of each RI (as labeled to the right of each well's electropherogram) are automatically generated by the Revelation software.
Primer sequences:
31bMAGI_65972L5: ACGAGAGCTGCAATCGAATC
31bMAGI_65972R4: TTAGGTGGTGGTTCGTCTCC

to increase the amount of RNA collected. LCM-454 transcriptome sequencing has been demonstrated in maize (Emrich et al. 2006), where cDNA extracted from developmentally important shoot apical meristem (SAM) cells (Baurle and Laux 2003; Guyomarc'h et al. 2005) from B73 and Mo17 maize inbred lines has been used for SNP discovery (W.B. Barbazuk et al., unpubl.).

SNP Mining

Putative SNPs are identified as mismatches between aligned sequences. Several computational tools are available for SNP identification (Nickerson et al. 1997; Marth et al. 1999; Manaster et al. 2005; Wang and Huang 2005; Weckx et al. 2005; Zhang et al. 2005). POLYBAYES has been used to identify SNPs in several systems, including maize (Useche et al. 2001), and is recommended for identifying SNPs in 454 sequences. POLYBAYES uses a Bayesian statistical model that considers depth of coverage, sequence quality, and an a priori expected polymorphism rate to determine the probability that polymorphic sites within a multiple sequence alignment (MSA) are SNPs rather than disagreements resulting from either sequencing errors or the alignment of paralogous (rather than allelic) sequences (Marth et al. 1999). Results from POLYBAYES can be obtained as easily parseable text files, or as .ace files that can be read and displayed graphically in CONSED (Gordon et al. 1998). In addition, POLYBAYES uses either genomic or EST sequence data as a template upon which all remaining sequences to be assessed are multiply aligned with CROSS_MATCH (P. Green, unpubl.) prior to scanning for polymorphism. Template-based MSAs are often correct, even in the presence of abundantly expressed or alternatively spliced transcripts (Marth et al. 1999), and therefore are more likely to overcome false polymorphism calls due to the higher sequence error rate associated with 454 sequence compared to Sanger sequenced reads.

A flowchart describing POLYBAYES-mediated SNP mining is illustrated in Figure 26-2, in which the 454 ESTs are derived from Mo17 maize, and the anchor sequence is from B73 maize. The MSAs provided by CROSS_MATCH can be central processing unit (CPU)- intensive, so large 454 sequence sets are best preprocessed using BLAST to address each 454 sequence to a template. CROSS_MATCH is subsequently run on small sequence subsets consisting of the template and the aligning 454 ESTs.

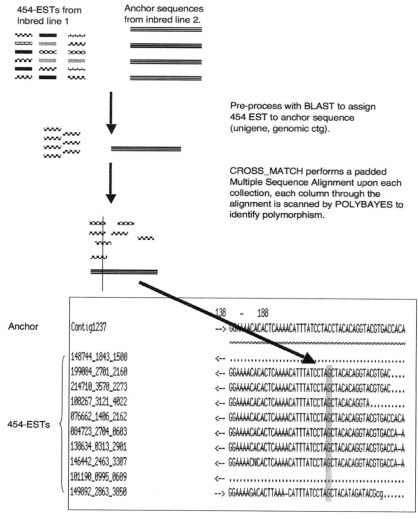

FIGURE 26-2. Flowchart for mining SNPs from 454 EST sequence with POLYBAYES. ESTs are associated with high-quality EST or genomic sequence assemblies that serve as anchors for template-driven MSAs. POLYBAYES surveys each column of the alignments, looking for well-supported polymorphisms between the 454 ESTs and the template. In this example, the 454 ESTs are derived from Mo17 maize, and the anchor sequence is from B73 maize. Position 166 is polymorphic between B73 (C) and Mo17 (G).

POLYBAYES evaluates the resulting MSA and calls SNPs. Sequence errors such as insertions or substitutions are not supported by a majority of sequences and are easily identified in deep alignments. At the time of writing, the maize genome sequencing project is less than 20% complete. In the interim, a collection of high quality (1 disagreement in 5000 bp) maize genomic sequence assemblies (Emrich et al. 2004; Fu et al. 2005) is available (http://magi.plantgenomics.iastate.edu/). It is composed of B73 genome survey sequences (GSSs) that are enriched for genes (Whitelaw et al. 2003).

Previous SNP analysis of maize 454 EST reads (W.B. Barbazuk, unpubl.) used default quality values of 18 for each base within the 454 reads. This corresponds to an error rate of approximately 1/65, which overcompensates for the error rate observed for current 454 sequencing (Margulies et al. 2005; Emrich et al. 2006). Although each base within the 454 sequence reads is assigned a quality score by the 454 sequencing machine, these scores are only reliable when confirmed within independent sequences covering the same region.

The strategy described here identified over 7000 potential SNPs between Mo17 and B73. The depth of coverage by each genotype influences confidence at each SNP site, and 90% of putative SNPs at sites sampled to at least 3X for both inbreds validate by sequencing. In general, SNPs with deep coverage are more likely to be valid. In the protocol, default quality values were assigned to 454 sequences based on an empirical evaluation of the base error rate rather than using the 454 quality scores. As a result, sequence depth and relative allele proportions have the greatest influence on polymorphism detection and, based on this observation, potential SNPs were filtered by examining these statistics at each polymorphic site. POLYBAYES consults base quality as well as sequence depth; hence, accurate base quality values will likely improve performance. As the 454 sequencing technology matures, and the 454 base quality scores increase in accuracy, incorporation of these scores will require no modification beyond making the quality files available to POLYBAYES. Furthermore, since POLYBAYES uses quality files when available and default quality values otherwise, POLYBAYES is well suited for SNP discovery from sequences obtained from mixed platforms (i.e., combining singleton 454 reads without quality scores, 454 assemblies with quality scores, and Sanger reads with quality scores).

The example here requires that both B73 and Mo17 inbreds be monoallelic at each nucleotide before accepting a given site as polymorphic. (Relatively rare exceptions are caused by nearly identical paralogs [NIPs; Emrich et al. 2004].) This is quite restrictive and has a high false negative rate. Consider a site that is well sampled and monoallelic in Mo17 and represented by an additional ten B73 sequences, nine of which are concordant (but different from the Mo17 allele) and one which is not. This site would have been filtered out, although the one disconcordant B73 call is likely a sequence error. Therefore, further parsing of the SNP output files can be used to represent quality more accurately.

SUMMARY

Because the rice genome sequence is now nearly complete, comparative sequence alignments between rice sequence and maize DNA-based markers anchored to the physical and genetic maps can identify potentially colinear regions. Such regions provide the basis for a rice–maize synteny map that can give maize researchers the benefits of an annotated genome well in advance of the completion of the maize sequencing effort. There are three major Web resources for rice–maize synteny maps:

1. Gramene (Jaiswal et al. 2006), http://www.gramene.org.

2. Arizona Genomics Computational Lab (AGCoL), http://www.agcol.arizona.edu.

3. TIGR, http://www.tigr.org/tdb/synteny/maize_IBMn/search_desc.shtml.

The increased density of markers on the maize genetic map, in combination with the synteny among grass genomes, facilitates positional cloning projects (Bortiri et al. 2006).

The high degree of genetic variation observed between two modern maize lines approximates that seen between humans and chimpanzees (Zhao et al. 2006). This exceptional within-species genetic diversity underlies the phenotypic diversity that permits maize to be adapted to diverse environments. Understanding the extent of genetic diversity and the mechanisms underlying it is critical to future plant breeding. Two public SNP discovery projects in maize are under way. The goal of the Panzea project led by John Doebley at the University of Wisconsin, Madison, is to examine the impact of past selection on molecular diversity in maize genes. This project is performing SNP discovery in approximately 5000 loci, and tests of selection are being used to identify genes showing positive, diversifying, and purifying selection. Wright et al. (2005) examined SNPs within sequences from 774 random genes from maize and teosinte, identified several candidate selected genes, and concluded that 2–4% of maize genes have undergone artificial selection. A subsequent study (Yamasaki et al. 2005) examined sequences from 1095 genes within 14 inbred maize lines, searching for genes that exhibit no sequence diversity and that may have been involved in domestication. Project data are available from http://www.panzea.org/.

The second public maize SNP discovery project led by Ed Buckler at Cornell University is developing methods to discover and score DNA polymorphisms at nearly 400,000 locations in the genome, which will provide an exceptional resource for future maize association mapping studies.

ACKNOWLEDGMENTS

We thank Mike Scanlon and current and former Schnable lab personnel Mikio Nakazono, Dave Skibbe, and Marianne Smith for their contributions to the protocols in this volume. This project was supported by competitive grants from the National Science Foundation Plant Genome Program to P.S.S. (DBI-0321711 and DBI-0321595) and W.B.B. (DBI-0501758); from the National Research Initiative of the USDA Cooperative State Research, Education and Extension Service (grant number 04-00913) to P.S.S.; by the Hatch Act and State of Iowa funds to P.S.S., and with funds from ISU's Plant Sciences Institute and the Donald Danforth Plant Science Center to P.S.S. and W.B.B., respectively.

REFERENCES

Asano T., Masumura T., Kusano H., Kikuchi S., Kurita A., Shimada H., and Kadowaki K. 2002. Construction of a specialized cDNA library from plant cells isolated by laser capture microdissection: Toward comprehensive analysis of the genes expressed in the rice phloem. *Plant J.* **32:** 401–408.

Bainbridge M.N., Warren R.L., Hirst M., Romanuik T., Zeng T., Go A., Delaney A., Griffith M., Hickenbotham M., Magrini V., et al. 2006. Analysis of the prostate cancer cell line LNCaP transcriptome using a sequencing-by-synthesis approach. *BMC Genomics* **7:** 246.

Barbazuk W.B., Bedell J.A., and Rabinowicz P.D. 2005. Reduced representation sequencing: A success in maize and a promise for other plant genomes. *BioEssays* **27:** 839–848.

Batley J., Barker G., O'Sullivan H., Edwards K.J., and Edwards D. 2003. Mining for single nucleotide polymorphisms and insertions/deletions in maize expressed sequence tag data. *Plant Physiol.* **132:** 84–91.

Baurle I. and Laux T. 2003. Apical meristems: The plant's fountain of youth. *BioEssays* **25:** 961–970.

Bortiri E., Jackson D., and Hake S. 2006. Advances in maize genomics: The emergence of positional cloning. *Curr. Opin. Plant Biol.* **9:** 164–171.

Bray N.J., Buckland P.R., Owen M.J., and O'Donovan M.C. 2003. *Cis*-acting variation in the expression of a high proportion of genes in human brain. *Hum. Genet.* **113:** 149–153.

Brendel V., Xing L., and Zhu W. 2004. Gene structure prediction from consensus spliced alignment of multiple ESTs matching the same genomic locus. *Bioinformatics* **20:** 1157–1169.

Burr B., Burr F.A., Thompson K.H., Albertson M.C., and Stuber C.W. 1988. Gene mapping with recombinant inbreds in maize. *Genetics* **118:** 519–526.

Ching A., Caldwell K.S., Jung M., Dolan M., Smith O.S., Tingey S., Morgante M., and Rafalski A.J. 2002. SNP frequency, haplotype structure and linkage disequilibrium in elite maize inbred lines. *BMC Genet.* **3:** 19.

Coe E., Cone K., McMullen M., Chen S.S., Davis G., Gardiner J., Liscum E., Polacco M., Paterson A., Sanchez-Villeda H., et al. 2002. Access to the maize genome: An integrated physical and genetic map. *Plant Physiol.* **128:** 9–12.

Cone K.C., McMullen M.D., Bi I.V., Davis G.L., Yim Y.S., Gardiner J.M., Polacco M.L., Sanchez-Villeda H., Fang Z., Schroeder S.G., et al. 2002. Genetic, physical, and informatics resources for maize. On the road to an integrated map. *Plant Physiol.* **130:** 1598–1605.

Cowles C.R., Hirschhorn J.N., Altshuler D., and Lander E.S. 2002. Detection of regulatory variation in mouse genes. *Nat. Genet.* **32:** 432–437.

Davis G.L., McMullen M.D., Baysdorfer C., Musket T., Grant D., Staebell M., Xu G., Polacco M., Koster L., Melia-Hancock S., et al. 1999. A maize map standard with sequenced core markers, grass genome reference points and 932 expressed sequence tagged sites (ESTs) in a 1736-locus map. *Genetics* **152:** 1137–1172.

Dietrich C.R., Cui F., Packila M.L., Li J., Ashlock D.A., Nikolau B.J., and Schnable P.S. 2002. Maize *Mu* transposons are targeted to the 5′ untranslated region of the *gl8* gene and sequences flanking *Mu* target-site duplications exhibit nonrandom nucleotide composition throughout the genome. *Genetics* **160:** 697–716.

Dong Q., Lawrence C.J., Schlueter S.D., Wilkerson M.D., Kurtz S., Lushbough C., and Brendel V. 2005. Comparative plant genomics resources at PlantGDB. *Plant Physiol.* **139:** 610–618.

Emerson R., Beadle G., and Fraser A. 1935. A summary of linkage studies in maize. *Cornell Univ. Agric. Exp. Stn. Memoir* **180:** 1–83.

Emrich S.J., Barbazuk W.B., Li L., and Schnable P.S. 2006. Gene discovery and annotation using LCM-454 transcriptome sequencing. *Genome Res.* **17:** 69–73.

Emrich S.J., Aluru S., Fu Y., Wen T.J., Narayanan M., Guo L., Ashlock D.A., and Schnable P.S. 2004. A strategy for assembling the maize (*Zea mays* L.) genome. *Bioinformatics* **20:** 140–147.

Fu Y., Emrich S.J., Guo L., Wen T.J., Ashlock D.A., Aluru S., and Schnable P.S. 2005. Quality assessment of maize assembled genomic islands (MAGIs) and large-scale experimental verification of predicted genes. *Proc. Natl. Acad. Sci.* **102:** 12282–12287.

Fu Y., Wen T.J., Ronin Y.I., Chen H.D., Gou L., Mester D.I., Yang Y., Lee M., Korol A.B., Ashlock D.A., and Schnable P.S. 2006. Genetic dissection of intermated recombinant inbred lines using a new genetic map of maize. *Genetics* **174:** 1671–1683.

Gordon D., Abajian C., and Green P. 1998. *Consed:* A graphical tool for sequence finishing. *Genome Res.* **8:** 195–202.

Gowda M., Li H., Alessi J., Chen F., Pratt R., and Wang G. 2006. Robust analysis of 5′-transcript ends (5′-RATE): A novel technique for transcriptome analysis and genome annotation. *Nucleic Acids Res.* **34:** e126.

Guo M., Rupe M.A., Zinselmeier C., Habben J., Bowen B.A., and Smith O.S. 2004. Allelic variation of gene expression in maize hybrids. *Plant Cell* **16:** 1707–1716.

Gut I.G. 2001. Automation in genotyping of single nucleotide polymorphisms. *Hum. Mutat.* **17:** 475–492.

Guyomarc'h S., Bertrand C., Delarue M., and Zhou D.X. 2005. Regulation of meristem activity by chromatin remodelling. *Trends Plant Sci.* **10:** 332–338.

Helentjaris T., Weber D.F., and Wright S. 1986. Use of monosomics to map cloned DNA fragments in maize. *Proc. Natl. Acad. Sci.* **83:** 6035–6039.

Hsia A.P., Wen T.J., Chen H.D., Liu Z., Yandeau-Nelson M.D., Wei Y., Guo L., and Schnable P.S. 2005. Temperature gradient capillary electrophoresis (TGCE)—A tool for the high-throughput discovery and mapping of SNPs and IDPs. *Theor. Appl. Genet.* **111:** 218–225.

Jaiswal P., Ni J., Yap I., Ware D., Spooner W., Youens-Clark K., Ren L., Liang C., Zhao W., Ratnapu K., et al. 2006. Gramene: A bird's eye view of cereal genomes. *Nucleic Acids Res.* **34**(Database issue)**:** D717–D723.

Jurinke C., Denissenko M.F., Oeth P., Ehrich M., van den Boom D., and Cantor C.R. 2005. A single nucleotide polymorphism based approach for the identification and characterization of gene expression modulation using MassARRAY. *Mutat. Res.* **573:** 83–95.

Kerk N.M., Ceserani T., Tausta S.L., Sussex I.M., and Nelson T.M. 2003. Laser capture microdissection of cells from plant tissues. *Plant Physiol.* **132:** 27–35.

Kristensen V.N., Kelefiotis D., Kristensen T., and Borresen-Dale A.L. 2001. High-throughput methods for detection of genetic variation. *BioTechniques* **30:** 318–326.

Kwok P.Y. 2001. Methods for genotyping single nucleotide polymorphisms. *Annu. Rev. Genomics Hum. Genet.* **2:** 235–258.

Lee M., Sharopova N., Beavis W.D., Grant D., Katt M., Blair D., and Hallauer A. 2002. Expanding the genetic map of maize with the intermated B73 × Mo17 (*IBM*) *population*. *Plant Mol. Biol.* **48:** 453–461.

Maher P.M., Chou H.H., Hahn E., Wen T.J., and Schnable P.S. 2006. GRAMA: Genetic mapping analysis of temperature gradient capillary electrophoresis data. *Theor. Appl. Genet.* **113:** 156–162.

Manaster C., Zheng W., Teuber M., Wachter S., Doring F., Schreiber S., and Hampe J. 2005. InSNP: A tool for automated detection and visualization of SNPs and InDels. *Hum. Mutat.* **26:** 11–19.

Margulies M., Egholm M., Altman W.E., Attiya S., Bader J.S., Bemben L.A., Berka J., Braverman M.S., Chen Y.J., Chen Z., et al. 2005. Genome sequencing in microfabricated high-density picolitre reactors. *Nature* **437:** 376–380.

Marth G.T., Korf I., Yandell M.D., Yeh R.T., Gu Z., Zakeri H., Stitziel N.O., Hillier L., Kwok P.Y., and Gish W.R. 1999. A general approach to single-nucleotide polymorphism discovery. *Nat. Genet.* **23:** 452–456.

Martienssen R.A., Rabinowicz P.D., O'Shaughnessy A., and McCombie W.R. 2004. Sequencing the maize genome. *Curr. Opin. Plant Biol.* **7:** 102–107.

Matsuoka Y., Vigouroux Y., Goodman M.M., Sanchez G.J., Buckler E., and Doebley J. 2002. A single domestication for maize shown by multilocus microsatellite genotyping. *Proc. Natl. Acad. Sci.* **99:** 6080–6084.

Nickerson D.A., Tobe V.O., and Taylor S.L. 1997. PolyPhred: Automating the detection and genotyping of single nucleotide substitutions using fluorescence-based resequencing. *Nucleic Acids Res.* **25:** 2745–2751.

Oeth P., Beaulieu M., Park C., Kosman D., del Mistro G., and Van den Boom D. 2005. iPLEX™ assay: Increased plexing efficiency and flexibility for Mass ARRAY® system through single base primer extension with mass-modified terminators. Sequenom Application Note No. 8876–006, Sequenom Inc., http://www.sequenom.com/Assets/pdfs/appnotes/8876-006.pdf

Pastinen T., Sladek R., Gurd S., Sammak A., Ge B., Lepage P., Lavergne K., Villeneuve A., Gaudin T., Brandstrom H., et al. 2004. A survey of genetic and epigenetic variation affecting human gene expression. *Physiol. Genomics* **16:** 184–193.

Piperno D.R. and Flannery K.V. 2001. The earliest archaeological maize (*Zea mays* L.) from highland Mexico: New accelerator mass spectrometry dates and their implications. *Proc. Natl. Acad. Sci.* **98:** 2101–2103.

Quackenbush J., Cho J., Lee D., Liang F., Holt I., Karamycheva S., Parvizi B., Pertea G., Sultana R., and White J. 2001. The TIGR Gene Indices: Analysis of gene transcript sequences in highly sampled eukaryotic species. *Nucleic Acids Res.* **29:** 159–164.

Rabinowicz P.D. and Bennetzen J.L. 2006. The maize genome as a model for efficient sequence analysis of large plant genomes. *Curr. Opin. Plant Biol.* **9:** 149–156.

Rafalski J.A. 2002. Novel genetic mapping tools in plants: SNPs and LD-based approaches. *Plant Sci.* **162:** 329–333.

Rogers S.O. and Blendich A.J. 1985. Extraction of DNA from milligram amounts of fresh herbarium and mummified plant tissues. *Plant Mol. Biol.* **5:** 69–76.

Schnable P.S., Hochholdinger F., and Nakazono M. 2004. Global expression profiling applied to plant development. *Curr. Opin. Plant Biol.* **7:** 50–56.

Sharopova N., McMullen M.D., Schultz L., Schroeder S., Sanchez-Villeda H., Gardiner J., Bergstrom D., Houchins K., Melia-Hancock S., Musket T., et al. 2002. Development and mapping of SSR markers for maize. *Plant Mol. Biol.* **48:** 463–481.

Stupar R.M. and Springer N.M. 2006. *Cis*-transcriptional variation in maize inbred lines B73 and Mo17 leads to additive expression patterns in the F1 hybrid. *Genetics* **173:** 2199–2210.

Tenaillon M.I., Sawkins M.C., Long A.D., Gaut R.L., Doebley J.F., and Gaut B.S. 2001. Patterns of DNA sequence polymorphism along chromosome 1 of maize (*Zea mays* ssp. *mays* L.). *Proc. Natl. Acad. Sci.* **98:** 9161–9166.

Useche F.J., Gao G., Harafey M., and Rafalski A. 2001. High-throughput identification, database storage and analysis of SNPs in EST sequences. *Genome Inform.* **12:** 194–203.

Wang J. and Huang X. 2005. A method for finding single-nucleotide polymorphisms with allele frequencies in sequences of deep coverage. *BMC Bioinformatics* **6:** 220.

Weckx S., Del-Favero J., Rademakers R., Claes L., Cruts M., De Jonghe P., Van Broeckhoven C., and De Rijk P. 2005. novoSNP, a novel computational tool for sequence variation discovery. *Genome Res.* **15:** 436–442.

Wei H., Fu Y., and Arora R. 2005. Intron-flanking EST-PCR markers: From genetic marker development to gene structure analysis in *Rhododendron*. *Theor. Appl. Genet.* **111:** 1347–1356.

Whitelaw C.A., Barbazuk W.B., Pertea G., Chan A.P., Cheung F., Lee Y., Zheng L., van Heeringen S., Karamycheva S., Bennetzen J.L., et al. 2003. Enrichment of gene-coding sequences in maize by genome filtration. *Science* **302:** 2118–2120.

Wright S.I., Bi I.V., Schroeder S.G., Yamasaki M., Doebley J.F., McMullen M.D., and Gaut B.S. 2005. The effects of artificial selection on the maize genome. *Science* **308:** 1310–1314.

Wu T.D. and Watanabe C.K. 2005. GMAP: A genomic mapping and alignment program for mRNA and EST sequences. *Bioinformatics* **21:** 1859–1875.

Yamasaki M., Tenaillon M.I., Bi I.V., Schroeder S.G., Sanchez-Villeda H., Doebley J.F., Gaut B.S., and McMullen M.D. 2005. A large-scale screen for artificial selection in maize identifies candidate agronomic loci for domestication and crop improvement. *Plant Cell* **17:** 2859–2872.

Zhang J., Wheeler D.A., Yakub I., Wei S., Sood R., Rowe W., Liu P.P., Gibbs R.A., and Buetow K.H. 2005. SNPdetector: A software tool for sensitive and accurate SNP detection. *PLoS Comput. Biol.* **1:** e53.

Zhao W., Canaran P., Jurkuta R., Fulton T., Glaubitz J., Buckler E., Doebley J., Gaut B., Goodman M., Holland J., et al. 2006. Panzea: A database and resource for molecular and functional diversity in the maize genome. *Nucleic Acids Res.* **34**(Database issue): D752–D757.

WWW RESOURCES

http://www.agcol.arizona.edu Arizona Genomics Computational Laboratory

http://www.complex.iastate.edu/download/GRAMA/index.html GRAMA (Genetic Recombinant Analysis and Mapping Assistant), Complex Computation Laboratory, Iowa State University.

http://www.gramene.org GRAMENE: A resource for comparative grass genomics

http://www.tigr.org/tdb/synteny/maize_IBMn/search_desc.shtml TIGR Rice Genome Annotation, Maize Genetic Markers Mapped to Rice Pseudomolecules

27 | Rice

Hei Leung,[1] Kenneth L. McNally,[2] and David Mackill[1]

[1]Plant Breeding, Genetics, and Biotechnology Division; [2]T.T. Chang Genetic Resources Center, International Rice Research Institute, Manila, Philippines

INTRODUCTION

Rice is one of the oldest cereals, with reported cultivation dating back 7000 years (Watanabe 1997). Its production is distributed over a broad geographic range, including almost all continents, with the biggest concentration in Asia (Khush 1997). Rice is cultivated in diverse ecologies, including dry upland, lowland with controlled irrigation, and deepwater environments. These ecosystems, which largely reflect the availability of water, have influenced the adaptation and natural evolution of rice since its domestication. Natural and artificial selection over this long period of cultivation have generated an amazing range of genetic diversity.

Cultivated rice (*Oryza sativa*) is diploid, with 12 chromosomes (2n = 24). Some consider domesticated rice to consist of two subspecies, *indica* and *japonica*, that are widely grown in most rice-growing regions, whereas others view these two

"subspecies" as ecotypes or cultigens. In terms of production, indica rice is the predominant type grown in tropical and subtropical areas, constituting about 80% of world rice production (Mackill 1995). On the other hand, the japonica type is important in temperate areas. Rice cultivation is heavily influenced by consumer preference and environmental requirements, and thus, characteristics among indica and japonica varieties overlap considerably, and mixing of their gene pools occurs frequently. Another domesticated diploid species is *Oryza glaberrima*, which is grown primarily in Africa. In addition to cultivated rice, a rich variety of wild *Oryza* relatives includes diploid and tetraploid species.

The rich genetic diversity present in rice not only provides a foundation for genetic improvement, but also serves as an excellent resource for understanding crop evolution and domestication history. Among all crop species, rice is exceptionally well endowed with extensive genetic resources supported by advanced genetic tools that make it a model genetic system. Rice is the first crop plant with a sequenced genome. International rice genome sequencing efforts have generated complete genome information for two varieties of indica and japonica rice. This, together with expanding genomic information from other plant species, presents a new paradigm for understanding, exploring, and using rice genetic resources (Leung and An 2004).

Under the new paradigm, genetic knowledge of rice can be integrated across species through comparative genomics analysis and can lead to accelerated discovery of gene functions. Furthermore, genome-wide analysis has the potential to reveal new insights into genetic pathways and create new opportunities to meet both anticipated and unforeseen challenges. To realize this potential, abundant genetic diversity, either natural or induced, must be available or generated. Conserved germ plasm maintained in national and international gene banks needs to be characterized; otherwise, this large store of genetic diversity will remain unused. Thus, conservation, generation, and characterization of genetic diversity, and discovery of gene functions, are interdependent activities that need to be integrated to achieve the desired outcomes. This chapter describes the current status of genetic resources available for probing the genetic and functional diversity of rice and how they are being used. We also discuss what additional resources should be generated to make best use of genomic information. We hope that this brief overview offers a glimpse of the exciting opportunities in the investigation of rice genetics that will eventually have an impact on rice improvement.

DEPTH OF GENETIC VARIABILITY IN RICE

The International Rice Genebank Collection (IRGC) at the International Rice Research Institute (IRRI) maintains the world's largest collection of rice germ plasm, including domesticated rice and its wild relatives. At present, there are >102,000 accessions from the Asian cultivated rice *O. sativa*, 1,650 accessions from the African cultivated rice *O. glaberrima*, and 4,510 accessions from 22 related wild relatives. Table 27-1 summarizes this wealth of germ plasm diversity maintained in the Genebank of IRRI. The collection represents an exceptionally deep gene pool that is largely uncharacterized. Efficient techniques are therefore much needed to understand, explore, and make use of this diversity.

Asian cultivated rice exists in a diverse range of types that have been classified into six groups based on isozyme analysis. In the early 1980s, Second (1982, 1985) hypothesized that the indica and japonica types of *O. sativa* arose from independent domestication events based on an analysis of isozyme alleles. Recently, further support for the separate domestication of the indica and japonica types has come from an analysis of the interspersion patterns of short interspersed nuclear elements (SINEs; Cheng et al. 2003), transposition histories of long terminal repeat (LTR) retrotransposons (Vitte et al. 2004), population structure analysis using simple sequence repeats (SSRs; Garris et al. 2005), and phylogenetic analysis of nuclear gene introns (Zhu and Ge 2005). The separate domestication of the two most important types of rice has important implications for the application of association genetics, since population structure must be taken into account. Figure 27-1 shows the overall relationships of major domesticated rice types and their hypothesized progenitors as determined by SSR genotyping. This diagram is based on work from a Generation Challenge Programme project (www.generationcp.org) involving IRRI, CIRAD, CIAT, EMBRAPA, WARDA, and Cornell University (K. McNally et al., unpubl.), and a similar analysis conducted by Garris et al. (2005). The structure shows five main groups corresponding to the isozyme variety groups of indica, aus, aromatic, and japonica, with the latter subdivided into temperate and tropical types. Wild AA genome species group between the aus/indica and aromatic/japonica groups. Individuals with a proportion of alleles derived from two or more major groups are indicated as admixed types.

Current taxonomy for the wild *Oryza* consists of 22 species that are mostly diploid, with five allotetraploid types (Table 27-1). These species have 10 diverse types of genomes. Assignments

TABLE 27-1. Domesticated and Wild Rice Germ Plasm in the International Rice Genebank

Genome	Species[a]	Distribution[b]	Number of accessions[c]
AA	O. sativa	worldwide	102,553
AA	O. glaberrima	West Africa	1650
AA	O. barthii	Africa	211
AA	O. nivara	tropical, subtropical Asia	1260
AA	O. rufipogon	tropical, subtropical Asia, tropical Australia	1048
AA	O. longistaminata	Africa	218
AA	O. glumaepatula	South, Central America	54
AA	O. meridionalis	tropical Australia	55
AA	hybrid or uncertain classification	worldwide	1002
BB, BBCC	O. punctata	Africa	75
BBCC	O. minuta	Philippines, Papua New Guinea	63
CC	O. officinalis	tropical, subtropical Asia, tropical Australia	279
CC	O. rhizomatis	Sri Lanka	20
CC	O. eichingeri	South Asia, East Africa	24
CCDD	O. alta	South, Central America	7
CCDD	O. latifolia	South, Central America	58
CCDD	O. grandiglumis	South, Central America	10
EE	O. australiensis	tropical Australia	36
FF	O. brachyantha	Africa	17
GG	O. granulata	South, Southeast Asia	24
GG	O. meyeriana	Southeast Asia	10
GG	O. neocaledonica	New Caledonia	1
HHJJ	O. longiglumis	Indonesia (Irian Jaya), Papua New Guinea	6
HHJJ	O. ridleyi	South Asia	14
HHKK	O. schlechteri	Papua New Guinea	1

[a]Modified from Leung and An (2004) and Lu (1999).
[b]Brar and Khush (1997).
[c]Information from IRGC database as of October 25, 2006. www.irgcis.irri.org:81/grc/irgcishome.html

following those of Khush (1997) are the AA genome (the Sativa complex consisting of the cultivated *O. sativa* and *glaberrima* and the wild *O. rufipogon, nivara, glumaepatula, meridionalis, barthii,* and *longistaminata*); the BB, CC, BBCC, and CCDD genomes (the Officinalis complex with *O. alta, punctata, eichingeri, minuta, officinalis, rhizomatis, grandiglumis* and *latifolia*); the EE genome (*O. australiensis*), the FF genome (*O. brachyantha*); the GG genome (the Meyeriana complex of *O. granulata, meyeriana,* and *neocalidonica*); the HHJJ genome (the Ridleyanae complex of *O. longiglumis* and *ridleyi*); and the HHKK genome (*O. schlechteri*). Figure 27-2 shows the phenotypic diversity of the representative *Oryza* species with AA genomes that are being actively used in rice breeding.

Thirteen representative accessions for the wild rice genomes were used to create bacterial artificial chromosome (BAC) libraries under the OMAP project (www.omap.org) (Ammiraju et al. 2006; Wing et al. 2007). BAC-end sequencing and fingerprinting of the clones have allowed anchoring to the Nipponbare genome and building of contigs. Based on these physical maps for the diverse genomes, clones corresponding to specific regions of interest can be selected for further study. These libraries and physical maps provide tools to understand the distant and diverse rice genomes. Wide hybridization between the diverse relatives and *O. sativa* has introgressed a number of useful traits for biotic and abiotic stress into modern varieties (Brar and Khush 1997). Some examples include grassy stunt virus resistance (from *O. nivara*), bacterial blight resistance (from *O. longistaminata*), tungro and acid sulfate tolerance (from *O. rufipogon*), and brown plant-hopper resistance (from *O. australiensis*).

Besides the International Rice Genebank, IRRI also hosts the International Network for the Genetic Evaluation of Rice (INGER), a program which involves screening for various traits in

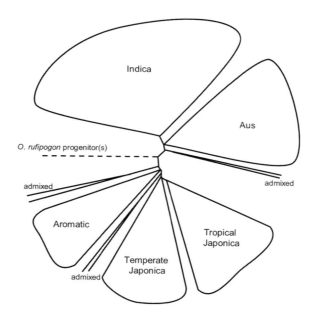

FIGURE 27-1. Representation of the genetic diversity relationships for the major variety groups of *O. sativa* based on SSR genotyping from a Generation Challenge Program project involving IRRI, CIRAD, CIAT, EMBRAPA, WARDA, and Cornell University (K. McNally et al., unpubl.) and a similar analysis conducted by Garris et al (2005). Position of the wild progenitor(s) is indicated by a broken line. Admixed refers to lines having a proportion of alleles derived from two or more major groups.

target environments. This program maintains a collection of over 40,000 rice varieties from breeding programs worldwide. Furthermore, the breeding pedigrees of all IRRI-bred and many non-IRRI-bred modern rice varieties, many of which are derived from the mating of traditional varieties, are maintained in the International Rice Information System (www.iris.irri.org). Besides being a pedigree database, IRIS includes information on variety characterization from IRRI and INGER nurseries.

PHYSICAL MAPS OF RICE

The sequencing of the whole rice genome was first discussed at the International Congress on Plant Molecular Biology held in Singapore in 1997. The International Rice Genome Sequencing Project (IRGSP) was formalized in 1998, with participation from multiple research institutions and laboratories from 10 countries. Less than a decade later, the scientific community had at least two genome sequence assemblies on two varieties of rice: a japonica variety, Nipponbare, and an indica variety, 93-11. The draft sequence of these two varieties was first produced by a whole-genome shotgun approach by the Beijing Genomics Institute (Yu et al. 2002) and Syngenta (Goff et al. 2002). With this approach, random genomic clones were sequenced and then assembled informatically. The IRGSP adopted a clone-by-clone sequencing approach in which large insert clones (BAC) were first aligned to produce a contiguous physical map. Then, each BAC was sequenced by sequencing libraries with small inserts. The map-based sequencing takes more time and effort, but it is considered the gold standard for accuracy in both chromosomal location of the sequence and the determination of nucleotides. The final description of all 12 chromosomes of Nipponbare was published in 2005 (IRGSP 2005). Prior to publication of the final genome sequence, the sequences of completed individual chromosomes were also published (Feng et al. 2002; Sasaki et al. 2002; The Rice Chromosome 10 Sequencing Consortium 2003; The Rice

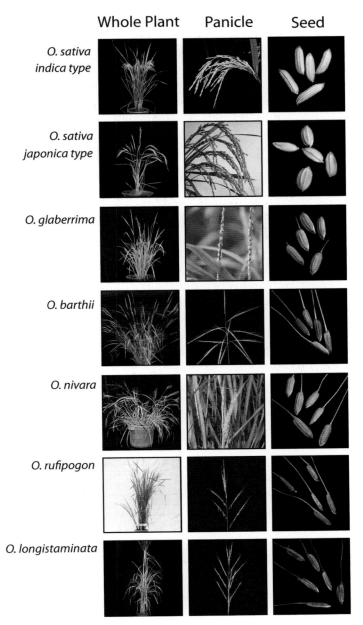

FIGURE 27-2. Phenotypic diversity of representative types of *Oryza* species with AA genomes. The three stages—whole plant, panicle, and seed—capture some of the distinguishing features seen in representative species. Because most AA genome species are inter-fertile, the variation present in the AA genomes is widely used in rice improvement programs.

Chromosome 3 Sequencing Consortium 2005). The early release of sequence information on individual chromosomes enabled the community to actively use the sequence information for cross-species analysis and accelerate the use of single-nucleotide polymorphism (SNP) and indel data as markers for selection (Feltus et al. 2004; Shen et al. 2004).

The IRGSP determined the sequence of 370 Mb of Nipponbare, representing 95% coverage of the 398-Mb genome, and nearly complete coverage of the euchromatic regions. Taking into account transposable-element-related sequences, and centromeric and telomeric regions, about 60% of the genome (240 Mb) consists of single- to low-copy sequences. A total of 37,544 protein-coding genes are found (not including transposable element-related genes). These represent a conservative estimate of protein-coding genes, which have direct experimental support from expressed sequence tags (ESTs) or full-length cDNAs. This yields a gene density of 1 gene per 9.9 kb (IRGSP 2005; Matsumoto et al. 2007). All sequences are searchable and publicly available through the Rice Annotation Project database (http://rapdb.lab.nig.ac.jp/) and The Institute for Genomic Research (TIGR) Web site (www.tigr.org/tdb/e2k1/osa1/). The number of functional elements is likely much greater, considering that many predicted genes are not yet supported by expression data. There is also a potentially large reservoir of noncoding RNA genes whose functions are still little studied (Sunkar et al. 2005).

WHOLE-GENOME SNP DISCOVERY

The published genome sequences of Nipponbare and 93-11 enabled analysis of SNPs between genotypes representing japonica and indica types. Based on these two genomes, Feltus et al. (2004) identified over 400,000 candidate SNP/indels after removing multiple-copy and low-quality sequences. Randomly selected loci (109) were validated by sequencing. Of these, 87 contained SNPs that matched those in the database, giving approximately 80% accuracy for their predictions. In another analysis, the IRGSP predicted 80,127 polymorphic sites between indica and japonica by comparing 12 Mb of indica Kasalath to japonica Nipponbare. Projecting this frequency to the whole genome would yield an estimate of 2.6 million SNPs between the two types, approximately 6 per kilobase. Yu et al. (2005) predicted a frequency of 3 SNPs/kb of coding sequence to 28 SNPs/kb for transposable elements.

Shen et al. (2004) surveyed indels and SNPs between Nipponbare and 93-11, and found 1 SNP per 268 bp and 1 indel per 953 bp. Thus, between these two cultivars, SNPs are approximately three times more frequent than indel polymorphisms. However, the distribution of SNPs is not uniform along a particular chromosome. Densities of polymorphism are lower around the centromeric regions for chromosomes 4, 5, 8, and 10, but polymorphisms are abundant around the centromere of chromosome 7. Regardless of the analytical methods, there is clearly an enormous amount of sequence polymorphism between one or two representative varieties of japonica and indica rice.

With an abundance of SNPs in the genome, it is possible to establish the causal relationships between SNP patterns and phenotypes. A key question is how these SNPs are distributed in multiple genomes representing the diverse gene pool of rice. Whole-genome SNP identification offers information on blocks of SNPs in physical proximity (SNP haplotypes) that reflect evolutionary and functional relationships. It also provides the "anchor" to relate other forms of polymorphisms (including biochemical, metabolic, and physiological) with phenotypic performance. A multi-varietal SNP database will be a significant advance beyond one or two versions of rice genome sequence in terms of practical utility because only through a comparative analysis can we relate allelic differences across diverse varieties with altered transcript levels, protein synthesis, subtle but critical differences in metabolism, and eventually, phenotypic performance.

Genome-wide SNP Comparative Analysis in 20 Diverse Lines

Recognizing the potential of having SNP data from diverse genotypes, the International Rice Functional Genomics Consortium, coordinated by IRRI, has led an effort to undertake genome-wide SNP discovery by resequencing diverse rice varieties through DNA–DNA hybridization on high-density oligonucleotide arrays (McNally et al. 2006). The project involves partnership with Perlegen Sciences to apply tiling arrays to determine sequence variation relative to a template (reference genome). In this approach of resequencing, all nonrepetitive regions of the genome greater than about 60 bases in length are tiled using 25-mer oligonucleotides with a sliding window offset by one base for both strands, where the middle base is fourfold degenerate. Hence, eight oligonucleotides are present on the array to interrogate every position contained in the nonrepetitive regions.

Currently, a collection of 25 diverse varieties/landraces has been assembled for seed increase and purification. Twenty of these lines were selected for use in the SNP discovery project. These lines have undergone one round of single seed descent and are being increased to amounts sufficient for phenotyping. Additionally, F_1 intercrosses for a half diallele mating scheme have been made for these lines in preparation for the production of recombinant inbred lines (RILs) and other populations.

Release 4 of the high-quality BAC-by-BAC japonica sequence of Nipponbare was masked for repetitive DNA by a pipeline developed at McGill University by Thomas Bureau. After masking, 95 Mb of Nipponbare was found to be unique by the strict criterion that each segment has no significant BLAST hit other than to itself. The same procedure was used to mask the whole-genome shotgun indica sequence of 93-11 (Beijing Genomics Institute). Both sequences were aligned to one another using the Mummer/Nucmer tools (TIGR). The masked sequence was compared to the TIGR gene models to define the 100 Mb of the genome that will be tiled onto the arrays. The tiled regions are dispersed throughout the genome, with at least one tiled region for every 100 kb of the complete genome. Hybridizations of the 20 diverse lines to these arrays will result in a collection of SNPs with whole-genome coverage.

A pilot experiment has been conducted involving arrays where 379 kb of unique sequence from a region of 684 kb on the long arm of chromosome 3 was interrogated. All 20 varieties were hybridized to the arrays, with a total of 2132 SNPs detected (on average 1 SNP/200 bp). In several contrasts between japonica lines, fewer than 10 SNPs were detected in the 379-kb region. It is not clear whether the low diversity between these lines is unique to this region or is general across these genomes, a question that can be resolved when the genome-wide data set becomes available. A recent study by Tang et al. (2006) showed that the degree of divergence may differ sharply between genomic regions of indica and japonica types. They found that at least 6% of the genomes are unusually divergent, whereas other reference regions show low divergence. These contrasting regions of divergence may reflect the history of domestication.

SNP Haplotyping

SNPs that are highly correlated with one another exhibit high linkage disequilibrium (LD). A particular combination of alleles along a chromosome determines a haplotype. Comparing haplotypes and the extent of high LD across individuals separated in time by many meioses allows the identification of segments or haplotype blocks that correspond to minimal units of recombination. Since an allele within a haplotype block is predictive of the other alleles in the block, genotyping of one or a few alleles per block is sufficient to establish the haplotype of a particular region. Such predictive SNPs are termed tag-SNPs. Whole-genome scans can therefore be accomplished by genotyping a collection of tag-SNPs that define the haplotype blocks throughout the genome. The anticipated SNP data sets will help establish the number of tag-SNPs that are needed to capture all haplotypic variation in the rice gene pool.

With the high-density oligomer resequencing approach, SNP haplotypes have been determined in both human and mouse genomes (Frazer et al. 2004; Hinds et al. 2005). In the case of humans (Hinds et al. 2005), a set of 1.6 million SNPs likely to be common in individuals of diverse ancestry was identified by array-based resequencing of a panel of 24 individuals. When this SNP collection was used to genotype a panel of 71 individuals, up to 73% of the common variation was detected. An international consortium has produced a haplotype map for the human genome (International HapMap Consortium 2005). For the Phase I map, over 1 million SNPs were genotyped on a panel of 269 individuals from 4 populations with 1 SNP in every 5-kb interval of the human genome. For the analysis of the mouse genome by Frazer et al. (2004), SNP screening was accomplished on 13 classic inbred and wild inbred mouse lines across 5 genomic regions totaling 4.6 Mb. This analysis revealed haplotypic blocks ranging from 12 kb to 608 kb that had simple and distinct phylogenetic relationships in the 13 classic inbred lines. This study suggested that for exhaustive SNP discovery in mouse, 97% of the common ancestral variation could be found within 12 inbred lines.

Two studies describing the extent of LD in rice have been published. Garris et al. (2003) studied the haplotype diversity and LD around the *xa5* locus on the short arm of chromosome 5. They found significant LD between sites 100 kb apart. Olsen et al. (2006) found that LD at the *Waxy* locus on chromosome 6 extended to 250 kb. Other diversity measurements indicated a selective sweep at this locus during domestication. Hence, it appears that the length for a haplotype block in rice is on the order of 100–250 kb. To extend these LD studies in rice, CIRAD and IRRI are determining the extent of haplotype structure that is indicative of their differentiation. A set of 1536 SNPs suitable for genome scans is being used to genotype 900 types predicted to cover the range of indica/japonica diversification and prospective natural hybrids between them.

The outcome from these genome-wide SNP analyses will be a fine-scale LD map for common SNP variation among the various types of rice. This will clarify the origin of the aus and aromatic varietal groups in relation to indica and japonica. Analysis of the SNP data with phenotypic data on a range of traits will identify loci involved in the differentiation among types. This may reveal patterns of relationships whose existence enables LD-based mapping of genes of agricultural interest and may inspire applications in other inbreeding species. However, it is important to note that, given the many options for genetic crosses in rice, one does not need to achieve exceedingly high resolution in LD-based mapping. At a certain point, genotype–phenotype relationship can be better determined using other data sets or experimental approaches that are cost-effective.

Identifying Allelic Variation

Because of LD structure, it is possible to comprehensively define the natural variation of rice germ plasm with SNP haplotypic blocks that are conserved or selected for during years of domestication. If association between SNP haplotypes and phenotypes is strong, this will increase the likelihood of finding useful allelic variation in candidate genes for specific functions. Thus, establishing genetic population structure through SNP haplotyping can provide a foundation for the effective exploration of allelic diversity in the rice gene pool. New and improved techniques are now available. One such technique is TILLING (Targeted Induced Lesions In the Genome), which was originally developed as a reverse genetic technique to screen induced point mutations (McCallum et al. 2000). EcoTILLING is an adaptation of the technique to probe the natural variation at individual loci in germ plasm (Comai et al. 2004).

As a first step, a subset of 1536 accessions/varieties of cultivated rice *O. sativa* from the total collection was assembled, and referred to as the minicore collection. This minicore collection represents the geographic range of rice germ plasm and includes material spanning the range of genetic diversity in Asian cultivated rice. The African cultivated species *O. glaberrima* is represented by 190 accessions. In addition to domesticated rice, we use a panel of 93 accessions of wild relatives representing all the genome types (AA, BB, CC, BBCC, CCDD, EE, FF, GG, HHJJ, HHKK) of the genus *Oryza*. For the AA genome types, we have a panel of 95 representative accessions for delineating the biosystematic relationships within the species most closely affiliated with *O. sativa*.

Our preliminary results from testing 10 genes on the *O. sativa* minicore collection and *O. glaberrima*, and five genes for the wild *Oryza* subset of 93 accessions, suggest that EcoTILLING is a promising approach for characterizing the rice collection. First, the frequency of SNPs detected is consistent with that expected from analysis. Second, primers designed for presumably conserved sequences (e.g., DREB2 coding region) based on Nipponbare sequence function across all AA genomes and in a majority of the genomes beyond AA. Although how well a primer pair works across genomes depends on the degree of sequence conservation, the ability to evaluate SNP or indel allelic variation across genomes opens up new opportunities. Finally, the simplified method of EcoTILLING on agarose gels, which is applicable to the cultivated and most of the AA genome *Oryza* species, allows a faster and more economical approach to detect and characterize variation (Raghavan et al. 2007).

GENETIC RESOURCES FOR MAPPING AND ASSIGNING GENE FUNCTION

Specialized genetic stocks and mapping populations are a prerequisite for determining gene function and must be made available or developed if we are to fully make use of rice sequence information. One can consider two categories of specialized genetic stocks. The first category is induced mutations, for which a chosen genotype, often selected for specific phenotypes, is mutagenized to produce variation at a locus. Mutagenesis is well developed in rice through the use of endogenous or heterologous transposons, insertional vectors, and chemical and irradiation treatments. The advantage of induced mutations is that the genetic background (barring transformation-induced somaclonal variation) is nearly identical, such that the mutant and wild type can be considered as isogenic lines, making comparative analysis simple. A good-sized collection of mutants is already available, although it is still far from achieving saturation of the genome (for review, see Hirochika et al. 2004). However, induced mutations are limited by allelic diversity at the loci of a chosen genotype (two alternative alleles at a locus).

The second category of specialized stocks captures the natural variation already present in the species. Novel genetic recombinants can be created by crosses designed to enable investigation of genetic factors in different backgrounds. Although many mapping populations have been produced in different genetic studies or from breeding programs, relatively few recombinant populations are systematically maintained and distributed for public use. Therefore, there is a demand for recombinant populations that are optimally designed for the analysis of gene functions.

Induced Variation

Because of multinational efforts to produce mutant collections for functional genomics studies, rice is particularly rich in induced genetic variation. Rice probably has the largest store of induced mutations in any crop plant (Hirochika et al. 2004). These collections include transposon or T-DNA insertion lines that disrupt or activate gene functions. As of November 2006, the OryGenes DB (http://orygenesdb.cirad.fr) has compiled about 140,000 insertion sequences with known flanking sequence tags (FSTs) in the rice genome (see Table 27-2). With this collection of insertion mutants, there is a 50% chance of finding a mutation tag for any given gene (A. Pereiera, pers. comm.). The use of multiple insertion vectors and transposons with different integration properties is helpful toward achieving genome coverage. However, saturation mutagenesis may prove difficult with insertional mutagenesis, as some genes or genomic regions may not be accessible to DNA integration. To saturate the genome with mutations, chemical and irradiation mutagenesis have proved highly complementary to insertional mutagenesis. Bhat et al. (2007) summarized the current status of chemical and irradiation mutagenesis in rice. Unlike transformation or tissue-culture-mediated mutagenesis, chemical and irradiation mutagenesis can be applied to any genotype. The nontransgenic nature of these mutants makes them particularly useful for wide distribution, so that biological traits can be evaluated in agronomically relevant environments.

IRRI has produced and maintained large collections of chemical and irradiation-induced mutants in several genetic backgrounds (Table 27-3). The largest mutant collection has been made in IR64, the most popular indica variety in tropical Asia. In this mutant collection, multiple mutagens were used to create different sizes of genetic lesions, making the mutants amenable to forward and reverse genetics (Wu et al. 2005). As of November 2006, more than 50,000 M_4 lines have

TABLE 27-2. Current Status of Insertion Rice Mutants with FSTs Produced by Different Research Institutions

Institution	Contact	Vector	No. of flanking sequences
CSIRO	narayana.upadhyaya@csiro.au	T-DNA	787
CIRAD-INRA-IRD-CNRS, Genoplante	emmanuel.guiderdoni@cirad.fr	T-DNA	7480
National Institute of Agrobiological Sciences	hirohiko@nias.affrc.go.jp	*Tos17*	18,024
CerealGene Tags, European Union	andy.pereira@wur.nl	*Ds*	1380
Gyeongsang National University	cdhan@nongae.gsnu.ac.kr	*Ds*	1040
Postech	genean@postech.ac.kr	T-DNA	80,259
National Center of Plant Gene Research (Wuhan)	swang@mail.hzau.edu.cn	T-DNA	15,727
National University of Singapore	sri@tll.org.sg	*Ds*	1469
Taiwan Rice Insertional Mutant Program	bohsing@gate.sinica.edu.tw	T-DNA	7053
University of California at Davis	sundar@ucdavis.edu	*Ds*	6878
Total			140,097

Downloaded November 5, 2006, from OryGenesDB (http://orygenesdb.cirad.fr/).

been produced. Sufficient seeds are produced for distribution and maintained as a permanent stock at IRRI's GeneBank for future use. Abundant genetic variation is observed in the mutant collection, including morphological variants (~ 8%) and conditional traits (e.g., gain and loss of resistance to biotic and abiotic stresses). With advances in whole-genome assay techniques, the utility of chemical-irradiation-induced mutants has also grown. Deletions can now be detected routinely by chip-based technology (Borevitz et al. 2003). In rice, the comprehensive oligonucleotide coverage (multiple oligonucleotides per gene) offered on the Affymetrix arrays makes them particularly useful for discovering deleted regions in rice mutants (M. Bruce et al., unpubl.). As mentioned earlier, TILLING is available for detecting point mutations and small deletions/insertions.

TABLE 27-3. Mutant Collections Induced by Irradiation and Chemical Mutagenesis at IRRI, as of February 2006

Genotype	Mutagen (dose)	Types of genetic lesions	Primary use	Number of lines advanced
IR64	fast neutron (60 Gy)	large deletions, translocations	forward genetics	8073
			deletion detection by arrays or PCR	
	gamma ray (250 and 500 Gy)	1–2 kb deletions, point mutations	forward genetics	15,295
			deletion detection by arrays or PCR	
	diepoxybutane	kilobase-range deletions, point mutations	forward genetics	16,520
			deletion detection by arrays or PCR	
	EMS	point mutations	forward genetics, TILLING	12,539
Total				52,427
Other genotypes				
Shan Huang Zhang No. 2	1.6% EMS		blast resistance	7387 M$_3$
Jinbubyeo	0.6% EMS		cold tolerance	3000 M$_3$
FL478	1% EMS		salinity tolerance	11,000 M$_1$
IRBB61	1% EMS		bacterial blight resistance	11,000 M$_2$

Creating Genotypic Diversity through Active Recombination

Natural or induced allelic variation at a locus provides the building blocks by which an enormous amount of genotypic diversity can be created through recombination. Whereas allelic diversity could be finite over a long time, the genotypic diversity that can be created by recombination within a few generations is enormous. Thus, special genetic stocks that maximize recombination are important resources for identifying gene function. Several types of segregating populations are particularly useful for genetic mapping as well as plant breeding. These include RILs, chromosomal segmental substitution lines, and advanced introgression lines. From these segregating materials, one can extract different specialized genetic stocks such as near-isogenic lines and heterogeneous inbred families (sister lines derived from an inbred line that is heterozygous at a target locus).

RILs have been the mainstay of mapping experiments. Advanced fixed lines can be maintained forever, providing many options for phenotyping in replicated experiments across locations. This is particularly useful for detecting small phenotypic effects. To further the community phenotyping concept and to maximize use of the SNP information generated in the multiple varieties, we have initiated a half-diallele crossing scheme to make pairwise crosses between all the "resequenced" varieties. Not all crosses will be successful due to fertility barriers, but the RILs derived from these densely genotyped lines will be particularly informative in quantitative trait locus (QTL) analysis.

TRANSCRIPTION MAPS

Because the rice genome sequence is complete, it is possible to generate transcription maps by describing expression patterns in a chromosomal context. Kikuchi et al. (2007) summarized the various platforms currently available for transcriptome analysis. The 22K Oligoarray (Agilent) has been used to examine transcriptional control of hormonal response and several stress treatments (K. Satoh and S. Kikuchi, NIAS; R. Muthurajan and R. Mauleon, IRRI, both unpubl.). It is expected to expand to 44K on a single chip (Kikuchi et al. 2007). Another 60K oligonucleotide array (on two slides) produced by the Beijing Genomics Institute and Yale University has been used to examine whole-genome expression profiles at different developmental stages (Ma et al. 2005). Third, a 45K oligonucleotide chip (version 3) supported by the U.S. National Science Foundation is produced and distributed by the University of California-Davis. Finally, a 50K Affymetrix chip is commercially available. Unlike the long-oligonucleotide arrays (60- to 70-mers) mentioned above, the Affymetrix chip contains 25-mer oligonucleotides, with about 11 oligonucleotides designed to constitute a probe set for each gene. These platforms differ in oligonucleotide design, genome coverage, and cost. One can select the appropriate platform by considering the experimental objectives and cost. Because of the near-genome coverage of rice oligonucleotide arrays, it is possible to examine gene expression in relation to gene order, an approach that is not possible unless complete and contiguous genome sequence data are available.

Unlike DNA-based maps that are constant with respect to genotypes, transcription maps are dynamic, depending on the developmental stages and environmental conditions under which observations are made. This makes transcriptome analysis more sensitive to perturbation by experimental conditions. Transcript maps provide unique "molecular phenotypes" that contribute to understanding genetic regulation at the genome level and help bridge the gap between genes and phenotypes. The RiceAtlas Project at Yale University has produced transcriptome data for 40 rice cell types representing various organs and developmental stages of representative indica and japonica varieties under normal growing conditions (http://bioinformatics.med.yale.edu/rc/overview.jspx). This rich database will provide a reference point to compare with transcriptome data derived from other studies.

There is emerging evidence that gene expression patterns are influenced by chromosomal locations. Results of transcriptome analysis in animals (Caron et al. 2000; Cohen et al. 2000; Spellman and Rubin 2002; Doss et al. 2005; Petkov et al. 2005) and plants (Williams and Bowles 2004; Jiao et al. 2005; Zhan et al. 2006) all suggest the presence of chromosome domains and evidence of correlated expression among neighboring genes. In rice, Ma et al. (2005) reported that about 10% of the genome exhibited correlated expression, which is in line with results reported in *Arabidopsis* (Zhan et al. 2006). Collaboration between the National Institute of Agrobiological Sciences (NIAS) and IRRI has focused on transcript maps specific to stress response. A rich gene expression data set on the 22K Oligoarray has been obtained from different rice genotypes in response to pathogen and water stresses. Of special interest are the observed relationships between expression patterns and chromosomal locations. For example, in a comparative analysis of gene expression in response to water stress during vegetative growth in two varieties, Apo (drought-tolerant) and IR64 (drought-sensitive), we found 14 regions of correlated expression (20-gene window) in Apo during drought stress (R. Muthurajan, K. Satoh, R. Mauleon, IRRI and NIAS, unpubl.). One region on chromosome 1 (38.5–39 Mb) colocalized with a reported drought tolerance QTL, and a second region on chromosome 8 (22.5–23.5 Mb) corresponded to another drought tolerance QTL. The use of the regions of correlated expression analysis enables us to identify discrete chromosomal regions (10–20 genes) that may underlie QTL effects even though they may not show a large magnitude of differential expression. Transcript maps showing differentially expressed genes and regions of correlated expression can be used to short-list genes of interest in regions contributing to QTLs. We are evaluating the possibility that coordinated expression of multiple adjacent genes may contribute to a single QTL region, a hypothesis with practical implications for selecting genotypes with the desired phenotypes.

In addition to transcriptome data derived from oligonucleotide arrays, condition-specific expression data have been produced using serial analysis of gene expression (SAGE) and massively parallel signature sequencing (MPSS) technologies (Gowda et al. 2004; Nakano et al. 2006). These data can be anchored onto the pseudomolecules to produce a rich source of gene expression information. Proteomics data can be analyzed in a similar way (Komatsu and Yano 2006) to gain additional insights into the role that chromosomal architecture may play in gene expression and eventually phenotypes.

SELECTION IN PRACTICE

Molecular markers for gene mapping in rice have been available for a long time, and this has led to the genetic mapping of many genes of economic importance. The application of these genes in marker-assisted selection (MAS), however, has not been extensive. Mapping genes and QTLs is only the first step toward practical MAS and this requires several additional steps before being useful in breeding. Validating and applying molecular markers for selection requires confirmation of the effect of the gene or QTL in a different genetic population, fine-scale mapping, and development of a toolkit of polymorphic markers tightly linked to and flanking the gene (Langridge et al. 2001; Collard and Mackill 2007).

For major genes and QTLs of large effect, it is quite easy to identify markers for MAS. These genes have been relatively easy to manipulate in breeding programs without resort to markers. However, MAS offers considerable advantages for the pyramiding of major genes for the same trait into a single plant. Very often, genes controlling the same trait may have epistatic interactions; thus, the presence of one gene may phenotypically mask the others. This has been pursued as a strategy for obtaining multiple resistance genes in a genotype with effective resistance to diseases such as bacterial blight (Huang et al. 1997; Joseph et al. 2004) and blast (Hittalmani et al. 2000; Tabien et al. 2000; Fjellstrom et al. 2004).

Another application of MAS for major genes is the enhancement of existing varieties by selective transfer of a major gene or QTL through marker-assisted backcrossing (MAB). The potential for using markers to transfer only a small chromosome segment of interest was recognized during the early development of molecular markers for plants (Tanksley et al. 1989). Simulation studies showed the potential of markers to accelerate backcross breeding and reduce linkage drag in the development of improved varieties (Frisch et al. 1999a,b; Hospital 2001; Servin and Hospital 2002). The use of flanking markers for recombinant selection can dramatically reduce the size of the chromosome fragment surrounding the introduced gene (Frisch et al. 1999b; Hospital 2001). Chen et al. (2000, 2001) applied the methods to transfer a small chromosomal segment surrounding the *Xa21* gene for bacterial blight resistance into two widely used hybrid parents in China. The method requires using larger than usual backcross populations and has not been widely taken up by rice breeders. However, recent success with transferring the submergence tolerance *Sub1* locus into widely grown varieties (Xu et al. 2006) may encourage more widespread use of this procedure.

QTLs present some difficulties in MAS because they are often difficult to fine-map and their effects are lower when manipulated as single factors. In addition, there are more likely to be genetic background effects and QTL by environment interactions that limit the usefulness of QTLs in breeding. For these reasons, QTLs of large effect are more amenable to MAS or MAB (Holland 2004; Mackill 2006). QTL mapping can be subject to many errors, and the accuracy of mapping is highly dependent on using a large population size (Beavis 1998). It is particularly important to validate QTL mapping studies using later generations of the populations or different mapping populations. However, the map position of verified QTLs is usually quite accurate, as suggested by a recent survey of QTLs that have been cloned (Price 2006).

QTLs of relatively large effect can be manipulated in the same way as major genes, although examples are limited in the rice literature. Steele et al. (2006) provide an example for root traits, but noted that most root QTLs transferred into a different genetic background were not expressed. For QTLs of smaller effects, it may not be possible to apply conventional MAS approaches. Fine-scale mapping of these QTLs may be quite difficult, and genetic background effects may be very high. For these types of traits, understanding the nature of the trait and identifying the genes underlying the QTLs will be very useful. Research on partial resistance to rice blast disease has shown that genes known to be involved in host defense often map near QTLs, and alleles at some of these are associated with the level of partial resistance (Liu et al. 2004; Wu et al. 2004). Combining transcriptome analysis with genetic mapping provides further evidence for the roles of candidate defense genes (R. Mauleon and B. Liu, IRRI, unpubl.). Finally, mutant analyses and RNA interference studies, wherein expression of the candidate genes is suppressed, have confirmed roles for some genes in QTL-governed resistance (P. Manosalva et al., unpubl.).

The use of background selection in backcrossing and the interest in manipulating a number of genes simultaneously have led to a need for more efficient whole-genome scans in MAS. With SSR markers, it is necessary to use at least four markers per chromosome arm to select against the donor chromosomes (Whittaker et al. 1996; Servin and Hospital 2002), and this still leaves the possibility of smaller regions where double recombinants are present. Multiplexing of SSR markers with automated sequencing equipment is one method for making background selection more efficient (Coburn et al. 2002). However, array-based methods that detect SNPs or SFPs (single-feature polymorphisms) are likely to be more suitable for performing whole-genome scans. Array-based methods for scanning SNPs or SFPs across the whole genome (Borevitz et al. 2003; Hazen and Kay 2003) will likely supersede the use of SSR markers for MAS because of their power to determine the whole-genome composition of plants in segregating populations.

Most recently, D. Galbraith and coworkers at the University of Arizona have designed oligonucleotides that detect SFPs across the rice genome. A unique feature of the SFP-detection microarrays is their relatively low cost, making it feasible to practice the technique in routine genetic studies which track breeding materials. Further cost reduction will make it economically feasible to have genetic "barcoding" of a major portion of the more than 102,000 accessions maintained in the International GeneBank at IRRI. As transcriptome analysis becomes less costly, transcript maps for selected genotypes under different conditions can also be produced, as illustrated in a recent study in *Arabidopsis* (Kliebenstein et al. 2006). Convergence of results from mapping, expression analysis, and selection response will provide a strong foundation for efficient use of both the allelic and genotypic diversity of rice.

SUMMARY

The available whole-genome sequence has opened up new ways to understand and use genetic variation. Knowing all gene sequences in their chromosomal context is essential for understanding genome regulation. Analysis of rice genetic diversity will go beyond the sampling of allelic variation to include an integrated analysis of variation at both the DNA and expression levels. This will enable us to understand how DNA sequence variation is eventually manifested in phenotypic variation through the response of the whole genome. This new dimension of understanding of genetic variation and diversity is unprecedented in a crop plant. It holds the promise of revealing the underlying control of many agronomic traits that have so far been difficult to identify or manipulate in breeding programs.

It is anticipated that with decreasing costs in genome sequencing and possibly expression analysis, genome variation and expression polymorphism can be documented comprehensively. The limiting factor in the next few years may be the availability of genetic resources designed for high-resolution mapping and gene identification. How can we use novel and efficient techniques to silence or activate target genes or gene clusters? How can we apply the power of recombination to generate desirable combinations of genes or groups of genes that serve a particular function at the whole-plant or crop level? These are just some of the questions that the new genomics tools and whole-genome analyses hope to address.

Finally, the potential of applying genomic information for the identification of functional natural or induced variants can only be realized when a system is in place to evaluate the traits. Many traits important for agriculture are conditional, such as tolerance of biotic and abiotic stresses. Phenotyping of minicore varieties is ongoing at IRRI, but a larger collective effort is needed to evaluate phenotypes in different environments and locations. The INGER provides such a mechanism, but a main challenge is to have effective coordination within the phenotyping network to produce high-quality and broadly accessible data. Another challenge is to have active exchange of genetic resources to enable independent testing and validation. Thus, sharing of data and free flow of genetic materials may hold the key for the future exploration of rice genetic diversity and use.

ACKNOWLEDGMENTS

We thank several anonymous colleagues for reading the manuscript and for providing unpublished information. The contribution of Ariel Javellana in photography and composing Figure 27-2 is gratefully acknowledged. Some unpublished work is cited from collaborative projects supported by the USDA NRI and the Generation Challenge Program.

REFERENCES

Ammiraju J.S.S., Luo M., Goicoechea J.L., Wang W., Kudrna D., Mueller C., Talag J., Kim H.R., Sisneros N.B., Blackmon B., et al. 2006. The *Oryza* bacterial artificial chromosome library resource: Construction and analysis of 12 deep-coverage large-insert BAC libraries that represent the 10 genome types of the genus *Oryza*. *Genome Res.* **16:** 140–147.

Beavis W.D. 1998. QTL analysis: Power, precision and accuracy. In *Molecular dissection of complex traits* (ed. A.H. Paterson), pp. 145–162. CRC Press, Boca Raton, Florida.

Bhat R., Upadhyaya N.M., Chaudhury A., Raghavan C., Qiu F., Wang H., Wu J., McNally K., Leung H., Till B., et al. 2007. Chemical and irradiation induced mutants and TILLING. In *Rice functional genomics: Challenges, progress and prospects* (ed. N.M. Upadhyaya), pp. 151–186. Springer, New York.

Borevitz J.O., Liang D., Plouffe D., Chan H.-S., Zhu T., Weigel D., Berry C.C., Winzeler E., and Chory J. 2003. Large-scale identification of single-feature polymorphisms in complex genomes. *Genome Res.* **13:** 513–523.

Brar D.S. and Khush G.S. 1997. Alien introgression in rice. *Plant Mol. Biol.* **35:** 35–47.

Caron H., van Schaik B., van der Mee M., Bass F., Riggins G., van Sluis P., Hermus M.-C., van Asperen R., Boon K., Vaûte P.A., et al. 2000. The human transcriptome map: Clustering of highly expressed genes in chromosomal domains. *Science* **291:** 1289–1292.

Chen S., Lin X.H., Xu C.G., and Zhang Q.F. 2000. Improvement of bacterial blight resistance of 'Minghui 63', an elite restorer line of hybrid rice, by molecular marker-assisted selection. *Crop Sci.* **40:** 239–244.

Chen S., Xu C.G., Lin X.H., and Zhang Q. 2001. Improving bacterial blight resistance of '6078', an elite restorer line of hybrid rice, by molecular marker-assisted selection. *Plant Breed.* **120:** 133–137.

Cheng C., Motohashi R., Tsuchimoto S., Fukuta Y., Ohtsubo H., and Ohtsubo E. 2003. Polyphyletic origin of cultivated rice: Based on interspersion pattern of SINES. *Mol. Biol. Evol.* **20:** 67–75.

Coburn J.R., Temnykh S.V., Paul E.M., and McCouch S.R. 2002. Design and application of microsatellite marker panels for semiautomated genotyping of rice (*Oryza sativa* L.). *Crop Sci.* **42:** 2092–2099.

Cohen B.A., Mitra R.D., Hughes J.D., and Church G.M. 2000. A computational analysis of whole-genome expression data reveals chromosomal domains of gene expression. *Nat. Genet.* **26:** 183–186.

Collard B.C.Y. and Mackill D.J. 2007. Marker-assisted selection: An approach for precision plant breeding in the 21st century. *Philos. Trans. R. Soc. B Rev.* (in press).

Comai L., Young K., Till B.J., Reynolds S.H., Greene E.A., Codomo C.A., Enns L.C., Johnson J.E., Burtner C., Odden A.R., and Henikoff S. 2004. Efficient discovery of DNA polymorphisms in natural populations by Ecotilling. *Plant J.* **37:** 778–786.

Doss S., Schadt E.E., Drake T.A., and Lusis A.J. 2005. Cis-acting expression of quantitative trait loci in mice. *Genome Res.* **15:** 681–691.

Feltus F.A., Wan J., Schulze S.R., Estill J.C., Jiang N., and Paterson A.H. 2004. An SNP resource for rice genetics and breeding based on subspecies *indica* and *japonica* genome alignments. *Genome Res.* **14:** 1812–1819.

Feng Q., Zhang Y., Hao P., Wang S., Fu G., Huang Y., Li Y., Zhu J., Liu Y., Hu X., et al. 2002. Sequence and analysis of rice chromosome 4. *Nature* **420:** 316–320.

Fjellstrom R., Conaway-Bormans C.A., McClung A.M., Marchetti M.A., Shank A.R., and Park W.D. 2004. Development of DNA markers suitable for marker assisted selection of three Pi genes conferring resistance to multiple *Pyricularia grisea* pathotypes. *Crop Sci.* **44:** 1790–1798.

Frazer K.A, Wade C.M., Hinds D.A., Patil N., Cox D.R., and Daly M.J. 2004. Segmental phylogenetic relationships of inbred mouse strains revealed by fine-scale analysis of sequence variation across 4.6 Mb of mouse genome. *Genome Res.* **14:** 1493–1500.

Frisch M., Bohn M., and Melchinger A.E. 1999a. Comparison of selection strategies for marker-assisted backcrossing of a gene. *Crop Sci.* **39:** 1295–1301.

———. 1999b. Minimum sample size and optimal positioning of flanking markers in marker-assisted backcrossing for transfer of a target gene. *Crop Sci.* **39:** 967–975.

Garris A.J., McCouch S.R., and Kresovich S. 2003. Population structure and its effect on haplotype diversity and linkage disequilibrium surrounding the *xa5* locus of rice (*Oryza sativa* L.). *Genetics* **165:** 759–769.

Garris A.J., Tai T.H., Coburn J., Kresovich S., and McCouch S.R. 2005. Genetic structure and diversity in *Oryza sativa* L. *Genetics* **169:** 1631–1638.

Goff S.A., Ricke D., Lan T.H., Presting G., Wang R., Dunn M., Glazebrook J., Sessions A., Oeller P., Varma H., et al. 2002. A draft sequence of the rice genome (*Oryza sativa* L. ssp. *japonica*). *Science* **296:** 92–100.

Gowda M., Jantasuriyarat C., Dean R.A., and Wang G.L. 2004. Robust-LongSAGE (RL-SAGE): A substantially improved LongSAGE method for gene discovery and transcriptome analysis. *Plant Physiol.* **134:** 890–897.

Hazen S.P. and Kay S.A. 2003. Gene arrays are not just for measuring gene expression. *Trends Plant Sci.* **8:** 413–416.

Hinds D.A., Stuve L.L., Nilsen G.B., Halperin E., Eskin E., Ballinger D.G., Frazer K.A., and Cox D.R. 2005. Whole-genome patterns of common DNA variation in three human populations. *Science* **307:** 1072–1079.

Hirochika H., Guiderdoni E., An G., Hsing Y., Eun M.Y., Han C.D., Upadhyaya N., Ramachandran S., Zhang Q., Pereira A., et al. 2004. Rice mutant resources for gene discovery. *Plant Mol. Biol.* **54:** 325–334.

Hittalmani S., Parco A., Mew T.V., Zeigler R.S., and Huang N. 2000. Fine mapping and DNA marker-assisted pyramiding of the three major genes for blast resistance in rice. *Theor. Appl. Genet.* **100:** 1121–1128.

Holland J.B. 2004. Implementation of molecular markers for quantitative traits in breeding programs–challenges and opportunities. In *New directions for a diverse planet: Proceedings of the 4th International Crop Science Congress*, Brisbane, Australia. The Regional Institute Ltd., Gosford, Australia (www.cropscience.org.au).

Hospital F. 2001. Size of donor chromosome segments around introgressed loci and reduction of linkage drag in marker-assisted backcross programs. *Genetics* **158:** 1363–1379.

Huang N., Angeles E.R., Domingo J., Magpantay G., Singh S., Zhang G., Kumaravadivel N., Bennett J., and Khush G.S. 1997. Pyramiding of bacterial blight resistance genes in rice: Marker-assisted selection using RFLP and PCR. *Theor. Appl. Genet.* **95:** 313–320.

International HapMap Consortium. 2005. A haplotype map of the human genome. *Nature* **437:** 1299–1320.

International Rice Genome Sequencing Project (IRGSP). 2005. The map-based sequence of the rice genome. *Nature* **436:** 793–800.

Joseph M., Gopalakrishnan S., Sharma R.K., Singh V.P., Singh A.K., Singh N.K., and Mohapatra T. 2004. Combining bacterial blight resistance and Basmati quality characteristics by phenotypic and molecular marker-assisted selection in rice. *Mol. Breed.* **13:** 377–387.

Jiao Y., Jia, P., Wang X., Su N., Yu S., Zhang D., Ma L., Feng Q., Jin Z., Li L., et al. 2005. A tiling microarray expression analysis of rice chromosome 4 suggests a chromosome-level regulation of transcription. *Plant Cell* **17:** 1641–1657.

Khush G.S. 1997. Origin, dispersal, cultivation and variation of rice. *Plant Mol. Biol.* **35:** 25–34.

Kikuchi S., Wang G.L., and Li L. 2007. Genome-wide RNA expression profiling in rice. In *Rice functional genomics: Challenges, progress and prospects* (ed. N.M. Upadhyaya), pp. 36–63. Springer, New York.

Kliebenstein D.J., West M.A.L., van Leeuwen H., Kim K., Doerge R.W., Michelmore R.W., and St. Claire D.A. 2006. Genomic survey of gene expression diversity in *Arabidopsis thaliana*. *Genetics* 172: 1179–1189.

Komatsu S. and Yano H. 2006. Update and challenges on proteomics in rice. *Proteomics* 6: 4057–4068.

Langridge P., Lagudah E.S., Holton T.A., Appels R., Sharp P.J., and Chalmers K.J. 2001. Trends in genetic and genome analyses in wheat: A review. *Aust. J. Agric. Res.* 52: 1043–1077.

Leung H. and An G. 2004. Rice functional genomics: Large scale gene discovery and applications for crop improvement. *Adv. Agron.* 82: 55–111.

Liu B., Zhang S.H., Zhu X.Y., Yang Q.Y., Wu S.Z., Mei M.T., Mauleon R., Leach J., Mew T., and Leung H. 2004. Candidate defense genes as predictors of quantitative blast resistance in rice. *Mol. Plant-Microbe Interact.* 17: 1146–1152.

Lu B.R. 1999. Taxonomy of the genus *Oryza* (Poaceae): Historical perspective and current status. *Int. Rice Res. Notes* 24: 4–8.

Ma L., Chen C., Liu X., Jiao Y., Su N., Li L., Wang X., Cao M., Sun N., Zhang X., et al. 2005. A microarray analysis of the rice transcriptome and its comparison to *Arabidopsis*. *Genome Res.* 15: 1274–1283.

Mackill D.J. 1995. Classifying japonica rice cultivars with RAPD markers. *Crop Sci.* 35: 889–894.

———. 2006. Breeding for resistance to abiotic stresses in rice: The value of quantitative trait loci. In *Plant breeding: The Arnel R. Hallauer International Symposium* (ed. K.R. Lamkey and M. Lee), pp. 201–212. Blackwell, Ames, Iowa.

Matsumoto T., Wing R.A., Han B., and Sasaki T. 2007. Rice genome sequence: The foundation for understanding the genetic systems. In *Rice functional genomics: Challenges, progress and prospects* (ed. N.M. Upadhyaya), pp. 5–21. Springer, New York.

McCallum C.M., Comai L., Greene E.A., and Henikoff S. 2000. Targeting induced local lesions IN genomes (TILLING) for plant functional genomics. *Plant Physiol.* 123: 439–442.

McNally K.L., Bruskiewich R., Mackill D., Leach J.E., Buell C.R., and Leung H. 2006. Sequencing multiple and diverse rice varieties: Connecting whole-genome variation with phenotypes. *Plant Physiol.* 141: 26–31.

Nakano M., Nobuta K., Vemaraju,K., Tej S.S., Skogen J.W., and Meyers B.C. 2006. Plant MPSS databases: Signature-based transcriptional resources for analyses of mRNA and small RNA. *Nucleic Acids Res.* 34: D731–D735.

Olsen K.M., Caicedo A.L., Polato N., McClung A., McCouch S., and Puruggan M. 2006. Selection under domestication: Evidence for a sweep in the rice *Waxy* genomic region. *Genetics* 173: 975–983.

Petkov P.M., Graber J.H., Churchill G.A., DiPetrillo K., King B.L. and Paigen K. 2005. Evidence of a large-scale functional organization of mammalian chromosomes. *PloS Genet.* 1: 312–322.

Price A.H. 2006. Believe it or not, QTLs are accurate. *Trends Plant Sci.* 11: 213–216.

Raghavan C., Naredo M.E.B., Wang H., Atienza G., Liu B., Qiu F., McNally K.L., and Leung H. 2007. Rapid method for detecting SNPs on agarose gels and its application in candidate gene mapping. *Mol. Breeding.* 19: 87–101.

The Rice Chromosome 10 Sequencing Consortium. 2003. In-depth view of structure, activity and evolution of rice chromosome 10. *Science* 300: 1566–1569.

The Rice Chromosome 3 Sequencing Consortium. 2005. Sequence, annotation, and analysis of synteny between rice chromosome 3 and diverse grass species. *Genome Res.* 15: 1284–1291.

Sasaki T., Matsumoto T., Yamamoto K., Sakata K., Baba T., Katayose Y., Wu J., Niimura Y., Cheng Z., Nagamura Y., et al. 2002. The genome sequence and structure of rice chromosome 1. *Nature* 420: 312–316.

Second G. 1982. Origin of the genetic diversity of cultivated rice (*Oryza sativa* L.): Study of the polymorphism scored at 40 isozyme loci. *Jpn. J. Genet.* 57: 25–57.

———. 1985. Evolutionary relationships in the *sativa* group of *Oryza* based on isozyme data. *Genet. Sel. Evol.* 17: 89–114.

Servin B. and Hospital F. 2002. Optimal positioning of markers to control genetic background in marker-assisted backcrossing. *J. Hered.* 93: 214–217.

Shen Y.J., Jiang H., Jin J.P., Zhang Z.B, Xi B., He Y.Y., Wang G., Wang C., Qian L., Li X., et al. 2004. Development of genome-wide DNA polymorphism database for map-based cloning of rice genes. *Plant Physiol.* 135: 1198–1205.

Spellman P.T. and Rubin G.M. 2002. Evidence for large domains of similarly expressed genes in the *Drosophila* genome. *J. Biol.* 1: 5.

Steele K.A., Price A.H., Shashidhar H.E., and Witcombe J.R. 2006. Marker-assisted selection to introgress rice QTLs controlling root traits into an Indian upland rice variety. *Theor. Appl. Genet.* 112: 208–221.

Sunkar R., Girke T., and Zhu J.K. 2005. Identification and characterization of endogenous small interfering RNAs from rice. *Nucleic Acids Res.* 14: 4443–4454.

Tang T., Lu J., Huang J., He J., McCouch S.R., Shen Y., Kai Z., Purugganan M.D., Shi S., and Wu C.I. 2006. Genomic variation in rice: Genesis of highly polymorphic linkage blocks during domestication. *PLoS Genet.* 2: 1824–1833.

Tabien R.E., Li Z., Paterson A.H., Marchetti M.A., Stansel J.W., and Pinson S.R.M. 2000. Mapping of four major rice blast resistance genes from 'Lemont' and 'Teqing' and evaluation of their combinatorial effect for field resistance. *Theor. Appl. Genet.* 101: 1215–1225.

Tanksley S.D., Young N.D., Paterson A.H., and Bonierbale M.W. 1989. RFLP mapping in plant breeding: New tools for an old science. *BioTechnology* 7: 257–264.

Vitte C., Ishii T., Lamy F., Brar D., and Panaud O. 2004. Genomic paleontology provides evidence for two distinct origins of Asian rice (*Oryza sativa* L.). *Mol. Gen. Genomics* 272: 504–511.

Watanabe Y. 1997. Phylogeny and geographical distribution of genus *Oryza*. In *Science of the rice plant* (ed. T. Matsuo et al.), pp. 29–39. Food and Agriculture Policy Research Center, Tokyo, Japan.

Whittaker J.C., Thompson R., and Visscher P.M. 1996. On the mapping of QTL by regression of phenotype on marker-type. *Heredity* 77: 23–32.

Williams E.J.B. and Bowles D.J. 2004. Coexpression of neighboring genes in *Arabidopsis thaliana*. *Genome Res.* 14: 1060–1067.

Wing R.A., Kim H.-R., Goicoechea J.L., Yu Y., Kudrna D., Zuccolo A., Ammiraju J.S.S., Luo M., Nelson W., Ma J., et al. 2007. The *Oryza* Map Alignment Project (OMAP): A new resource for comparative genome studies within *Oryza*. In *Rice functional genomics: Challenges, progress and prospects* (ed. N.M. Upadhyaya), pp. 408–421. Springer, New York.

Wu J.L., Sinha P.K., Variar M., Zheng K.L., Leach J.E., Courtois B., and Leung H. 2004. Association between molecular markers and blast resistance in an advanced backcross population of rice. *Theor. Appl. Genet.* 108: 1024–1032.

Wu J.L., Wu C., Lei C., Baraoidan M., Boredos A., Madamba R.S., Ramos-Pamplona M., Mauleon R., Portugal A., Ulat V., et al. 2005. Chemical- and irradiation-induced mutants of *indica* rice IR64 for forward and reverse genetics. *Plant Mol. Biol.* 59: 85–97.

Xu K., Xia X., Fukao T., Canlas P., Maghirang-Rodriguez R., Heuer S., Ismail A.I., Bailey-Serres J., Ronald P.C., and Mackill D.J. 2006. *Sub1A* is an ethylene response factor-like gene that confers submergence tolerance to rice. *Nature* 442: 705–708.

Yu J., Hu S., Wang J., Wong G.K., Li S., Liu B., Deng Y., Dai L., Zhou Y., Zhang X., et al. 2002. A draft sequence of the rice genome (*Oryza sativa* L. ssp. *indica*). *Science* **296:** 79–92.

Yu J., Wang J., Lin W., Li S., Li H., Zhou J., Ni P., Dong W., Hu S., Zeng C., et al. 2005. The genomes of *Oryza sativa*: A history of duplications. *PLoS Biol.* **3:** e38.

Zhan S., Horrocks J., and Lukens L.N. 2006. Islands of co-expressed neighboring genes in *Arabidopsis thaliana* suggest high-order chromosome domains. *Plant J.* **45:** 347–357.

Zhu Q. and Ge S. 2005. Phylogenetic relationships among A-genome species of the genus *Oryza* revealed by intron sequences of four nuclear genes. *New Phytol.* **167:** 249–265.

WWW RESOURCES

http://bioinformatics.med.yale.edu/rc/overview.jspx Yale Virtual Center for Cellular Expression Profiling of Rice. T. Nelson and X.-W. Deng, Yale Department of Molecular, Cellular & Developmental Biology (MCDB); H. Zhao, Department of Epidemiology & Public Health, Yale School of Medicine

http://orygenesdb.cirad.fr OryGenesDB, an interactive tool for rice reverse genetics. Centre de coopération internationale en recherche agronomique pour le développement

http://rapdb.lab.nig.ac.jp/ Rice Annotation Project DataBase. Center for Information Biology and DNA Data Bank of Japan, National Institute of Genetics, National Institute of Agrobiological Sciences, National Institute of Advanced Industrial Science and Technology, Japan Biological Informatics Consortium

www.generationcp.org Generation Challenge Programme, Science for Better Crops

www.irgcis.irri.org:81/grc/irgcishome.html

www.iris.irri.org International Rice Information System (IRIS) is the rice implementation of the International Crop Information System (ICIS) which is a database system that provides integrated management of global information on genetic resources and crop cultivars. This includes germplasm pedigrees, field evaluations, structural and functional genomic data (including links to external plant databases) and environmental (GIS) data.

www.tigr.org/tdb/e2k1/osa1/ TIGR (The Institute for Genomic Research) Rice Genome Annotation Database

www.omap.org The Oryza Map Alignment Project. Principal investigators R.A. Wing, University of Arizona, Arizona Genomics Institute; S.A. Jackson, Purdue University; L.D. Stein, Cold Spring Harbor Laboratory; C. Soderlund, Unviersity of Arizona, Arizona Genomics Computation Laboratory

28 | The Mouse

Claire M. Wade and Mark J. Daly

*Center for Human Genetic Research, Massachusetts General Hospital, Boston, Massachusetts 02114;
Broad Institute of MIT and Harvard, Cambridge, Massachusetts 02142*

INTRODUCTION

Inbred mammals represent a subset of sequenced organisms which have played a special role in our understanding of genetics and genomics. The inbred mouse, in particular, has been a popular subject of mammalian research for many years because mice are inexpensive, are relatively easy to house and breed, and can acclimatize to human touch. Additionally, well-described breeding strategies and methods for genome manipulation are available. The Jackson Laboratory in Bar Harbor, Maine, alone provides researchers

with mice from over 2,700 strains, a testament to the popularity of this species. With the advent of the sequencing of the mouse genome in 2002, a vast array of genomic resources has been developed to assist the mouse research community. Whereas directed crosses and quantitative trait locus (QTL) mapping are standard tools in the mouse geneticist's arsenal, it has recently been suggested that the special genetic architecture of the inbred laboratory mouse strains might make further strides in gene mapping possible using single-nucleotide polymorphism (SNP) haplotypes. In this chapter, we discuss the implications of mouse history and mouse genomic resources for these newer approaches to gene mapping in the mouse.

Effect of Mouse History on the Inbred Laboratory Mouse Genome

The common house mouse, *Mus musculus*, originated in what is today India and Southeast Asia (Silver 1995) and, in modern times, opportunistically expanded with human migrations and agricultural practice. Although most, if not all, house mice are of the species *Mus musculus*, this species is itself divided into a number of subspecies that have arisen in different geographical regions of the world. The most important subspecies from a scientific point of view are *Mus musculus musculus*, which is dispersed through Eastern Europe and Asia; *Mus musculus domesticus*, which is distributed throughout Western Europe, and as a result of colonization by Western

353

Europeans, the Americas and Australia; *Mus musculus castaneus* from Southeast Asia; and *Mus musculus molossinus* from Japan, although other subspecies (e.g., *Mus musculus bactrianus*) are known to exist.

The mice that later became the inbred mouse lines originated as hybrids developed as pet mice and show mice (more commonly referred to as "fancy" mice because of their interesting coat colors and behaviors). Such mice were first kept by wealthy individuals in both Europe and Asia, who bred them for diversion. People who enjoyed breeding and keeping these mice strove to create new and different phenotypes. As people began to travel the world, the mice were exchanged, and this enabled the different subspecies of *Mus musculus* to be bred together.

The inbred mouse strains commonly used in scientific research have varied origins. Many of the so-called classic inbred laboratory mouse strains originated in the laboratories of Clarence Cook Little and William Castle, who derived their mouse stocks predominantly from a single fancy mouse breeder, Abbie Lathrop (Silver 1995). Nearly all laboratory mice are inbred. This provides exceptional genetic stability within an inbred line and omits any requirement for chromosomal phasing within the parental strains, since all loci are homozygous. Most importantly, completely homozygous genomes provide a renewable animal resource for experimentation such that replication within experiments and across laboratories and time can be performed. Other major inbred strains not derived from the stocks of Castle and Little include the Swiss-derived mouse strains and the wild-derived inbred mouse strains. Wild-derived inbred mouse strains may be representatives of the different *Mus musculus* subspecies (particularly *domesticus*, *musculus*, and *molossinus*), or they may be other closely related *Mus* species such as *Mus spretus*. These mice have much higher levels of divergence than that seen among the Swiss mice and the Castle–Little mice (Beck et al. 2000). In general, they should be thought of as representing a single copy of each chromosome selected from an outbred population.

Observations of Genomic Variation

The distribution of polymorphisms in comparisons between pairs of mouse strains is not random across the genome. Rather, SNPs are "clumped" into regions of high diversity and low diversity. This has been documented statistically by Lindblad-Toh et al. (2000), who observed that SNPs occurring in expressed sequence tags (ESTs) did not follow the expected Poisson distribution. There were significant numbers of ESTs with either fewer or more SNPs than expected, given the number of bases examined.

As part of the International Mouse Genome Sequencing Consortium mouse genome sequencing project, SNP discovery was conducted in a handful of inbred laboratory mice (129S1/SvImJ, BALBc/ByJ, C3H/HeJ) by comparing light whole-genome shotgun sequences (50,000 reads) of these strains against the fully sequenced strain (C57BL6/J). The findings of Lindblad-Toh et al. (2000) were confirmed in the analysis of the genomic patterning of the ascertained SNP (Wade et al. 2002). At the same time, Celera Discovery Corporation was sequencing the mouse by using a light coverage (1.5×) of four mouse strains (129X1/SvJ, DBA/2J, A/J, C57BL6/J) plus a yet lighter coverage of 129S1/SvImJ (Mural et al. 2002). Richard Mural (pers. comm.) noticed an equivalent uneven dispersion of SNPs among the strains. It was proposed by Wade et al. (2002) and earlier researchers such as Bonhomme et al. (1987) that this uneven dispersal of SNPs was due to the hybrid history of the mouse strains. A region of low diversity observed in a pair of strains would indicate a common ancestral history in that region, whereas a region of high diversity would indicate a disparate subspecific ancestry at the locus in question. The predominant ancestral source in the strain was *Mus musculus domesticus*, and the most common second ancestral type was *Mus musculus musculus*.

Copy number polymorphisms and insertion/deletion events have not yet been extensively studied in the mouse. The most extensive published analyses include those using comparative genome hybridization techniques to study both copy number alterations between strains and insertion/deletion events (Chung et al. 2004; Adams et al. 2005; Snijders et al. 2005). Some of these studies have detected large discrepancies in copy number between strains. In particular, regions on chromosome 7 and chromosome 14 have been cited, although some of these studies have analyzed the same cell lines, rather than live mice, and it is possible that some results may be cell-line passage events rather than found in all members of that strain. Insertion/deletion events on a smaller scale have been studied over 26 kb (Ideraabdullah et al. 2004), with the conclusions that insertion/deletion events among strains tend to be small (40% were single base events) and their origin appears to be inter-subspecific (i.e., between subspecies) rather than intra-subspecific.

Impact of Mouse Comparative Genomic Organization on Trait Mapping Practices

Genes with significant influence on quantitative traits (i.e., traits with continuous rather than all-or-none phenotypes, such as obesity, and predisposition to common health problems such as heart disease or diabetes) have traditionally been difficult to identify. In the past, mapping of such traits has been carried out by identifying lines or families of individuals with disparate predisposition to the phenotype and then intercrossing F_2 progeny to isolate a genomic segment harboring factors influencing the trait, using microsatellites or SNP markers at low density. Although this is frequently successful at associating a marker with the phenotype, the low levels of recombination over the few generations of breeding typically make it difficult to narrow the intervals of association to less than several megabases.

The patterns of SNP variation in the mouse suggest the possibility of using these polymorphisms for the mapping of QTL. The mouse genetics community until recently has remained relatively unconvinced that there exists a direct correlation between strain phenotype and inter-strain SNP variation (first proposed by Grupe et al. 2001). The reluctance to embrace "in silico mapping" is due to valid concerns about the potential statistical power (Chesler et al. 2001) and the limited number of phenotyped strains available for comparison (Pletcher et al. 2004), which might make the resolution of QTLs problematic and which limit the power of detection to cases where the genetic variation explains a meaningful proportion of the overall phenotypic variance. Some researchers have not wanted to accept methods that suggest the exclusion of regions of concordant ancestry from further consideration in the analysis (Yalcin et al. 2004), as occasional sequence differences can be found between otherwise concordant strains. Such methods have been proposed by Wade and Daly (2005) and Cervino et al. (2006), because the overwhelming majority of variation will be found in easily recognizable segments covering only about 50% of any genomic interval. Researchers using in silico mapping methods have reported good success with the mapping of Mendelian loci. However, it is nearly certain that few alleles in complex phenotypes will explain enough of the phenotypic variance for a limited panel of strains (such as the 50 mouse Haplotype Map strains) to identify a cause under a genome-wide association testing paradigm. This problem is magnified because there are few statistical methods that effectively test the strength of association between known phenotypes and SNP haplotypes while allowing for the dependencies among adjacent markers and the relationships between the strains themselves. Because many of the inbred laboratory strains share common origins, one cannot assume that all strains are equally unrelated on a genome-wide basis (Fig. 28-1). These problems can be substantially reduced in many cases by examining polymorphisms only within a genomic region already defined by a known QTL. This greatly reduces the analytical testing burden, thereby increasing the power of the study to detect a significant association. The strain interrelationship issue can largely be resolved by phenotyping/analyzing strains that have a relatively long inter-strain phylogenetic branch length and only analyzing one strain from a group known to share close ancestry (such as the C57/C58 groups of strains). Examples of recent success with this paradigm are described below.

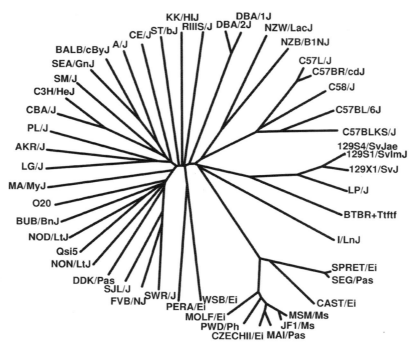

FIGURE 28-1. Genome-wide phylogeny of 49 mouse strains genotyped for the Mouse Haplotype Map project, showing clustering of closely related strains. Branch lengths are exaggerated among the inbred strains relative to the wild-derived strains, because SNPs are ascertained for polymorphism among the classic strains.

Alternative Mouse Mapping Resources

Fortunately for geneticists working with laboratory mice, there are many resources to aid in the quest for the often elusive genetic contributions to complex phenotypes. The International Mouse Genome Sequencing Consortium draft mouse genome sequence was released in February 2002, and the accompanying paper was published in December of the same year (Waterston et al. 2002). Celera Discovery Systems released their independent analysis of mouse chromosome 16 earlier that year (Mural et al. 2002). Since that time, the mouse genome has been taken to the final stages of genome finishing. This means that all known gaps in the assembly are filled using a variety of more costly but accurate techniques. The paper to accompany the mouse genome finishing process is now in preparation.

During the mouse genome sequencing project, close to 400,000 SNPs were submitted to dbSNP(http://www.ncbi.nlm.nih.gov/entrez/query.fcgi?CMD=search&DB=SNP&term=txid10090%5Borgn%5D). Soon thereafter, SNPs were added by several other groups, including, but not limited to, Celera Discovery Systems, Roche Biosciences, and The Sanger Institute. Recently, the number of SNPs in the public domain has grown enormously to 6.5 million, predominantly due to recent strain resequencing undertaken by Perlegen Sciences as a research contract with the National Institute of Environmental and Health Sciences (K. Frazer, pers. comm.). This sequencing effort involved the tiling of 15 inbred mouse strains on microarrays. The sequence included on the arrays represents the nonrepetitive portions of the mouse genome. The strains included were 129S1/SvImJ, A/J, AKR/J, BALBc/ByJ, BTBR T+tf/J, C3H/HeJ, DBA/2J FVB/NJ, KK/HIJ, NZW/LacJ, NOD/LtJ, CAST/EiJ, MOLF/EiJ, PWD/PhJ, and WSB/EiJ. The last four of these (wild-derived inbreds) were included as representatives of putative ancestral strains to study mouse ancestry, as well as being strains used in experimentation.

The availability of large numbers of SNPs in the public domain has encouraged significant genotyping efforts internationally. One focus of this genotyping effort has been to characterize the

TABLE 28-1. Mouse Strains Included in Haplotype Map (www.broad.mit.edu/personal/mjdaly/mousehapmap.html)

Inbred laboratory strains		Wild-derived strains
129S1/SvImJ	I/LnJ	*Mus m. castaneus*
129X1/SvJ	KK/HlJ	CAST/Ei
A/J	LG/J	
AKR/J	LP/J	*Mus m. musculus*
BALB/cByJ	MA/MyJ	CZECHII/Ei
BTBR+Ttftf	NOD/LtJ	PWD/Ph
BUB/BnJ	NON/LtJ	
C3H/HeJ	NZB/B1NJ	*Mus m. molossinus*
C57BL/6J	NZW/LacJ	JF1/Ms
C57BLKS/J	O20	MAI/Pas
C57BR/cdJ	PL/J	MOLF/Ei
C57L/J	Qsi5	MSM/Ms
C58/J	RIIIS/J	
CBA/J	SEA/GnJ	*Mus m. domesticus*
CE/J	SJL/J	PERA/Ei
DBA/1J	SM/J	WSB/Ei
DBA/2J	ST/bJ	
DDK/Pas	SWR/J	*Mus spretus*
FVB/NJ		SEG/Pas
		SPRET/Ei

genomes of the commonly used inbred laboratory mouse strains. The Genomics Institute of the Novartis Research Foundation released genotypes for 10,990 polymorphic sites in 48 strains to collaborators in the mouse community (Pletcher et al. 2004). The genotyped strains were the high-priority mouse strains defined by The Jackson Laboratories in consultation with the mouse community for their Mouse Phenome Database (Bogue 2003; Bogue and Grubb 2004). In 2005, the Wellcome Trust released a large set of genotypes (see below) that included many inbred laboratory strains. During 2006, The Broad Institute of Harvard and MIT released genotypes for 49 inbred mouse strains over 148,000 polymorphic sites segregating among the classic (as opposed to wild-derived) inbred strains (www.broad.mit.edu/personal/mjdaly/mousehapmap.html). The resequencing efforts of Perlegen Sciences resulted in the identification of potential polymorphic sites in their represented strain group, and simultaneously provided genotypes for the smaller set of strains examined in that experiment. Strains genotyped for the mouse haplotype map are summarized in Table 28-1.

Genotypes for a large number of recombinant inbred mouse strains have been publicly distributed through the Wellcome-CTC Mouse Strain SNP Genotype Set Web site (http://www.well.ox.ac.uk/ mouse/INBREDS). The 480 strains that have been genotyped for their 13,377 SNP set include the recombinant inbred strain panels; 2300 Heterogeneous Stock mice; and many laboratory inbred strains and wild-derived inbred strains. The Heterogeneous Stock is derived from later generations of intercrossed F_2 mice from multiple pairs of contributor strains.

In part to overcome limitations with respect to the number and genetic diversity of current inbred strains, important new mouse resources are being developed which will synergize well with these genetic resources. Mouse chromosome substitution strains have recently been developed for strains A/J and 129S1/SvImJ on a C57BL6/J background. The A/J chromosome substitution strains are available (Nadeau et al. 2000; Singer et al. 2004), but the 129 set of strains is still in development. These strains typically consist of a C57BL6/J genetic background, with a single chromosome included from the substitution strain. Such strains enable the dissection of complex traits on a chromosome-by-chromosome basis (Petryshen et al. 2005). Their development requires many generations of breeding and genotyping of crossbred mice.

The Complex Trait Consortium is an organization designed to promote the development of resources that can be used to understand human diseases. As part of this initiative, the group is developing "The Collaborative Cross" in the mouse (Churchill et al. 2004). This is intended to

TABLE 28-2. Mouse World Wide Web Resources

Type of resource	Strains included	URL	Principal investigator
Strain genotypes—148,000 loci plus assay	50 inbred laboratory strains	http://www.broad.mit.edu/personal/mjdaly/mousehapmap.html	Mark Daly
Strain genotypes ~10,990 loci; also mapping methods	recombinant inbred strains plus laboratory inbred strains	http://www.well.ox.ac.uk/mouse/	Richard Mott
Strain genotypes—high density	15 inbred laboratory strains	http://mouse.perlegen.com	Kelly Frazer
Strain and mapping resources	most commercial strains	http://www.jax.org	Ken Paigen
SNP database	all genotyped to date	http://www.ncbi.nlm.nih.gov/SNP/	National Institutes of Health
Mouse Genome Resources	C57BL/6J	http://www.ncbi.nlm.nih.gov/genome/guide/mouse/	National Institutes of Health
Phenotype database	50 common strains	http://www.jax.org/phenome	Molly Bogue
Genome browser	C57BL/6J	http://genome.ucsc.edu	Jim Kent
Genome browser	C57BL/6J	http://www.ensembl.org	Ewan Birney
Genome browser and mapping tools	all strains	http://www.informatics.jax.org	Jackson Laboratory
Complex trait mapping resources	all laboratory strains	http://www.complextrait.org	Complex Trait Consortium
Identity by descent mapping	genotyped strains	http://mouseibd.florida.scripps.edu	Allesandra Cervino
Association mapping	phenome strains	http://snp.ucsd.edu/mouse	Eleazar Eskin

provide a common reference panel of diverse individuals for the integrative analysis of complex diseases in mammals. The consortium aims to produce a large common set of genetically defined mice. Approximately 1000 strains should be available when strain generation is complete. The mice will be genotyped once, and the individual lines will be maintained as stocks that can be purchased and phenotyped by mouse genetic researchers.

A final critical part of these resources is the mouse phenome database project being curated by The Jackson Laboratories in Bar Harbor, Maine (Bogue 2003; Bogue and Grubb 2004). This project involves the standardized phenotyping of a large number of laboratory inbred mice. The results of the phenotyping experiments are shared by the mouse community. Data are available not only for commonly used inbred strains, but also for a number of recombinant inbred mouse strains. Mouse trait mapping resources available to the public through the World Wide Web are summarized in Table 28-2.

Approaches to Planning Mouse Complex Trait Mapping Experiments

The approach to trait mapping taken by the researcher depends on the quantity of existing information on the trait and the expected complexity of the genetic architecture of the trait to be mapped.

Case 1: Known QTL

Where QTLs have been detected previously by linkage, an in silico mapping method might be used to narrow the interval of association. To do this, the researcher should access all available genotypes for the strains in which the QTL was discovered. This process is greatly enhanced if the same QTL has been observed in multiple crosses. Within the interval, the researcher would then identify regions where the strains that are expected to differ for the trait have different ancestral haplotypes (i.e., are genetically diverged). The relatively uncontroversial assumption is that one might further evaluate those genes known to bear genetic differences between the mapping strains. For one pair of strains, this process alone could reduce the interval by up to 50–60% (Wade et al. 2002). The density of genotypes available within the region of initial linkage will be influenced heavily by the strains used for the initial QTL mapping. Strains represented in the Perlegen resequencing data will have the greatest density of coverage, followed by the 48 high-priority strains.

Once the narrowest possible set of intervals has been established by this process, the researcher might resequence exons and non-exonic sites within the regions that have high mammalian sequence conservation. Alternatively, the researcher might perform expression analyses (Mehrabian et al. 2005; Drake et al. 2006) to identify genes within the interval for which an expression difference exists between the mapping strains.

Should no genes within the greater interval be identified by these methods, the trait being mapped may be a recent mutation on a common ancestral background or may be caused by insertion/deletion polymorphisms. Recent studies (see, e.g., McCarroll et al. 2006) have determined that insertion/deletion polymorphisms commonly segregate in high-linkage disequilibrium with ancestral haplotypes. Thus, this type of polymorphism would very likely be embedded in an interval in which the SNP variation suggested many differences. When there is haplotype similarity over the association interval, it might be useful to phenotype other strains with existing dense genotyping data to identify the endpoints for the region of association. Across strains, the length of linkage disequilibrium is approximately 100 kb (Frazer et al. 2004), relatively shorter than between strain pairs (closer to 1 Mb of disequilibrium). Several recent studies suggest that the use of in silico association following QTL mapping may be an effective positional cloning technique. Liu et al. (2006) have identified a novel lung cancer susceptibility gene via association to a coding SNP across 21 strains. Zheng et al. (2006) have identified a 1.3-Mb haplotype unique to strains with severe doxorubicin-induced nephropathy, embedded in a significantly linked QTL. This greatly reduces the burden of following up the linkage result.

Case 2: No Preexisting Information

Where there are no significant known regions of linkage for the trait, a first approach might be to phenotype multiple representatives of strains for which dense genotype data exist. Where possible, wild-derived inbred strains should be avoided, because they do not offer the same mosaic pattern of limited variation shared by the inbred laboratory strains from the laboratories of Castle and Little and the Swiss mice. Within the classic inbred strains, it may be preferable to include only one representative from a close family (such as the C57, C58 mouse families), since these mice are often genetically and phenotypically similar. Once a phenotyping scan has identified mice that have divergent phenotypes in the trait, it might be possible to employ a combination of candidate gene analysis, in silico mapping, and expression analyses on these strains to identify intervals for further study.

SUMMARY

Although the number of mammals being sequenced continues to rise, few offer the same advantages for genetic research as the laboratory mouse. The large number of resources available to the mouse research community provides an unsurpassed opportunity to explore functional mammalian variation. Methods are available to better exploit existing genetic mapping results, and to locate new intervals of influence on complex traits. The most effective methods for identifying regions of significant association will employ a combination of techniques, including haplotype analysis and expression analysis.

REFERENCES

Adams D.J., Dermitzakis E.T., Cox T., Smith J., Davies R., Banerjee R., Bonfield J., Mullikin J.C., Chung Y.J., Rogers J., and Bradley A. 2005. Complex haplotypes, copy number polymorphisms and coding variation in two recently divergent mouse strains. *Nat. Genet.* **37:** 532–536.

Beck J.A., Lloyd S., Hafezparast M., Lennon-Pierce M., Eppig J.T., Festing M.F., and Fisher E.M. 2000. Genealogies of mouse inbred strains. *Nat. Genet.* **24:** 23–25.

Bogue M. 2003. Mouse Phenome Project: Understanding human biology through mouse genetics and genomics. *J. Appl. Physiol.* **95:** 1335–1337.

Bogue M.A. and Grubb S.C. 2004. The Mouse Phenome Project. *Genetica* **122:** 71–74.

Bonhomme F., Guenet J.-L., Dod B., Moriwaki K., and Bulfield G. 1987. The polyphyletic origin of laboratory inbred mice and their rate of evolution. *J. Linn. Soc.* **30:** 51–58.

Cervino A.C., Gosink M., Fallahi M., Pascal B., Mader C., and Tsinoremas N.F. 2006. A comprehensive mouse IBD database for the efficient localization of quantitative trait loci. *Mamm. Genome* **17:** 565–574.

Chesler E.J., Rodriguez-Zas S.L., and Mogil J.S. 2001. In silico mapping of mouse quantitative trait loci. *Science* **294:** 2423.

Chung Y.J., Jonkers J., Kitson H., Fiegler H., Humphray S., Scott C., Hunt S., Yu Y., Nishijima I., Velds A., et al. 2004. A whole-genome mouse BAC microarray with 1-Mb resolution for analysis of DNA copy number changes by array comparative genomic hybridization. *Genome Res.* **14:** 188–196.

Churchill G.A., Airey D.C., Allayee H., Angel J.M., Attie A.D., Beatty J., Beavis W.D., Belknap J.K., Bennett B., Berrettini W., et al. 2004. The Collaborative Cross, a community resource for the genetic analysis of complex traits. *Nat. Genet.* **36:** 1133–1137.

Drake T.A., Schadt E.E., and Lusis A.J. 2006. Integrating genetic and gene expression data: Application to cardiovascular and metabolic traits in mice. *Mamm. Genome* **17:** 466–479.

Frazer K.A., Wade C.M., Hinds D.A., Patil N., Cox D.R., and Daly M.J. 2004. Segmental phylogenetic relationships of inbred mouse strains revealed by fine-scale analysis of sequence variation across 4.6 Mb of mouse genome. *Genome Res.* **14:** 1493–1500.

Grupe A., Germer S., Usuka J., Aud D., Belknap J.K., Klein R.F., Ahluwalia M.K., Higuchi R., and Peltz G. 2001. In silico mapping of complex disease-related traits in mice. *Science* **292:** 1915–1918.

Ideraabdullah F.Y., de la Casa-Esperon E., Bell T.A., Detwiler D.A., Magnuson T., Sapienza C., and de Villena F.P. 2004. Genetic and haplotype diversity among wild-derived mouse inbred strains. *Genome Res.* **14:** 1880–1887.

Lindblad-Toh K., Winchester E., Daly M.J., Wang D.G., Hirschhorn J.N., Laviolette J.P., Ardlie K., Reich D.E., Robinson E., Sklar P., et al. 2000. Large-scale discovery and genotyping of single-nucleotide polymorphisms in the mouse. *Nat. Genet.* **24:** 381–386.

Liu P., Wang Y., Vikis H., Maciag A., Wang D., Lu Y., Liu Y., and You M. 2006. Candidate lung tumor susceptibility genes identified through whole-genome association analyses in inbred mice. *Nat. Genet.* **38:** 888–895.

McCarroll S.A., Hadnott T.N., Perry G.H., Sabeti P.C., Zody M.C., Barrett J.C., Dallaire S., Gabriel S.B., Lee C., Daly M.J., and Altshuler D.M. (International HapMap Consortium). 2006. Common deletion polymorphisms in the human genome. *Nat. Genet.* **38:** 86–92.

Mehrabian M., Allayee H., Stockton J., Lum P.Y., Drake T.A., Castellani L.W., Suh M., Armour C., Edwards S., Lamb J., et al. 2005. Integrating genotypic and expression data in a segregating mouse population to identify 5-lipoxygenase as a susceptibility gene for obesity and bone traits. *Nat. Genet.* **37:** 1224–1233.

Mural R.J., Adams M.D., Myers E.W., Smith H.O., Miklos G.L., Wides R., Halpern A., Li P.W., Sutton G.G., Nadeau J., et al. 2002. A comparison of whole-genome shotgun-derived mouse chromosome 16 and the human genome. *Science* **296:** 1661–1671.

Nadeau J.H., Singer J.B., Matin A., and Lander E.S. 2000. Analysing complex genetic traits with chromosome substitution strains. *Nat. Genet.* **24:** 221–225.

Petryshen T.L., Kirby A., Hammer R.P., Jr., Purcell S., O'Leary S.B., Singer J.B., Hill A.E., Nadeau J.H., Daly M.J., and Sklar P. 2005. Two quantitative trait loci for prepulse inhibition of startle identified on mouse chromosome 16 using chromosome substitution strains. *Genetics* **171:** 1895–1904.

Pletcher M.T., McClurg P., Batalov S., Su A.I., Barnes S.W., Lagler E., Korstanje R., Wang X., Nusskern D., Bogue M.A., et al. 2004. Use of a dense single nucleotide polymorphism map for in silico mapping in the mouse. *PLoS Biol.* **2:** e393.

Silver L.M. 1995. *Mouse genetics: Concepts and applications.* Oxford University Press, New York.

Singer J.B., Hill A.E., Burrage L.C., Olszens K.R., Song J., Justice M., O'Brien W.E., Conti D.V., Witte J.S., Lander E.S., and Nadeau J.H. 2004. Genetic dissection of complex traits with chromosome substitution strains of mice. *Science* **304:** 445–448.

Snijders A.M., Nowak N.J., Huey B., Fridlyand J., Law S., Conroy J., Tokuyasu T., Demir K., Chiu R., Mao J.H., et al. 2005. Mapping segmental and sequence variations among laboratory mice using BAC array CGH. *Genome Res.* **15:** 302–311.

Wade C.M. and Daly M.J. 2005. Genetic variation in laboratory mice. *Nat. Genet.* **37:** 1175–1180.

Wade C.M., Kulbokas E.J., III, Kirby A.W., Zody M.C., Mullikin J.C., Lander E.S., Lindblad-Toh K., and Daly M.J. 2002. The mosaic structure of variation in the laboratory mouse genome. *Nature* **420:** 574–578.

Waterston R.H., Lindblad-Toh K., Birney E., Rogers J., Abril J.F., Agarwal P., Agarwala R., Ainscough R., Alexandersson M., An P., et al. (Mouse Genome Sequencing Consortium) 2002. Initial sequencing and comparative analysis of the mouse genome. *Nature* **420:** 520–562.

Yalcin B., Fullerton J., Miller S., Keays D.A., Brady S., Bhomra A., Jefferson A., Volpi E., Copley R.R., Flint J., and Mott R. 2004. Unexpected complexity in the haplotypes of commonly used inbred strains of laboratory mice. *Proc. Natl. Acad. Sci.* **101:** 9734–9739.

Zheng Z., Pavlidis P., Chua S., D'Agati V.D., and Gharavi A.G. 2006. An ancestral haplotype defines susceptibility to doxorubicin nephropathy in the laboratory mouse. *J. Am. Soc. Nephrol.* **17:** 1796–800.

29 | The Rat

Edwin Cuppen,[1] Norbert Hübner,[2] Howard J. Jacob,[3] and Anne E. Kwitek[3]

[1]*Hubrecht Laboratory, Utrecht, The Netherlands;* [2]*Max-Delbrück-Center for Molecular Medicine, Berlin-Buch, Germany;* [3]*Medical College of Wisconsin, Milwaukee, Wisconsin 53226*

INTRODUCTION

The laboratory rat (*Rattus norvegicus*) is one of the most extensively studied model organisms in fields such as physiology, toxicology, and neurobiology. It is also the major animal model in the initial stages of drug development. The power of the laboratory rat is the biological characterization of the over 500 strains (http://rgd.mcw.edu/strains/), most of which were developed as models for complex, common diseases. Although the rat is primarily known as a physiological and neurobiological model, there has been a steady increase in the use of the rat for genomic and genetic studies over the last decades. Given the need to annotate the human genome with function, linking the rat into this process via its own genome project, in combination with genetic and genomics approaches, is a logical and necessary requirement for accelerating improvements in health care.

RESOURCES

Genetic and Genomic Data

In 1987, there were ten identified rat linkage groups and four named chromosomes, constructed with 39 phenotypes (coat color, eye color, growth, tumors, teeth, etc.) and 33 electrophoretic and coat color markers (Robinson 1987). A major contribution to the genetic mapping revolution in the human genome was Weber and May's use of simple sequence length polymorphisms (SSLPs), also known as CA-repeats and microsatellites (Weber and May 1989). This new class of genetic markers was also deployed in the rat. Since then, the rat genome project has yielded a tremendous wealth of genomic resources, including genetic maps; radiation hybrid (RH) cell lines and the associated RH maps (over 5,000 genetic markers and 19,500 genes and expressed sequence tags [ESTs] mapped); cDNA libraries generating more than 683,500 ESTs (with more being generated) clustered into over 40,000 UniGenes; over 10,033 genetic markers; and a published draft (~6.8×) sequence of the genome based on the inbred BN/NHsdMcwi (Brown Norway) strain (Gibbs et al. 2004). The novel sequencing strategy combined whole-genome shotgun (WGS) with bacterial artificial chromosome (BAC) sequencing and covered 90% of the rat genome. Moreover, Celera has released a 1.5× draft sequence from the Sprague-Dawley rat (Kaiser 2005). The sequenced rat genome is estimated to be 2.75 Gb, distributed across 21 of the 22 chromosomes (the Y chromosome is not yet complete), and is predicted to encode approximately 20,973 genes, with 28,516 transcripts and 205,623 exons (Gibbs et al. 2004). The exact number of genes and transcripts will take several more years to resolve, but the bulk of the data is available for investigators to use now. Because of the success of the rat sequencing project and the value of the rat for functional genomics, the Mammalian Gene Collection (Gerhard et al. 2004) (full-length cDNA project) decided to sequence 6,000 full-length genes from the same BN strain that was sequenced, with over 4,500 nonredundant genes completed. Most of these resources are publicly available through NCBI, the Rat Genome Database (RGD), RatMap, UCSC, Ensembl, and other genome databases (Table 29-1).

TABLE 29-1. List of Major Rat Resources

Database	Data type	References	URL
Rat Genome Database (RGD)	several	Twigger et al. 2005	http://rgd.mcw.edu
RatMap	several	Petersen et al. 2005	http://ratmap.gen.gu.se
NCBI Rat Genome Resources	several		http://www.ncbi.nlm.nih.gov/ genome/guide/rat/index.html
UCSC Rat Browser	several	Karolchik et al. 2003	http://genome.brc.mcw.edu/
Ensembl Rat Browser	several	Hubbard et al. 2005	http://www.ensembl.org/Rattus_norvegicus/ index.html
Baylor College of Medicine Rat Resources	sequence, contigs, assembly	Gibbs et al. 2004	http://www.hgsc.bcm.tmc.edu/projects/rat/
Rat EST Project at the University of Iowa	rat ESTs	Scheetz et al. 2004	http://ratest.uiowa.edu/
PhysGen Program for Genomics Applications (PGA)	strains, phenotypes, genotypes	Jacob and Kwitek 2002	http://pga.mcw.edu
National Bio Resource Project Japan	strains, phenotypes, genotypes	Mashimo et al. 2005	http://www.anim.med.kyoto-u.ac.jp/nbr/
TIGR Program for Genomics Applications (TREX)	microarray		http://pga.tigr.org/
NIAMS: ARB Rat Genetic Database	strains, maps, markers	Dracheva et al. 2000	http://www.niams.nih.gov/rtbc/ratgbase/
Wellcome Trust Centre: Rat Mapping Resources	maps, markers	Wilder et al. 2004	http://www.well.ox.ac.uk/rat_mapping_resources/
RRRC: Rat Resource and Research Center	strains		http://www.nrrrc.missouri.edu/
TIGR Gene Index	genes	Lee et al. 2005	http://www.tigr.org/tigr-scripts/tgi/T_index.cgi? species=rat
ECR Comparative Genome Browser	comparative genomics	Ovcharenko et al. 2004	http://ecrbrowser.dcode.org/index.php?db=rn3
VISTA: Comparative Sequence Alignment Browser	comparative genomics	Frazer et al. 2004a	http://pipeline.lbl.gov/cgi-bin/gateway2? bg=rn3&selector=vista
LONI Rat Atlas Image Database	anatomy	Toga et al. 1995	http://www.loni.ucla.edu/Research/Atlases/ RatAtlas.html
EU Rat SNP and haplotype Map Project (STAR)	strains, genotypes		http://www.snp-star.eu
Rat candidate SNP database	strains, genotypes	Guryev et al. 2005	http://cascad.niob.knaw.nl/
European Rat Tools for Functional Genomics (Euratools)	strains, phenotypes, genotypes		http://euratools.csc.mrc.ac.uk

The data from the rat genome sequence provide researchers with a precise knowledge of the rat gene content, essential for the advance of biomedical research. The data also improve physical and genetic map resolution, since chromosomal position no longer depends on recombination rates and statistical analysis. However, it must be noted that the rat sequence is a draft sequence, and that other lines of evidence may be required, i.e., genetic linkage analysis or other forms of mapping, to ensure that local regions of the genome under investigation have been assembled correctly. The genomic toolbox is now nearly complete and, as outlined in this review, is having an important impact on research using the rat.

Phenotypic Data

Many rat strains have been selectively bred for multifactorial disease (polygenic with environmental influence) and then bred to isogeneity. Figure 29-1 shows the phylogenetic relationship of rat strains. Currently, there are 1,015 rat strains in the RGD, over 50% of which are inbred strains for complex traits (538 strains). In these strains, there are 168 different diseases and 393 phenotypes, as defined by RGD's strain disease and phenotype ontologies. These include arthritis, cancer, hypertension, multiple sclerosis (MS), and seizures. In some cases, there are multiple inbred strains for a single multifactorial disease. For example, five different rat strains (BUF, DA,

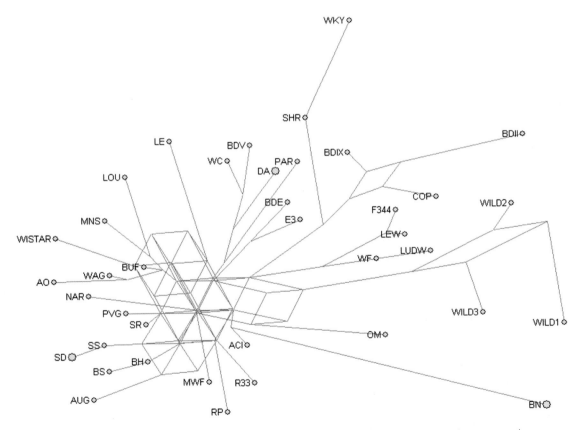

FIGURE 29-1. Phylogenetic relationship for rat strains. Rat strain relationships presented in a network structure. Because many strains are highly related and there are potential complex gene flows, ancestral nodes cannot be determined with high confidence. Thus, the data are not well-suited to traditional phylogenetic trees. The diagram is based on 861 SNP markers in 36 commonly used rat inbred strains and three wild rat individuals. End nodes (yellow dots) represent strains. Some end nodes are double-sized, meaning that they are supported by two samples. Interconnecting nodes where lines come together represent possible precursors. (Reprinted from Smits et al. 2005 [© Bio Med Central Ltd.].)

F344, LEW, and PVG) have an increased risk of MS. Crosses between two of these disease strains (DA and LEW) and resistant control strains have resulted in the identification of 18 quantitative trait loci (QTLs) involved in experimental allergic encephalomyelitis, an animal model of MS (Dahlman et al. 1999a,b; Roth et al. 1999; Bergsteinsdottir et al. 2000). Overlapping QTL confidence intervals for the same trait in multiple strains (e.g., Eae2 and Eae11) may then allow identification of shared haplotypes between the disease strains, which can facilitate positional cloning of the disease allele.

As inbred strains are developed, multiple genes conferring disease may be concurrently fixed, resulting in multiple disease models within a single inbred strain. However, some of these traits may remain unidentified. Because of this, strains need to be better characterized, at both the phenotypic and the genomic levels. Major efforts are focusing on generating a rat phenome. Mashimo et al., from the National Bio Resource Project for the Rat (NBRP), have characterized 109 traits in 54 inbred rats (Mashimo et al. 2005), and PhysGen (http://pga.mcw.edu) has characterized 11 different strains (9 inbred and 2 outbred) for over 280 different traits, and has generated and characterized two chromosome substitution panels (44 strains derived from the sequenced BN and the FHH and SS hypertensive strains) (Jacob and Kwitek 2002; Kwitek et al. 2006). Importantly, all these experiments are performed using the same methodology.

APPROACHES AND TOOLS

Novel animal models for complex disease allow one to map phenotypes to the genome, and facilitate gene identification, by narrowing the chromosomal region where linkage to a phenotype resides and by fixing the effect of a disease locus in a homogeneous genetic background. Here, we describe the development of new or "designer" rat models to follow up genetic linkage or QTL studies.

QTL Mapping

QTL mapping is a proven method to assign the biology of the rat to the genomic sequence by identifying chromosomal regions that contain genes affecting complex phenotypes. Although a QTL is a rather large genetic locus, the gene(s) within this interval is responsible for a component of the trait variation, enabling the genome to be annotated with physiology. Importantly, most rat models reflect a clinical phenotype, and several comparative mapping studies have determined that common phenotypes often map to conserved genomic regions between rat and human (outlined in detail below). The ultimate goal of QTL mapping is to use positional cloning to identify the genes that underlie complex phenotypes and diseases and to gain a better understanding of their physiology and pathophysiology.

To date, there have been 536 QTL papers published with over 1,000 QTLs reported for different physiological and pathophysiological traits. These include investigations of the genetic basis of blood pressure (Rapp 2000), diabetes (Jacob et al. 1992; Galli et al. 1996; Pravenec et al. 1996), cardiovascular disease (Stoll et al. 2001; Moreno et al. 2003), stroke (Rubattu et al. 1996), ethanol preference (Murphy et al. 2002), behavioral conditioning and anxiety (Fernandez-Teruel et al. 2002; Flint 2003), fat accumulation (Tanomura et al. 2002), arthritis (Olofsson et al. 2003a), copper metabolism (de Wolf et al. 2002), pituitary tumor growth (Wendell and Gorski 1997), aerobic capacity (Ways et al. 2002), and chemical carcinogenesis (De Miglio et al. 2002). Most of the QTLs have been mapped in the last 3 years, mainly due to advances in technologies that allow high-throughput genotyping and an accelerated development of genetically modified strains.

QTL mapping is often followed by confirmation of the loci by the development of congenic lines, in order to evaluate the QTLs in the absence of other mapped QTLs, and as a step in positional cloning (Flint et al. 2005). To date, 118 QTLs mapped in rat have been confirmed by congenic lines, many of which have narrowed the critical genomic interval to a handful of candidate genes.

More than 50% of these congenics (59 strains) were developed for studying blood pressure control, followed by congenics for non-insulin-dependent diabetes mellitus (29 strains). From the 118 congenic lines developed following QTL mapping, 61 congenic lines have been published since 2002. Following this trend, acceleration in the rate of gene discovery in the rat is expected, reflecting the availability of the rat sequence, the accessibility of high-throughput sequencing to search for sequence variants, as well as microarray technologies for gene identification, pathway analysis, and mapping of *cis* and *trans* regulatory elements. These resources will greatly facilitate the identification of genes underlying the hundreds of QTLs mapped for complex diseases and phenotypes.

Comparative Mapping

The primary motivation for the rat genome project was to leverage the deep biological history to annotate the human sequence with common complex diseases (Jacob and Kwitek 2002). Most rat research is ultimately translational, aimed at improving human health through the understanding of key genetic and physiological factors in common disease pathways. Using the evolutionarily conserved regions between genomes (Brudno et al. 2004; Gibbs et al. 2004; Wilder et al. 2004) to map disease-causing genes or regions from one organism to another has begun to bear fruit in humans.

The genomic sequence of the rat and many other species allows comparative genome analysis at a nucleotide level rather than the identification of conserved synteny based on low-resolution ordering of orthologous genes. Comparative analysis is based on the hypothesis that functionally important sequences will be conserved across species. In 2000, Stoll et al. reported that QTLs in the rat could be used to predict the likely locations of human QTLs (Stoll et al. 2000). Since then, numerous other studies have demonstrated evolutionarily conserved regions in human, mouse, and rat that are linked to the same phenotype in all three species (Stoll et al. 2000; Sugiyama et al. 2001; Jacob and Kwitek 2002; Korstanje and DiPetrillo 2004). Over the next 5 years, a large increase can be expected in the number of studies using cross-species comparisons to find causes of common complex diseases (Glazier et al. 2002; Korstanje and Paigen 2002). Approximately 100 papers report that a particular disease trait maps to the same conserved region in rat and human, illustrating the near-term benefits of the rat genome project. However, one must keep in mind that QTLs are large and numerous for multifactorial traits, such as behavior and metabolic syndromes. Therefore, overlapping QTLs may sometimes be merely a chance event. To address this issue, the rat can be extensively evaluated for disease sub-phenotypes to better match QTLs by intermediate phenotype. Furthermore, with the promise of a rat SNP map, fine-resolution mapping may reduce the size of a QTL. Finally, comparative mapping data from additional species such as dog, cow, or other models might be used to confirm conserved QTLs. Integration of the genome sequence with existing mapping data and the biological data attached to those maps, plus the creation and annotation of a comprehensive catalog of gene products, will increase the use of such comparative studies and the impact the rat has on translational research.

"Designer" Strains

Congenic strains

To validate the functional importance of a genomic region, initially identified by genetic linkage analysis, congenic techniques were originally developed to study the MHC in the mouse by Nobel Prize-winner Dr. Snell (Snell 1948) at the Jackson Lab. This strategy remains a common way to study genes nearly 60 years later. Since congenic strains differ only in a short chromosomal segment from their background strain, it is possible to investigate the phenotypic effect of the locus, isolated from other effects caused by other loci on the original genetic background (Fig. 29-2). The development of congenic strains can be accelerated by genotyping the whole genome and selecting the breeders that, besides containing the target region of the donor strain, have a greater proportion of alleles from the recipient strain throughout the genome. This process of whole-genome

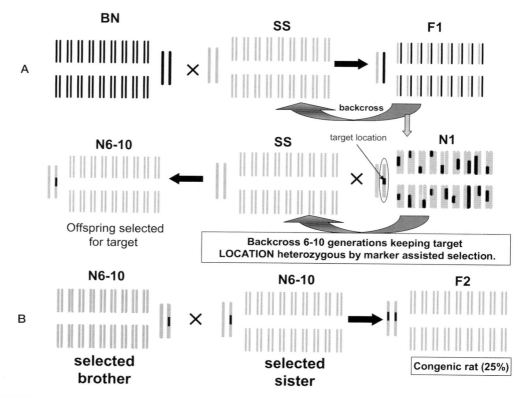

FIGURE 29-2. Schematic representation of the generation of a congenic strain from two genetically different rat strains. (*A*) Parental strains Brown Norway (BN) and Dahl salt-sensitive (SS) are intercrossed for the generation of a heterozygous F$_1$ population. The F$_1$ is then crossed with the parental background of interest (in this example, the SS) to generate an N1 population. The N1 rats are then backcrossed 6–10 generations using marker-assisted selection of the offspring, in order to substitute a selected genomic region from the BN rat. (*B*) A male and a female rat, selected by genotyping for this specific target region containing the phenotype of interest, are then mated. 25% of the offspring from this cross will be homozygous for this region. These rats are then inbred to produce a stable inbred congenic strain. (Reprinted from Cowley et al. 2004 [© Blackwell].)

marker-assisted selection, also called "speed congenics," can reduce the breeding time by half (Visscher 1999). With this method, generation of a congenic strain is reduced from 4–5 years to 2–3 years. In complex diseases, with multiple QTLs determining a trait, generation of double or triple congenic strains is sometimes necessary in order to confirm a causative locus. These multiple congenics are typically constructed one at a time and then assembled onto multi-congenics.

Consomic strains

Consomic strains are rat strains in which a whole chromosome from one strain is transferred to the genomic background of another strain, by a method similar to congenic generation. The Medical College of Wisconsin has assembled two complete panels of consomic strains, using the BN strain that was sequenced as the donor strain. In these consomic strains, a chromosome from the BN/NHsdMcwi rat was substituted, one at a time, into the genetic background of the SS/JrHsdMcwi (Dahl salt-sensitive; SS) or FHH/EurMcwi (Fawn Hooded Hypertensive; FHH) rats. The SS rat is a model for salt-sensitive hypertension (Rapp 1982), insulin resistance (Kotchen et al. 1991), hyperlipidemia (Reaven et al. 1991), endothelial dysfunction (Luscher et al. 1987), cardiac hypertrophy (Ganguli et al. 1979), and glomerulosclerosis (Roman and Kaldunski 1991). The FHH rat is a model for systolic hypertension, renal disease, pulmonary hypertension, a bleeding disorder, alcoholism, and depression (Provoost 1994). The two consomic panels capture nearly 50% of the genetic variation in the rat (Steen et al. 1999) and provide a foundation of genetic resources for the study of disorders of heart, kidney, lung, and vasculature. Given the 50% genetic variability, it is reasonable to assume that a similar level of biological variability can be expected, making the consomic strains a powerful tool for mapping additional complex traits.

One major advantage of using consomic strains is that congenic lines can be generated rapidly. Generation of congenic rats from consomic rat strains takes at most three generations of breeding, following an intercross with the consomic and parental strain (Fig. 29-3). Consomic rat strains can be used to assess whether background effects modify gene function and to develop polygenic models to study gene–gene interactions. The contribution of genes on each chromosome to the observed traits can be assessed both by phenotyping and by expression profiling. Comparisons between the consomic and parental strains provide valuable insights into the genomic pathways (clustered gene expression patterns) that differ between strains and how these differences might be connected to a particular pathologic phenotype. The advantage of studying a consomic rather than the donor parental strain is that all of the genome, except the replaced chromosome, is genetically identical to the recipient strain, significantly reducing heterogeneity.

Recombinant inbred strains

Recombinant inbred (RI) strains provide an additional tool for mapping phenotypes to the genome. This strategy is based on the generation of a panel of inbred strains derived from an F_2 population (Pravenec et al. 1996). Lines that contain more than one QTL are generated, permitting the analysis of gene interaction and detection of weak loci. However, the presence of a unique genome background in each strain prevents their use in rapid generation of congenic animals.

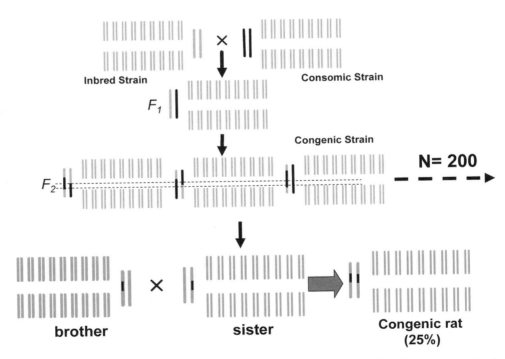

FIGURE 29-3. Generation of congenic rats from consomic strains. The parental strain is crossed with the consomic strain, to generate an F_1 population with identical genetic background and a heterozygous target chromosome. These F_1 rats are intercrossed to generate an F_2 population of rats whose target chromosome will be congenic, due to recombination events. Two similar F_2 rats are selected (by genotype) and mated to fix the region of interest. (Reprinted, with permission, from Cowley et al. 2004 [© Blackwell].)

One of the largest rodent recombinant inbred panels is the reciprocal HXB/BXH recombinant inbred strains, derived from the Spontaneously Hypertensive (SHR) and the BN rat strains (Pravenec et al. 1989, 1999). These strains are a great resource for genetic analysis of cardiovascular and metabolic phenotypes. The use of this RI strain panel has facilitated the mapping of several traits, including blood pressure (Pravenec et al. 1995), reproductive traits (Zidek et al. 1999), metabolic traits (Pravenec et al. 2002), behavior (Conti et al. 2004), and susceptibility to cancer (Bila and Kren 1996). Other RIs have also been developed, such as the LEXF (Shisa et al. 1997) and the SWXJ (Svenson et al. 1995) panels, although phenotypic characterization of these panels has not been as extensive as that for the HXB/BXH.

Heterogeneous stocks

Heterogeneous stocks (HS) are derived from crossing eight inbred strains followed by continuous outbreeding for several generations (Hansen and Spuhler 1984). Although this strategy was developed before the human and rat genome projects, the resources of the rat genome project make this collection of rat strains extremely powerful. The chromosomes of the HS progeny represent a random mosaic of the founding animals with an average distance between recombination events close to a single centiMorgan (cM). This high degree of recombination enables the fine mapping of QTLs into sub-cM intervals and the identification of multiple QTLs within what was previously identified as a single QTL (e.g., Ariyarajah et al. 2004; Stylianou et al. 2004). The HS rat colony was derived by the NIH in 1984 for alcohol studies (Pandey et al. 2002). Fine mapping of QTL to sub-cM intervals has been successful for traits of anxiety (Mott et al. 2000), ethanol-induced locomotor activity (Demarest et al. 2001), and conditioned fear (Mott et al. 2000) in HS mice. Studies are currently under way to detect QTLs for multiple traits, including behavior and diabetic traits, using HS rats.

Positionally Cloned Genes

Traditional positional cloning in the rat has been coming to fruition in identifying many disease genes over the past 2 years. Numerous genes have now been identified in the rat by positional cloning, concurrent with the great increase in rat genomic resources. These include genes for cancer (*BHD, Tsc2*) (Yeung et al. 1994; Okimoto et al. 2004), type 1 diabetes (*Gimap5, Cblb*) (MacMurray et al. 2002; Yokoi et al. 2002), type 2 diabetes (*Cd36*) (Aitman et al. 1999), neurological disorders (*Cct4, Reln, Unc5h3*) (Lee et al. 2003; Yokoi et al. 2003; Kuramoto et al. 2004), arthritis (*Ncf1*) (Olofsson et al. 2003b), renal disease (*Pkhd1, Rab38*) (Ward et al. 2002), glomerulonephritis (*Fcgr3*) (Aitman et al. 2006), bleeding disorders (*Rab38, VKOR*) (Oiso et al. 2004; Rost et al. 2004), retinal degeneration (*Mertk*) (Gal et al. 2000), and hypotrichosis (*Dsg4, Whn*) (Segre et al. 1995; Jahoda et al. 2004). Many of these genes were cloned from spontaneous mutants with Mendelian inheritance of disease; e.g., the *Pkdh1* mutation in the PCK rat causes autosomal recessive polycystic kidney disease (ARPKD). However, the number of identified genes involved in complex traits is on the rise.

Among the most challenging tasks in genomics are the prediction of gene function and the study of the interactions between genes, known as functional genomics. DNA microarrays can be used to study the genes and pathways involved in the pathogenesis of diseases and in the physiological responses to physiological stressors, drugs, and environmental stimuli. Microarray studies in rat have been used in conjunction with other genetic strategies, like QTL analysis, congenic mapping, or transgenic techniques to accelerate the search for genes underlying various phenotypes (Aitman et al. 1999; Monti et al. 2001; Liang et al. 2003; Vitt et al. 2004). The cloning of the *Cd36* gene is one of the first examples of cloning a complex trait gene in the rat using a combined approach of introducing a QTL to generate congenic strains and profiling their expression patterns compared to the parental strain. A more recent study looked at gene expression in a panel of BXH/HXB recombinant inbred (RI) rat strains (Hubner et al. 2005) to identify eQTL (expression

QTL) in the rat genome. eQTLs that overlap with previously identified QTLs for metabolic syndromes have provided nearly 76 candidate genes to be evaluated.

Target Validation

Once a gene has been positionally cloned or implicated to be causal, the causality must be proved, a phase of gene discovery termed target validation. The gold standard for proving a gene is causal is to knock in the particular allele or mutation responsible for the trait, or to replace the defect with a "normal" allele to demonstrate that this specific substitution changes the phenotype. Although this can provide conclusive proof, it is an onerous and expensive process that is not likely to be feasible for all 30,000+ genes. The rat genome sequence has facilitated an alternative approach, transgenic rescue, whereby the phenotype is normalized via a transgene, particularly when the trait shows a recessive mode of inheritance (Pravenec et al. 2001; Jacob and Kwitek 2002). Recently, cloning of fertile adult rats has been achieved by nuclear transfer (Zhou et al. 2003), which opens the door for targeted gene manipulation, such as knock-out technology. However, it will be some time before this can be done routinely, as its efficiency is too low to be used as a general method.

Transgenic Rats

For more than 15 years, genetics studies have followed two tracks to unravel gene function: positional cloning from genetic mapping to gene identification, and transgenesis (random insertion of genes, knock-outs, knock-ins, and conditional knock-outs). Traditional rat transgenesis by pronuclear injection was established in 1990 (Hammer et al. 1990; Mullins et al. 1990). However, because the rat lacks viable ES cell lines, traditional knock-out and knock-in technology is unavailable, somewhat limiting the use of rat for gene-manipulation studies. Nonetheless, over 200 transgenic rats have been generated.

As in the mouse, the major purpose of generating rats via transgenesis was to study a particular gene of interest. It is now relatively straightforward to alter the expression of specific rat genes as well as to use rats as surrogate hosts for expression of genes from other species. Many transgenic rat strains have been "humanized" by using a gene from human, providing a bridge between genetic linkage studies in humans, and functional association of a (mutant) gene with particular pathological features. For example, humanized rats were used to dissect complex diseases such as heart hypertrophy (Tian et al. 2004), end-organ damage (Hocher et al. 1996), and hypertension (for review, see Pinto-Sietsma and Paul 1997; Liefeldt et al. 1999; Bohlender et al. 2000). These examples provide proof of principle that human disease modeling in rats is valuable, and one can expect that the etiology of other diseases will be similarly illuminated through sequence knowledge and transgenesis. Furthermore, transgenic rats expressing human genes can be used to follow up disease progression in longitudinal in vivo studies, and to monitor the effects of long-term treatments, cell implantation, or antisense approaches on the course of disease. Finally, the need to validate a gene cloned by position via transgenic rescue will further increase the use of transgenic technologies.

N-Ethyl-N-Nitrosourea Mutagenesis

For many years, the genetics community relied on spontaneous mutations as a source of rat models, which is a major limiting factor in the use of the rat for structure–function studies, particularly for disease gene validation. An alternative involves inducing mutation through the use of chemical mutagens such as *N*-ethyl-*N*-nitrosourea (ENU). ENU mutagenesis has been utilized for many years, but the last 10 years have seen an increase in its use (Guenet 2004). The typical strategy is to treat males with ENU, inducing mutations in the spermatogonial stem cells (predominantly loss of function); breed them to untreated females; and screen the offspring for phenotypic effects. As the ENU approach has advanced, phenotyping is now emerging as the rate-limiting step in such large-scale studies. There has been an increasing emphasis on large-scale ENU screens for systematic and comprehensive gene function analyses of the mouse genome (Hrabe de Angelis et al. 2000; Nadeau 2000; Nolan et al. 2000; Brown and Balling 2001). The mutagenesis programs are phenotype-driven, employing appropriate screens of mutagenized animals to identify novel mutant phenotypes, particularly those that model human disease, followed by mapping and isolation of the underlying causal genes. However, large-scale phenotypic screens are very expensive and consume a tremendous amount of animal per diem charges. As such, there has not been a similar attempt at this type of screen using rats.

In 2003, two groups published on the use of ENU in combination with a gene screen to generate rat gene knock-outs. The major difference in this strategy is that the gene(s) of interest is screened and only the animals that have a mutation in specific genes of interest are kept (Zan et al. 2003; Smits et al. 2004a). In this way, per diem costs are minimized, although gene screening also involves a cost. This strategy has been used to knock out several genes in the rat (Zan et al. 2003; Homberg et al. 2005; Smits et al. 2006). One other advantage is that ENU mutagenesis can be done in the strain of choice (after determination of the appropriate ENU dose). This prevents problems associated with genome background effects that are relatively common when knock-outs are generated on a limited number of ES-cell lines. This method will likely gain in popularity until there are ES cells for rats, and will remain an alternative approach when genome backgrounds affect a phenotype. There are currently 26 ENU-induced mutant strains registered in the RGD.

GENETIC VARIATION

Although mapping QTLs provides valuable genomic information, there remains a need for improving the tools used in identifying genes that affect complex traits. Currently, QTLs can be localized to specific genetic intervals of about 2–10 cM, but further characterization is difficult. Single-nucleotide polymorphisms (SNPs) and haplotype maps are powerful tools for reducing the size of QTL intervals. SNP discovery and a SNP-based haplotype map will enable this reduction via two general avenues. The first is the advancement of correlations between phenotypic data and ancestral sequence origin across many existing inbred strains. This will immediately identify short genomic regions most likely to harbor the responsible genes. The second is the identification of segments that would be shared by rat strains used for simple intercross/backcross experiments (Cuppen 2005).

SNPs and Haplotypes

Variation at the DNA level is a major factor underlying the phenotypic diversity between individuals in a population. The most common type of genetic variation is SNPs. Although the majority of SNPs do not have a functional effect, others may affect chromosome organization, gene expression, or protein function. SNPs and their individual states (alleles) are not randomly distributed throughout the genome or within a population. Recombination and mutation events, in combination with selection processes and population history, have resulted in common block-like structures in genomes. These structures are characterized by a common combination of SNP alleles, a so-called haplotype. Selection for specific haplotypes within a population is primarily driven by the advantageous effect of an individual polymorphism in the haplotype block.

This mosaic block-wise pattern can be observed in many species and, unlike the variation between individuals of an outbred population (e.g., humans), there is relatively little variation among the commonly used inbred mice strains (Lindblad-Toh et al. 2000; Wade et al. 2002; Wiltshire et al. 2003). In addition, pair-wise comparison of genome-wide SNP data, sampled at relatively low resolution, revealed that the variation is a mosaic of regions with either extremely low (<1 polymorphism per 10 kb) or high (>40 per 10 kb) levels of polymorphisms. Such regions are 10–120 Mb in size and are assumed to be the result of a recent genetic bottleneck (Wade et al. 2002; Wiltshire et al. 2003). Although wild mice and rats were domesticated in ancient China and Japan, commonly used laboratory strains originate from just a small selection of animals.

Haplotype information can potentially accelerate genetic mapping and cloning procedures in laboratory animals in two ways. First, inbred strains are considered to have no genetic variation between individuals, thereby fixing the haplotype-block structure and organization. As a result of common ancestry, patterns of allelic similarities and differences among strains (strain-distribution patterns or SDPs) can be discerned for every variable locus (Grupe et al. 2001). Theoretically, mutations can be mapped by correlating phenotype and genotype SDPs, because a common phenotypic trait is most likely caused by a common ancestral polymorphism instead of independent newly acquired mutations in different strains. Several recent reports show a proof of principle for this approach (Grupe et al. 2001; Liao et al. 2004; Pletcher et al. 2004; Wang et al. 2004), although general applicability remains to be shown.

Second, genotyping a limited number of carefully chosen SNPs per haplotype block (tag-SNPs) is sufficient to assess the information of all of the other polymorphisms in the block (Cardon and Abecasis 2003; Sebastiani et al. 2003). This approach reduces the number of SNPs to be genotyped but has the disadvantage that the genomic region with the polymorphism of interest cannot be narrowed further than the smallest common haplotype block structure or SDP. Owing to the common ancestry of the strains, these regions can extend across many ancestral segments and harbor hundreds of characterized and uncharacterized polymorphisms. Introducing an additional group of carefully chosen strains (i.e., taking into account their phylogenetic relationships) is expected to increase the number of blocks and to provide the high resolution required for gene identification.

Rat SNP Data

Current information on genomic variation and SDPs in the rat is based on minimal SNP data inventories. dbSNP (build 126) contains 43,229 rat RefSNPs. This data set encompasses SNPs within cDNA and nontranscribed genomic regions and provides important information on the expected frequency of SNPs in the rat genome. Zimdahl et al. (2004) sequenced cDNA libraries from the SHRSP, BN, WKY, and SD strains, identifying 12,395 polymorphic sites in an interstrain comparison. Based on comparisons between the BN and other strains, their estimated discovery rate was at 1 SNP per 1100 base pairs (bp) of cDNA. Smits et al. (2004b) screened 55 genes in 96 strains, finding a total of 103 SNPs. Considering only intronic SNPs, and grouping closely linked polymorphisms as a single SNP, they calculate a frequency of 1 SNP per 367 bp, significantly higher than the previous study due to the large number of strains examined.

Guryev et al. (2004) used publicly available sequences consisting of WGS, EST, and mRNA data for in silico identification of 33,305 high-quality candidate SNPs in gene-coding regions. Experimental verification of 471 candidate SNPs using a limited set of rat strains revealed a confirmation rate of approximately 50%. Although the majority of SNPs were identified between Sprague-Dawley (EST data) and Brown Norway (WGS data) strains, 66% of the verified variations were common among different rat strains (minor allele frequency of >20%). Based on the confirmation data, a SNP frequency of 1 per 226 bp was obtained. This frequency is much higher than the previous number, as about half of the polymorphisms are intronic.

Because most SNP discovery approaches are biased by the choice of strain, Smits et al. (2005) used wild rat strains in a shotgun sequencing approach and found 485 SNPs in 814 kbp of sequence, in a comparison with the BN genome sequence. Interestingly, genotyping 36 commonly used inbred rat strains showed that 84% of these alleles are also polymorphic in a representative set of laboratory strains. As genotyping was performed by dideoxy resequencing, an additional 358 SNPs were discovered. (Such sequencing data can be analyzed using a variety of packages, e.g.,

Staden [Bonfield et al. 1998] or Polyphred [Stephens et al. 2006].) Based on the genotyping results, the SNP rate between BN and the wild rat is about 1 SNP per 190 bp. The SNP rate within the 36 rat strains, including BN, is 1 per 158 bp.

Finally, Celera produced genomic shotgun sequences for the Sprague-Dawley rat strain at about 1.5x coverage. Preliminary analysis indicates that about 2–4 million SNPs can be mined from these data by comparison to the BN genome, resulting in an estimated average SNP frequency between these two strains of 1 per 400–800 bp.

Rat Haplotype Data

More extensive, fine-scale analyses in larger sets of strains by Yalcin et al. (2004a), Frazer et al. (2004b), and recently, Guryev et al. (2006) have shed new light on rodent haplotype-block structure and its utility in genetic studies in mice and rats. Frazer and coworkers analyzed five genomic intervals, totaling 4.6 Mb (0.17% of the mouse genome), in 13 commonly used inbred mouse strains and 2 inbred strains derived from wild mice. Using a high-density oligonucleotide array, they resequenced 3 Mb of nonrepetitive sequence in these regions and discovered 18,366 SNPs. However, only 4,065 of the SNPs were polymorphic within the 13 commonly used inbred strains. Analysis using a hidden Markov model, assuming a two-state model fitted for low and high SNP rate and bins of 10,000 nucleotides (Wade et al. 2002), defined approximately 50 haplotype blocks ranging from 12 kb to 608 kb in size. Most cases represented the contribution of the different ancestral subspecies *Mus musculus musculus* and *Mus musculus domesticus*. On average, only 1–3 SNPs are needed to distinguish a specific haplotype block. Extrapolation of these results to the entire 2.5-Gb mouse genome suggests the need for an estimated 50,000 tag-SNPs to define the phylogenetic relationship for any segment in these inbred strains.

Although these results are consistent with previous data (Wade et al. 2002; Wiltshire et al. 2003), this study also shows that detailed sequence information from at least 12 inbred strains is needed to capture the majority (>95%) of variable sites present in commonly used mouse inbred strains. This implies that the resequencing of at least 12 inbred rat strains is required for reliable genome-wide haplotype-block determination and subsequent tag-SNP identification. Furthermore, the estimated false-negative rate of >50% for the microarray-based resequencing approach, the strain selection, and the computational algorithm used remain as variables associated with the proposed haplotype-block structure.

Yalcin and coworkers analyzed a contiguous 4.8-Mb region of mouse containing a QTL influencing anxiety (Yalcin et al. 2004b). This was a test case for the claimed usefulness of correlating haplotype blocks and SDPs in QTL mapping. High-resolution SNP data were obtained by resequencing 1- to 2-kb segments at 8-kb intervals in eight inbred mouse strains. In total, 0.58 Mb of sequence was sampled, resulting in the identification of 1,720 variants, 1,325 of which were SNPs. The analysis of SNP density confirms the mosaic pattern of segments with regions of extremely

FIGURE 29-4. Patterns of LD for orthologous genomic segments of approximately 5 Mb in rat, human, and mouse. LD plots for orthologous genomic segments in rat (1,351 SNPs), human (1,479 SNPs), and mouse (311 SNPs) are shown. For each panel, the following information is shown: LD plot (*top*), haplotype blocks in SNP coordinates (*middle*), and physical map and haplotype blocks in physical coordinates (*bottom*). The haplotype map has a gradient representation for D′ values that assists visual comparison of haplotype structure. Haplotype blocks were built with stringent criteria, sometimes resulting in splitting of visually recognized blocks. LD patterns in the rat and mouse inbred strains have common features. Both organisms exhibit extended blocks of increased LD corresponding to the following genomic segments: (1) the cluster of 5 genes: B3galt2, Cdc73 (Hrpt2), Glrx2, Trove2 (Ssa2), and Uchl5; (2) the large Fam5C (Brinp3) gene, and (3) regions flanking the Rgs18 gene. Although the human haplotype structure is characterized by much smaller blocks, the most extensive human regions displaying high LD, and thus extended haplotype blocks, include the cluster of 5 genes mentioned above, the coding part of Fam5C (Brinp3), and the region flanking Rgs18. Three characteristic haplotype blocks that are conserved cross-species have been color-coded. (Reprinted from Guryev et al. 2006.)

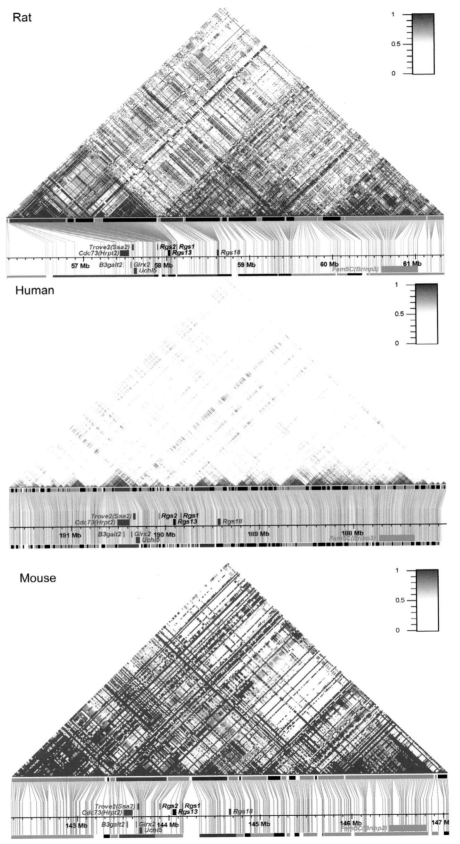

FIGURE 29-4. (*See facing page for legend.*)

low or extremely high SNP density. Segments with an intermediate SNP density are also observed. However, with the current SNP sampling density, it remains unclear whether these consist of a mixture of smaller segments of either type.

In a recent study, Guryev et al. (2006) compared the rat, mouse, and human haplotype structures of a 5-Mb region from rat chromosome 1 and its conserved syntenic regions in mouse and human. They show that haplotype block structure is conserved across mammals, most prominently in genic regions, suggesting the existence of an evolutionary selection process that drives the conservation of long-range allele combinations (Fig. 29-4) (Guryev et al. 2006). Indeed, genome-wide, gene-centric analysis of the human HapMap data revealed that equally spaced polymorphic positions in genic regions and their upstream regulatory regions are genetically more tightly linked than in non-genic regions. These findings may complicate the identification of causal polymorphisms underlying phenotypic traits, because in regions where haplotype structure is conserved, combinations of tightly linked polymorphisms (rather than a single causal SNP) might contribute to the phenotypic difference. On the other hand, evolutionary conservation of haplotype structure may actually be utilized for the identification and characterization of functionally important genomic regions. Figure 29-5 compares rat chromosome 10 with the human sequence.

SNP Projects in Progress

Haplotype-based rat genetics is currently gaining momentum. To fully realize the power of the rat genome sequence, two initiatives are under way to dissect the ancestral segments making up the most commonly used inbred lines. One initiative, named STAR, is funded by the European Union and aims to generate a haplotype map for the rat (http://www.snp-star.eu and http://euratools.csc.mrc.ac.uk). The main goal is to genotype about 100,000 SNPs in at least 200 different inbred strains. These genotyping data will be useful to correlate phenotypes with the ancestral origin of a strain, allowing the identification and fine mapping of the critical regions responsible for specific traits. Through the inclusion of essentially all important rat strains that are used around the world for QTL mapping experiments, researchers will be able to reduce the critical interval by linkage analysis using shared segments between their strains of interest, or to select ideal strain combinations for intercross/backcross experiments. The second initiative is supported by the NIH. The rat genome sequencing consortium will initiate a large SNP discovery effort by shotgun sequencing eight commonly used inbred rat strains, aiming at the discovery of more than 280,000 SNPs for each of the eight strains.

This level of coverage should address genetic mapping needs and allow an in-depth evaluation of which additional resources will be required to maximize disease gene discovery. The strains used in both efforts were selected to cover the "evolutionary" range of the known genetic variation in the lab rat which, in addition to maximizing diversity, also includes the strains where large numbers of QTLs have been mapped and other genetic tools such as congenics, consomics, and recombinant inbred strains exist.

SNP data obtained in any haplotype map project will undoubtedly be informative for a variety of purposes, including genetic mapping and cloning approaches that do not depend on haplotype-block definition (Liao et al. 2004; Pletcher et al. 2004). Rapid development of high-throughput SNP-typing strategies, combinations of mapping gene expression data and physiological QTLs (Hubner et al. 2005; Malek et al. 2006), and the increased efficiency of resequencing candidate regions might overcome many of the bottlenecks that are associated with the traditional mapping and cloning approaches. The combination of the genomic sequence and a larger SNP database will further enhance the toolkit for investigators using rats to study human disease.

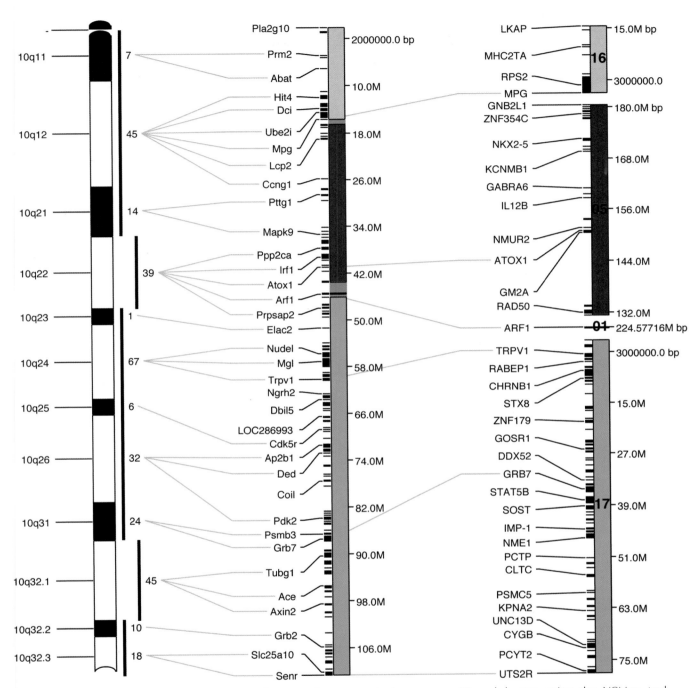

FIGURE 29-5. Comparative map between rat chromosome 10 and human using the VCMap tool (http://rgd.mcw.edu/VCMAP). On the left is the cytogenetic map of RNO10. The bars directly to the right of the cytogenetic map depict the cytogenetic bins to which genes map. The numbers depict the number of genes in each bin. In the center is the genome map of RNO10, measured in megabase pairs of DNA (Mb). Gene symbols are noted on the left-hand side of the map. The colored regions indicate the conserved synteny with the human genome. To the far right are the human conserved segments, annotated with the human chromosome number and measured in Mb along each chromosome. Chromosome 16 is blue, chromosome 5 is red, and chromosome 17 is turquoise.

REFERENCES

Aitman T.J., Glazier A.M., Wallace C.A., Cooper L.D., Norsworthy P.J., Wahid F.N., Al-Majali K.M., Trembling P.M., Mann C.J., Shoulders C.C., et al. 1999. Identification of Cd36 (Fat) as an insulin-resistance gene causing defective fatty acid and glucose metabolism in hypertensive rats. Nat. Genet. 21: 76–83.

Aitman T.J., Dong R., Vyse T.J., Norsworthy P.J., Johnson M.D., Smith J., Mangion J., Roberton-Lowe C., Marshall A.J., Petretto E., et al. 2006. Copy number polymorphism in Fcgr3 predisposes to glomerulonephritis in rats and humans. Nature 439: 851–855.

Ariyarajah A., Palijan A., Dutil J., Prithiviraj K., Deng Y., and Deng A.Y. 2004. Dissecting quantitative trait loci into opposite blood pressure effects on Dahl rat chromosome 8 by congenic strains. J. Hypertens. 22: 1495–1502.

Bergsteinsdottir K., Yang H.T., Pettersson U., and Holmdahl R. 2000. Evidence for common autoimmune disease genes controlling onset, severity, and chronicity based on experimental models for multiple sclerosis and rheumatoid arthritis. J. Immunol. 164: 1564–1568.

Bila V. and Kren V. 1996. The teratogenic action of retinoic acid in rat congenic and recombinant inbred strains. Folia Biol. 42: 167–173.

Bohlender J., Ganten D., and Luft F.C. 2000. Rats transgenic for human renin and human angiotensinogen as a model for gestational hypertension. J. Am. Soc. Nephrol. 11: 2056–2061.

Bonfield J.K., Rada C., and Staden R. 1998. Automated detection of point mutations using fluorescent sequence trace subtraction. Nucleic Acids Res. 26: 3404–3409.

Brown S.D. and Balling R. 2001. Systematic approaches to mouse mutagenesis. Curr. Opin. Genet. Dev. 11: 268–273.

Brudno M., Poliakov A., Salamov A., Cooper G.M., Sidow A., Rubin E.M., Solovyev V., Batzoglou S., and Dubchak I. 2004. Automated whole-genome multiple alignment of rat, mouse, and human. Genome Res. 14: 685–692.

Cardon L.R. and Abecasis G.R. 2003. Using haplotype blocks to map human complex trait loci. Trends Genet. 19: 135–140.

Conti L.H., Jirout M., Breen L., Vanella J.J., Schork N.J., and Printz M.P. 2004. Identification of quantitative trait loci for anxiety and locomotion phenotypes in rat recombinant inbred strains. Behav. Genet. 34: 93–103.

Cowley A.W., Jr., Roman R.J., and Jacob H.J. 2004. Application of chromosomal substitution techniques in gene-function discovery. J. Physiol. 554: 46–55.

Cuppen E. 2005. Haplotype-based genetics in mice and rats. Trends Genet. 21: 318–322.

Dahlman I., Jacobsson L., Glaser A., Lorentzen J.C., Andersson M., Luthman H., and Olsson T. 1999a. Genome-wide linkage analysis of chronic relapsing experimental autoimmune encephalomyelitis in the rat identifies a major susceptibility locus on chromosome 9. J. Immunol. 162: 2581–2588.

Dahlman I., Wallstrom E., Weissert R., Storch M., Kornek B., Jacobsson L., Linington C., Luthman H., Lassmann H., and Olsson T. 1999b. Linkage analysis of myelin oligodendrocyte glycoprotein-induced experimental autoimmune encephalomyelitis in the rat identifies a locus controlling demyelination on chromosome 18. Hum. Mol. Genet. 8: 2183–2190.

De Miglio M.R., Pascale R.M., Simile M.M., Muroni M.R., Calvisi D.F., Virdis P., Bosinco G.M., Frau M., Seddaiu M.A., Ladu S., and Feo F. 2002. Chromosome mapping of multiple loci affecting the genetic predisposition to rat liver carcinogenesis. Cancer Res. 62: 4459–4463.

de Wolf I.D., Bonne A.C., Fielmich-Bouman X.M., van Oost B.A., Beynen A.C., van Zutphen L.F., and van Lith H.A. 2002. Quantitative trait loci influencing hepatic copper in rats. Exp. Biol. Med. 227: 529–534.

Demarest K., Koyner J., McCaughran J., Jr., Cipp L., and Hitzemann R. 2001. Further characterization and high-resolution mapping of quantitative trait loci for ethanol-induced locomotor activity. Behav. Genet. 31: 79–91.

Dracheva S.V., Remmers E.F., Chen S., Chang L., Gulko P.S., Kawahito Y., Longman R.E., Wang J., Du Y., and Shepard J. 2000. An integrated genetic linkage map with 1137 markers constructed from five F2 crosses of autoimmune disease-prone and -resistant inbred rat strains. Genomics 63: 202–226.

Fernandez-Teruel A., Escorihuela R.M., Gray J.A., Aguilar R., Gil L., Gimenez-Llort L., Tobena A., Bhomra A., Nicod A., Mott R., et al. 2002. A quantitative trait locus influencing anxiety in the laboratory rat. Genome Res. 12: 618–626.

Flint J. 2003. Analysis of quantitative trait loci that influence animal behavior. J. Neurobiol. 54: 46–77.

Flint J., Valdar W., Shifman S., and Mott R. 2005. Strategies for mapping and cloning quantitative trait genes in rodents. Nat. Rev. Genet. 6: 271–286.

Frazer K.A., Pachter L., Poliakov A., Rubin E.M., and Dubchak I. 2004a. VISTA: Computational tools for comparative genomics. Nucleic Acids Res. 32: W273–279.

Frazer K.A., Wade C.M., Hinds D.A., Patil N., Cox D.R., and Daly M.J. 2004b. Segmental phylogenetic relationships of inbred mouse strains revealed by fine-scale analysis of sequence variation across 4.6 mb of mouse genome. Genome Res. 14: 1493–1500.

Gal A., Li Y., Thompson D.A., Weir J., Orth U., Jacobson S.G., Apfelstedt-Sylla E., and Vollrath D. 2000. Mutations in MERTK, the human orthologue of the RCS rat retinal dystrophy gene, cause retinitis pigmentosa. Nat. Genet. 26: 270–271.

Galli J., Li L.S., Glaser A., Ostenson C.G., Jiao H., Fakhrai-Rad H., Jacob H.J., Lander E.S., and Luthman H. 1996. Genetic analysis of non-insulin dependent diabetes mellitus in the GK rat. Nat. Genet. 12: 31–37.

Ganguli M., Tobian L., and Iwai J. 1979. Cardiac output and peripheral resistance in strains of rats sensitive and resistant to NaCl hypertension. Hypertension 1: 3–7.

Gerhard D.S., Wagner L., Feingold E.A., Shenmen C.M., Grouse L.H., Schuler G., Klein S.L., Old S., Rasooly R., Good P., et al. 2004. The status, quality, and expansion of the NIH full-length cDNA project: The Mammalian Gene Collection (MGC). Genome Res. 14: 2121–2127.

Gibbs R.A., Weinstock G.M., Metzker M.L., Muzny D.M., Sodergren E.J., Scherer S., Scott G., Steffen D., Worley K.C., Burch P.E., et al. 2004. Genome sequence of the Brown Norway rat yields insights into mammalian evolution. Nature 428: 493–521.

Glazier A.M., Nadeau J.H., and Aitman T.J. 2002. Finding genes that underlie complex traits. Science 298: 2345–2349.

Grupe A., Germer S., Usuka J., Aud D., Belknap J.K., Klein R.F., Ahluwalia M.K., Higuchi R., and Peltz G. 2001. In silico mapping of complex disease-related traits in mice. Science 292: 1915–1918.

Guenet J.L. 2004. Chemical mutagenesis of the mouse genome: An overview. Genetica 122: 9–24.

Guryev V., Berezikov E., and Cuppen E. 2005. CASCAD: A database of annotated candidate single nucleotide polymorphisms associated with expressed sequences. BMC Genomics 6: 10.

Guryev V., Berezikov E., Malik R., Plasterk R.H.A., and Cuppen E. 2004. Single nucleotide polymorphisms associated with rat expressed sequences. Genome Res. 14: 1438–1443.

Guryev V., Smits B.M.G., van de Belt J., Verheul M., Hubner N., and Cuppen E. 2006. Haplotype block structure is conserved across mammals. PLoS Genet. 2: e121.

Hammer R.E., Maika S.D., Richardson J.A., Tang J.P., and Taurog J.D. 1990. Spontaneous inflammatory disease in transgenic rats expressing HLA-B27 and human β_2m: An animal model of HLA-B27-associated human disorders. *Cell* **63:** 1099–1112.

Hansen C. and Spuhler K. 1984. Development of the National Institutes of Health genetically heterogeneous rat stock. *Alcohol Clin. Exp. Res.* **8:** 477–479.

Hocher B., Liefeldt L., Thone-Reineke C., Orzechowski H.D., Distler A., Bauer C., and Paul M. 1996. Characterization of the renal phenotype of transgenic rats expressing the human endothelin-2 gene. *Hypertension* **28:** 196–201.

Homberg J.R., Olivier J.D., Smits B., Mudde J., Cools A.R., Ellenbroek B.A., and Cuppen E. 2005. O9 phenotyping of the serotonin transporter knockout rat. *Behav. Pharmacol.* (suppl. 1) **16:** S21.

Hrabe de Angelis M.H., Flaswinkel H., Fuchs H., Rathkolb B., Soewarto D., Marschall S., Heffner S., Pargent W., Wuensch K., Jung M., et al. 2000. Genome-wide, large-scale production of mutant mice by ENU mutagenesis. *Nat. Genet.* **25:** 444–447.

Hubbard T., Andrews D., Caccamo M., Cameron G., Chen Y., Clamp M., Clarke L., Coates G., Cox T., Cunningham F., et al. 2005. Ensembl 2005. *Nucleic Acids Res.* **33:** D447–D453.

Hubner N., Wallace C.A., Zimdahl H., Petretto E., Schulz H., Maciver F., Mueller M., Hummel O., Monti J., Zidek V., et al. 2005. Integrated transcriptional profiling and linkage analysis for identification of genes underlying disease. *Nat. Genet.* **37:** 243–253.

Jacob H.J. and Kwitek A.E. 2002. Rat genetics: Attaching physiology and pharmacology to the genome. *Nat. Rev. Genet.* **3:** 33–42.

Jacob H.J., Pettersson A., Wilson D., Mao Y., Lernmark A., and Lander E.S. 1992. Genetic dissection of autoimmune type I diabetes in the BB rat. *Nat. Genet.* **2:** 56–60.

Jahoda C.A., Kljuic A., O'Shaughnessy R., Crossley N., Whitehouse C.J., Robinson M., Reynolds A.J., Demarchez M., Porter R.M., Shapiro L., and Christiano A.M. 2004. The *lanceolate hair* rat phenotype results from a missense mutation in a calcium coordinating site of the *desmoglein 4* gene. *Genomics* **83:** 747–756.

Kaiser J. 2005. Genomics: Celera to end subscriptions and give data to public GenBank. *Science* **308:** 775.

Karolchik D., Baertsch R., Diekhans M., Furey T.S., Hinrichs A., Lu Y.T., Roskin K.M., Schwartz M., Sugnet C.W., Thomas D.J., et al. 2003. The UCSC Genome Browser Database. *Nucleic Acids Res.* **31:** 51–54.

Korstanje R. and DiPetrillo K. 2004. Unraveling the genetics of chronic kidney disease using animal models. *Am. J. Physiol. Renal Physiol.* **287:** F347–F352.

Korstanje R. and Paigen B. 2002. From QTL to gene: The harvest begins. *Nat. Genet.* **31:** 235–236.

Kotchen T.A., Zhang H.Y., Covelli M., and Blehschmidt N. 1991. Insulin resistance and blood pressure in Dahl rats and in one-kidney, one-clip hypertensive rats. *Am. J. Physiol.* **261:** E692–E697.

Kuramoto T., Kuwamura M., and Serikawa T. 2004. Rat neurological mutations *cerebellar vermis defect* and *hobble* are caused by mutations in the netrin-1 receptor gene *Unc5h3*. *Brain Res. Mol. Brain Res.* **122:** 103–108.

Kwitek A.E., Jacob H.J., Baker J.E., Dwinell M.R., Forster H.V., Greene A.S., Kunert M.P., Lombard J.H., Mattson D.L., Pritchard K.A., Jr., et al. 2006. BN phenome: Detailed characterization of the cardiovascular, renal, and pulmonary systems of the sequenced rat. *Physiol. Genomics* **25:** 303–313.

Lee M.J., Stephenson D.A., Groves M.J., Sweeney M.G., Davis M.B., An S.F., Houlden H., Salih M.A., Timmerman V., de Jonghe P., et al. 2003. Hereditary sensory neuropathy is caused by a mutation in the delta subunit of the cytosolic chaperonin-containing t-complex peptide-1 (*Cct4*) gene. *Hum. Mol. Genet.* **12:** 1917–1925.

Lee Y., Tsai J., Sunkara S., Karamycheva S., Pertea G., Sultana R., Antonescu V., Chan A., Cheung F., and Quackenbush J. 2005. The TIGR Gene Indices: Clustering and assembling EST and known genes and integration with eukaryotic genomes. *Nucleic Acids Res.* **33:** D71–D74.

Liang M., Yuan B., Rute E., Greene A.S., Olivier M., and Cowley A.W., Jr. 2003. Insights into Dahl salt-sensitive hypertension revealed by temporal patterns of renal medullary gene expression. *Physiol. Genomics* **12:** 229–237.

Liao G., Wang J., Guo J., Allard J., Cheng J., Ng A., Shafer S., Puech A., McPherson J.D., Foernzler D., Peltz G., and Usuka J. 2004. In silico genetics: Identification of a functional element regulating H2-Eα gene expression. *Science* **306:** 690–695.

Liefeldt L., Schonfelder G., Bocker W., Hocher B., Talsness C.E., Rettig R., and Paul M. 1999. Transgenic rats expressing the human ET-2 gene: A model for the study of endothelin actions in vivo. *J. Mol. Med.* **77:** 565–574.

Lindblad-Toh K., Winchester E., Daly M.J., Wang D.G., Hirschhorn J.N., Laviolette J.P., Ardlie K., Reich D.E., Robinson E., Sklar P., et al. 2000. Large-scale discovery and genotyping of single-nucleotide polymorphisms in the mouse. *Nat. Genet.* **24:** 381–386.

Luscher T.F., Raij L., and Vanhoutte P.M. 1987. Endothelium-dependent vascular responses in normotensive and hypertensive Dahl rats. *Hypertension* **9:** 157–163.

MacMurray A.J., Moralejo D.H., Kwitek A.E., Rutledge E.A., Van Yserloo B., Gohlke P., Speros S.J., Snyder B., Schaefer J., Bieg S., et al. 2002. Lymphopenia in the BB rat model of type 1 diabetes is due to a mutation in a novel immune-associated nucleotide (*Ian*)-related gene. *Genome Res.* **12:** 1029–1039.

Malek R.L., Wang H.Y., Kwitek A.E., Greene A.S., Bhagabati N., Borchardt G., Cahill L., Currier T., Frank B., Fu X., et al. 2006. Physiogenomic resources for rat models of heart, lung and blood disorders. *Nat. Genet.* **38:** 234–239.

Mashimo T., Birger V., Kuramoto T., and Serikawa T. 2005. Rat Phenome Project: The untapped potential of existing rat strains. *J. Appl. Physiol.* **98:** 371–379.

Monti J., Gross V., Luft F.C., Franca Milia A., Schulz H., Dietz R., Sharma A.M., and Hubner N. 2001. Expression analysis using oligonucleotide microarrays in mice lacking bradykinin type 2 receptors. *Hypertension* **38:** E1–E3.

Moreno C., Dumas P., Kaldunski M.L., Tonellato P.J., Greene A.S., Roman R.J., Cheng Q., Wang Z., Jacob H.J., and Cowley A.W., Jr. 2003. Genomic map of cardiovascular phenotypes of hypertension in female Dahl S rats. *Physiol. Genomics* **15:** 243–257.

Mott R., Talbot C.J., Turri M.G., Collins A.C., and Flint J. 2000. A method for fine mapping quantitative trait loci in outbred animal stocks. *Proc. Natl. Acad. Sci.* **97:** 12649–12654.

Mullins J.J., Peters J., and Ganten D. 1990. Fulminant hypertension in transgenic rats harbouring the mouse Ren-2 gene. *Nature* **344:** 541–544.

Murphy J.M., Stewart R.B., Bell R.L., Badia-Elder N.E., Carr L.G., McBride W.J., Lumeng L., and Li T.K. 2002. Phenotypic and genotypic characterization of the Indiana University rat lines selectively bred for high and low alcohol preference. *Behav. Genet.* **32:** 363–388.

Nadeau J.H. 2000. Muta-genetics or muta-genomics: The feasibility of large-scale mutagenesis and phenotyping programs. *Mamm. Genome* **11:** 603–607.

Nolan P.M., Peters J., Strivens M., Rogers D., Hagan J., Spurr N., Gray I.C., Vizor L., Brooker D., Whitehill E., et al. 2000. A systematic, genome-wide, phenotype-driven mutagenesis programme for gene function studies in the mouse. *Nat. Genet.* **25:** 440–443.

Oiso N., Riddle S.R., Serikawa T., Kuramoto T., and Spritz R.A. 2004. The rat Ruby (*R*) locus is *Rab38*: Identical mutations in Fawn-hooded and Tester-Moriyama rats derived from an ancestral Long Evans rat sub-strain. *Mamm. Genome* **15**: 307–314.

Okimoto K., Sakurai J., Kobayashi T., Mitani H., Hirayama Y., Nickerson M.L., Warren M.B., Zbar B., Schmidt L.S., and Hino O. 2004. A germ-line insertion in the Birt-Hogg-Dubé (*BHD*) gene gives rise to the Nihon rat model of inherited renal cancer. *Proc. Natl. Acad. Sci.* **101**: 2023–2027.

Olofsson P., Holmberg J., Pettersson U., and Holmdahl R. 2003a. Identification and isolation of dominant susceptibility loci for pristane-induced arthritis. *J. Immunol.* **171**: 407–416.

Olofsson P., Holmberg J., Tordsson J., Lu S., Akerstrom B., and Holmdahl R. 2003b. Positional identification of *Ncf1* as a gene that regulates arthritis severity in rats. *Nat. Genet.* **33**: 25–32.

Ovcharenko I., Nobrega M.A., Loots G.G., and Stubbs L. 2004. ECR Browser: A tool for visualizing and accessing data from comparisons of multiple vertebrate genomes. *Nucleic Acids Res.* **32**: W280–W286.

Pandey J., Cracchiolo D., Hansen F.M., and Wendell D.L. 2002. Strain differences and inheritance of angiogenic versus angiostatic activity in oestrogen-induced rat pituitary tumours. *Angiogenesis* **5**: 53–66.

Petersen G., Johnson P., Andersson L., Klinga-Levan K., Gómez-Fabre P.M., and Ståhl F. 2005. RatMap—Rat genome tools and data. *Nucleic Acids Res.* **33**: D492–D494.

Pinto-Sietsma S.J. and Paul M. 1997. Transgenic rats as models for hypertension. *J. Hum. Hypertens.* **11**: 577–581.

Pletcher M.T., McClurg P., Batalov S., Su A.I., Barnes S.W., Lagler E., Korstanje R., Wang X., Nusskern D., Bogue M.A., et al. 2004. Use of a dense single nucleotide polymorphism map for in silico mapping in the mouse. *PLoS Biol.* **2**: e393.

Pravenec M., Klir P., Kren V., Zicha J., and Kunes J. 1989. An analysis of spontaneous hypertension in spontaneously hypertensive rats by means of new recombinant inbred strains. *J. Hypertens.* **7**: 217–221.

Pravenec M., Kren V., Krenova D., Bila V., Zidek V., Simakova M., Musilova A., van Lith H.A., and van Zutphen L.F. 1999. HXB/Ipcv and BXH/Cub recombinant inbred strains of the rat: Strain distribution patterns of 632 alleles. *Folia Biol.* **45**: 203–215.

Pravenec M., Gauguier D., Schott J.J., Buard J., Kren V., Bila V., Szpirer C., Szpirer J., Wang J.M., Huang H., et al. 1995. Mapping of quantitative trait loci for blood pressure and cardiac mass in the rat by genome scanning of recombinant inbred strains. *J. Clin. Invest.* **96**: 1973–1978.

Pravenec M., Gauguier D., Schott J.J., Buard J., Kren V., Bila V., Szpirer C., Szpirer J., Wang J.M., Huang H., et al. 1996. A genetic linkage map of the rat derived from recombinant inbred strains. *Mamm. Genome* **7**: 117–127.

Pravenec M., Landa V., Zidek V., Musilova A., Kren V., Kazdova L., Aitman T.J., Glazier A.M., Ibrahimi A., Abumrad N.A., et al. 2001. Transgenic rescue of defective Cd36 ameliorates insulin resistance in spontaneously hypertensive rats. *Nat. Genet.* **27**: 156–158.

Pravenec M., Zidek V., Musilova A., Simakova M., Kostka V., Mlejnek P., Kren V., Krenova D., Bila V., Mikova B., et al. 2002. Genetic analysis of metabolic defects in the spontaneously hypertensive rat. *Mamm. Genome* **13**: 253–258.

Provoost A.P. 1994. Spontaneous glomerulosclerosis: Insights from the fawn-hooded rat. *Kidney Int. Suppl.* **45**: S2–S5.

Rapp J.P. 1982. Dahl salt-susceptible and salt-resistant rats. A review. *Hypertension* **4**: 753–763.

———. 2000. Genetic analysis of inherited hypertension in the rat. *Physiol. Rev.* **80**: 135–172.

Reaven G.M., Twersky J., and Chang H. 1991. Abnormalities of carbohydrate and lipid metabolism in Dahl rats. *Hypertension* **18**: 630–635.

Robinson R. 1987. Genetic linkage in the Norway rat. *Genetica* **74**: 137–142.

Roman R.J. and Kaldunski M. 1991. Pressure natriuresis and cortical and papillary blood flow in inbred Dahl rats. *Am. J. Physiol.* **261**: R595–R602.

Rost S., Fregin A., Ivaskevicius V., Conzelmann E., Hortnagel K., Pelz H.J., Lappegard K., Seifried E., Scharrer I., Tuddenham E.G., et al. 2004. Mutations in *VKORC1* cause warfarin resistance and multiple coagulation factor deficiency type 2. *Nature* **427**: 537–541.

Roth M.P., Viratelle C., Dolbois L., Delverdier M., Borot N., Pelletier L., Druet P., Clanet M., and Coppin H. 1999. A genome-wide search identifies two susceptibility loci for experimental autoimmune encephalomyelitis on rat chromosomes 4 and 10. *J. Immunol.* **162**: 1917–1922.

Rubattu S., Volpe M., Kreutz R., Ganten U., Ganten D., and Lindpaintner K. 1996. Chromosomal mapping of quantitative trait loci contributing to stroke in a rat model of complex human disease. *Nat. Genet.* **13**: 429–434.

Scheetz T.E., Laffin J.J., Berger B., Holte S., Baumes S.A., Brown R., II, Chang S., Coco J., Conklin J., Crouch K., et al. 2004. High-throughput gene discovery in the rat. *Genome Res.* **14**: 733–741.

Sebastiani P., Lazarus R., Weiss S.T., Kunkel L.M., Kohane I.S., and Ramoni M.F. 2003. Minimal haplotype tagging. *Proc. Natl. Acad. Sci.* **100**: 9900–9905.

Segre J.A., Nemhauser J.L., Taylor B.A., Nadeau J.H., and Lander E.S. 1995. Positional cloning of the nude locus: Genetic, physical, and transcription maps of the region and mutations in the mouse and rat. *Genomics* **28**: 549–559.

Shisa H., Lu L., Katoh H., Kawarai A., Tanuma J., Matsushima Y., and Hiai H. 1997. The LEXF: A new set of rat recombinant inbred strains between LE/Stm and F344. *Mamm. Genome* **8**: 324–327.

Smits B.M., Mudde J., Plasterk R.H., and Cuppen E. 2004a. Target-selected mutagenesis of the rat. *Genomics* **83**: 332–334.

Smits B.M.G., van Zutphen B.F.M., Plasterk R.H.A., and Cuppen E. 2004b. Genetic variation in coding regions between and within commonly used inbred rat strains. *Genome Res.* **14**: 1285–1290.

Smits B.M.G., Guryev V., Zeegers D., Wedekind D., Hedrich H.J., and Cuppen E. 2005. Efficient single nucleotide polymorphism discovery in laboratory rat strains using wild rat-derived SNP candidates. *BMC Genomics* **6**: 170.

Smits B.M.G., Mudde J.B., van de Belt J., Verheul M., Olivier J., Homberg J., Guryev V., Cools A.R., Ellenbroek B.A., Plasterk R.H.A., and Cuppen E. 2006. Generation of gene knockouts and mutant models in the laboratory rat by ENU-driven Target-selected mutagenesis. *Pharmacogenet. Genomics* **16**: 159–169.

Snell G. 1948. Methods for the study of histocompatibility genes. *J. Genet.* **49**: 87–108.

Steen R.G., Kwitek-Black A.E., Glenn C., Gullings-Handley J., Van Etten W., Atkinson O.S., Appel D., Twigger S., Muir M., Mull T., et al. 1999. A high-density integrated genetic linkage and radiation hybrid map of the laboratory rat. *Genome Res.* **9**: AP1–8, insert.

Stephens M., Sloan J.S., Robertson P.D., Scheet P., and Nickerson D.A. 2006. Automating sequence-based detection and genotyping of SNPs from diploid samples. *Nat. Genet.* **38**: 375–381.

Stoll M., Cowley A.W., Jr., Tonellato P.J., Greene A.S., Kaldunski M.L., Roman R.J., Dumas P., Schork N.J., Wang Z., and Jacob H.J. 2001. A genomic-systems biology map for cardiovascular function. *Science* **294**: 1723–1726.

Stoll M., Kwitek-Black A.E., Cowley A.W., Jr., Harris E.L., Harrap S.B., Krieger J.E., Printz M.P., Provoost A.P., Sassard J., and Jacob H.J. 2000. New target regions for human hypertension via comparative genomics. *Genome Res.* **10:** 473–482.

Stylianou I.M., Christians J.K., Keightley P.D., Bünger L., Clinton M., Bulfield G., and Horvat S. 2004. Genetic complexity of an obesity QTL (*Fob3*) revealed by detailed genetic mapping. *Mamm. Genome* **15:** 472–481.

Sugiyama F., Churchill G.A., Higgins D.C., Johns C., Makaritsis K.P., Gavras H., and Paigen B. 2001. Concordance of murine quantitative trait loci for salt-induced hypertension with rat and human loci. *Genomics* **71:** 70–77.

Svenson K.L., Cheah Y.C., Shultz K.L., Mu J.L., Paigen B., and Beamer W.G. 1995. Strain distribution pattern for SSLP markers in the SWXJ recombinant inbred strain set: Chromosomes 1 to 6. *Mamm. Genome* **6:** 867–872.

Tanomura H., Miyake T., Taniguchi Y., Manabe N., Kose H., Matsumoto K., Yamada T., and Sasaki Y. 2002. Detection of a quantitative trait locus for intramuscular fat accumulation using the OLETF rat. *J. Vet. Med. Sci.* **64:** 45–50.

Tian X.L., Pinto Y.M., Costerousse O., Franz W.M., Lippoldt A., Hoffmann S., Unger T., and Paul M. 2004. Over-expression of angiotensin converting enzyme-1 augments cardiac hypertrophy in transgenic rats. *Hum. Mol. Genet.* **13:** 1441–1450.

Toga A.W., Santori E.M., Hazani R., and Ambach K. 1995. A 3D digital map of rat brain. *Brain Res. Bull.* **38:** 77–85.

Twigger S.N., Pasko D., Nie J., Shimoyama M., Bromberg S., Campbell D., Chen J., Dela Cruz N., Fan C., Foote C., et al. 2005. Tools and strategies for physiological genomics: The Rat Genome Database. *Physiol. Genomics* **23:** 246–256.

Visscher P.M. 1999. Speed congenics: Accelerated genome recovery using genetic markers. *Genet. Res.* **74:** 81–85.

Vitt U., Gietzen D., Stevens K., Wingrove J., Becha S., Bulloch S., Burrill J., Chawla N., Chien J., Crawford M., et al. 2004. Identification of candidate disease genes by EST alignments, synteny, and expression and verification of Ensembl genes on rat chromosome 1q43-54. *Genome Res.* **14:** 640–650.

Wade C.M., Kulbokas E.J., III, Kirby A.W., Zody M.C., Mullikin J.C., Lander E.S., Lindblad-Toh K., and Daly M.J. 2002. The mosaic structure of variation in the laboratory mouse genome. *Nature* **420:** 574–578.

Wang X., Korstanje R., Higgins D., and Paigen B. 2004. Haplotype analysis in multiple crosses to identify a QTL gene. *Genome Res.* **14:** 1767–1772.

Ward C.J., Hogan M.C., Rossetti S., Walker D., Sneddon T., Wang X., Kubly V., Cunningham J.M., Bacallao R., Ishibashi M., et al. 2002. The gene mutated in autosomal recessive polycystic kidney disease encodes a large, receptor-like protein. *Nat. Genet.* **30:** 259–269.

Ways J.A., Cicila G.T., Garrett M.R., and Koch L.G. 2002. A genome scan for loci associated with aerobic running capacity in rats. *Genomics* **80:** 13–20.

Weber J.L. and P.E. May. 1989. Abundant class of human DNA polymorphisms which can be typed using the polymerase chain reaction. *Am. J. Hum. Genet.* **44:** 388–396.

Wendell D.L. and Gorski J. 1997. Quantitative trait loci for estrogen-dependent pituitary tumor growth in the rat. *Mamm. Genome* **8:** 823–829.

Wilder S.P., Bihoreau M.-T., Argoud K., Watanabe T.K., Lathrop M., and Gauguier D. 2004. Integration of the rat recombination and EST maps in the rat genomic sequence and comparative mapping analysis with the mouse genome. *Genome Res.* **14:** 758–765.

Wiltshire T., Pletcher M.T., Batalov S., Barnes S.W., Tarantino L.M., Cooke M.P., Wu H., Smylie K., Santrosyan A., Copeland N.G., et al. 2003. Genome-wide single-nucleotide polymorphism analysis defines haplotype patterns in mouse. *Proc. Natl. Acad. Sci.* **100:** 3380–3385.

Yalcin B., Fullerton J., Miller S., Keays D.A., Brady S., Bhomra A., Jefferson A., Volpi E., Copley R.R., Flint J., and Mott R. 2004a. Unexpected complexity in the haplotypes of commonly used inbred strains of laboratory mice. *Proc. Natl. Acad. Sci.* **101:** 9734–9739.

Yalcin B., Willis-Owen S.A., Fullerton J., Meesaq A., Deacon R.M., Rawlins J.N., Copley R.R., Morris A.P., Flint J., and Mott R. 2004b. Genetic dissection of a behavioral quantitative trait locus shows that Rgs2 modulates anxiety in mice. *Nat. Genet.* **36:** 1197–1202.

Yeung R.S., Xiao G.H., Jin F., Lee W.C., Testa J.R., and Knudson A.G. 1994. Predisposition to renal carcinoma in the Eker rat is determined by germ-line mutation of the tuberous sclerosis 2 (*TSC2*) gene. *Proc. Natl. Acad. Sci.* **91:** 11413–11416.

Yokoi N., Namae M., Wang H.Y., Kojima K., Fuse M., Yasuda K., Serikawa T., Seino S., and Komeda K. 2003. Rat neurological disease *creeping* is caused by a mutation in the *reelin* gene. *Brain Res. Mol. Brain Res.* **112:** 1–7.

Yokoi N., Komeda K., Wang H.Y., Yano H., Kitada K., Saitoh Y., Seino Y., Yasuda K., Serikawa T., and Seino S. 2002. *Cblb* is a major susceptibility gene for rat type 1 diabetes mellitus. *Nat. Genet.* **31:** 391–394.

Zan Y., Haag J.D., Chen K.S., Shepel L.A., Wigington D., Wang Y.R., Hu R., Lopez-Guajardo C.C., Brose H.L., Porter K.I., et al. 2003. Production of knockout rats using ENU mutagenesis and a yeast-based screening assay. *Nat. Biotechnol.* **21:** 645–651.

Zhou Q., Renard J.P., Le Friec G., Brochard V., Beaujean N., Cherifi Y., Fraichard A., and Cozzi J. 2003. Generation of fertile cloned rats by regulating oocyte activation. *Science* **302:** 1179.

Zidek V., Pintir J., Musilova A., Bila V., Kren V., and Pravenec M. 1999. Mapping of quantitative trait loci for seminal vesicle mass and litter size to rat chromosome 8. *J. Reprod. Fertil.* **116:** 329–333.

Zimdahl H., Nyakatura G., Brandt P., Schulz H., Hummel O., Fartmann B., Brett D., Droege M., Monti J., Lee Y.A., et al. 2004. A SNP map of the rat genome generated from cDNA sequences. *Science* **303:** 807.

30 | The Cat

Monika J. Lipinski, Nicholas Billings, and Leslie A. Lyons

Department of Population Health and Reproduction, School of Veterinary Medicine, University of California, Davis, California 95616

INTRODUCTION

The domestic cat has many secrets hidden in its genome; these same secrets have already been revealed for other species. Domestic cat origins are a mystery, and the causes for much of the cat's splendor and beauty are yet to be discovered. Domestic cat genomics have benefited from the improved technologies of the Human Genome Project, and the cat now represents the second carnivore species with a focused genome sequencing effort. Currently, a low-coverage, 2X sequence is

available for the domestic cat, but deeper coverage and resequencing in other breeds and individuals are on the horizon. Regardless of the sequence coverage, cat genetics and genomics are already leaping forward. Many cat diseases and phenotypic mutations have been identified within the last few years. The unique breed development for the cat and the differences in genome organization suggest that the cat is not just a small dog. In this chapter, we present the nuances of feline breed development and dynamics, and the common and unique biological and physiological aspects that can be the focus of feline-specific studies. Such studies will undoubtedly lead to health improvements in felines, humans, and other species.

CAT ORIGINS

The domestic cat is one of 36 extant species of felids (Kitchener 1991; Seidensticker and Lumpkin 1991; Sunquist and Sunquist 2002), grouping with four other Old World wildcat species in the domestic cat "*Felis*" lineage. This lineage includes *F. bieti*, the Chinese desert cat; *F. margarita*, the sand cat; *F. nigripes*, the black-footed cat; and *F. chaus*, the jungle cat (Johnson et al. 2006). Two additional wildcats, the European wildcat, *F. silvestris*, and the African wildcat, *F. libyca*, are usually considered to be separate species and distinct from the domestic cat. However, these two wildcats produce fertile hybrids with domestic cats, and the speciation of the small wildcats and domestic cats remains a mystery, particularly since genetic studies have not yet been able to focus

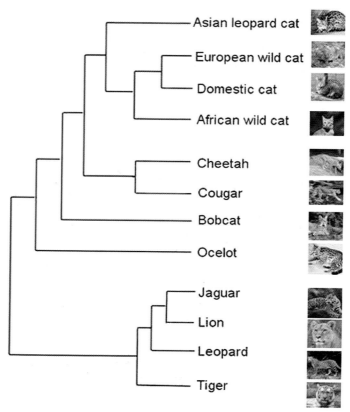

FIGURE 30-1. Phylogenetic relationship of small wildcats. (Based on Johnson et al. 2006.)

on their recent divergence (see Fig. 30-1). Many other subspecies of European, Asian, and African wildcats have been described. However, the true origins and the number of domestication events that led to the domestication of the cat have not been resolved and remain mysterious.

The focal points for cat domestication likely occurred at sites of agricultural development, and the linkage of the domestication process to grain storage appears evident. Various independent sites of agricultural development are known, including the Fertile Crescent region, particularly in Syria around the 10th and 9th millennia B.C. Agriculture appeared in the Nile Valley of Egypt some 5000 years after the earliest farming settlements in the upper Fertile Crescent. Additionally, crop assemblages had already migrated to Asia and the India subcontinent a few thousand years earlier (Zohary and Hopf 2002), implying that Egypt may have been one of the later regions for cat domestication. The Egyptians were prolific in documenting their culture, and the presence of the "domestic cat" is highly evident in their society. Hence, many assume that the Egyptians were the first to domesticate cats. Some of the earliest artist representations of cats occur around 2500 B.C., but definitive breeding programs to produce cats for mummification were not established until the late Ptolemic period, from 500 B.C. to 100 A.D. (Malek 2006).

Several small wildcat species, namely African, Asian, and European wildcats, were available in all the early farming areas as the potential progenitors of the domestic cat. As humans made the transition from hunter–gatherers to the more sedentary lifestyle of the farmer, permanent settlements developed. Villages produced refuse piles and grain stores, attracting mice and rats, primary prey species for the small wildcat. Thus, it is likely that cats actively participated in their owners' domestication, with both humans and felines developing a symbiotic, mutual tolerance. Cats that accepted human presence obtained a more consistent food supply. Human acceptance of cats initially supported the control of pestilence and of contamination of grain stores, thereby controlling zoonotic disease transmission from rodents. But when did the cat begin to seek human affection and companionship, and when did man develop the first controlled cat breeding programs? The answers to these questions remain to be discovered. Regardless of where or when cat breeding developed, the cat is one of the most recently domesticated of our companion animals, distinct from the much earlier domestication of the other favorite companion animal, the domestic dog. Thus, the dynamics of genetic variation across cat breeds is likely significantly different from species that have more ancient domestication events and longer breed histories.

CAT BREEDS

Cat breed dynamics are significantly different from those of other companion and agricultural species. These nuances are important for the development of the appropriate genetic tools, resources, and techniques that will be the most beneficial and efficient for cat genetic research. Random-bred and feral cats, not fancy breed populations, represent the overwhelming majority of cats throughout the world. Considering the worldwide distribution of cats, the U.S. has the highest proportion of purebred cats. The University of California Davis Veterinary Teaching Hospital has over 20,000 feline clients per year; however, only 10–15% are purebred cats (Louwerens et al. 2005). Thus, population studies for complex traits will likely require markers and linkage disequilibrium (LD) estimates in a variety of random-bred cat populations.

The first documented cat show that judged cats on their aesthetic value occurred in London at the Crystal Palace in 1871. This first competition presented a handful of breeds, including the Persian, Abyssinian, and Siamese. The first cat registry in the U.S., the Cat Fanciers Association (CFA), was developed by 1905, with the Maine coon being an additional American breed. Various encyclopedic volumes pertaining to the domestic cat list approximately 50–80 cat breeds worldwide (Fogle 1997; Morris 1999). However, a majority of breeds were developed in the past 50 years. Many listed breeds have not been developed into viable populations and, hence, have been lost to posterity. The most pertinent cat breeds are listed in Table 30-1. Most worldwide cat fancy associations recognize approximately 35–40 cat breeds. However, only a few breeds overwhelmingly dominate the breed populations. Persian cats and their related breeds, such as Exotics, a shorthaired Persian variety, are the most popular cat breeds worldwide and represent an overwhelming majority of purebred cats. Although perhaps only 20–30% of cats produced by breeders are registered, the CFA (one of the largest cat registries worldwide) generally registers approximately 40,000 total purebreds annually. Approximately 16,000–20,000 are Persians, and approximately 3,000 are Exotics. Thus, the Persian group of cats represents over 50% of the cat fancy population. Common breeds that generally have at least 1,000 annual registrants are Abyssinians, Maine coon cats, and Siamese. Other popular breeds include Birman and Burmese (Cat Fanciers' Almanac). Most of these popular breeds also represent the oldest and most established cat breeds worldwide. Thus, genetic tools and single-nucleotide polymorphism (SNP) studies should primarily focus on domestic cats and a handful of fancy cat breeds.

Additionally, substructuring of the breeds may need to be evaluated for genetic applications. Many breeds are derived from an older breed, forming breed families. Approximately 17 breeds can be considered "foundation" or "natural" breeds (Table 30-1), implying that many other breeds have been derived from these foundation cats. Derived breeds are often single-gene variants, such as longhaired and shorthaired varieties, or even a no-haired variety, as found in the Devon rex and Sphynx grouping. Color variants also tend to demarcate breeds, such as the "pointed" variety of the Persian, known as the Himalayan by many cat enthusiasts and as a separate breed by some associations. Many cat breeds originated from single-gene traits (such as folded ears of the Scottish fold and dorsally curled pinnea of the American curl) and then later developed into a more conformationally unique breed. The newly identified spontaneous mutations are recognized often in random-bred cat populations, followed by morphological molding with various desired breed combinations. Thus, many new breeds and some established breeds have allowable outcrosses to influence the "type" and to support genetic diversity in the breed foundation. Persians have a highly desired brachycephalic head type and thus tend to influence many breeds. Breeders

TABLE 30-1. Traditional Cat Breeds and Breed Families

Breed	Origin	Est'd. date	Derived breeds	Head type
Abyssinian[a]	India	1868	Somali[b]	mesaticephalic
American bobtail	mutation–U.S.	1960		mesaticephalic
American curl	mutation–U.S.	1981		mesaticephalic
American shorthair	U.S.	1966		brachycephalic
American wirehair	mutation–U.S.	1966		mesaticephalic
Australian mist	mix–Australia	1990s		mesaticephalic
Birman[a]	Burma	<1868	Snowshoe[b]	mesaticephalic
British shorthair[a]	England	1870s		brachycephalic
Burmese[a]	Burma	1350–1767	(Asian) Bombay, Tiffanie,[b] Malayan, Burmilla	brachycephalic
Cornish rex	mutation–U.K.	1950		dolichocephalic
Chartreux[a]	France	1300		mesaticephalic
Devon rex	mutation–U.K.	1960	Sphynx (1966)	mesaticephalic
Egyptian Mau[a]	Egypt	1953		mesaticephalic
European shorthair	Europe			brachycephalic
Japanese bobtail[a]	Japan	500–1100		mesaticephalic
Korat[a]	Thailand	1350–1767		mesaticephalic
LaPerm	mutation–U.S.	1986		mesaticephalic
Maine coon[a]	U.S.	1860s		mesaticephalic
Manx	Isle of Man	<1868	Cymric[b]	mesaticephalic
Munchkin	U.S.	1990s		mesaticephalic
Norwegian forest[a]	Norway	<1868		mesaticephalic
Ocicat	crossbred	1964	Siamese × Abyssinian	mesaticephalic
Ojos Azules	mutation	1980s		mesaticephalic
Persian[a]	Persia	<1868	Exotic,[b] Kashmir, Himalayan, Peke-faced, Burmilla	brachycephalic
Russian blue[a]	Russia	<1868	Nebelung[b]	mesaticephalic
Ragdoll	selection	1960s	Ragamuffin	
Scottish Fold	mutation	1961	Highland fold[b] (Coupari)	brachycephalic
Selkirk rex	mutation–U.S.	1980s		mesaticephalic
Siamese[a]	Thailand	1350–1767	Colorpoint,[b] Javanese,[b] Balinese,[b] Oriental,[b] Havana brown, Don Sphynx	dolichocephalic
Siberian[a]	Russia	<1868		mesaticephalic
Sokoke[a]	Africa			mesaticephalic
Tonkinese	crossbred	1950s	Siamese × Burmese	brachycephalic
Turkish Angora[a]	Ankara	1400		mesaticephalic
Turkish Van[a]	Van Lake	<1868		mesaticephalic

[a]Denotes foundation or natural cat breeds.

[b]Many derived breeds are long- or shorthaired varieties of the foundation breed but have different breed names; others are delineated by longhair or shorthair in the breed name. At least 10 additional rex-coated cat populations have not developed into viable populations or are extinct.

desiring the dolichocephalic type often outcross with the Siamese family of cats. The outcrosses that are valid for any breed can vary between cat registries, and the same breed may have a different name depending on the country. For example, the Burmese registered by the Governing Council of the Cat Fancy (GCCF) in the United Kingdom and the Feline International Federations (FIFe) in Europe are known as the Foreign Burmese breed in the U.S., and these cat "breeds" have significantly different craniofacial types between the countries. The Havana brown has developed into a distinctive breed in the U.S., also with a significantly different craniofacial structure than its foundation breeds, the Siamese and Oriental shorthairs. However, in Europe, the chestnut color variety of the Oriental shorthair is similar to the Havana brown. Some breeds, such as Korats and Turkish Vans, have very similar standards across most or all countries and registries. Oddly, some cat breeds are actually hybrids between clearly different species of cats and the domestic cat. Asian leopard cats are part of the foundation of the Bengal breed, which is a highly popular breed worldwide, but is not registered by the CFA. Serval hybrids, known as Savannahs, and jungle cat hybrids, known as Chaussies, are also growing in popularity. Hence, genomic tools should give some attention to these three cat species to support disease studies within these hybrid cat breeds.

Thus, from a phylogenetic point of view, cat breeders are "splitters" instead of "lumpers," and a few foundation breeds encompass most variation of cat breeds. An evaluation of 39 microsatellite markers in 19 cat breeds, 12 random-bred populations, and 3 wildcat subspecies has revealed some of the basal relationships of the foundation cat breeds (see Fig. 30-2) by neighbor joining (N-J) trees constructed using the software Phylip: http://evolution.genetics.washington.edu/phylip.html (Felsenstein 1989). Both Cavalli-Sforza's chord measure (Cavalli-Sforza and Edwards 1967) and Nei's genetic distance (Nei 1972) produce similar results, but the N-J tree produced by chord distance (Fig. 30-2) had stronger support, as indicated by higher bootstrap values. (Bootstrapping is a resampling [pseudoreplication of data] method used to estimate the reliability of a tree. The numbers correspond to the percentage of times that a particular branching pattern occurred when the data was resampled and trees were constructed.) The relationships show that breeds which are documented derivatives (such as the Singapura and the Burmese or the Havana brown and the Siamese) strongly cluster. Newly developing breeds from eastern Africa (such as the Sokoke) adhere to their feral relatives, cats from the Kenyan islands of Lamu and Pate. Additionally, three "clusterings" of cats appear to be evident: cats from the Far East, such as the Siamese, Burmese, Havana brown, Singapura, Korats, and Birmans; cats with Arabic influence, such as the wildcats and Kenyan cats; and cats from the Mediterranean, which include almost all other breeds and populations.

Additional structure of the cat breeds is evident using a Bayesian clustering analysis of the 19 breeds with the software Structure (Pritchard et al. 2000). Assuming that the 19 breeds were genet-

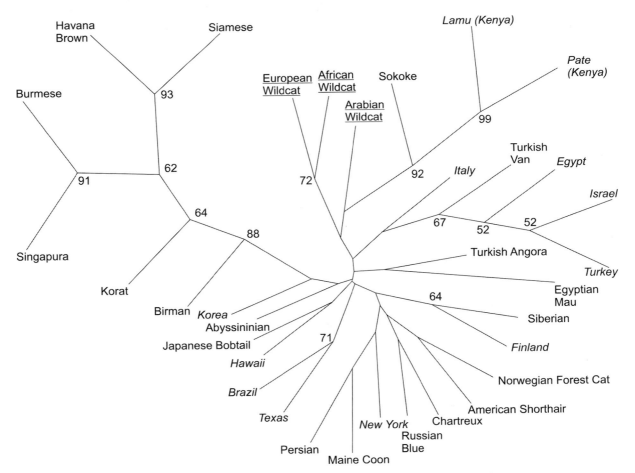

FIGURE 30-2. Phylogenetic analysis of cat breeds. Neighbor-joining trees were constructed using the software Phylip (Felsenstein 1989). Presented is the tree produced by chord distance. The numbers at the nodes represent the highest bootstrap values. Random-bred populations are in italics, and wildcat populations are underlined.

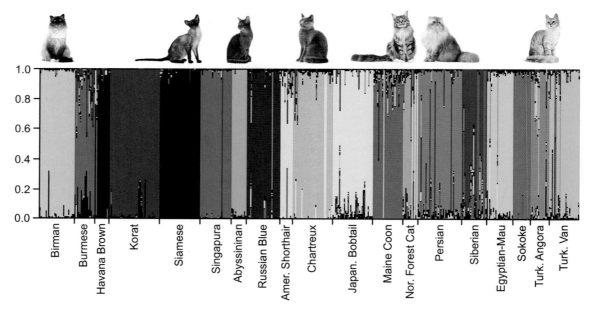

FIGURE 30-3. Structure of foundation domestic cat breeds. Individuals from 19 cat breeds were genotyped for 39 microsatellite loci. Using the software Structure, the breeds can be clearly distinguished assuming the breeds represent the true number of populations (K = 19). Cat breeds represented by the same color, such as Havana brown and Siamese or Burmese and Singapura, could not be resolved.

ically defined populations, the number of inferred populations was set to 10 (K = 19). This method separated the cats into inferred clusters that corresponded to their breed labels with high posterior probability of assignment and correctly assigned individuals to their respective breeds with 98% accuracy (see Fig. 30-3). A majority of the analyzed breeds form strong groupings with some allele sharing between breeds developed from similar regions. The Far Eastern breeds, Birmans to Singapuras, are clearly defined but do share some genetic components. Four of the Far Eastern breeds remained unresolved. The Bayesian model was unable to separate the Singapura and Burmese breeds, suggesting a recent common origin of the two breeds. Cat-breeding folklore suggests that both breeds are indigenous to ancient Burma. Alternately, popular belief also suggests that Burmese cats from the U.S. were taken to Singapore, bred with native cats, and later returned to the U.S. as the new Singapura breed. Additionally, the Havana brown and Siamese separation was inconsistent. Abyssinians appear to be a strongly demarcated breed, whereas the more European breeds are slightly less distinct. In recent years, random-bred or feral populations of cats have been developed into newer, region-specific breeds, such as the Siberian from Russia and the Sokoke from Africa. The Siberian breed has genetic variation comparable to random-bred populations, as seen in the Structure analysis (see Fig. 30-3).

CAT PHENOTYPIC VARIATION

Single-gene mutations established many of the cat breeds, and many breed populations segregate for the variant and normal alleles. Over two dozen single-gene phenotypic variants have been identified in domestic cats and associated breeds, including dominant, codominant, recessive, sex-linked, and homozygote-lethal traits (Table 30-2). The genetic heterogeneity, pleiotrophic, and epistatic effects of cat coat colorations allow cats to be an excellent primer for basic and advanced genetic instruction. Although one of the first traits to be mapped to a particular chromosome in any species was the sex-linked orange coloration of the cat (Ibsen 1916), the first autosomal trait linkage in the cat was not recognized until 70 years later, the polymorphism for

TABLE 30-2. Phenotypic Variation in the Domestic Cat

Locus	Alleles	Phenotype	Variant	Inheritance
Agouti (ASIP)	$A > a$	banded fur	solid, self	recessive
Black (TYRP1)	$B > b > b^l$	black pigment	brown pigments	allelic series
Color (TYR)	$C > c^b > c^s > c$	normal color	temperature-sensitive colors, albino	allelic series
Curl	$Cu > cu$	normal pinnea	curled pinnea	dominant
Dwarfism	not designated	normal	shortened legs	dominant
Dilute (MLPH)	$D > d$	dense pigment	dilute pigment	recessive
Fold	$Fd > fd$	normal pinnea		dominant
Gloves	$G > g$	normal color	white feet	recessive
Hairless	$Hr > hr$	normal fur	no hair	recessive
Inhibitor	$I > i$	normal color		dominant
Long (FGF5)	$L > l$	short fur	long fur	recessive
Manx (tailless)	$M > m$	normal tail	no (short) tail	dominant
Orange	X^o, x^o	normal color	orangish pigments	sex-linked
Peterbald	not designated	hairless	normal hair	dominant
Polydactyla	$Pd > pd$	extra toes	normal toes	dominant
Rex (Cornish)	$R > r$	normal hair	curly hair	recessive
Rex (Devon)	$R^e > r^e$	normal hair	curly hair	recessive
Rex (LaPerm)	not designated	curly hair	normal fur	dominant
Rex (Selkirk)	not designated	curly hair	normal fur	dominant
Rex (wirehair)	not designated	normal hair	curly hair	incomplete
Spotting (KIT?)	S, s	normal color	ventral white	additive
Tabby	$T^u > T^m > t^b$	no pattern	pattern: stripes	allelic series
White	$W > w$	all white	normal color	dominant

Loci with known gene mutations have the gene symbol in parentheses. The phenotype represents the dominant allele and the variants are the recessive alleles.

hemoglobin and the Siamese pattern caused by mutations in *tyrosinase* (O'Brien et al. 1986). Now that genomic tools for the cat have become effective, the genetic mechanisms for many cat phenotypes are being deciphered, including the Siamese pattern, which is also called "points." The points coloration found in Siamese and Himalayans was easily identified because the same gene, *tyrosinase*, causes similar phenotypes in other species (Lyons et al. 2005b). To date, most phenotypic mutations in the cat have been identified by a candidate gene approach because an obvious enzymatic mechanism has been disrupted or the same coat coloration has been documented by other species. Family-based linkage studies also have implicated candidate genes, such as *KIT* and white spotting (Cooper et al. 2006), but other linkage analyses have implicated genetic regions with no known candidates, such as chromosome B1 and the *Tabby* locus (Lyons et al. 2006). To date, all the coloration mutations in cats have been shown to be the same in different breeds and are likely identical by descent; however, the several mutations likely confer long fur in the domestic cat (Drogemuller et al. 2007).

CAT DISEASE MUTATIONS

Although many of the phenotypic traits of the cat seem to have mutations that are identical by descent across the breeds, disease mutations are proving to be more distinct across breeds. Documented in the cat are at least 277 disorders or "phenes" with a heritable component in other species; at least 46 are suggested to be single-gene traits in cats (http://omia.angis.org.au/). Eighteen different genes account for the 24 known mutations causing diseases in cats (Table 30-3). However, only 10 disease-causing mutations are segregating in cat breeds. Several cat diseases are maintained only in research colonies. Although several of the disease mutations found in cats have been identified in a breed, not all are of concern to the breed, because the disease has been eliminated or was a sporadic mutation.

TABLE 30-3. Cat Traits and Diseases with Known Mutations

Disease/Coat color	Gene	Mutation	Breeds	Ref.
Agouti	*ASIP*	del122–123	all breeds	Eizirik et al. (2003)
Brown	*TYRP1*	b = C8G	all breeds	Lyons et al. (2005a)
		b[l] = C298T		
Dilution	*MLPH*	T83del	all breeds	Ishida et al. (2006)
Color	*TYR*	c[b] = G715T	all breeds	Lyons et al. (2005b);
		c[s] = G940A		Imes et al. (2006)
		c = C975del		
AB blood type (type B)	*CMAH*	18indel-53	all breeds	Bighignoli et al. (2007)
Gangliosidosis 1[a]	*GBL1*	G1457C	Korat, Siamese	De Maria et al. (1998)
Gangliosidosis 2[a]	*HEXB*	15 bp del (intron)	Burmese	(unpubl.)
Gangliosidosis 2	*HEXB*	inv1467–1491	DSH	Martin et al. (2004)
Gangliosidosis 2	*HEXB*	C667T	DSH (Japan)	Kanae et al. (2006)
Gangliosidosis 2[a]	*HEXB*	C39del	Korat	Muldoon et al. (1994)
Gangliosidosis 2	*GM2A*	del390–393	DSH	Martin et al. (2005)
Glycogen storage disease IV[a]	*GBE1*	230 bp ins 5′–6 kb del	Norwegian forest	Fyfe et al. (2007)
Hemophilia B	*F9*	G247A	DSH	Goree et al.(2005)
Hemophilia B	*F9*	C1014T	DSH	Goree et al. (2005)
Hypertrophic cardiomyopathy[a]	*MYBPC*	G93C	Maine coon	Meurs et al. (2005a)
Hypertrophic cardiomyopathy	*MYBPC*	C2458T	Ragdolls	Meurs et al. (2007)
Lipoprotein lipase deficiency	*LPL*	G1234A	DSH	Ginzinger et al. (1996)
Alpha mannosidosis	*LAMAN*	del1748–1751	Persian	Berg et al. (1997)
Mucolipidosis II	*GNPTA*	C2655T	DSH	Giger et al. (2006)
Mucopolysaccharidosis I	*IDUA*	del1047–1049	DSH	He et al. (1999)
Mucopolysaccharidosis VI	*ARSB*	T1427C	Siamese	Yogalingam et al. (1996)
Mucopolysaccharidosis VI	*ARSB*	G1558A	Siamese	Yogalingam et al. (1998)
Mucopolysaccharidosis VII	*GUSB*	A1052G	DSH	Fyfe et al. (1999)
Muscular dystrophy	*DMD*	900 bp del M promoter–exon 1	DSH	Winand et al. (1994)
Niemann-Pick C	*NPC*	G2864C	Persian	Somers et al. (2003)
Polycystic kidney disease[a]	*PKD1*	C10063A	Persian	Lyons et al. (2004)
Progressive retinal atrophy[a]	*CEP290*	IVS50 + 9T>G	Abyssinian	Menotti-Raymond et al. (2007)
Pyruvate kinase deficiency[a]	*PKLR*	13 bp del in exon 6	Abyssinian	(unpubl.)[b]
Spinal muscular atrophy[a]	*LIX1*	140 kb del, exons 4–6	Maine coon	Fyfe et al. (2006)

[a]Diseases that are currently segregating within the breed.
[b]Mutations that are unpublished to date.

Most of the earliest disease mutations identified in the cat also have benefited from the candidate gene approach, because similar traits exist in other species with known genes and mutations. The first disease mutations found in the cat were a promoter mutation in Duchenne muscular dystrophy (DMD), *dystrophin* (Winand et al. 1994), and a mutation in the beta subunit of *beta-hexosaminidase A* (*HEXB*), the well-known lysosomal storage disease (LSD) in humans, Sandhoff disease (Muldoon et al. 1994). Neither condition is prevalent in breeds, and hence, neither constitutes a breed predisposition. Several additional inborn errors of metabolism or LSDs have been identified (Table 30-3) in Persians and Siamese and are maintained as research colonies, but again, these conditions do not segregate in the breed at large. Korats and Persians have more than one metabolism defect, and the MPS VI group of cats were found to be compound heterozygotes for mutations in *ARSB* (Crawley et al. 1998). Korats are a small population breed, and this breed continues to segregate for two gangliosidoses, GM1 and GM2, and the pointed phenotype, all of which are undesirable traits in the breed.

Two of the most prevalent disorder mutations in the cat cause cardiac and renal disease. Hypertrophic cardiomyopathy (HCM) is a heterogeneous cardiac disease that has been recognized in several breeds, including American shorthair, Bengals, Ragdolls, and Sphynx. However, HCM is scientifically documented as heritable in only the Maine coon cat (Kittleson et al. 1999). Mutations in *myosin C binding protein* (*MYCBP*) are highly correlated with clinical presentation of HCM in the Maine coon cat and Ragdolls (Meurs et al. 2005, 2007) and are considered the causative mutations.

Although the frequency of the disease has not been clearly established for the Maine coon breed, an active genetic typing program, combined with echocardiogram evaluations, is assisting in the reduction of this disease within the Maine coon breed. The same mutation has been evaluated in other breeds with HCM, but as in humans, HCM appears to be heterogenic in cats, and the mutation is not correlated with disease in other breeds (K. Meurs, pers. comm.; L. Lyons, data not shown).

Other diseases found in the cat, such as polycystic kidney disease (PKD) in Persian cats, have benefited from a combined genetic linkage analysis and candidate gene approach (Lyons et al. 2004; Young et al. 2005). Feline PKD was first clinically described as an autosomal dominant inherited trait in 1990 (Biller et al. 1990, 1996). Approximately 38% of Persians worldwide have PKD (Beck and Lavelle 2001; Barrs et al. 2001; Cannon et al. 2001; Barthez et al. 2003), making PKD the most common inherited disease in the domestic cat. PKD affects 600,000 people in the U.S. alone, and 12.5 million worldwide. There are more people afflicted with PKD than with cystic fibrosis, muscular dystrophy, hemophilia, Down syndrome, and sickle cell anemia combined. Over 90% of PKD is heritable. More than 60% of the individuals with PKD develop kidney failure, or end-stage renal disease (ESRD). Over 95% of cats with PKD will develop renal cysts by 8 months of age, which can be accurately determined by ultrasound. Depending on disease severity, cats can live a normal lifespan of 10–14 years with PKD, or succumb within a few years of onset. Similar disease variation and progression are also seen in humans. Ultrasonographic identification of renal cysts allowed the accurate ascertainment of large pedigrees of Persian cats with PKD. In humans, two genes, *PKD1* and *PKD2*, are responsible for a majority of PKD. The two major genes produce very large transcripts and, because limited domestic cat genetic sequence was available, a linkage analysis was first conducted to implicate a candidate gene for feline PKD prior to a candidate gene approach (Young et al. 2005). Once a cat microsatellite marker showed significant linkage to the region of the cat genome with *PKD1*, a gene scan identified a C10063A transversion that changes an arginine to an OPA stop codon, which should disrupt approximately 25% of the polycystin-1 protein (Lyons et al. 2004).

Most cat diseases and their mutations are specific to particular breeds. However, any disease found in the Persian or the Siamese families can spread to other breeds due to the influence of these foundation breeds on modifying morphological structures of new and developing breeds. For example, PKD has also been documented in Scottish folds, Selkirk rex, and British shorthairs, all brachycephalic breeds that have used Persians to modify structure (Lyons et al. 2004). HCM is considered highly prevalent in Maine coons, which is a large population breed, but GM1 and GM2 are at low frequencies in a very small population breed. Thus, breeders are now enthusiastically using DNA testing to identify carriers, but wrestling with the disease-management decisions for their breed.

Overall, 23 mutations cause diseases in the cat, four coat-color loci represent an additional 7 mutations, and a blood group mutation is associated with the type B blood group in cats. Cats have even been shown to lack a "sweet tooth" due to the pseudogenization of *Tas1r2* sweet receptor gene (Li et al. 2005, 2006). In addition to the more than 277 traits that could be heritable conditions in the cat, the identification of several additional mutations should be on the horizon. Active linkage studies have implicated genes or genetic regions for white spotting, tabby, orange, a progressive retinal atrophy in the Persian, and a craniofacial defect in the Burmese. Improved genetic resources will expedite mutation discoveries for the single-gene traits of the cat and allow cats to be of assistance for studying complex traits and health conditions.

FELINE GENOMICS

Early chromosome-banding studies of the domestic cat revealed an easily distinguishable karyotype consisting of 18 autosomal pairs and the XY sex chromosome pair, resulting in a 2N complement of 38 cat chromosomes (Wurster-Hill and Gray 1973). The traditional grouping of chromosomes into alphabetic groups based on size and centromeric position has only relatively recently been renamed to more standard nomenclature for the cat (Cho et al. 1997). Basic light microscopy and Giemsa banding also showed that domestic cats have a chromosomal architecture that is highly representative for all felids and even carnivores (Modi and O'Brien 1988). Only minor chromosomal rearrangements are noted among the 36 extant felids. Most noticeably, a Robertsonian fusion in the South American ocelot lineage leads to a reduced complement, 2N = 36 (Wurster-Hill and Gray 1973). The variation of chromosomal sizes allowed the easy development of chromosome paints by flow sorting (Wienberg et al. 1997). Chromosome-painting techniques supported early somatic-cell hybrid maps in that the cat appeared to be highly conserved in chromosomal arrangement to humans, specifically as compared to mice (Stanyon et al. 1999). Hence, chromosome painting gave an excellent overview of cat genome organization, which greatly facilitates candidate gene approaches, since the location of particular genes can be anticipated in cats from comparison with the genetic map of humans.

Genetic and radiation hybrid maps of the cat augment the low-resolution genetic comparisons provided by chromosomal studies. The Bengal, a hybrid between domestic cats (primarily Abyssinians and Egyptian or Indian Maus) and different subspecies of Asian leopard cat (*Prionailurus [Felis] bengalensis*), has been in production since the late 1960s. It is currently a very popular breed with unique coloration and coat patterns, although not all registries recognize these cats. The evolutionary distance between the parental type cats of the Bengal breed is significant. Thus, a Bengal pedigree was the basis of the first recombination map for the cat (Menotti-Raymond et al. 1999). The Bengal breed has various health concerns and predispositions, such as HCM, retinal degeneration, and chronic inflammatory bowel disease. These conditions may be excellent candidates for LD and admixture mapping in the hybrid cat populations. The interspecies hybrid-based linkage map contains approximately 250 microsatellite markers (Menotti-Raymond et al. 1999, 2003b) that are effective for the initiation of linkage studies in families segregating for phenotypic traits. The linkage map is due for updating, likely in the coming year. Already, linkage maps have assisted targeted candidate gene approaches, as seen for *PKD* (Young et al. 2005), and linkage analyses for *Tabby* (Lyons et al. 2006), *white spotting* (Cooper et al. 2006), and *Orange* (Grahn et al. 2005). The genetic map has also led to the first disease gene isolated by positional cloning, *LIX1*, which causes spinal muscular atrophy in the Maine coon cat (He et al. 2005; Fyfe et al. 2006). The current 5000 Rad radiation hybrid map of the cat consists of 1784 markers (Murphy et al. 1999, 2000, 2006; Menotti-Raymond et al. 2003a), supporting the conserved genomic organization as compared to humans and assisting with sequence contig construction. The cat's importance in human health, comparative genomics, and evolutionary studies supported the National Institutes of Health–National Human Genomics Research Institute (NIH-NHGRI) decision to produce a low-coverage, 2X, sequence of the cat genome (http://www.genome.gov/Pages/Research/Sequencing/SeqProposals/CatSEQ.pdf). Led by the Broad Institute and AgenCourt, approximately 327,037 SNPs have been identified in the sequence from the solitary, highly inbred Abyssinian cat. An additional 7X coverage of the same cat has been scheduled for completion (http://www.genome.gov/19517271), which will provide a deeper coverage draft sequence of the cat in the coming years.

LD estimates in the cat are under evaluation but not yet published. Knowing the breed dynamics of the domestic cat, it is likely that LD will be less extensive than in the dog, but more extensive than in humans. An internationally tested, microsatellite-based DNA profiling panel developed for parentage and individual identification in domestic cats requires fewer markers than in other species, because most cat breeds have adequate variation at all markers (Lipinski et al. 2007). Nineteen microsatellite markers were included in the panel development and were genotyped in a variety of domestic cat DNA test samples. Most markers consisted of dinucleotide repeats. In addition to the autosomal markers, the panel included two gender-specific markers, *amelogenin* and *zinc-finger XY*, which produce genotypes for both the X and Y chromosome. The international cat DNA profiling panel has a power of exclusion comparable to panels used in other species, ranging from 90.08% to 99.79% across breeds and 99.47% to 99.87% in random-bred cat populations. However, only 10 markers were required to obtain adequate exclusion probabilities (Table 30-4). Dog breeds and other species generally require 15 or more markers for adequate exclusions across all breeds.

TABLE 30-4. Genetic Markers Selected as a "Core" Panel for Cat Parentage and Identification Testing

Marker	Chr.	Repeat	Forward primer 5'–3' Reverse primer 5'–3'	Label	µM	PE (Min-Max) (breeds)	PE (Min-Max) (random)
FCA069	B4	AC	AATCACTCATGCACGAATGC AATTTAACGTTAGGCTTTTTGCC	VIC	0.20	0.1324–0.5336	0.3958–0.5948
FCA075	E2	TG	ATGCTAATCAGTGGCATTTGG GAACAAAATTCCAGACGTGC	NED	0.10	0.1442–0.5771	0.4240–0.5992
FCA105	A2	TG	TTGACCCTCATACCTTCTTTGG TGGGAGAATAAATTTGCAAAGC	PET	0.20	0.2221–0.5585	0.6110–0.7101
FCA149[a]	B1	TG	CCTATCAAAGTTCTCACCAAATCA GTCTCACCATGTGTGGGATG	PET	0.18	0.1783–0.5995	0.3586–0.5767
FCA220	F2	CA	CGATGGAAATTGTATCCATGG GAATGAAGGCAGTCACAAACTG	FAM	0.30	0.0000–0.3383	0.1851–0.4221
FCA229	A1	GT	CAAACTGACAAGCTTAGAGGGC GCAGAAGTCCAATCTCAAAGTC	NED	0.25	0.0452–0.5131	0.3927–0.5813
FCA310[a]	C2	(CA)₅TA(CA)₇ TA(CA)₈	TTAATTGTATCCCAAGTGGTCA TAATGCTGCAATGTAGGGCA	FAM	0.30	0.1196–0.5256	0.3417–0.5611
FCA441[b]	D3	TAGA	ATCGGTAGGTAGGTAGATATAG GCTTGCTTCAAAATTTTCAC	VIC	0.15	0.2061–0.5774	0.3388–0.5505
FCA678[c]	A1	AC	TCCCTCAGCAATCTCCAGAA GAGGGAGCTAGCTGAAATTGTT	NED	0.25	0.0415–0.4908	0.3016–0.5715
AMEL[d]	XY	—	CGAGGTAATTTTTCTGTTTACT GAAACTGAGTCAGAGAGGC			n.a.	n.a.
ZFXY[d]	XY	—	AAGTTTACACAACCACCTGG CACAGAATTTACACTTGTGCA	PET	0.20	n.a.	n.a.
				Total PE		0.9008–0.9979	0.9947–0.9987

n.a. indicates not applicable.

[a] Markers that are of the first ten published feline microsatellites (Menotti-Raymond and O'Brien 1995).

[b] A marker that is currently included in the feline forensic panel (Menotti-Raymond et al. 2005).

[c] The two markers on the X and Y chromosomes were added to the panel after the comparison test (Pilgrim et al. 2005).

[d] Newly designed primers presented herein for *FCA678* generate a product 30 bp less than originally published primers.

SUMMARY AND CONCLUSIONS

Genetic studies and resources in the cat have improved significantly in the past decade, and additional leaps are on the horizon. The cat is slowly revealing its mysteries, but cat origins and the mutations causing many diseases and traits are yet to be identified. Updated maps, deeper genomic sequences, and the RPCI-85 cat BAC (bacterial artificial chromosome; http://bacpac.chori.org/) and other libraries should provide sufficient and highly efficient resources for future genetic studies in felids. Cats are excellent models of infectious and acquired diseases, and morphological and behavior traits segregate in breeds and populations. Improved veterinary facilities and medicine support the study of new models in the domestic cat, and existing models are proving useful for gene therapy and stem cell studies, opening the door for advances in human and feline health and medicine.

ACKNOWLEDGMENTS

Funding for this project was provided to L. A. Lyons from NIH-NCRR grant RR016094, the Winn Feline Foundation, and the George and Phyllis Miller Feline Health Fund, Center for Companion Animal Health, and the Koret Center for Veterinary Genetics, School of Veterinary Medicine, University of California, Davis.

REFERENCES

Barrs V.R., Gunew M., Foster S.F., Beatty J.A., and Malik R. 2001. Prevalence of autosomal dominant polycystic kidney disease in Persian cats and related-breeds in Sydney and Brisbane. *Aust. Vet. J.* **79:** 257–259.

Barthez P.Y., Rivier P., and Begon D. 2003. Prevalence of polycystic kidney disease in Persian and Persian related cats in France. *J. Feline Med. Surg.* **5:** 345–347.

Beck C. and Lavelle R.B. 2001. Feline polycystic kidney disease in Persian and other cats: A prospective study using ultrasonography. *Aust. Vet. J.* **79:** 181–184.

Berg T., Tollersrud O.K., Walkley S.U., Siegel D., and Nilssen O. 1997. Purification of feline lysosomal α-mannosidase, determination of its cDNA sequence and identification of a mutation causing α-mannosidosis in Persian cats. *Biochem. J.* **328:** 863–870.

Bighignoli B., Grahn R.A., Millon L.V., Longeri M., Polli M., and Lyons L.A. 2007. Genetic mutations for the feline AB blood group identified in *CMAH*. *BMC Genet.* **8:** 27.

Biller D.S., Chew D.J., and DiBartola S.P. 1990. Polycystic kidney disease in a family of Persian cats. *J. Am. Vet. Med. Assoc.* **196:** 1288–1290.

Biller D.S., DiBartola S.P., Eaton K.A., Pflueger S., Wellman M.L., and Radin M.J. 1996. Inheritance of polycystic kidney disease in Persian cats. *J. Hered.* **87:** 1–5.

Cannon M.J., MacKay A.D., Barr F.J., Rudorf H., Bradley K.J., and Gruffydd-Jones T.J. 2001. Prevalence of polycystic kidney disease in Persian cats in the United Kingdom. *Vet. Rec.* **149:** 409–411.

Cavalli-Sforza L.L. and Edwards A.W.F. 1967. Phylogenetic analysis: Models and estimation procedures. *Evolution* **21:** 550–570.

Cho K.W., Youn H.Y., Watari T., Tsujimoto H., Hasegawa A., and Satoh H. 1997. A proposed nomenclature of the domestic cat karyotype. *Cytogenet. Cell Genet.* **79:** 71–78.

Cooper M.P., Fretwell N., Bailey S.J., and Lyons L.A. 2006. White spotting in the domestic cat (*Felis catus*) maps near *KIT* on feline chromosome B1. *Anim. Genet.* **37:** 163–165.

Crawley A.C., Yogalingam G., Muller V.J., and Hopwood J.J. 1998. Two mutations within a feline mucopolysaccharidosis type VI colony cause three different clinical phenotypes. *J. Clin. Invest.* **101:** 109–119.

Drogemuller C., Rufenacht S., Wichert B., and Leeb T. 2007. Mutations within the FGF5 gene are associated with hair length in cats. *Anim. Genet.* **38:** 218–221.

De Maria R., Divari S., Bo S., Sonnio S., Lotti D., Capucchio M.T., and Castagnaro M. 1998. β-galactosidase deficiency in a Korat cat: A new form of feline G_{M1}-gangliosidosis. *Acta Neuropathol.* **96:** 307–314.

Eizirik E., Yuhki N., Johnson W.E., Menotti-Raymond M., Hannah S.S., and O'Brien S.J. 2003. Molecular genetics and evolution of melanism in the cat family. *Curr. Biol.* **13:** 448–453.

Felsenstein J. 1989. PHYLIP—Phylogeny Inference Package (Version 3.2). *Cladistics* **5:** 164–166.

Fogle B. 1997. *The encyclopedia of the cat.* DK Publishing, New York.

Fyfe J.C., Kurzhals R.L., Lassaline M.E., Henthorn P.S., Alur P.R., Wang P., Wolfe J.H., Giger U., Haskins M.E., Patterson D.F, et al. 1999. Molecular basis of feline β-glucuronidase deficiency: An animal model of mucopolysaccharidosis VII. *Genomics* **58:** 121–128.

Fyfe J.C., Menotti-Raymond M., David V.A., Brichta L., Schaffer A.A., Agarwala R., Murphy W.J., Wedemeyer W.J., Gregory B.L., Buzzell B.G., et al. 2006. An ~140-kb deletion associated with feline spinal muscular atrophy implies an essential *LIX1* function for motor neuron survival. *Genome Res.* **16:** 1084–1090.

Fyfe J.C., Kurzhals R.L., Hawkins M.G., Wang P., Yuhki N., Giger U., Van Winkle T.J., Haskins M.E., Patterson D.F., and Henthorn P.S. 2007. A complex rearrangement in GBE1 causes both perinatal hypoglycemic collapse and late-juvenile-onset neuromuscular degeneration in glycogen storage disease type IV of Norwegian forest cats. *Mol. Genet. Metab.* **90:** 383–392.

Giger U., Tcherneva E., Caverly J., Seng A., Huff A. M., Cullen K., Van Hoeven M., Mazrier H., and Haskin M.E. 2006. A missense point mutation in *N*-acetylglucosamine-1-phosphotransferase causes mucolipidosis II in domestic shorthair cats. *J. Vet. Intern. Med.* **20**: 781.

Ginzinger D.G., Lewis M.E., Ma Y., Jones B.R., Liu G., and Jones S.D. 1996. A mutation in the lipoprotein lipase gene is the molecular basis of chylomicronemia in a colony of domestic cats. *J. Clin. Invest.* **97**: 1257–1266.

Goree M., Catalfamo J.L., Aber S., and Boudreaux M.K. 2005. Characterization of the mutations causing hemophilia B in 2 domestic cats. *J. Vet. Intern. Med.* **19**: 200–204.

Grahn R.A., Lemesch B.M., Millon L.V., Matise T., Rogers Q.R., Morris J.G., Fretwell N., Bailey S.J., Batt R.M., and Lyons L.A. 2005. Localizing the X-linked orange colour phenotype using feline resource families. *Anim. Genet.* **36**: 67–70.

He Q., Lowrie C., Shelton G.D., Castellani R.J., Menotti-Raymond M., Murphy W., O'Brien S.J., Swanson W.F., and Fyfe J.C. 2005. Inherited motor neuron disease in domestic cats: A model of spinal muscular atrophy. *Pediatr. Res.* **57**: 324–330.

He X., Li C.M., Simonaro C.M., Wan Q., Haskins M.E., Desnick R.J., and Schuchman E.H. 1999. Identification and characterization of the molecular lesion causing mucopolysaccharidosis type I in cats. *Mol. Genet. Metab.* **67**: 106–112.

Ibsen H.L. 1916. Tricolor inheritance. III. Tortoiseshell cats. *Genetics* **1**: 377–386.

Imes D.L., Geary L.A., Grahn R.A., and Lyons L.A. 2006. Albinism in the domestic cat (*Felis catus*) is associated with a *tyrosinase* (*TYR*) mutation. *Anim. Genet.* **37**: 175–178.

Ishida Y., David V.A., Eizirik E., Schaffer A.A., Neelam B.A., Roelke M.E., Hannah S.S., O'Brien S.J., and Menotti-Raymond M. 2006. A homozygous single-base deletion in *MLPH* causes the *dilute* coat color phenotype in the domestic cat. *Genomics* **88**: 698–705.

Johnson W.E., Eizirik E., Pecon-Slattery J., Murphy W.J., Antunes A., Teeling E., and O'Brien S.J. 2006. The late Miocene radiation of modern Felidae: A genetic assessment. *Science* **311**: 73–77.

Kanae Y., Endoh D., Yamato O., Hayashi D., Matsunaga S., Ogawa H., Maede Y., and Hayashi M. 2006. Nonsense mutation of feline β-hexosaminidase β-subunit (*HEXB*) gene causing Sandhoff disease in a family of Japanese domestic cats. *Res. Vet. Sci.* **82**: 54–60.

Kitchener A. 1991. *The natural history of wild cats*. Cornell University Press, New York.

Kittleson M.D., Meurs K.M., Munro M.J., Kittleson J.A., Liu S.K., Pion P.D., and Towbin J.A. 1999. Familial hypertrophic cardiomyopathy in Maine coon cats: An animal model of human disease. *Circulation* **99**: 3172–3180.

Li X., Li W., Wang H., Bayley D.L., Cao J., Reed D.R., Bachmanov A.A., Huang L., Legrand-Defretin V., Beauchamp G.K., and Brand J.G. 2006. Cats lack a sweet taste receptor. *J. Nutr.* **136**: 1932S–1934S.

Li X., Li W., Wang H., Cao J., Maehashi K., Huang L., Bachmanov A.A., Reed D.R., Legrand-Defretin V., Beauchamp G.K., and Brand J.G. 2005. Pseudogenization of a sweet-receptor gene accounts for cats' indifference toward sugar. *PLoS Genet.* **1**: 27–35.

Lipinski M.J., Amigues Y., Blasi M., Broad T.E., Cherbonnel C., Cho G.J., Corley S., Daftari P., Delattre D.R., Dileanis S., et al. 2007. An international microsatellite-based DNA profiling panel for the domestic cat (*Felis catus*). *Anim. Genet.* (in press).

Louwerens M., London C.A., Pedersen N.C., and Lyons L.A. 2005. Feline lymphoma in the post-feline leukemia virus era. *J. Vet. Intern. Med.* **19**: 329–335.

Lyons L.A., Foe I.T., Rah H.C., and Grahn R.A. 2005a. Chocolate coated cats: *TYRP1* mutations for brown color in domestic cats. *Mamm. Genome* **16**: 356–366.

Lyons L.A., Imes D.L., Rah H.C., and Grahn R.A. 2005b. Tyrosinase mutations associated with Siamese and Burmese patterns in the domestic cat (*Felis catus*). *Anim. Genet.* **36**: 119–126.

Lyons L.A., Biller D.S., Erdman C.A., Lipinski M.J., Young A.E., Roe B.A., Qin B., and Grahn R.A. 2004. Feline polycystic kidney disease mutation identified in PKD1. *J. Am. Soc. Nephrol.* **15**: 2548–2555.

Lyons L.A., Bailey S.J., Baysac K.C., Byrns G., Erdman C.A., Fretwell N., Froenicke L., Gazlay K.W., Geary L.A., Grahn J.C., et al. 2006. The Tabby cat locus maps to feline chromosome B1. *Anim. Genet.* **37**: 383–386.

Malek J. 2006. *The cat in ancient Egypt*. British Museum Press, London.

Martin D.R., Krum B.K., Varadarajan G.S., Hathcock T.L., Smith B.F., and Baker H.J. 2004. An inversion of 25 base pairs causes feline G_{M2} gangliosidosis variant 0. *Exp. Neurol.* **187**: 30–37.

Martin D.R., Cox N.R., Morrison N.E., Kennamer D.M., Peck S.L., Dodson A.N., Gentry A.S., Griffin B., Rolsma M.D., and Baker H.J. 2005. Mutation of the G_{M2} activator protein in a feline model of G_{M2} gangliosidosis. *Acta Neuropathol.* **110**: 443–550.

Menotti-Raymond M.A. and O'Brien S.J. 1995. Evolutionary conservation of ten microsatellite loci in four species of Felidae. *J. Hered.* **86**: 319–322.

Menotti-Raymond M.A., David V.A., Wachter L.L., Butler J.M., and O'Brien S.J. 2005. An STR forensic typing system for genetic individualization of domestic cat (*Felis catus*) samples. *J. Forensic Sci.* **50**: 1061–1070.

Menotti-Raymond M., David V.A., Agarwala R., Schaffer A.A., Stephens R., O'Brien S.J., and Murphy W.J. 2003a. Radiation hybrid mapping of 304 novel microsatellites in the domestic cat genome. *Cytogenet. Genome Res.* **102**: 272–276.

Menotti-Raymond M., David V.A., Lyons L.A., Schaffer A.A., Tomlin J.F., Hutton M.K., and O'Brien S.J. 1999. A genetic linkage map of microsatellites in the domestic cat (*Felis catus*). *Genomics* **57**: 9–23.

Menotti-Raymond M., David V.A., Schaffer A.A., Stephens R., Wells D., Kumar-Singh R., O'Brien S.J., and Narfstrom K. 2007. Mutation in CEP290 discovered for cat model of human retinal degeneration. *J. Hered.* (in press).

Menotti-Raymond M., David V.A., Chen Z.Q., Menotti K.A., Sun S., Schaffer A.A., Agarwala R., Tomlin J.F., O'Brien S.J., and Murphy W.J. 2003b. Second-generation integrated genetic linkage/radiation hybrid maps of the domestic cat (*Felis catus*). *J. Hered.* **94**: 95–106.

Meurs K.M., Norgard M.M., Ederer M.M., Hendrix K.P., and Kittleson M.D. 2007. A substitution mutation in the myosin binding protein C gene in ragdoll hypertrophic cardiomyopathy. *Genomics* (in press).

Meurs K.M., Sanchez X., David R.M., Bowles N.E., Towbin J.A., Reiser P.J., Kittleson J.A., Munro M.J., Dryburgh K., Macdonald K.A., and Kittleson M.D. 2005. A cardiac myosin binding protein C mutation in the Maine Coon cat with familial hypertrophic cardiomyopathy. *Hum. Mol. Genet.* **14**: 3587–3593.

Modi W.S. and O'Brien S.J. 1988. Quantitative cladistic analysis of chromosomal banding data among species in three orders of mammals: Hominoid primates, felids, and arvicolid rodents. *In Chromosome structure and function* (Eds. J. Perry Gustafson and R. Appels). Plenum Publishing, New York.

Morris D. 1999. Cat breeds of the world: A complete illustrated encyclopedia. Viking Penguin, New York.

Muldoon L.L., Neuwelt E.A., Pagel M.A., and Weiss D.L. 1994. Characterization of the molecular defect in a feline model for type II G_{M2}-gangliosidosis (Sandhoff disease). *Am. J. Pathol.* **144**: 1109–1118.

Murphy W.J., Menotti-Raymond M., Lyons L.A., Thompson M.A., and O'Brien S.J. 1999. Development of a feline whole genome radiation hybrid panel and comparative mapping of human chromosome 12 and 22 loci. *Genomics* **57:** 1–8.

Murphy W.J., Sun S., Chen Z., Yuhki N., Hirschmann D., Menotti-Raymond M., and O'Brien S.J. 2000. A radiation hybrid map of the cat genome: Implications for comparative mapping. *Genome Res.* **10:** 691–702.

Murphy W.J., Davis B., David V.A., Agarwala R., Schaffer A.A., Pearks Wilkerson A.J., Neelam B., O'Brien S J., and Menotti-Raymond M. 2006. A 1.5-Mb-resolution radiation hybrid map of the cat genome and comparative analysis with the canine and human genomes. *Genomics* **89:** 189–196.

Nei M. 1972. Genetic distance between populations. *Am. Naturalist* **106:** 283–292.

O'Brien S.J., Haskins M.E., Winkler C.A., Nash W.G., and Patterson D.F. 1986. Chromosomal mapping of beta-globin and albino loci in the domestic cat. A conserved mammalian chromosome group. *J. Hered.* **77:** 374–378.

Pilgrim K.L., McKelvey K.S., Riddle A.E., and Schwartz M.K. 2005. Felid sex identification based on noninvasive genetic samples. *Mol. Biol. Notes* **5:** 60–61.

Pritchard J.K., Stephens M., and Donnelly P. 2000. Inference of population structure using multilocus genotype data. *Genetics* **155:** 945–959.

Seidensticker J. and Lumpkin S. 1991. *Great cats: Majestic creatures of the wild.* Rodale Press, Emmaus, Pennsylvania.

Somers K.L., Royals M.A., Carstea E.D., Rafi M.A., Wenger D.A., and Thrall M.A. 2003. Mutation analysis of feline Niemann-Pick C1 disease. *Mol. Genet. Metab.* **79:** 99–103.

Stanyon R., Yang F., Cavagna P., O'Brien P.C., Bagga M., Ferguson-Smith M.A., and Wienberg J. 1999. Reciprocal chromosome painting shows that genomic rearrangement between rat and mouse proceeds ten times faster than between humans and cats. *Cytogenet. Cell Genet.* **84:** 150–155.

Sunquist M. and Sunquist F. 2002. *Wild cats of the world.* University of Chicago Press, Illinois.

Wienberg J., Stanyon R., Nash W.G., O'Brien P.C., Yang F., O'Brien S.J., and Ferguson-Smith M.A. 1997. Conservation of human vs. feline genome organization revealed by reciprocal chromosome painting. *Cytogenet. Cell Genet.* **77:** 211–217.

Winand N.J., Edwards M., Pradhan D., Berian C.A., and Cooper B.J. 1994. Deletion of the dystrophin muscle promoter in feline muscular dystrophy. *Neuromuscul. Disord.* **4:** 433–445.

Wurster-Hill D.H. and Gray C.W. 1973. Giemsa banding patterns in the chromosomes of twelve species of cats (Felidae). *Cytogenet. Cell Genet.* **12:** 388–397.

Yogalingam G., Hopwood J.J., Crawley A., and Anson D.S. 1998. Mild feline mucopolysaccharidosis type VI. Identification of an *N*-acetylgalactosamine-4-sulfatase mutation causing instability and increased specific activity. *J. Biol. Chem.* **273:** 13421–13429.

Yogalingam G., Litjens T., Bielicki J., Crawley A.C., Muller V., Anson D.S., and Hopwood J.J. 1996. Feline mucopolysaccharidosis type VI. Characterization of recombinant *N*-acetyl-galactosamine 4-sulfatase and identification of a mutation causing the disease. *J. Biol. Chem.* **271:** 27259–27265.

Young A.E., Biller D.S., Herrgesell E.J., Roberts H.R., and Lyons L.A. 2005. Feline polycystic kidney disease is linked to the *PKD1* region. *Mamm. Genome* **16:** 59–65.

Zohary D. and Hopf M. 2002. Domestication of plants in the old world. 3 ed. Oxford University Press, Oxford.

31 | The Dog

Kerstin Lindblad-Toh[1] and Elaine A. Ostrander[2]

[1]*Broad Institute of MIT and Harvard, Cambridge, Massachusetts 02142;* [2]*National Human Genome Research Institute, National Institutes of Health, Bethesda, Maryland 20892*

INTRODUCTION

In the past 5 years, the field of canine genetics and genomics has advanced tremendously. The unique breeding history of the domestic dog, including hundreds of distinct breeds created in the last few hundred years, combined with a new suite of genomic tools, has put canines in the forefront of the genomics landscape. Resources now include an almost complete genome sequence, a 2.5 million single-nucleotide polymorphism (SNP) map, and a growing understanding of the haplotype structure and breed relationships as well as both simple sequence length polymorphism (SSLP) and SNP mapping tools. In this chapter, we discuss these recent advances and the role of genetic variation in controlling phenotypes of interest, including both Mendelian disease and complex traits. We predict that a myriad of diseases and other complex traits will be mapped in the next few years and that many will inform human health and biology.

Dog History and Breeds

The domestic dog is the most recently evolved species from the family Canidae (Wayne et al. 1987a,b, 1997), sharing a clade with the wolf-like canids such as the gray wolf, coyote, and jackals (Fig. 31-1). Dogs are believed to have arisen as recently as 30,000 years ago, with the initial domestication events occurring in eastern Asia (Vilà et al. 1997; Savolainen et al. 2002; Lindblad-Toh et al. 2005). Following these major domestication events, most dog breeds are believed to have been created in the last 200–300 years. As a result, there are currently over 400 domestic dog breeds demonstrating enormous levels of physical and behavioral variation. Of these, 155 breeds are registered by the American Kennel Club (AKC) in the United States (American Kennel Club 1998).

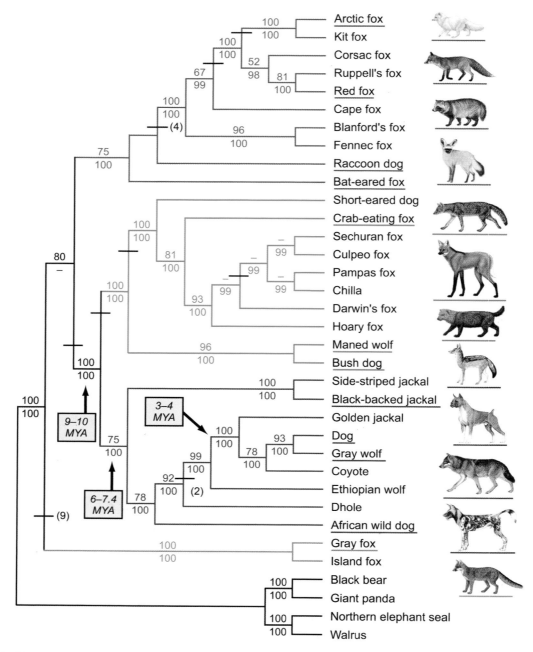

FIGURE 31-1. Canid tree. The canid phylogenetic tree is based on ~15 kb of exon and intron sequence. It was constructed using maximum parsimony as the optimality criterion and is the single most parsimonious tree. Colors on branches mark the red fox-like clade (*red*), the South American clade (*green*), the wolf-like clade (*blue*), and the gray and island fox clade (*orange*). Illustrated species are underlined. The tree was rooted with the black bear (*Ursus americanus*), giant panda (*Ailuropoda melanoleuca*), elephant seal (*Mirounga augustirostris*), and walrus (*Odobenus rosmarus*). (Reprinted, with permission, from Lindblad-Toh et al. 2005.)

The "breed" nomenclature has been defined by both dog fanciers and geneticists. To be a registered member of a breed, the AKC requires that both of a dog's parents were documented members of the same breed. Modern dog breeds have thus become closed breeding populations with infrequent introduction of new alleles. As a result, dog breeds are characterized by a lower level of genetic heterogeneity (1 SNP/1600 bp) than is typically observed in mixed-breed dogs (1SNP/900 bp) (Lindblad-Toh et al. 2005). This is due to a number of factors, including the small numbers of founders that defined many breeds, population bottlenecks, and the overrepresentation of some males (popular sires) in the breeding pool (Parker et al. 2004; Parker and Ostrander 2005). The current population of approximately 10 million purebred dogs in the United States therefore represents an ideal group in which to study population genetics and the role of genetic variation in controlling both simple and complex traits.

Several studies have tackled the issue of breed definition (Koskinen and Bredbacka 2000; Koskinen 2003; Parker et al. 2004). In the most expansive study of dog breeds, Parker et al. (2004) utilized data from 96 (CA)n repeat-based SSLPs, spanning all dog autosomes, on 414 dogs to determine the degree to which dogs could be assigned to their appropriate breed using a clustering algorithm (Fig. 31-2) (Parker et al. 2004). Eighty-five breeds were ordered into four clusters based on similar patterns of alleles, presumably representing shared ancestral pools (Ostrander and Wayne 2005). Ongoing studies are under way to include more breeds. These results are interesting in light of studies on genetic diversity in human populations. The overall level of nucleotide diversity is 8 × 10^{-4} (Parker et al. 2004; Lindblad-Toh et al. 2005), which is not too dissimilar from what is reported for humans (Cavelli-Sforza et al. 1994; Rosenberg et al. 2002). However, genetic variation between breeds is much greater than the observed variation between human populations (27.5% versus 5.4% by ANOVA) (Parker et al. 2004). Thus, dog "breeds" are statistically meaningful entities.

The Dog as a Model for Human Diseases

The population structure of the domestic dog lends itself to the study of both monogenic and complex traits, particularly those associated with disease susceptibility (Ostrander and Kruglyak 2000). Diseases that occur in both humans and companion animals and that have been difficult to tackle in human populations are particularly well suited for canine mapping studies. Good candidates include cancer, diabetes, autoimmune disease, motor neuron disease, deafness, epilepsy, and heart disease (Ostrander and Friedrichsen 2004; Ostrander et al. 2004). As a consequence of the tight population bottlenecks accompanying breed creation, and subsequent inbreeding, many breeds suffer from high rates of specific diseases, suggesting enrichment of genetic risk factors (Patterson et al. 1988; Ostrander and Kruglyak 2000; Sargan 2004; Parker and Ostrander 2005).

The low level of variation within a breed may also lead to fixation of certain risk factors, thereby reducing the overall level of noise and making the effects of other risk factors more easily observable. In addition, family history is well recorded and dogs are second only to humans in terms of medical scrutiny and the number of dollars spent on health care annually (Patterson 2000; American Veterinary Medical Association 2002). Thus, several hundred genetic diseases have been characterized in the dog, a number surpassed only by studies of human populations (Patterson et al. 1982; Patterson 2000; Sargan 2004). These are collated in an on-line database called IDID (Inherited Disease in Dogs), which is organized in a fashion similar to the Online Mendelian Inheritance of Man (OMIM) database (Sargan 2004). Furthermore, dogs share a large part of their environment with humans. In aggregate, these factors make the dog an ideal system in which to study the genetics of both simple and complex traits that are of interest to those studying both human genetics and companion animal health. Table 31-1 summarizes the major diseases of domestic dogs and breeds which are at risk.

Prior to the availability of the canine genome sequence, geneticists used large canine pedigrees to overcome many of the disadvantages faced by human geneticists. Early successful genetic mapping studies in the dog include those for progressive retinal atrophy (PRA) (Acland et al. 1994, 1998, 1999; Sidjanin et al. 2002), copper toxicosis (Yuzbasiyan-Gurkan et al. 1997), renal cancer

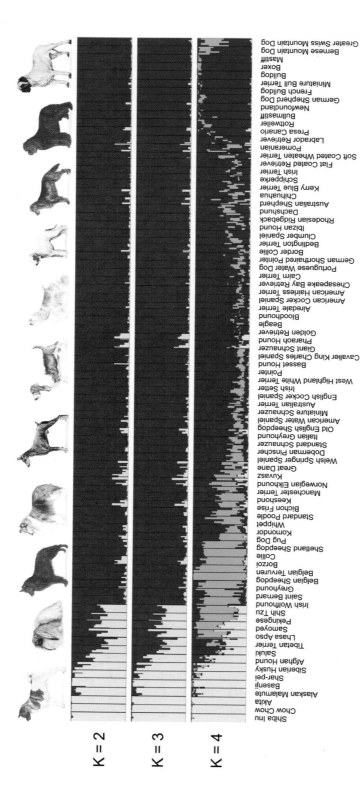

FIGURE 31-2. Population structure of the domestic dog. Figure is derived from the work of Parker and colleagues (2004). Five dogs from each of 85 breeds were genotyped using 85 (CA)n repeat-based microsatellites. Markers spanned all autosomes at 30 Mb density. Analysis was performed using the computer program *structure*. The program was asked to divide the samples into optimal groups, assuming, initially, that only two groups existed (K = 2). At that level, a group of breeds that were largely Asian or Ancient in origin were selected as being most unrelated to the remainder. This included the Lhasa apso, shar-pei, and Akita. One additional degree of freedom (K = 3) separated out the mastiff group and includes, for example, the boxer, bulldog, and Presa Canario. At K = 4, a group of largely working breeds, such as Saint Bernards and Belgian sheepdogs, was identified. The remainder of dogs, so called group four, is enriched for sight and scent hounds and includes breeds such as the spaniels and retrievers. Each group of dogs is more closely related one to another than to any other grouping of dogs. The addition of more dogs per breed, and more breeds total, should allow a refinement of this analysis.

TABLE 31-1. Major Diseases of Domestic Dogs and Breeds at Risk

Disease	Breeds at higher risk for disease
Cancers	Airedale Terrier, Akita, American Eskimo Dog, Belgian Malinois, Bloodhound, Boxer, Briard, Canaan Dog, Curley Coated Retriever, Dandie Dinmont Terrier, English Foxhound, English Setter, Flat Coated Retriever, German Wirehaired Pointer, Great Dane, Greyhound, Irish Water Spaniel, Irish Wolfhound, Japanese Chin, Kuvasz, Otterhound, Pembroke Welsh Corgi, Portuguese Water Dog, Rhodesian Ridgeback, Rottweiler, Scottish Deerhound, Scottish Terrier, Skye Terrier, Soft Coated Wheaten Terrier, Staffordshire Terrier, Tibetan Terrier, Vizsla
Epilepsy	Australian Terrier, Belgian Malinois, Belgian Tervuren, Boston Terrier, Canaan Dog, Chesapeake Bay Retriever, Clumber Spaniel, Collie, Curley Coated Retriever, Dachshund, Dalmatian, English Foxhound, English Springer Spaniel, English Toy Spaniel, Field Spaniel, Giant Schnauzer, Great Pyrenees, Greater Swiss Mountain Dog, Harrier, Irish Setter, Irish Water Spaniel, Italian Greyhound, Japanese Chin, Miniature Pinscher, Newfoundland, Otterhound, Petit Basset Griffon Vendeen, Poodle, Portuguese Water Dog, Pug, Schipperke, Sealyham Terrier, Shetland Sheepdog, Siberian Husky, Vizsla, Welsh Springer Spaniel, Welsh Terrier
Hip dysplasia	Airedale Terrier, American Eskimo Dog, American Water Spaniel, Basset Hound, Belgian Malinois, Black and Tan Coonhound, Bloodhound, Chesapeake Bay Retriever, Clumber Spaniel, Curly Coated Retriever, English Foxhound, English Setter, Field Spaniel, Flat Coated Retriever, German Wirehaired Pointer, Giant Schnauzer, Great Dane, Greyhound, Harrier, Mastiff, Newfoundland, Otterhound, Pembroke Welsh Corgi, Pug, Rhodesian Ridgeback, Rottweiler, Shetland Sheepdog, Staffordshire Bull Terrier, Staffordshire Terrier, Tibetan Terrier, Weimaraner, Welsh Springer Spaniel
Thyroid disease	Akita, American Eskimo Dog, Australian Terrier, Basset Hound, Belgian Tervuren, Canaan Dog, Chesapeake Bay Retriever, Dachshund, Dandie Dinmont Terrier, English Foxhound, English Terrier, English Springer Spaniel, Field Spaniel, German Wirehaired Pointer, Giant Schnauzer, Greyhound, Irish Water Spaniel, Italian Greyhound, Kuvasz, Maltese, Miniature Pinscher, Norwegian Elkhound, Rhodesian Ridgeback, Scottish Deerhound, Shetland Sheepdog, Siberian Husky, Sussex Spaniel, Tibetan Terrier, Vizsla, Welsh Springer Spaniel
Allergies	Airedale Terrier, American Water Spaniel, Australian Terrier, Bichon Frise, Black and Tan Coonhound, Boston Terrier, Bull Terrier, Chinese Shar-Pei, Dalmatian, English Springer Spaniel, French Bulldog, Irish Setter, Irish Water Spaniel, Kuvasz, Otterhound, Scottish Deerhound, Staffordshire Bull Terrier
Bloat	Akita, Bloodhound, Briard, Collie, Curly Coated Retriever, English Foxhound, Great Dane, Greater Swiss Mountain Dog, Greyhound, Irish Setter, Irish Wolfhound, Komondor, Poodle, Scottish Deerhound, Sussex Spaniel, Weimaraner
Heart disease	American Water Spaniel, Belgian Malinois, Bloodhound, Briard, Bull Terrier, English Toy Spaniel, Field Spaniel, Irish Wolfhound, Japanese Chin, Maltese, Mastiff, Miniature Bull Terrier, Staffordshire Terrier, Sussex Spaniel, West Highland White Terrier
Autoimmune diseases	Airedale Terrier, Akita, Belgian Tervuren, Briard, Canaan Dog, English Cocker Spaniel, German Wirehaired Pointer, Italian Greyhound, Maltese, Pembroke Welsh Corgi, Petit Basset Griffon Vendeen, Skye Terrier, Staffordshire Terrier
Progressive retinal atrophy	Airedale Terrier, American Eskimo Dog, Basenji, Belgian Tervuren, Chesapeake Bay Retriever, Dachshund, English Springer Spaniel, Italian Greyhound, Mastiff, Papillon, Portuguese Water Dog, Tibetan Terrier
Cataracts	Bedlington Terrier, Bichon Frise, Black and Tan Coonhound, Boston Terrier, Cairn Terrier, Chesapeake Bay Retriever, Havanese, Norwegian Elkhound, Siberian Husky, Staffordshire Bull Terrier, Tibetan Terrier

Information provided by the American Kennel Club Canine Health Foundation.

(Jonasdottir et al. 2000), and narcolepsy (Mignot et al. 1991; Lin et al. 1999). All of these studies involved large, multigenerational families of highly penetrant Mendelian traits, making them amenable to linkage analysis with SSLPs (Sargan et al. 2007).

The Canine Genome Sequence

Although canine genetics has demonstrated significant progress in the past few years, the availability of a 7.5x assembly covering approximately 99% of the euchromatic genome of a female boxer

TABLE 31-2. Canine Genome and Mapping Resources

Genome assembly and browser:	
NCBI	www.ncbi.nih.gov/Genbank
UCSC browser	www.genome.ucsc.edu
Ensembl	www.ensembl.org/index.html
Broad Institute	www.broad.mit.edu/ftp/pub/assemblies/mammals/dog/canFam2/
Dog paper Supp Info	www.broad.mit.edu/ftp/pub/papers/dog_genome/suppinfo/
Maps:	
RH map	www-recomgen.univ-rennes1.fr/Dogs/paper/FISH-RHmap.html
FISH map	www.cvm.ncsu.edu/mbs/breen_matthew.htm
SNP map	www.broad.mit.edu/ftp/pub/papers/dog_genome/snps_canfam2/
Breeds, relationships, and diseases:	
American Kennel Club	www.akc.org
Parker breed relationships	www.fhcrc.org/science/dog_genome/dog.html
IDID	www.vet.cam.ac.uk/idid/
Mapping tools:	
SSLPs	www.cvm.tamu.edu/cgr/multiplex.html
Affymetrix array	www.affymetrix.com/index.affx
Illumina array	www.illumina.com/
Haploview software	www.broad.mit.edu/mpg/haploview

(Lindblad-Toh et al. 2005) has dramatically increased the rate at which canine geneticists can answer questions of interest. The dog euchromatic genome is approximately 2.4 billion bases and comprises about 243 segments of conserved synteny when compared to the human genome. There is a 1-1-1 correspondence between orthologs of the human, mouse, and dog for 75% of the 19,000 canine genes. The full genome sequence can be accessed at a number of Web sites including http://www. genome.ucsc.edu; http://www.ncbi.nih.gov; and http://www.ensembl.org. See Table 31-2 for a more complete list of canine genome and mapping resources. A primer for mining the canine genome sequence was recently published (O'Rourke 2005).

Although this genome sequence (CanFam2.0) is highly accurate and contiguous (half of the sequence resides in continuous sequence blocks of at least 180 kb in size), a small percentage of the genome remains in gaps or regions of lower structural integrity (marked as "uncertified regions," available at http://www.broad.mit.edu/ftp/pub/assemblies/mammals/dog/canFam2/). A genome improvement effort is ongoing at the Broad Institute of MIT and Harvard. It aims to bring a small portion of the genome to finished standard (International Human Genome Sequencing Consortium 2004), to resolve the majority of larger gaps and uncertified regions, and to resolve errors in gene sequence. The territory to be finished includes the ENCODE regions (http://www.genome.gov/12513456), forty-four 0.5–2.0 Mb regions representing a mostly random 1% of the genome, that are being used to pilot annotation of the human genome. This genome improvement effort should result in a near-finished quality genome in late 2007.

Canine Genomic Variation

A comprehensive SNP map containing 2.5 million SNPs was generated by identification of SNPs within the boxer assembly and by comparison to the 1.5X partial sequence of the standard poodle (Kirkness et al. 2003). In addition, 100,000 sequence reads from each of nine dogs of unrelated breeds were compared to the boxer assembly. This SNP map contains on average a SNP every 1 kb and is publicly available (http://www.broad.mit.edu/ftp/pub/papers/dog_genome/snps_canfam2/ and http://www.broad. mit.edu/mammals/dog/) (Lindblad-Toh et al. 2005). The SNPs have a mean validation rate of 98%, with a slightly lower rate for SNP discovery through comparison of the poodle and boxer sequences. The general utility across breeds is approximately 70%, with an essentially random distribution across breeds.

Haplotype Structure and Association Mapping Strategies

Canine population history includes two population bottlenecks, the most recent of which led to breed creation in the last few hundred years. This established fact has long allowed geneticists to speculate that linkage disequilibrium (LD) in dogs would be long and therefore highly amenable to genome-wide association mapping. This has now been proven through two studies. Initially, Sutter and colleagues examined the extent of LD in five breeds with distinct breed histories and reported that the average length of LD in these five breeds is approximately 2 Mb (Sutter et al. 2004). This is 40–100 times further than LD typically extends in the human genome. As part of the genome sequencing project, the extent and number of haplotypes within 10 individual dog breeds and across the dog population as a whole were carefully examined in ten random 15-Mb regions. Haplotypes and LD extend for long distances (several megabases) within breeds, and typically three to six breed-specific haplotypes are present at each locus (Lindblad-Toh et al. 2005; Wade et al. 2006). In the whole dog population, haplotypes and LD are much shorter. Only three to five ancestral haplotypes are observed at each locus, with many haplotypes shared even between distantly related breeds. This suggests that many disease mutations are likely to be ancestral, and may be shared not only by closely related breeds, but by diverse sets of breeds as well.

Whereas human association studies require approximately 500,000 evenly spaced SNPs (Kruglyak 1999; International HapMap Consortium 2003), the fact that LD extends over approximately 50-fold greater distances in dog suggests that dog association studies would require perhaps 10,000 evenly spaced SNPs. To estimate this, sequence for 1-Mb-sized regions was generated using a population coalescence simulation, calibrated to produce sequence with SNP rates and LD curves equivalent to real data (Fig. 31-3) (Lindblad-Toh 2005). Different marker densities were modeled by selecting, at random, an appropriate number of evenly spaced SNPs. Individual SNPs with minor allele frequencies (MAF) <20% were selected as "disease alleles," and the ability to map

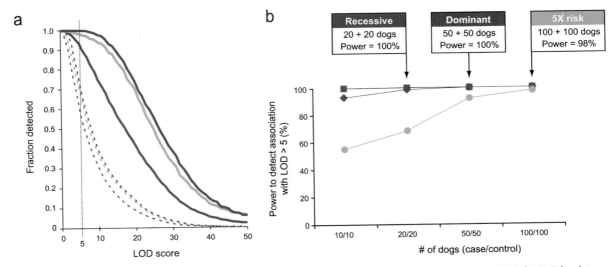

FIGURE 31-3. Power to detect a disease locus by genome-wide association mapping. (a) Multi-SNP haplotypes (*solid lines*) with SNP densities equivalent to a genome-wide map with a total of 7,500 (*red*), 15,000 (*green*), and 30,000 (*blue*) SNPs are more sensitive for detecting association than single SNPs (*dotted lines*). One SNP was designated as a disease allele under a fivefold multiplicative increase in risk model. SNP genotypes across surrounding chromosomal regions of 1 Mb were simulated, by using the coalescent model corresponding to observed within-breed variation. Diploid genotypes across the chromosomal region were then generated for 100 affected and 100 unaffected dogs, based on the genotype at the disease locus and the genetic model. Association analysis was performed to detect the presence of a disease allele (Lindblad-Toh et al. 2005). (b) One SNP was designated as a disease allele under one of three genetic models: simple Mendelian recessive (*blue*), simple Mendelian dominant (*red*), and fivefold multiplicative increase in risk (*green*).

them by association analysis with different marker densities was tested. Analysis using 15,000 or 30,000 genome-wide SNPs gave equivalent results. Somewhat lower power was observed for 7,500 SNPs (Fig. 31-3a).

The number of dogs needed for each study has also been examined in detail (Lindblad-Toh et al. 2005). For disease alleles causing a simple Mendelian trait with high penetrance and no phenocopies, there is overwhelming power to map a recessive locus, using about 20 affected cases and 20 unaffected controls. In the case of a dominant trait, 50 affected and 50 unaffected dogs are required. For a multigenic trait, the power to detect disease alleles depends on several factors, including the relative risk conferred by the allele, the allele frequency, and the interaction with other alleles. The simple model of an allele that increases risk by a multiplicative factor (λ) of two- or fivefold has been investigated (Lindblad-Toh et al. 2005). Using the same SNP density and significance threshold as proposed above, the power to detect a locus with a sample of 100 affected and 100 unaffected dogs is 98% for $\lambda = 5$, and 50% for $\lambda = 2$. In fact, approximately 500 affected and 500 unaffected dogs should provide sufficient power to map a locus carrying an allele that confers a twofold increased risk, given a reasonable population frequency (<20% allele) (Fig. 31-3b).

Two-Tiered Mapping Strategy

The above data suggest that mapping of both Mendelian and complex traits should be possible with small numbers of samples as long as a single breed is used for the initial genome-wide scan and phenotyping is accurate. However, the sharing of short "ancestral" haplotype segments suggests that a two-tiered mapping strategy, where the fine mapping effort is performed using multiple breeds showing the same phenotype, would permit the localization of the causative mutation to a very discrete region (Fig. 31-4a). An excellent example is the mapping of a gene for progressive rod–cone degeneration (prcd). The disease was initially mapped to canine chromosome 9 (CFA9) using a conventional microsatellite genome scan (Acland et al. 1998). Fine mapping using a combination of breeds was then used to reduce the disease interval to 106 kb (Goldstein et al. 2006), where the single missense mutation accounting for the canine disease was located. The same missense mutation was found in an autosomal recessive form of the disease in a patient from Bangladesh (Zangerl et al. 2006).

Tools for Canine Genome-wide Association Mapping

To permit genome-wide association mapping using the predicted approximately 15,000 SNPs, the Broad Institute of Harvard and MIT generated a genome-wide SNP genotyping array in collaboration with Affymetrix. The final quality-controlled (QCed) array contains approximately 26,000 functional SNPs covering the whole genome and representing the diversity of the dog population. A second-generation array is expected to contain closer to 50,000 functional SNPs. In addition, an Illumina array containing essentially the same approximately 26,000 SNPs is being generated and should be available in early 2007. These arrays constitute a resource for the canine research community.

The original Affymetrix array contains a set of approximately 65,000 SNPs covering the whole genome and representing the diversity of breeds. Roughly 70% of SNPs were expected to be polymorphic within any specific breed (Lindblad-Toh et al. 2005). A final set of approximately 15,000–30,000 functional SNPs was expected to pass initial QC based on the 25–50% success rate inherent to the technology. The principle of the array is similar to that of the human 500K Affymetrix SNP array (Matsuzaki et al. 2004) where DNA is digested and the approximately 10% of genomic DNA falling within a certain size fraction is hybridized to the oligonucleotide array. The lower complexity of DNA serves to reduce cross hybridization. (It is important that the starting DNA have a high molecular weight to avoid degraded DNA falling within the selected size range. Therefore, we recommend the use of DNA prepared from blood rather than from cheek

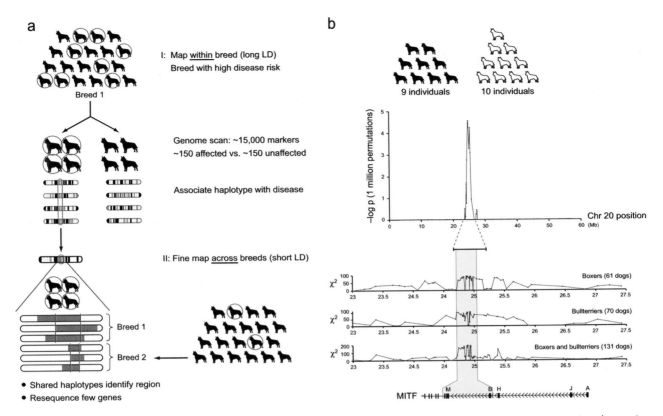

FIGURE 31-4. Two-tiered association mapping strategy identifies <0.5-Mb regions. (a) A two-tiered mapping strategy has been proposed. In the first step, whole-genome-wide association mapping using ~10,000–20,000 SNPs is employed in a breed with a high disease risk. Second, fine mapping of the disease-associated region is performed in the initial breed together with several related breeds, thus taking advantage of the ancestrally shared disease haplotypes. This permits rapid narrowing of the region to enable mutation screening of only part of a gene or a few genes. (b) Mapping of the white coat color locus in boxers employed 10 white dogs and 9 solid-colored dogs in the genome-wide association step, thereby identifying a ~1-Mb region on chromosome 20. Fine mapping utilized a second breed (white bullterriers) that refined the associated interval to ~100 kb in size. This region contains only four exons of the MITF gene. (a, Reprinted, with permission, from Wade et al. 2006.)

swabs for the arrays.) To further reduce noise, alleles A and B are represented by multiple randomly placed 25-mers. A set of highly performing SNPs was selected based on an original data set of several hundred dogs from multiple breeds.

The genome coverage of the selected approximately 26,000 SNPs is considered quite good. Ninety-seven percent of 1-Mb bins across the dog's 38 autosomes contain five or more SNPs, and 100% contain at least two. Coverage of chromosome X is less dense, likely attributable to its lower genetic diversity and higher repeat content, with five or more SNPs in just 42% of 1-Mb bins, whereas 88% contain at least one SNP (Hillbertz et al. 2007; Karlsson et al. 2007). On average, 93% of SNPs are called in each dog, with an accuracy of >99.5%. The fraction of SNPs that are polymorphic within a breed is >70% (MAF >5%), as expected. Within breeds, the average haplotype block size is estimated at approximately 600 kb, if a 4-gamete rule and a 5% MAF are applied. However, the higher number of SNPs on the next generation array may refine this number. Within each block, 3.8 haplotypes are observed at a frequency >5% (Fig. 31-5).

Analysis of about 10 dogs per breed using the full approximately 26,000-SNP set easily clusters the dogs into breeds with an average drift time of 0.3–0.4 using Eigenstrat (Price et al. 2006). Average drift time is similar to FST (F-statistics, a measure of genetic variance among populations), but is a genome-wide measure. This method is based on principal components analysis and explicitly models ancestry differences within a population. Sixteen breeds were immediately separated based on 15

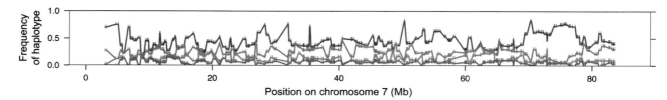

FIGURE 31-5. Frequency and size of haplotypes along chromosome 7 in the rottweiler breed. Haplotypes average ~600 kb in size, and a typical locus shows 3–4 haplotypes with the most common haplotype at a frequency of 40–80%. The most common haplotype is marked in red. Additional haplotypes are marked in blue, green, and purple.

components (Price et al. 2006). Such a clear breed separation is consistent with previous data (Parker et al. 2004) and is a reflection of the strong bottlenecks that occurred at breed creation. In contrast, a comparison of American and Dutch golden retrievers shows a drift time of only 0.11, similar to what is seen when comparing Caucasian and Asian human populations (International HapMap Consortium 2003). Thus, dog breeds are very distinct entities as compared to human populations, and even within breeds, strong stratifications exist due to physical location or possibly popular sire effects. It is therefore important to carefully consider the choices made when pairing cases and controls from a population or a pedigree for a whole-genome association analysis.

At the time of writing, the array has been used to map several monogenic traits and provided preliminary results for complex traits. In all cases, loci have been mapped at genome-wide significance when using the sample numbers predicted in the power calculations described above. Figure 4b shows how white coat color in boxers, a co-dominant trait, was mapped using 9 solid and 10 white boxer dogs. The identified region was <1 Mb in size and contained the MITF (microphthalmia-associated transcription factor) gene known to cause pigmentation disorders in both human and mouse (Udono et al. 2000; Steingrimsson et al. 2004; Karlsson et al. 2007).

Fine Mapping in Multiple Breeds May Help Precisely Identify Mutations

Whereas the majority of genome-wide association projects will identify a region about 1 Mb in size, often containing only a few genes, fine mapping in multiple breeds may provide even more exact information. For identification of the white coat color gene in boxers, a second breed segregating the trait (white bullterriers) was included in the fine mapping step leading to the identification of an approximately 100-kb recombination-free interval containing four exons, several conserved noncoding elements, and the disease mutation (Fig. 31-4b) (Karlsson et al. 2007). This shared haplotype is very similar in size to that seen in the previously described prcd study where multiple breeds were also used (Goldstein et al. 2006).

It seems likely, then, that the two-stage mapping strategy will work also for complex traits, assuming phenotypes can be sufficiently well-defined to suggest shared risk factors. Naturally, more closely related breeds are more likely to share the same risk alleles, but based on the regular occurrence of shared haplotypes also across distantly related breeds (Lindblad-Toh et al. 2005; Hillbertz et al. 2007), it is still worth considering distantly related breeds that share a consistent phenotype.

Variation Underlying Complex Traits

Few actual variants underlying complex traits have been found either in humans or in dogs. However, one can hypothesize that many such mutations will be regulatory in nature. This is particularly likely in dogs, where strong selection has been applied and the deleterious mutations resulting from coding mutations are less likely to have been tolerated by dog breeders.

Of critical importance has been the knowledge gained about types of variants responsible for genetically simple canine diseases. Several phenotypes are associated with insertion of a canine-specific short interspersed nuclear element (SINE) (Minnick et al. 1992; Bentolila et al. 1999; Vassetzky and Kramerov 2002). These retrotransposons are derived from a tRNA-Lys and occur frequently throughout the canine genome (Coltman and Wright 1994; Bentolila et al. 1999; Kirkness et al. 2003). In addition to inherited narcolepsy, aberrant insertion of SINEC_Cf elements is associated with centronuclear myopathy in the Labrador retriever (Pelé et al. 2005) and gray or "merle" coat coloring in several breeds (Clark et al. 2006).

Another interesting example of a novel variant associated with disease is a form of canine epilepsy similar to human Lafora disease. Studying the miniature wire-haired dachshund, Lohi and collaborators have shown that the disease is caused by expansion of an unstable dodecamer repeat in the *Epm2b* (*Nhlrc1*) gene (Lohi et al. 2005). Although trinucleotide repeat expansion has been reported in association with several human neurologic disorders (Kazemi-Esfarjani et al. 1995), this is the first report of a repeat expansion causing a disease in a species other than human.

These types of mutations may play a role also in complex traits in dogs (Kirkness 2006). Thus, mapping traits in dogs may prove simple, but finding the underlying germ-line variant may not be. Identifying the region containing the mutation with high precision, using multiple breeds when available, will help identify a number of variants that must be tested for functional consequence.

Identification of Regions under Selection

Many traits have been strongly selected for under the influence of required breed standards. Many of these are morphological or other physical traits, but behavior is also included. One would hypothesize that such selection would also leave footprints in the genome, with particular alleles being fixed or highly overrepresented. Analysis of genome-wide array data from multiple individuals within a breed should identify genomic regions under selection. However, current evidence suggests that such loci are frequently shorter than 1 Mb (Hillbertz et al. 2007; Karlsson et al. 2007). This is the same size observed for random haplotypes that have homozygosed through genetic drift alone (Karlsson et al. 2007). Thus, identification of such regions will require careful analysis to remove statistical noise. The most promising method seems to be comparison of the homozygosed regions in multiple distantly related breeds that share the same selected phenotype.

SUMMARY

The dog system is at a crossroads. All the genetic tools and resources the community has craved for so long are now available. Maps, sequences, markers, and chips are all at the disposal of those trying to identify genes important in disease susceptibility, progression, and outcomes using the canine system. It is now the task of the clinicians and geneticists to define and collect the family and population resources needed to maximize the power of these advances. Such efforts are under way now in laboratories across the world. For the canine genetics community, both the quality and quantity of information we can generate have achieved a new level. Our task now is to make the best use of those resources, judiciously choosing traits and diseases to map that will continue to expand our knowledge about ourselves; our closest companion, the dog; and mammalian biology as a whole.

ACKNOWLEDGMENTS

We thank the many dog owners, breeders, and supporters who continue to provide us with samples and information about their pets. This work is supported by the Intramural Program of the National Human Genome Research Institute, National Human Genome Research Institute Grants HG03067 and HG003069, and The American Kennel Club Canine Health Foundation.

REFERENCES

Acland G.M., Blanton S.H., Hershfield B., and Aguiree G.D. 1994. XLPRA: A canine retinal degeneration inherited as an X-linked trait. *Am. J. Med. Genet.* **52:** 27–33.

Acland G.M., Ray K., Mellersh C.S., Gu W., Langston A.A., Rine J., Ostrander E.A., and Aguirre G.D. 1998. Linkage analysis and comparative mapping of canine progressive rod-cone degeneration (*prcd*) establishes potential locus homology with retinitis pigmentosa (RP17) in humans. *Proc. Natl. Acad. Sci.* **95:** 3048–3053.

———. 1999. A novel retinal degeneration locus identified by linkage and comparative mapping of canine early retinal degeneration. *Genomics* **59:** 134–142.

American Kennel Club. 1998. *The complete dog book*, 19th edition revised. Howell Book House, New York, p. 790.

American Veterinary Medical Association. 2002. *U.S. pet ownership and demographics sourcebook.* American Veterinary Medical Association, Schaumburg, Illinois.

Bentolila S., Bach J.M., Kessler J.L., Bordelais I., Cruaud C., Weissenbach J., and Panthier J.J. 1999. Analysis of major repetitive DNA sequences in the dog (*Canis familiaris*) genome. *Mamm. Genome* **10:** 699–705.

Cavelli-Sforza L.L., Menozzi P., and Piazza A. 1994. *The history and geography of human genes.* Princeton University Press, Princeton, New Jersey.

Clark L.A., Wahl J.M., Rees C.A., and Murphy K.E. 2006. Retrotransposon insertion in *SILV* is responsible for merle patterning of the domestic dog. *Proc. Natl. Acad. Sci.* **103:** 1376–1381.

Coltman D.W. and Wright J.M. 1994. *Can* SINEs: A family of tRNA-derived retroposons specific to the superfamily Canoidea. *Nucleic Acids Res.* **22:** 2726–2730.

Goldstein O., Zangerl B., Pearce-Kelling S., Sidjanin D.J., Kijas J.W., Felix J., Acland G.M., and Aguirre G.D. 2006. Linkage disequilibrium mapping in domestic dog breeds narrows the progressive rod–cone degeneration interval and identifies ancestral disease-transmitting chromosome. *Genomics* **18:** 541–550.

Hillbertz S.H., Isaksson M., Karlsson E.K., Hellmén E., Rosengren Pielberg G., Savolainen P., Wade C.M., von Euler H., Gustafson U., Hedhammar A., et al. 2007. A duplication of FGF3, FGF4, FGF9 and ORAOV1 causes the hair ridge and predisposes to dermoid sinus in Ridgeback dogs. *Nat. Genet.* (in press).

International HapMap Consortium. 2003. The International HapMap Project. *Nature* **426:** 789–796.

International Human Genome Sequencing Consortium. 2004. Finishing the euchromatic sequence of the human genome. *Nature* **431:** 931–945.

Jonasdottir T.J., Mellersh C.S., Moe L., Heggebo R., Gamlem H., Ostrander E.A., and Lingaas F. 2000. Genetic mapping of a naturally occurring hereditary renal cancer syndrome in dogs. *Proc. Natl. Acad. Sci.* **97:** 4132–4137.

Karlsson E.K., Baranowska I., Wade C.M., Nicolette H.C., Hillbertz S., Zody M.C., Anderson N., Biagi T.M., Patterson N., Rosengren Pielberg G., et al. 2007. Genome-wide association mapping in dogs—A powerful approach for gene discovery. *Nat. Genet.* (in press).

Kazemi-Esfarjani P., Trifiro M.A., and Pinsky L. 1995. Evidence for a repressive function of the long polyglutamine tract in the human androgen receptor: Possible pathogenetic relevance for the $(CAG)_n$-expanded neuronopathies. *Hum. Mol. Genet.* **4:** 523–527.

Kirkness E.F. 2006. SINEs of canine genomic diversity. In *The dog and its genome* (ed. E.A. Ostrander et al.), pp. 209–219. Cold Spring Harbor Laboratory Press, Cold Spring Harbor, New York.

Kirkness E.F., Bafna V., Halpern A.L., Levy S., Remington K., Rusch D.B., Delcher A.L., Pop M., Wang W., Fraser C.M., and Venter J.C. 2003. The dog genome: Survey sequencing and comparative analysis. *Science* **301:** 1898–1903.

Koskinen M.T. 2003. Individual assignment using microsatellite DNA reveals unambiguous breed identification in the domestic dog. *Anim. Genet.* **34:** 297–301.

Koskinen M.T. and Bredbacka P. 2000. Assessment of the population structure of five Finnish dog breeds with microsatellites. *Anim. Genet.* **31:** 310–317.

Kruglyak L. 1999. Prospects for whole-genome linkage disequilibrium mapping of common disease genes. *Nat. Genet.* **22:** 139–144.

Lin L., Faraco J., Li R., Kadotani H., Rogers W., Lin X., Qiu X., de Jong P.J., Nishino S., and Mignot E. 1999. The sleep disorder canine narcolepsy is caused by a mutation in the *hypocretin (orexin) receptor 2* gene. *Cell* **98:** 365–376.

Lindblad-Toh K., Wade C.M., Mikkelsen T.S., Karlsson E.K., Jaffe D.B., Kamal M., Clamp M., Chang J.L., Kulbokas E.J., III, Zody M.C., et al. 2005. Genome sequence, comparative analysis and haplotype structure of the domestic dog. *Nature* **438:** 803–819.

Lohi H., Young E.J., Fitzmaurice S.N., Rusbridge C., Chan E.M., Vervoort M., Turnbull J., Zhao X.C., Ianzano L., Paterson A.D., et al. 2005. Expanded repeat in canine epilepsy. *Science* **307:** 81.

Matsuzaki H., Dong S., Loi H., Di X., Liu G., Hubbell E., Law J., Berntsen T., Chadha M., Hui H., et al. 2004. Genotyping over 100,000 SNPs on a pair of oligonucleotide arrays. *Nat. Methods* **1:** 109–111.

Mignot E., Wang C., Rattazzi C., Gaiser C., Lovett M., Guilleminault C., Dement W.C., and Grumet F.C. 1991. Genetic linkage of autosomal recessive canine narcolepsy with a mu immunoglobulin heavy-chain switch-like segment. *Proc. Natl. Acad. Sci.* **88:** 3475–3478.

Minnick M.F., Stillwell L.C., Heineman J.M., and Stiegler G.L. 1992. A highly repetitive DNA sequence possibly unique to canids. *Gene* **110:** 235–238.

O'Rourke K. 2005. Mining the canine genome. Identification of genes helps breeders and researchers. *J. Am. Vet. Med. Assoc.* **226:** 863–864.

Ostrander E.A. and Friedrichsen D.M. 2004. Genetic factors: Finding cancer susceptibility genes. In *Clinical oncology*, 3rd edition (ed. M.D. Abeloff et al.), pp. 253–267. Elsevier Churchill Livingstone, Philadelphia, Pennsylvania.

Ostrander E.A. and Kruglyak L. 2000. Unleashing the canine genome. *Genome Res.* **10:** 1271–1274.

Ostrander E.A. and Wayne R.K. 2005. The canine genome. *Genome Res.* **15:** 1706–1716.

Ostrander E.A., Markianos K., and Stanford J.L. 2004. Finding prostate cancer susceptibility genes. *Annu. Rev. Genomics Hum. Genet.* **5:** 151–175.

Parker H.G. and Ostrander E.A. 2005. Canine genomics and genetics: Running with the pack. *PLoS Genet.* **1:** e58.

Parker H.G., Kim L.V., Sutter N.B., Carlson S., Lorentzen T.D., Malek T.B., Johnson G.S., DeFrance H.B., Ostrander E.A., and Kruglyak L. 2004. Genetic structure of the purebred domestic dog. *Science* **304:** 1160–1164.

Patterson D. 2000. Companion animal medicine in the age of medical genetics. *J. Vet. Intern. Med.* **14:** 1–9.

Patterson D.F., Haskins M.E., and Jezyk P.F. 1982. Models of human genetic disease in domestic animals. *Adv. Hum. Genet.* **12:** 263–339.

Patterson D.F., Haskins M.E., Jezyk P.F., Giger U., Meyers-Wallen V.N., Aguirre G., Fyfe J.C., and Wolfe J.H. 1988. Research on genetic diseases: Reciprocal benefits to animals and man. *J. Am. Vet. Med. Assoc.* **193:** 1131–1144.

Pelé M., Tiret L., Kessler J.L., Blot S., and Panthier J.J. 2005. SINE exonic insertion in the *PTPLA* gene leads to multiple splicing defects and segregates with the autosomal recessive centronuclear myopathy in dogs. *Hum. Mol. Genet.* **14:** 1417–1427.

Price A.L., Patterson N.J., Plenge R.M., Weinblatt M.E., Shadick N.A., and Reich D. 2006. Principal components analysis corrects for stratification in genome-wide association studies. *Nat. Genet.* **38:** 904–909.

Rosenberg N.A., Pritchard J.K., Weber J.L., Cann H.M., Kidd K.K., Zhivotovsky L.A., and Feldman M.W. 2002. Genetic structure of human populations. *Science* **298:** 2381–2385.

Sargan D.R. 2004. IDID: Inherited diseases in dogs: Web-based information for canine inherited disease genetics. *Mamm. Genome* **15:** 503–506.

Sargan D., Aguirre-Hernandez J., Galibert F., and Ostrander E.A. 2007. An extended microsatellite set for linkage mapping in the domestic dog. *J. Hered.* (in press).

Savolainen P., Zhang Y.P., Luo J., Lundeberg J., and Leitner T. 2002. Genetic evidence for an East Asian origin of domestic dogs. *Science* **298:** 1610–1613.

Sidjanin D.J., Lowe J.K., McElwee J.L., Milne B.S., Phippen T.M., Sargan D.R., Aguirre G.D., Acland G.M., and Ostrander E.A. 2002. Canine CNGB3 mutations establish cone degeneration as orthologous to the human achromatopsia locus ACHM3. *Hum. Mol. Genet.* **11:** 1823–1833.

Steingrimsson E., Copeland N.G., and Jenkins N.A. 2004. Melanocytes and the microphthalmia transcription factor network. *Annu. Rev. Genet.* **38:** 365–411.

Sutter N.B., Eberle M.A., Parker H.G., Pullar B.J., Kirkness E.F., Kruglyak L., and Ostrander E.A. 2004. Extensive and breed-specific linkage disequilibrium in *Canis familiaris*. *Genome Res.* **14:** 2388–2396.

Udono T., Yasumoto K., Takeda K., Amae S., Watanabe K., Saito H., Fuse N., Tachibana M., Takahashi K., Tamai M., and Shibahara S. 2000. Structural organization of the human microphthalmia-associated transcription factor gene containing four alternative promoters. *Biochim. Biophys. Acta* **1491:** 205–219.

Vassetzky N.S. and Kramerov D.A. 2002. CAN—A pan-carnivore SINE family. *Mamm. Genome* **13:** 50–57.

Vilà C., Savolainen P., Maldonado J.E., Amorim I.R., Rice J.E., Honeycutt R.L., Crandall K.A., Lundeberg J., and Wayne R.K. 1997. Multiple and ancient origins of the domestic dog (see comments). *Science* **276:** 1687–1689.

Wade C., Karlsson E.K., Mikkelsen T.S., Zody M.C., and Lindblad-Toh K. 2006. The dog genome: Sequence, evolution and haplotype structure. In *The dog and its genome* (ed. E.A. Ostrander et al.), pp. 179–207. Cold Spring Harbor Laboratory Press, Cold Spring Harbor, New York.

Wayne R.K., Nash W.G., and O'Brien, S. J. 1987a. Chromosomal evolution of the Canidae. I. Species with high diploid numbers. *Cytogenet. Cell Genet.* **44:** 123–133.

———. 1987b. Chromosomal evolution of the Canidae. II. Divergence from the primitive carnivore karyotype. *Cytogenet. Cell Genet.* **44:** 134–141.

Wayne R.K., Geffen E., Girman D.J., Koeppfli K.P., Lau L.M., and Marshall C.R. 1997. Molecular systematics of the Canidae. *Syst. Biol.* **46:** 622–653.

Yuzbasiyan-Gurkan V., Blanton S.H., Cao V., Ferguson P., Li J., Venta P.J., and Brewer G.J. 1997. Linkage of a microsatellite marker to the canine copper toxicosis locus in Bedlington terriers. *Am. J. Vet. Res.* **58:** 23–27.

Zangerl B., Goldstein O., Philp A.R., Lindauer S.J., Pearce-Kelling S.E., Mullins R.F., Graphodatsky A.S., Ripoll D., Felix J.S., Stone E.M., et al. 2006. Identical mutation in a novel retinal gene causes progressive rod–cone degeneration in dogs and retinitis pigmentosa in humans. *Genomics* **26:** 551–563.

WWW RESOURCE

http://www.genome.gov/125134 ENCODE, Project Background, National Human Genome Research Institute.

32 | The Chimpanzee

Tarjei S. Mikkelsen, Michael C. Zody, and Kerstin Lindblad-Toh

Broad Institute of MIT and Harvard, Cambridge, Massachusetts 02142

INTRODUCTION

As our closest extant evolutionary relative, the chimpanzee offers a unique perspective on the human species and its history. All heritable, biological traits unique to our species, such as distinct anatomy, cognitive capacities, and some disease susceptibilities, are ultimately caused by one or more discrete differences between the human and chimpanzee genomes. Comparative analysis of our genome sequences can help elucidate these differences, as well as the mutational processes and selective pressures that have generated them. In addition, as the closest outgroup to the modern human population, the chimpanzee has a special role in informing studies of human genetic variation.

A draft sequence of the genome of the common chimpanzee (*Pan troglodytes*) was initially generated by a U.S.-based consortium from 4-fold whole-genome shotgun (WGS) coverage of a single individual (Chimpanzee Sequencing and Analysis Consortium 2005). At the time of writing, further sequencing has been completed to yield a total of 6.5-fold WGS coverage, as well as a bacterial artificial chromosome (BAC)-based physical map (Warren et al. 2006). A growing number of additional genomic resources are also publicly available, including a BAC-based assembly of chromosome 21 from a different individual (Watanabe et al. 2004); two chromosome Y sequences (Hughes et al. 2005; Kuroki et al. 2006); PCR-amplified exons from over 13,000 known genes (Nielsen et al. 2005); cDNA sequences (Hellmann et al. 2003); and light WGS coverage of additional West African and Central African chimpanzees (Chimpanzee Sequencing and Analysis Consortium 2005). Pre-computed gene and genome alignments are also available from several research groups (http://genome.ucsc.edu/; http://www.ensembl.org; http://www.ncbi.nih. gov). Table 32-1 summarizes the available chimpanzee genome resources.

Here we describe practical aspects of, and insights from, comparative analyses of the human and chimpanzee genomes. Unless otherwise stated, supporting data can be found in the manuscript and supplementary materials published with the initial chimpanzee genome sequence (Chimpanzee Sequencing and Analysis Consortium 2005).

TABLE 32-1. Chimpanzee Genome Resources

Genome sequence and annotations:
http://www.ncbi.nlm.nih.gov/genome/guide/chimp/
http://www.ensembl.org/Pan_troglodytes/
http://genome.ucsc.edu

SNPs:
http://www.broad.mit.edu/mammals/chimp/SNP/

Overview of chimpanzee genome research:
http://www.nature.com/nature/focus/chimpgenome/

Practical Considerations for the Draft Genome Sequence

Prior to utilizing the chimpanzee genome sequence for a particular analysis, it is advisable to consider how to minimize the influence of the inevitable imperfections of a draft genome assembly. This is particularly important for the comparison of closely related species, such as humans and chimpanzees, where even modest error rates can be significant relative to the magnitude of genetic divergence.

Sequence quality scores provide an essential tool for identifying or ignoring putative errors in a genome assembly at the single-nucleotide level, and can be obtained from the same sources as the genome sequence. Genome assembly software uses quality scores from individual sequence reads to weigh the evidence provided by conflicting reads when constructing a consensus sequence, and reports derived quality scores for each nucleotide in this consensus. Currently used quality scores follow the PHRED scoring scheme (Ewing and Green 1998), which is logarithmically related to the estimated accuracy of the corresponding nucleotide (10 equals 90% accuracy, 20 equals 99%, 30 equals 99.9%, and so on). According to this scheme, the low divergence rate between humans and chimpanzees would mandate a very high quality score cutoff. However, the observation that most sequence errors occur adjacent to easily detected artifacts has led to the development of the neighborhood quality score (NQS) filter (Altshuler et al. 2000), which accepts lower-quality nucleotides, provided they are not flanked by obvious errors. Various NQS criteria have been devised for human–chimpanzee comparisons. We find that requiring a quality score of 30 at the central nucleotide and 25 at the five flanking nucleotides on each side, with no restrictions on the number of flanking substitutions, is sufficient to bring the 4-fold WGS assembly to a level of accuracy indistinguishable from "finished" sequences (>99.99% accuracy). Nevertheless, it is advisable to verify that the qualitative outcome of any particular analysis is robust to these parameters.

A potential bias in quality score assignments stems from the fact that the WGS approach employed to sequence the chimpanzee genome does not separate the two haplotypes present in the donor individual. Variant alleles result in conflicts between overlapping reads, and it may in some cases be difficult for genome assembly algorithms to distinguish between sequencing errors and genuine polymorphic sites. As a consequence, the current chimpanzee genome assembly is somewhat biased toward assigning low-quality scores to heterozygous positions relative to homozygous positions in the sequenced individual. In contrast, the human reference genome sequence originates from shotgun sequencing of large, single-haplotype clones (International Human Genome Sequencing Consortium 2004). Although this bias will have a negligible impact on most comparative analyses, it may be important to consider for any approach that is sensitive to small variations in lineage-specific divergence rates (see below).

Missed sequence overlaps are another common artifact of draft assemblies, or more accurately, of how they are typically presented. WGS assembly algorithms generate contiguous sequences (contigs) by merging overlapping reads. Incomplete coverage results in a gap between two contigs, and the length of missing sequence can often be estimated from paired-end reads spanning the gap. In some cases, two contigs may contain redundant regions that were not merged due to conflicting sequence content, often due to low-quality reads. Such situations are usually recognized by the software and flagged with negative gap length estimates, but when contigs are compiled into one continuous sequence per chromosome, as is often done for post-assembly analysis and presentation, negative gaps are by convention converted to positive gaps 100 bases long (filled by the arbitrary base "N"). If this continuous sequence is aligned to the human genome, the redundant regions flanking such gaps will appear as artificial "insertions" in the chimpanzee sequence. A mapping from the continuous sequence representation to the original WGS contigs is available with each draft assembly, and is a useful tool for avoiding such artifacts when inferring the presence of large-scale insertions and deletions (indels).

For some applications, researchers have elected to align individual chimpanzee sequence reads onto the human genome, rather than start from the assembled genome sequence (Hellmann et al. 2005; Patterson et al. 2006). Reads are typically aligned using software such as BLASTZ (Schwartz et al. 2003) or ARACHNE (Jaffe et al. 2003), and only unambiguously placed reads with high-quality scores are kept for downstream analysis. This approach has the advantage that it can be used to compute both divergence (from read–assembly differences) and diversity rates (from read–read differences), from multiple species or populations, without introducing unknown biases from potentially different assembly algorithms. Read alignments have also been used to discover large-scale rearrangements between the human and chimpanzee genomes from discordant placement of paired-end sequences from large-insert clones (Newman et al. 2005).

Genetic Divergence between Humans and Chimpanzees

Most nucleotides in the human and chimpanzee genomes are identical by descent, and an observed difference nearly always represents a single mutation. Enumeration of all differences between the human and chimpanzee genomes, as inferred from sequence alignments, can therefore yield a nearly complete catalog of the genetic alterations that took place during the evolution of the human and chimpanzee lineages.

Single-nucleotide substitutions are the most abundant type of differences. Overall, 1.23% of orthologous nucleotides differ, but these substitutions are not distributed uniformly throughout the genomes, in large part due to context-dependent and regional variation in mutation rates. For example, although CpG dinucleotides constitute only 2% of each genome, they account for approximately 25% of all nucleotide substitutions. On a larger scale, the divergence rate fluctuates from 0.99% to 1.54% across 1-Mb segments (25th–75th quartile range; Fig. 32-1). Orthologous sequences situated within 10 Mb of a telomere have, on average, accumulated 15% more substitutions than the rest of the genome, potentially reflecting recombination-mediated mutations (Hellmann et al. 2005).

Nucleotide indels are less abundant than substitutions, but they affect significantly more sequence overall. In total, 5–6 million indels have resulted in the human and chimpanzee genomes each containing 40–45 Mb of euchromatic sequence not present in the other. The vast majority of indels are small (98.6% are shorter than 80 bp), but the largest few contain most of the affected sequence (~70,000 indels longer than 80 bp constitute 75% of the lineage-specific sequences).

Transposable elements are the source of a distinct class of indels in both the human and chimpanzee genomes. The most striking difference is the emergence of large, subterminal caps of satellite repeats on chimpanzee chromosomes (Yunis and Parkash 1982). The euchromatic sequences also show evidence of lineage-specific insertions of all major classes of transposable elements. The short, primate-specific Alu element has been threefold more active in the human lineage, and more than 7000 human-specific Alu insertions can be found throughout our genome. The longer SVA and LINE-1 elements have been inserted at similar rates in humans and chimpanzees, with approximately 1000 and 2000 new insertions in each genome, respectively. Endogenous retroviruses have become all but extinct in the human lineage, with only a single retrovirus (HERV-K) having contributed less than 100 human-specific insertions. In contrast, the chimpanzee genome carries several hundred insertions that appear to originate from multiple, independent germ-line invasions of novel retroviruses (Yohn et al. 2005).

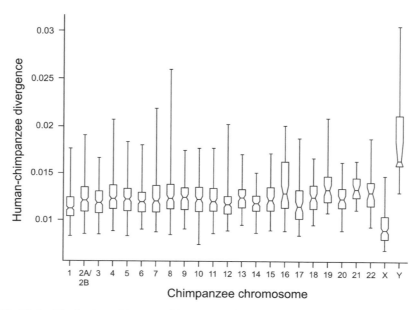

FIGURE 32-1. Distribution of human–chimpanzee divergence in 1-Mb segments, shown as a box plot. The edges of the box correspond to quartiles; the notches to the standard error of the mean; and the vertical bars to the range. (Reprinted, with permission, from Chimpanzee Sequencing and Analysis Consortium 2005.)

Large-scale chromosomal rearrangements make up the least abundant, but most dramatic, differences between the two genomes. Early cytogenetic characterization revealed nine pericentric inversions between human and chimpanzee chromosomes and a fusion of two ancestral chromosomes in the human lineage (Yunis and Prakash 1982). Surveys of structural variation using paired-end sequence mapping and the WGS assembly have refined the localization of these rearrangements and revealed several hundred additional chromosomal inversions, segmental duplications, and deletions (Newman et al. 2005). Figure 32-2 shows a synteny plot between human and chimpanzee chromosomes.

Impact of Extant and Ancestral Genetic Variation on Observed Divergence

Because the sequenced human and chimpanzee genomes represent only one extant allele at each nucleotide position, an observed difference is not necessarily a fixed difference between all humans and chimpanzees, but may instead be a variant allele in the human population or a variant allele in the chimpanzee population (Fig. 32-3).

The genetic divergence between two orthologous sequences is proportional to the time to the last common ancestor (TLCA) from which they originate, under the assumption that differences have accumulated at a constant rate as a result of new mutations (Zuckerkandl and Pauling 1965). Assuming no selection, the proportion of observed divergent nucleotides that are non-polymorphic in both the human and chimpanzee populations is $1 - (T_H + T_C)/(2 \times T_{HC})$, where T_H is the mean TLCA of a chromosomal segment in the human population, T_C is the TLCA in the chimpanzee population, and T_{HC} is the TLCA of humans and chimpanzees. From coalescence theory (Rosenberg and Feldmann 2002), the expected TLCA in the human and chimpanzee populations is $4 \times N_e \times g$, where N_e is the effective population size (estimated as 10,000 for humans, 10,000–20,000 for chimpanzees) and g is the generation time (assumed to be 25 years), giving T_H = 1 million years (Myr) and T_C = 1–2 Myr. Assuming T_{HC} = 7 Myr (Glazko and Nei 2003), the genome-wide proportion of fixed differences can be estimated as 0.78–0.86. It may therefore be important to genotype multiple individuals of both species before drawing conclusions about the implications of a particular substitution on human evolution.

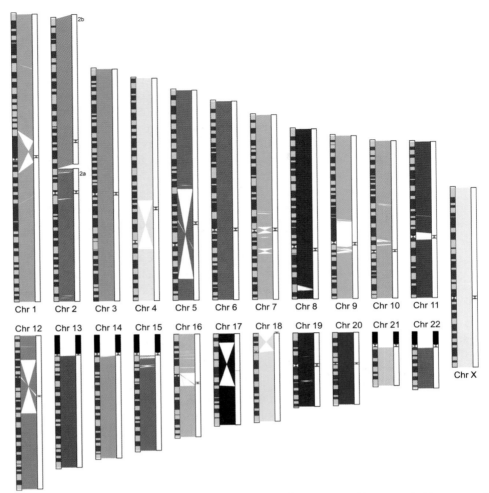

FIGURE 32-2. Synteny plot between human and chimpanzee chromosomes. Human and chimpanzee chromosomes are largely colinear. Exceptions include human chromosome 2, which corresponds to two chimpanzee chromosomes (2a and 2b), and large pericentric inversions on chromosomes 1, 4, 5, 7, 9, 12, 15, 16, 17, and 18. The inversions on chromosomes 9 and 16 have many duplications, making determination of synteny difficult. The inversions on chromosomes 7 and 15 are limited mostly to heterochromatic regions. Synteny maps were generated by clustering PatternHunter (http://www.bioinformaticssolutions.com) alignments at a resolution of >200 kb. Human chromosomes, with G-banding patterns, are shown to the left and chimpanzee to the right for each pair.

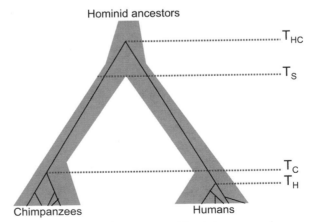

FIGURE 32-3. Simplified phylogeny of a single, non-recombined chromosomal segment present in both humans and chimpanzees. T_{HC} is the time to the last common ancestor (TLCA) of humans and chimpanzees from which the segment originated. T_H and T_C are the TLCA of humans and chimpanzees, respectively, from which every extant copy of the segment originated. T_S is the time since speciation of humans and chimpanzees.

For other types of genetic divergence, the impact of extant variation may be even greater. Based on initial surveys of structural variation, it has been estimated that for apparent deletions of length >12 kb in the chimpanzee genome relative to the human genome, the proportion of fixed differences may be as low as 0.67 (Newman et al. 2005). This likely reflects a high rate of recurrent structural rearrangements in certain regions of our genomes.

Although variation in mutation rates can explain a substantial fraction of the regional variation in divergence rates, another important force is genetic drift in the ancestral population. For pairs of diploid organisms, the TLCA for orthologous sequences is necessarily larger than the time since speciation, and is estimated to average approximately 7 Myr in humans and chimpanzees (Fig. 32-3). Due to recombination, the TLCA is not constant across a chromosome, but varies stochastically between segments that are, on average, shorter than 10 kb (the length of linkage disequilibrium in modern African populations and a likely upper bound for the ancestral hominid population; Reich et al. 2001). Based on the observed divergence rates, the range of the TLCA of human and chimpanzee chromosomal segments has been estimated to span more than 4 Myr. This range is so large that some researchers have proposed that one or more hybridization events may have taken place after the initial split of the human and chimpanzee lineages (Patterson et al. 2006).

Identification of Ancestral Alleles

The *ancestral allele* at a polymorphic locus is defined as the allele that was carried by the last common ancestor (LCA) of all humans. The complementary allele must originate from mutations postdating the LCA, and is referred to as the *derived allele*.

The chimpanzee genome sequence is an excellent tool for ancestral allele classification, particularly for single-nucleotide polymorphisms (SNPs) within the human population. Because 98.8% of orthologous nucleotides are identical by descent, one can generally assume that a human allele matching the chimpanzee genome sequence is ancestral. Depending on the stringency of the sequence alignments used, we could assign ancestral and derived status to 80–90% of the human variants cataloged in dbSNP this way using the initial chimpanzee genome sequence. Additional sequence data from chimpanzees, as well as from close outgroups such as gorillas and orangutans, facilitate refinement and extension of these assignments.

It is, of course, important to keep in mind that ancestral allele assignments using the chimpanzee genome are heuristic. The error rate of these assignments can be estimated as the probability that the orthologous chimpanzee nucleotide matches one of the two human alleles, but not the one that is ancestral in the human population. There are two simple cases in which this could happen: first, where the chimpanzee has mutated to the same nucleotide as the derived human allele, and second, where a derived human allele was fixed at some point in the past and then followed by a reversion mutation that is still segregating. Cases involving more mutations are possible but are at least two orders of magnitude less likely than these.

The estimated error rate for a typical SNP is 0.5%. The exceptions are those SNPs for which the human alleles are CpG and TpG and the chimpanzee sequence is TpG. For these, a nonnegligible fraction may have arisen by two independent deamination events within an ancestral CpG dinucleotide, which is a well-known mutational hot spot (see above, Genetic Divergence between Humans and Chimpanzees). Human SNPs in a CpG context for which the orthologous chimpanzee sequence is TpG account for 12% of all human variants and have an estimated error rate of 9.8%. Across all SNPs, the average error rate is thus estimated to be about 1.6%.

Classification of human alleles as ancestral or derived has a number of applications in the study of genetic variation, including inferences about demography, natural selection, and ancient phenotypes. For example, an elegant result in population genetics states that, for a randomly interbreeding population of constant size, the probability that an allele is ancestral is equal to its frequency (Watterson and Guess 1977). We explored the extent to which this simple theoretical expectation fits empirical data from the human population using about 120,000 SNPs genotyped by Affymetrix in 54 diverse individuals. We tabulated the proportion $p_a(x)$ of ancestral alleles for various frequencies of x and compared this with the prediction $p_a(x) = x$.

The data lie near the predicted line, but the observed slope (0.83) is significantly less than 1 (Fig. 32-4). One explanation for this deviation is that some ancestral alleles are incorrectly assigned (an error rate of ε would artificially decrease the slope by a factor of $1-2\varepsilon$). However, with ε estimated to be only 1.6%, errors can only explain a small part of the deviation. The most likely explanation is the presence of bottlenecks during human history, which flatten the distribution of allele frequencies. Theoretical calculations indicate that a recent bottleneck would decrease the slope by a factor of $(1 - b)$, where b is the inbreeding coefficient induced by the bottleneck. This suggests that measurements of the slope in different human groups may shed light on population-specific bottlenecks. Initial analyses of allele frequencies in several ENCODE regions (International HapMap Consortium 2005) suggest that the slope is lower than 1 in European and Asian samples, and close to 1 in an African sample, consistent with the "out-of-Africa" model of human migrations.

Signatures of Natural Selection

A major motivation for sequencing the chimpanzee genome was to elucidate the impact of natural selection on the evolutionary history of the human species. Evidence of past natural selection can be inferred from patterns of sequence divergence or variation that differ from those expected from neutral genetic drift. The impact of natural selection is generally divided into two major forms: *negative*, or purifying, selection leads to removal of deleterious alleles from a population; *positive*, or adaptive, selection leads to rapid fixation of advantageous alleles.

There are myriad proposed approaches to detecting signatures of both negative and positive natural selection. A key differentiating characteristic is the evolutionary time frame in which they can detect such signatures. The chimpanzee genome sequence is particularly helpful for detecting natural selection that occurred during the 5–7 Myr that separate the LCA of the human and

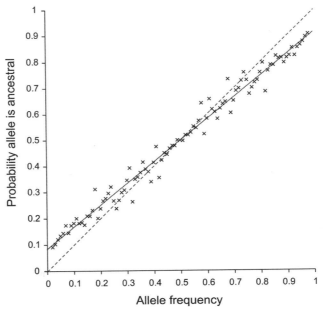

FIGURE 32-4. The observed fraction of ancestral alleles in 1% bins of observed frequency. The solid line shows the theoretical relationship $p_a(x) = x$. Note that because each variant yields a derived and an ancestral allele, the data are necessarily symmetrical about 0.5. (Modified, with permission, from Chimpanzee Sequencing and Analysis Consortium 2005.)

chimpanzee species and the LCA of anatomically modern humans (we refer to selection during this time span as *ancient*). Combined with human genetic variation data, the chimpanzee sequence can also provide a valuable baseline for inferences about natural selection within the last 250,000 years (*recent*). PAML (http://abacus.gene.ucl.ac.uk/software/paml.html) and MEGA (http://www.megasoftware.net) are two popular software packages that implement several of the methods described here.

Signatures of ancient, negative selection can be recognized from a reduced rate of divergence in functional, relative to neutrally evolving, sequences. For example, the average protein-coding gene has accumulated only two amino acid substitutions since our divergence, and the mean ratio of the rates of amino acid-altering (nonsynonymous) and synonymous substitutions (often denoted K_A/K_S or d_N/d_S) across all orthologs is 0.23, compared to the expected value of 1 in the absence of selection. Under the common assumption that synonymous mutations are selectively neutral, this implies that 77% of amino acid alterations are sufficiently deleterious as to be eliminated by natural selection during hominid evolution. This is likely to be a lower bound, as there is growing evidence that synonymous mutations do experience some negative selection in humans and other mammals (Chamary et al. 2006). Using the divergence rate of shared transposable elements as a proxy for the neutral rate of evolution, evidence of negative selection can also be detected in functional noncoding sequences, such as 5′ and 3′ untranslated regions, and *cis*-regulatory elements.

Signatures of recent, negative selection can be recognized from the patterns of human genetic variation within protein-coding regions. Examining HapMap SNPs with derived allele frequencies greater than 15%, we found that the ratio of amino acid-altering to synonymous polymorphisms was approximately 0.23, similar to the ratio for fixed differences. This implies that negative natural selection has removed most novel, deleterious alleles from the modern human population before they could reach a high frequency. However, polymorphisms in protein-coding genes with derived allele frequencies from 1 to 15% do have an elevated K_A/K_S ratio (estimated as ~0.3–0.4 from HapMap data). This suggests that a subset of deleterious amino acid alterations may temporarily attain readily detectable frequencies and thus contribute significantly to the human genetic load (Fay et al. 2001). Combining derived allele frequency data with predictions of functionally relevant substitutions from software such as Polyphen (Ramensky et al. 2002) and SIFT (Ng and Henikoff 2003) may assist in identifying deleterious alleles in a gene of interest.

It is notable that the K_A/K_S ratio of approximately 0.23 observed for human–chimpanzee divergence and human polymorphisms is significantly higher than the ratios observed in equivalent comparisons between other well-studied mammals. The nearly neutral theory (Ohta 1998) predicts that this elevation indicates a general relaxation of evolutionary constraints due to smaller effective population sizes and, consequently, a greater impact of genetic drift in human and chimpanzee evolution, relative to many other mammals.

Signatures of ancient, positive selection are more challenging to detect than negative selection, but are of even greater interest. Although effectively neutral alleles may well have phenotypic effects, it is generally thought that signatures of positive selection can help pinpoint the genetic changes most critical to the evolutionary history of hominids.

The low divergence rate between humans and chimpanzees effectively eliminates the possibility of detecting conclusive evidence of positive selection in most genes or other functional elements by statistical means alone, but several methods have been utilized to identify and rank candidate genes for follow-up analysis, and to elucidate large-scale patterns of adaptive evolution (Fay and Wu 2003). These methods all rely on detection of either an excess of amino acid-altering versus synonymous substitutions (K_A/K_S tests), accelerated substitution rates in one lineage relative to the other (relative rate tests), or an excess of fixed substitutions relative to the amount of polymorphism in the human population (the McDonald-Kreitman test). Development of analogous methods for the detection of positive selection in noncoding functional elements is a more recent, but active, area of research (Rockman et al. 2005).

The most stringent signature of positive selection is a significant excess of amino acid substitutions over synonymous substitutions in protein-coding genes ($K_A/K_S > 1$), which cannot be explained by neutral evolution. The K_A/K_S test can be applied to a single lineage, by using outgroups to infer lineage-specific substitutions, or to a pair of lineages. The latter approach will often have more statistical power if a gene has been under positive selection in both lineages. In our initial survey, we identified 585 orthologs with an excess of amino acid-altering substitutions between humans and chimpanzees, approximately 50% more than would be expected under the null hypothesis of no positive selection. This set of orthologs was highly enriched for genes involved in the immune system. This is a common observation in higher organisms studied so far, reflecting sustained selective pressures on host-defense mechanisms throughout evolution.

A major limitation of the standard K_A/K_S test is that it does not account for heterogeneous selection pressures acting on different regions of a gene. For protein sequences, it may be helpful to restrict the test to particular functional domains. For example, positive selection on the BRCA1 tumor suppressor in humans and chimpanzees appears to have been focused on its RAD51 interaction domain (Huttley et al. 2000). Variations of the K_A/K_S test have been developed to detect positive selection limited to individual codons (known as "heterogeneous site" and "branch-site" tests; Yang et al. 2000; Zhang et al. 2005), but some controversy remains over the applicability of these methods to comparisons of closely related species.

Lineage-specific acceleration of substitution rates, as inferred from the comparison of human, chimpanzee, and one or more additional species, can reveal more subtle signatures of selection. The limitations are that this signature cannot formally distinguish positive selection and relaxation of evolutionary constraints, and it also differs depending on the outgroups used. For example, by comparing the K_A/K_S ratios of orthologs between human and chimpanzee with matched ratios from mouse and rat, we found that genes involved in spermatogenesis and the male reproductive system show a highly significant hominid-specific acceleration of amino acid substitutions, potentially reflecting a particularly strong influence of sexual selection on hominid evolution. In contrast, there is no evidence of lineage-specific acceleration in these genes in humans relative to chimpanzees when the murids are used as outgroups, reflecting more similar selection pressures within the two hominid lineages.

Notably, transcription factor genes do show acceleration of amino acid substitution rates in humans relative to chimpanzees. These genes have a significant excess of amino acid substitutions between humans and chimpanzees, relative to the rate of polymorphic amino acids in modern humans (Bustamante et al. 2005). This supports the hypothesis that changes in gene regulation may have been a key factor underlying the rapid anatomical evolution of our hominid ancestors (King and Wilson 1975).

Signatures of recent, positive selection are also of great interest because they may help pinpoint genes critical to the evolution of anatomically modern humans, as well as identify genetic responses to new pathogens, diets, and other environmental conditions (Sabeti et al. 2006). The pattern of human genetic variation holds substantial information about selection events. Strong positive selection creates the distinctive signature of a "selective sweep," whereby a rare allele rapidly rises to fixation and carries the haplotype on which it occurs to high frequency (the "hitchhiking" effect). The surrounding region should show two distinctive signatures: a significant reduction of overall diversity, and an excess of derived alleles with high frequency in the population owing to hitchhiking of derived alleles on the selected haplotype (Przeworski 2002).

The chimpanzee genome provides crucial baseline information required for accurate assessment of both signatures: Human–chimpanzee divergence can be used to control for regional variation in mutation rate when searching for regions of low diversity, and the chimpanzee genome can be used to classify alleles as ancestral or derived (as discussed above). For example, in our initial survey, we identified six megabase-wide regions with exceptionally low human diversity relative to human–chimpanzee divergence, and an excess of higher frequency derived alleles, making them prime candidates for strong selective sweeps in recent history. In a different study (Nielsen et al. 2005), a subset of genes first identified as candidates for selection based on a K_A/K_S test were also found to show an excess of high-frequency, derived, amino acid-altering alleles, suggesting recent episodes of adaptive evolution.

SUMMARY

Although the chimpanzee is of limited use as a model organism in the traditional sense, due to both ethical restrictions and high costs of maintenance, its genome sequence constitutes an important resource for human population genetics and evolutionary studies of the human lineage.

ACKNOWLEDGMENTS

We thank all members of the Chimpanzee Sequencing Consortium for their contributions to the first comprehensive comparative analysis of the human and chimpanzee genomes. We also thank Manuel Garber for generating the synteny figure.

REFERENCES

Altshuler D., Pollara V.J., Cowles C.R., Van Etten W.J., Baldwin J., Linton L., and Lander E.S. 2000. An SNP map of the human genome generated by reduced representation shotgun sequencing. *Nature* 407: 513–516.

Bustamante C.D., Fledel-Alon A., Williamson S., Nielsen R., Hubisz M.T., Glanowski S., Tanenbaum D.M., White T.J., Sninsky J.J., Hernandez R.D., et al. 2005. Natural selection on protein-coding genes in the human genome. *Nature* 437: 1153–1157.

Chamary J.V., Parmley J.L., and Hurst L.D. 2006. Hearing silence: Non-neutral evolution at synonymous sites in mammals. *Nat. Rev. Genet.* 7: 98–108.

Chimpanzee Sequencing and Analysis Consortium. 2005. Initial sequence of the chimpanzee genome and comparison with the human genome. *Nature* 437: 69–87.

Ewing B. and Green P. 1998. Base-calling of automated sequencer traces using PHRED. II. Error probabilities. *Genome Res.* 8: 186–194.

Fay J.C. and Wu C.-I. 2003. Sequence divergence, functional constraint, and selection in protein evolution. *Annu. Rev. Genomics Hum. Genet.* 4: 213–235.

Fay J.C., Wyckoff G.J., and Wu C.-I. 2001. Positive and negative selection on the human genome. *Genetics* 158: 1227–1234.

Glazko G.V. and Nei M. 2003. Estimation of divergence times for major lineages of primate species. *Mol. Biol. Evol.* 20: 424–434.

Hellmann I., Prüfer K., Ji H., Zody M.C., Pääbo S., and Ptak S.E. 2005. Why do human diversity levels vary at a megabase scale? *Genome Res.* 15: 1222–1231.

Hellmann I., Zollner S., Enard W., Ebersberger I., Nickel B., and Pääbo S. 2003. Selection on human genes as revealed by comparisons to chimpanzee cDNA. *Genome Res.* 13: 831–837.

Hughes J.F., Skaletsky H., Pyntikova T., Minx P.J., Graves T., Rozen S., Wilson R.K., and Page D.C. 2005. Conservation of Y-linked genes during human evolution revealed by comparative sequencing in chimpanzee. *Nature* 437: 100–103.

Huttley G.A., Easteal S., Southey M.C., Tesoriero A., Giles G.G., McCredie M.R., Hopper J.L., and Venter D.J. 2000. Adaptive evolution of the tumor suppressor BRCA1 in humans and chimpanzees. Australian Breast Cancer Family Study. *Nat. Genet.* 26: 131–132.

International HapMap Consortium. 2005. A haplotypes map of the human genome. *Nature* 437: 1299–1320.

International Human Genome Sequencing Consortium. 2004. Finishing the euchromatic sequence of the human genome. *Nature* 431: 931–945.

Jaffe D.B., Butler J., Gnerre S., Mauceli E., Lindblad-Toh K., Mesirov J.P., Zody M.C. and Lander E.S. 2003. Whole-genome sequence assembly for mammalian genomes: Arachne 2. *Genome Res.* 13: 91–96.

King M.C. and Wilson A.C. 1975. Evolution at two levels in humans and chimpanzees. *Science* 188: 107–116.

Kuroki Y., Toyoda A., Noguchi H., Taylor T.D., Itoh T., Kim D.S., Kim D.W., Choi S.H., Kim I.C., Choi H.H., et al. 2006. Comparative analysis of chimpanzee and human Y chromosomes unveils complex evolutionary pathway. *Nat. Genet.* 38: 158–167.

Newman T.L., Tuzun E., Morrison V.A., Hayden K.E., Ventura M., McGrath S.D., Rocchi M., and Eichler E.E. 2005. A genome-wide survey of structural variation between human and chimpanzee. *Genome Res.* 15: 1344–1356.

Ng P.C. and Henikoff S. 2003. SIFT: Predicting amino acid changes that affect protein function. *Nucleic Acids Res.* 31: 3812–3814.

Nielsen R., Bustamante C., Clark A.G., Glanowski S., Sackton T.B., Hubisz M.J., Fledel-Alon A., Tanenbaum D.M., Civello D., White T.J., et al. 2005. A scan for positively selected genes in the genomes of humans and chimpanzees. *PLoS Biol.* 3: e170.

Ohta T. 1998. Evolution by nearly-neutral mutations. *Genetica* 102–103: 83–90.

Patterson N.P., Richter D.J., Gnerre S., Lander E.S., and Reich D. 2006. Genetic evidence for complex speciation of humans and chimpanzees. *Nature* 441: 1103–1108.

Przeworski M. 2002. The signature of positive selection at randomly chosen loci. *Genetics* **160:** 1179–1189.

Ramensky V., Bork P., and Sunyaev S. 2002. Human non-synonymous SNPs: Server and survey. *Nucleic Acids Res.* **30:** 3894–3900.

Reich D.E., Cargill M., Bolk S., Ireland J., Sabeti P.C., Dichter D.J., Lavery T., Kouyoumjian R., Farhadian S.F., Ward R., and Lander E.S. 2001. Linkage disequilibrium in the human genome. *Nature* **411:** 199–204.

Rockman M.V., Hahn M.W., Soranzo N., Zimprich F., Goldstein D.B. and Wray G.A. 2005. Ancient and recent positive selection transformed opioid *cis*-regulation in humans. *PLoS Biol.* **3:** e387.

Rosenberg H.F. and Feldmann M.W. 2002. The relationship between coalescence times and population divergence times. In *Modern developments in theoretical population genetics: The legacy of Gustave Malécot* (ed. M. Slatkin and M. Veuille), pp. 130–164. Oxford University Press, Oxford, United Kingdom.

Sabeti P.C., Schaffner S.F., Fry B., Lohmueller J., Varilly P., Shamovsky O., Palma A., Mikkelsen T.S., Altshuler D., and Lander E.S. 2006. Positive natural selection in the human lineage. *Science* **312:** 1614–1620.

Schwartz S., Kent W. J., Smit A., Zhang Z., Baertsch R., Hardison R.C., Haussler D., and Miller W. 2003. Human-mouse alignments with BLASTZ. *Genome Res.* **13:** 103–107.

Warren R.L., Varabei D., Platt D., Huang X., Messina D., Yang S.P., Kronstad J.W., Krzywinski M., Warren W.C., Wallis J.W., et al. 2006. Physical map-assisted whole-genome sequence assemblies. *Genome Res.* **16:** 768–775.

Watanabe H., Fujiyama A., Hattori M., Taylor T.D., Toyoda A., Kuroki Y., Noguchi H., BenKahla A., Lehrach H., Sudbrak R., et al. 2004. DNA sequence and comparative analysis of chimpanzee chromosome 22. *Nature* **429:** 382–388.

Watterson G.A. and Guess H.A. 1977. Is the most frequent allele the oldest? *Theor. Popul. Biol.* **11:** 141–160.

Yang Z., Nielsen R., Goldman N., and Pedersen A.M. 2000. Codon-substitution models for heterogeneous selection pressure at amino acid sites. *Genetics* **155:** 431–449.

Yohn C.T., Jiang Z., McGrath S.D., Hayden K.E., Khaitovich P., Johnson M.E., Eichler M.Y., McPherson J.D., Zhao S., Pääbo S., and Eichler E.E. 2005. Lineage-specific expansions of retroviral insertions within the genomes of African great apes but not humans and orangutans. *PLoS Biol.* **3:** e110.

Yunis J.J. and Prakash O. 1982. The origin of man: A chromosomal pictorial legacy. *Science* **215:** 1525–1530.

Zhang J., Nielsen R., and Yang Z. 2005. Evaluation of an improved branch-site likelihood method for detecting positive selection at the molecular level. *Mol. Biol. Evol.* **22:** 2472–2479.

Zuckerkandl E. and Pauling L. 1965. Molecules as documents of evolutionary history. *J. Theor. Biol.* **8:** 357–366.

33 | Genealogical Markers: mtDNA and the Y Chromosome

Mark Stoneking[1] and Manfred Kayser[2]

[1]Max Planck Institute for Evolutionary Anthropology, D-04103 Leipzig, Germany; [2]Department of Forensic Molecular Biology, Erasmus University Medical Centre, 3000 CA Rotterdam, The Netherlands

INTRODUCTION

Because they are haploid and inherited from one parent without any recombination, mitochondrial DNA (mtDNA) and the nonrecombining portion of the Y chromosome (NRY) have provided, and continue to provide, important insights into the genealogical (i.e., maternal and paternal) history of human populations or individuals. In addition, because it is present in many copies per cell, mtDNA is the genome of choice for analyzing ancient DNA. In this chapter, we first discuss some general issues that arise in analyzing human mtDNA and NRY variation. We then draw upon our own work to present some examples of the kinds of insights that have resulted from comparative analyses of mtDNA and NRY variation in human populations. Finally, we briefly summarize the use of mtDNA and NRY markers for forensic and genealogical purposes.

mtDNA VARIATION

The properties of mtDNA that make it valuable for population and evolutionary genetic studies have recently been reviewed in detail elsewhere (Pakendorf and Stoneking 2005) and include its high copy number, rapid rate of evolution, and haploid, uniparental, nonrecombining mode of inheritance. Prior to the advent of PCR, studies of human mtDNA variation utilized either low-resolution restriction-site mapping using Southern blots of whole genomic DNA (e.g., Johnson et al. 1983) or high-resolution mapping using purified, end-labeled mtDNA (e.g., Cann et al. 1987). For both types of studies, the early availability (in 1981) of the complete sequence of the human mtDNA genome (Anderson et al. 1981) made it possible to infer the mutations responsible for the observed restriction fragment length polymorphisms (RFLPs). With the development of PCR, some investigators utilized high-resolution mapping of PCR fragments to investigate human mtDNA variation (see, e.g., Torroni et al. 1992).

However, sequencing studies revealed that the most variable, and hence informative, portions of the mtDNA genome are located in the noncoding control region, with one hypervariable segment (HV1) harboring most of the variation, and at present, most studies of human mtDNA variation focus on sequence analysis of HV1. Tens of thousands of HV1 sequences have been published, and in addition to the usual databases of DNA sequences such as Genbank, a useful compilation of HV1 sequences is Hvrbase++ (www.hvrbase.org). More recently, as the time and cost of sequencing have continued to decrease, studies have turned to sequencing the entire mtDNA genome (Ingman et al. 2000), which thus provides the maximum amount of information that can be extracted from mtDNA. Such studies have provided important and interesting insights into the history of particular mtDNA lineages (Thangaraj et al. 2005; Trejaut et al. 2005), but at present, cost and time limitations preclude implementation of whole-genome mtDNA sequencing on the scale required for population genetic studies of many groups.

Although HV1 sequencing is the standard approach for analyzing variation in human mtDNA, utilized by dozens of laboratories, HV1 sequencing is not without idiosyncrasies. HV1 itself comprises an approximately 380-bp region beginning at nucleotide position (np)16024 (the start of the control region; Anderson et al. 1981), which is a convenient size for amplification and sequencing. However, a T→C mutation (or deletion) at np 16189 results in 9–10 consecutive cytosines known as the "C-stretch," which is too long to sequence through, most likely because of slippage during the PCR amplification (Bendall and Sykes 1995). Therefore, in order to have all positions sequenced twice, it is necessary to either sequence each strand twice, or to use internal primers that bracket the C-stretch.

In the past few years, there has been much discussion about errors in HV1 sequence studies (Bandelt et al. 2002; Forster 2003; Salas et al. 2005), and indeed, a "cottage industry" has arisen around the detection of errors in published studies and databases. The detection of errors relies on methods of statistical analyses, based on patterns of variation observed in sequences obtained to date (Bandelt et al. 2002), and hence is not without potential problems because new sequences may not conform to previous patterns (Barbujani et al. 2004). Moreover, such statistical analyses are not expected to capture all potential errors and are not an adequate substitute for careful laboratory practice. In this vein, we note with dismay an increasing tendency for published studies to obtain sequence information from only one strand of the HV1 PCR product. In our view, it is absolutely necessary to sequence both strands routinely, because in our experience this is the best way to detect errors (in particular, sample mix-ups, which are by far the most common error). Although not foolproof, comparison of new HV1 sequences to existing sequences for discrepant patterns is an important and useful method for detecting sequence errors and should be required of every study.

An outcome of the early studies of mtDNA RFLP variation was that individual mtDNA types could be grouped on the basis of shared diagnostic sites into haplogroups, which show some degree of regional geographic specificity. Letters of the alphabet were used to designate the major haplogroups, but since no consistent nomenclature was adopted for naming new haplogroups, the haplogroup tree is rather haphazard. An up-to-date version of the haplogroup tree, as well as a map indicating the geographic distribution of the major haplogroups, can be found at the Mitomap Web site (www.mitomap.org). Some haplogroups can be inferred from the HV1 sequences with a high degree of confidence, but some cannot; diagnostic sites must then be typed to assign haplogroups to the latter.

In analyzing mtDNA sequence and/or haplogroup data (or, for that matter, NRY haplotype and/or haplogroup data, discussed below), two general approaches are used: lineage-based and population-based (Pakendorf and Stoneking 2005). The lineage-based approach analyzes each distinct mtDNA lineage (i.e., haplogroup) in a phylogeographic approach, usually defining the geographic spread and time of origin of each lineage. This approach results in a detailed understanding of the history of the mtDNA lineages but does not directly reveal anything about the history of the populations carrying those lineages. Unfortunately, this limitation of lineage-based analyses was not widely appreciated initially, and there was a naïve tendency to equate the age of a haplogroup with the age of the population and to assume that each haplogroup in a population reflected a separate migration. Fortunately, it is now generally appreciated that population genetic methods, in which the population and not the individual lineages are the unit of analysis, are required to make inferences about population history from mtDNA and/or NRY data. Further on, as an example of how this is done from our own work, we present the insights gained from mtDNA and NRY data into the colonization of Polynesia.

Two issues have been raised that question the validity of applications of mtDNA analyses to human population variation and evolution: (1) recombination and (2) selection. With respect to recombination, conventional thinking for many years, based on a rather limited number of pedigree studies (Giles et al. 1980), was that mtDNA was strictly maternally inherited with no recombination. This view was challenged by studies claiming evidence for recombination based on either unusual patterns of variation observed in some mtDNA sequences (Hagelberg et al. 1999) or statistical analyses of patterns of variation (Awadalla et al. 1999). However, the unusual mtDNA sequences were later shown to be artifacts (Hagelberg et al. 2000), and the statistical analyses were flawed (Kumar et al. 2000). Even though paternal inheritance of human mtDNA has now been demonstrated in one individual from a family exhibiting an mtDNA disease (Schwartz and Vissing 2002), and recombination has been demonstrated in somatic tissue from this same individual (Kraytsberg et al. 2004), thousands of mother–offspring comparisons have failed to yield any evidence of recombination (for review, see Pakendorf and Stoneking 2005), and the general consensus is that parallel and back mutations at hypervariable sites are a more likely explanation for observed patterns of variation that might otherwise suggest recombination (Piganeau and Eyre-Walker 2004).

Similarly, it has been suggested that natural selection acting on particular mtDNA mutations, possibly in relation to climatic adaptation, could account for the distribution of mtDNA types in certain populations (Excoffier 1990; Mishmar et al. 2003; Ruiz-Pesini et al. 2004), casting doubt on inferences about population history using mtDNA. However, although non-neutral evolution in the form of purifying selection (Nachman et al. 1996) and population growth (Rogers and Harpending 1992) has influenced patterns of human mtDNA variation, claims of adaptive evolution have not been substantiated (Kivisild et al. 2006; Sun et al. 2006). In any event, any inferences about population history based on mtDNA should, at some point, be verified with analyses of additional loci, as there is always the risk that any analysis based on a single locus is yielding incorrect insights into population history, because of selection or chance events.

NRY VARIATION

Representing the largest haploid part of our genome with about 23,000,000 bp escaping interchromosomal recombination, the human Y chromosome should provide a rich resource of male-specific markers for genealogical studies. However, for a long time it was considered to be short of useful DNA polymorphisms, and the first NRY marker, 12f2 or DYS11, only became available in 1985 (Casanova et al. 1985). After more than two decades of intensive search for useful markers, a battery of well-described NRY markers is now available. A total of 245 NRY single-nucleotide polymorphisms (Y-SNPs) were analyzed in a worldwide panel of DNA samples to establish a comprehensive phylogenetic tree as well as a commonly used nomenclature of Y-SNP-based haplogroups, an initiative taken by the Y Chromosome Consortium, YCC (2002), with more and more markers being discovered and added (Jobling and Tyler-Smith 2003). Most of these markers were discovered by applying denaturing high performance liquid chromatography (dHPLC) to short fragments of amplified Y-chromosome DNA from worldwide individuals in combination with a reference DNA, usually from Africa (Underhill et al. 2000). This approach introduces a strong ascertainment bias, due to the limited number and choice of individuals used in the marker screening process, which needs to be considered especially when Y-SNPs are used for diversity estimates in geographic regions that were underrepresented in the original screening set.

In addition to Y-SNPs, hundreds of NRY short tandem repeat polymorphisms (Y-STRs) or microsatellites have been discovered and are available to complement Y-SNPs in male genealogical studies. The first Y-STR, 27H39LR or DYS19, was found in 1992 (Roewer et al. 1992) using traditional molecular techniques, i.e., repetitive probe hybridization to Y-chromosomal clones. Several additional Y-STRs were identified and characterized (Kayser et al. 1997). However, many more Y-STRs were discovered more recently by taking advantage of the almost complete DNA sequence of the human Y chromosome, established as part of the International Human Genome Project. Using this resource, computer algorithms able to recognize simple repetitive sequence structures were employed, followed by marker verification through molecular techniques, and 166 useful and previously unknown Y-STRs were identified (Kayser et al. 2004).

One complication that arises with Y-STRs, in comparison to autosomal STRs, is that, due to the highly duplicated and palindromic nature of the Y chromosome (Skaletsky et al. 2003), some Y-STR loci are present in multiple copies. This means that two (or more) different male-specific products can be observed after PCR amplification with a single primer pair. In the usual case of duplicate loci, which are not infrequent (Kayser et al. 2004), these are assigned to two loci, e.g., DYS385a/b. However, sometimes a single product is observed at a duplicate locus, in which case the usual practice is to assume that the alleles at the two loci have identical lengths, although in principle it is possible that there has been a deletion of one locus. In assigning alleles to individual loci of a duplicate pair, it is customary to assume that the larger allele always belongs to one locus and the smaller allele to the other, even though this does not hold for all cases, as was shown when an assay to separate the duplicate DYS385a/b loci was developed (Kittler et al. 2003). Another complication that sometimes arises is that new duplications can occur, resulting in polymorphisms for the presence/absence of a duplicated locus. For example, a duplication of DYS19 is quite common in Mongols and groups of Mongolian origin, such as Kalmyks (Zerjal et al. 2003; Nasidze et al. 2005).

Conventionally, the term "haplogroup" is used to refer to the Y-SNP status of an individual, whereas "haplotype" refers to Y-STRs (de Knijff 2000). It should be emphasized that this terminology differs from that employed for autosomal variation, in which haplotypes can consist of any combination of STRs and/or SNPs. The reason for this distinction is that Y-SNPs represent unique events which, in the absence of recombination, mark a unique Y-chromosome lineage; recombination along the autosomes means that individual autosomal SNPs do not mark unique lineages. Although in principle the same should hold true for Y-STRs as for Y-SNPs, in practice the high rate of parallel and back mutations at Y-STR loci means that individuals with the same Y-STR haplotype may belong to different Y-SNP haplogroups. Normally, Y-SNPs are used to characterize genealogical events further back in time, i.e., many thousands of years, whereas Y-STRs are applied for more recent events in population or individual genetic history. This is because Y-STRs have a mutation rate about 100,000 times higher than that for Y-SNPs (Kayser et al. 2000b; Thomson et al. 2000). Hence, ancient events are usually masked by recurrent Y-STR mutations, making their identification more difficult when only Y-STRs are analyzed (Kayser et al. 2001). Therefore, NRY-based genealogical studies as performed today usually employ both types of Y-chromosomal polymorphisms to take full advantage of the potential information provided by the Y chromosome: Y-SNPs for the broad classification of individuals or populations based on their paternal lineages (i.e., NRY haplogroups) and Y-STRs to define such lineages in respect to place and time of origins. Such combination allows deep insights into male genealogies and can provide detailed information about (male) human population history (Zerjal et al. 1997, 2003; Kayser et al. 2000a; Seielstad et al. 2003). In some cases, haplogroups and haplotypes provide similar conclusions (Rosser et al. 2000; Semino et al. 2000; Roewer et al. 2005) and (rarely) almost identical information (Kayser et al. 2005).

Due to the complete linkage of NRY-DNA, natural selection acting on the human Y chromosome would clearly influence genetic variation all across the chromosome. This issue has been discussed elsewhere (Jobling and Tyler-Smith 2003), and it has been concluded that, based on existing knowledge, there is no evidence for differential selection acting on the human Y chromosome. However, certain cultural factors that affect reproductive success are known to influence Y chromosome diversity, one of which is patrilocal residence, as discussed below. Other cultural factors influencing Y chromosome diversity are polygyny (where one man sires children with more than one woman), which was/is part of many human societies before/outside Christianity, and warfare, involving preferential killing of men over women as was practiced in certain regions (e.g., New Guinea) to obtain access to women. Patrilocality, polygyny, and male-biased warfare all lead to reduced Y chromosome diversity and a biased Y chromosome distribution. Hence, they have been suggested to explain the observed reduced Y chromosome, but not mtDNA, diversity in human populations from West New Guinea that until recently existed, and partly still today continues to exist, under traditional conditions (Kayser et al. 2003). Extreme male-biased migrations, e.g., as a consequence of war-like invasions, can also be reflected in Y chromosome diversity. An example is the relatively high frequency (about 8%) and peculiar distribution of a closely related cluster of Y lineages, found all over Central Asia, reflecting the genetic footprints of Genghis Khan and his descendants (Zerjal et al. 2003).

HUMAN EVOLUTION

mtDNA variation in contemporary human populations was originally interpreted as supporting a recent African origin of human mtDNA (Cann et al. 1987; Vigilant et al. 1991), and further work has strengthened the support for this interpretation (for review, see Pakendorf and Stoneking 2005). This is evidenced by the following: (1) African populations have the most mtDNA variation (in general, the population harboring the greatest number of mutations is likely to be the ancestral population); (2) phylogenetic trees of mtDNA types invariably include two primary branches, one consisting of exclusively African mtDNA types, the other of mtDNA types from Africa and the rest of the world (a pattern that is most easily explained if the common ancestor were African); and (3) an estimated age of the human mtDNA ancestor of about 150,000 years ago, with further indications in the patterns of mtDNA variation of expansion out of Africa around 50,000 years ago (Ingman et al. 2000). These conclusions, based on contemporary populations, are further supported by mtDNA sequences retrieved from Neandertals (Krings et al. 1997) and early modern humans (Serre et al. 2004): Neandertal mtDNA sequences fall outside the range of modern human variation, and there is no indication of Neandertal mtDNA in either ancient or contemporary modern humans.

For many years, discussion about genetic evidence concerning modern human origins focused on the mtDNA data. It was encouraging when data on the NRY became available and indicated a recent African origin of human NRY variation (Underhill et al. 2000). However, the age of the NRY ancestor appears to be somewhat younger than the age of the mtDNA ancestor, between about 60,000 and 100,000 years (Macpherson et al. 2004). This apparent discrepancy is actually to be expected, given the overall general tendency for a lower proportion of human males than human females to contribute offspring to subsequent generations.

It should be kept in mind that a recent African origin of human mtDNA and NRY variation need not necessarily mean a recent African origin of our entire genome, and indeed, some analyses do indicate that some small portion of our genome may be the result of admixture with archaic humans (Wall and Hammer 2006). Sequencing of genomic DNA from Neandertals (and other archaic humans) should provide a definitive answer to this question (Green et al. 2006; Noonan et al. 2006).

COMPARISONS OF mtDNA AND NRY VARIATION IN HUMAN POPULATIONS

An early observation of comparative analyses of mtDNA and NRY variation across a variety of human populations was that, on average, genetic distances between populations tend to be larger for the NRY than for mtDNA (Seielstad et al. 1998). Genetic differentiation between populations reflects a balance between genetic drift (random changes in allele frequencies from generation to generation), causing populations to become dissimilar genetically, and migration or gene flow between populations, causing them to remain similar genetically. The larger genetic distances for the NRY therefore were interpreted as indicating a lower rate of male than female migration among human populations. The extent to which a lower rate of male than female migration holds on a global scale has recently been challenged (Wilder et al. 2004), but nonetheless, on a local scale, genetic distances between populations are invariably larger for the NRY than for mtDNA (Kayser et al. 2003; Nasidze et al. 2004; Pakendorf et al. 2006).

The idea that the rate of male migration may be lower than the rate of female migration may seem counterintuitive, because we are accustomed to thinking of migrations as primarily male-dominated invasions and conquests. However, the mtDNA and NRY data may instead reflect an effect of residence pattern (Seielstad et al. 1998): Most human populations are patrilocal, and upon marrying, the female moves to the residence of the male. Patrilocality would be predicted to lead to the observed larger genetic distances between populations for the NRY, since males would be moving between groups at a lower rate than females. A test of this hypothesis would be to examine patterns of mtDNA and NRY variation in matrilocal groups, in which the males move to the residence of their female partner. The prediction therefore would be that matrilocal groups should show the opposite pattern; namely, larger genetic distances for mtDNA than for the NRY, and indeed, a study of matrilocal and patrilocal groups among the hill tribes of Thailand (Oota et al. 2001) showed that, rather astonishingly, this prediction was fulfilled quite nicely (Fig. 33-1).

This example illustrates that comparative analyses of mtDNA and NRY variation are particularly informative when one is investigating social/cultural situations in which males and females behave differently. Another example concerns the former caste system in India, in which different social restrictions on the mate choices of males versus females led to predictions about how mtDNA versus NRY variation should be structured within and between castes. Again, these predictions are fulfilled exactly (Bamshad et al. 1998). The insights into the impact of these and other social/cultural phenomena on the genetic structure of human populations

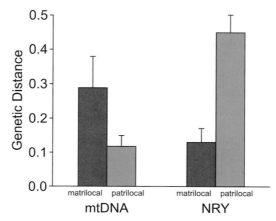

FIGURE 33-1. Genetic distances for matrilocal and patrilocal groups among the hill tribes of Thailand, showing that the usual pattern of greater genetic differentiation for the NRY than for mtDNA (seen here for the patrilocal groups) is reversed in the matrilocal groups. Data from Oota et al. (2001).

would not be possible without mtDNA and NRY analyses, and, for this reason, they are among the most significant contributions of these genealogical markers to human population genetics.

CASE STUDY: THE ORIGINS OF POLYNESIANS

In most instances, analyses of mtDNA and NRY variation in particular populations tend to agree, indicating that, in general, the maternal and paternal histories of human populations tend to be the same. However, in some cases they differ, and perhaps the most dramatic example concerns the origins of Polynesians.

Polynesians are the Pacific Islanders living on the many islands (Polynesia) in the vast triangle area between Tuvalu in the west, Hawaii in the north, Easter Island in the east, and New Zealand in the south. Due to the remoteness of these islands (thousands of kilometers distant from the continental mainland), the origin and migration history of Polynesians has fascinated scientists as well as lay people at least from the time of their "discovery" by the first European sailors in the 16th century. On the basis of archaeological data, it is known that Polynesia was first occupied by humans between 3000 and 800 years ago, earlier in Western than in Eastern Polynesia, and most recently in New Zealand (Kirch 2000). This is all surprisingly recent, considering that Australia and New Guinea were first occupied about 40,000–50,000 years ago (Groube et al. 1986; Roberts et al. 1990). There are two extreme hypotheses for the origin of Polynesians. The "Express train" model assumes an origin in East Asia about 6000 years ago and a fast migration eastward into the Pacific, without considerable pauses that could have given rise to admixture with, e.g., New Guinea aborigines (Diamond 1988). This model predicts that most (if not all) features of Polynesians should be traceable to Asians. In contrast, the "Entangled bank" model assumes a long period of interaction between New Guinea and surrounding regions, including Polynesia, from the time of the original settlement of New Guinea during the Pleistocene (Terrell 1988). Hence, this model predicts that most (if not all) features of Polynesians should be traceable to New Guineans or Melanesians (meant here as the geographic region including mainland New Guinea and surrounding islands). Other, more "intermediate" models such as the "Out-of-Taiwan" model (Bellwood 2004) or "Triple-I" model (for intrusion, innovation, and integration; Green 1991) do postulate an East Asian origin of Polynesian ancestors and allow for some (generally unspecified) amount of contact and admixture between Polynesian ancestors and Melanesians. According to this intermediate model, Polynesians should exhibit both Asian and Melanesian features.

Distinguishing among these various hypotheses concerning Polynesian origins thus requires assessing the relative genetic contributions of Asian and Melanesian groups to Polynesians. There are two features of mtDNA and NRY haplogroups that can be used to infer their geographic origin: (1) geographic distribution and (2) amount of associated mutational diversity. In general, one expects the origin of a haplogroup to have been in that group which exhibits the greatest amount of associated mutational diversity, since in the absence of recombination, mutational diversity is primarily a function of time. mtDNA and NRY haplogroups in Polynesians can be readily classified into two types: those that are also found widely throughout East and Southeast Asia, coastal (but not highland) New Guinea, and island Melanesia; and those that are found in island Melanesia, coastal and highland New Guinea, and eastern Indonesia, but not elsewhere in Southeast or East Asia (Kayser et al. 2006). The inference that mtDNA and NRY haplogroups of the first type are of Asian origin, whereas those of the second type are of Melanesian origin, is confirmed by the mutational diversity associated with these haplogroups. Asian groups have the highest mutational diversity associated with mtDNA and NRY haplogroups of the first type, whereas Melanesian groups have the highest mutational diversity associated with haplogroups of the second type (Kayser et al. 2006). We illustrate this by showing the frequency distributions across

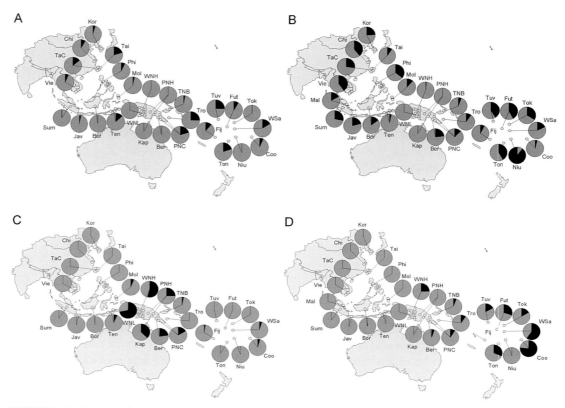

FIGURE 33-2. Examples of Asian and Melanesian haplogroups and their frequency distribution throughout Asia/Oceania: (*A*) Asian mtDNA haplogroup B4a, (*B*) Asian NRY haplogroup O-M122, (*C*) Melanesian mtDNA haplogroup Q1, and (*D*) Melanesian NRY haplogroup C-M208. The black portion of the pie charts represents the frequency of the respective haplogroup in the population sample analyzed. Population abbreviations are as follows: Cook (Coo), Niue (Niu), Tokelau (Tok), Western Samoa (WSa), Tonga (Ton), East Futuna (Fut), Tuvalu (Tuv), Fiji (Fij), New Britain (TNB), Trobriand PNG (Tro), Bereina PNG (Ber), PNG Coast (PNC), PNG Highlands (PNH), Kapuna PNG (Kap), WNG Highland (WNH), WNG Lowland (WNL), Moluccas (Mol), Nusa Tengarras (Ten), Philippines (Phi), Southern Borneo (Bor), Java (Jav), Sumatra (Sum), Malaysia (Mal), Korea (Kor), Taiwan Aborigines (Tai), Taiwan Chinese (TaC), Vietnam (Vie). Data from Kayser et al. (2006).

Asia/Oceania (Fig. 33-2) and the regional diversity differences (Fig. 33-3) of the Asian haplogroups B4a (mtDNA) and O-M122 (NRY), as well as the Melanesian haplogroups Q1 (mtDNA) and C-M208 (NRY).

Remarkably, the estimated Asian contribution for mtDNA haplogroups is much higher than the estimated Asian contribution for NRY haplogroups, and vice versa for the Melanesian contribution to Polynesia. By combining the evidence from five mtDNA and seven NRY haplogroups of Asian origin, as well as four mtDNA and seven NRY haplogroups of Melanesian origin, we can trace 94% of Polynesian mtDNAs but only 28% of Polynesian Y chromosomes to Asia, whereas 66% of Polynesian Y chromosome but only 6% of Polynesian mtDNAs can be traced to Melanesia (Kayser et al. 2006). One explanation for these genetic results is a sex-mediated genetic admixture bias between Asian migrants arriving in New Guinea and New Guinea aborigines with more Asian women (and less men) and more Melanesian men (and less women). Such a scenario would be supported by the assumed matrilineal descent and matrilocal residence of the ancient Polynesian society (Hage 1998).

The question as to whether Asian and Melanesian haplogroups arrived in Polynesia as a result of one or several waves of migration could in principle be addressed by dating the arrival of mtDNA and NRY haplogroups in Polynesia from the associated mtDNA HV1 sequence and

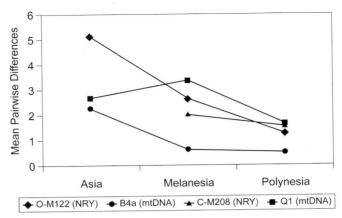

FIGURE 33-3. Regional comparisons of associated Y-STR diversity of NRY haplogroups C-M208 and O-M122, and associated HV1 sequence haplotype diversity for mtDNA haplogroups B4a and Q1. The diversity measure is the mean number of pairwise differences. Data from Kayser et al. (2006).

Y-STR diversity. Resulting dates for different haplogroups observed in Polynesia overlap between those of Asian and those of Melanesian origin, due to the large confidence intervals that are inevitable with the dating methods available today (Kayser et al. 2006). Taking into account this caveat, no distinction between the arrival dates of Asian and Melanesian haplogroups in Polynesia can be made based on the current data.

mtDNA and NRY markers allow two ways of analyzing data: population-based and lineage-based. In population-based analyses, the population is the unit of analysis, and the interest is in the relationships of populations to one another. When population-based analyses are performed, such as multidimensional scaling of Fst (fixation index resulting from comparing subpopulations to the total population) values established from NRY haplogroup frequencies or mtDNA sequences, Polynesian populations cluster together and are somewhat separate from Melanesian and East as well as Southeast Asian populations for both NRY and mtDNA markers (Fig. 33-4). Population clustering according to geographic regions is also observed for East/Southeast Asian groups and Melanesian groups with Island Melanesians/coastal New Guineans separated from New Guinea Highland groups/West New Guineans. Fijians are the closest of the Polynesian cluster to Melanesians. In the NRY plot, East Indonesian populations (Moluccas and Tenggaras islands) cluster together with island Melanesians and coastal New Guineans due to the high frequency of Melanesian NRY haplogroups in Eastern Indonesia, which is higher than Melanesian mtDNA haplogroups in those samples. There is a better fit between the genetic distances and the multidimensional scaling (MDS) plot based on mtDNA than that based on the NRY, as indicated by the stress values (mtDNA: 0.051 vs. NRY: 0.157). Stress values provide a measure of the goodness-of-fit of the plot to the observed distances between groups, and are calculated by the program used to generate the MDS plots.

In lineage-based analyses, the individual lineage is the unit of analysis, and the interest is in how different lineages are related to one another. Often, the time of origin of a lineage is estimated from the amount of sequence variation (for mtDNA lineages) or Y-STR haplotype variation (for Y-SNP haplogroups). However, although lineage-based analyses can be very informative, it must be kept in mind that additional inferences must be made in order to gain insights into population history. For example, the age of a lineage is not the same as the age of a population. Moreover, it is seldom the case that a single lineage will denote a migration, because when people migrate, it is expected that many lineages will migrate with them and hence should show the signature of the migration.

Lineage-based analyses for the Polynesian data, such as network analysis performed separately for each haplogroup using associated Y-STR haplotype diversity for NRY haplogroups and HV1 sequence diversity for mtDNA haplogroups, revealed a similar picture for the major Polynesian

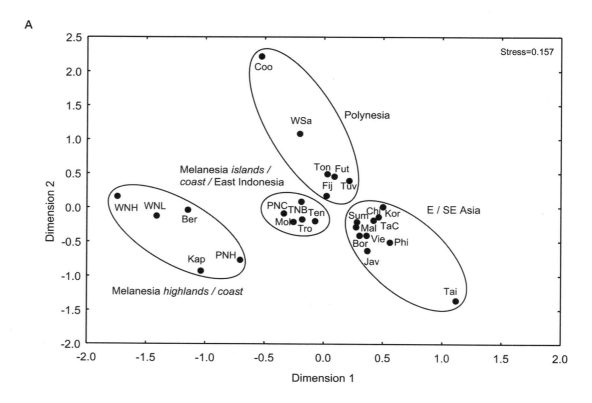

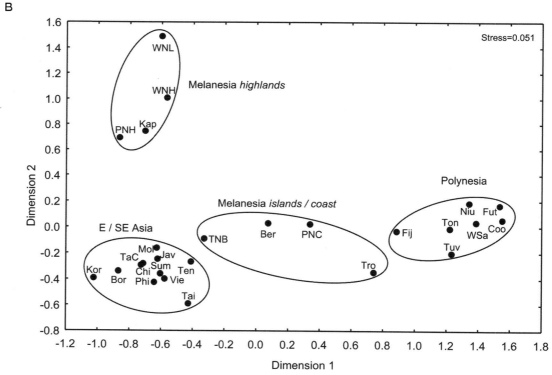

FIGURE 33-4. Two-dimensional plots from multidimensional scaling analyses using Fst values from pair-wise population comparisons based on (A) NRY haplogroups and (B) mtDNA HV1 sequence data. Population abbreviations are as in Fig. 33-2 (some populations with small sample sizes are omitted). The geographic regions of the population samples are highlighted. Data from Kayser et al. (2006).

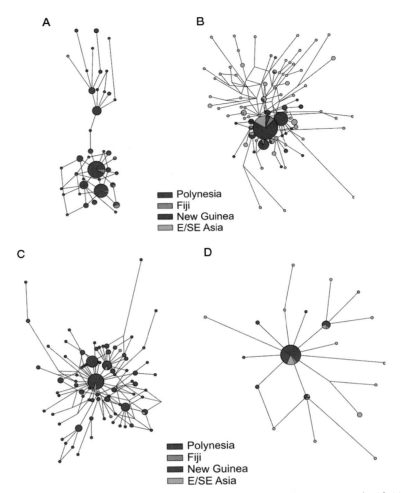

FIGURE 33-5. Median-Joining networks of haplotypes based on 7 Y-STR loci, associated with NRY haplogroups (*A*) C-M208 and (*B*) O-M122, and HV1 sequences associated with mtDNA haplogroups (*C*) B4a and (*D*) Q1. Circles denote haplotypes, with the area of the circle proportional to the number of individuals carrying the particular haplotype. Lines denote mutation steps. The geographic regions of the individual samples are indicated. Data from Kayser et al. (2006).

haplogroups (Fig. 33-5) as that from the population-based analyses. However, the distinction between geographic regions is not as evident as in the population-based analyses (with the exception of C-M208, which shows an almost complete separation of Polynesian and Melanesian haplotypes). Moreover, the networks illustrate an important point; namely, that frequency information alone can give misleading indications about the origin of a haplogroup. This can be seen in the network for mtDNA haplogroup B4a (Fig. 33-5C): Although B4a is more frequent in New Guinea than in Asia, which might then suggest an origin for B4a in New Guinea, the diversity in HV1 sequences associated with this haplogroup is highest in Asia (Fig. 33-3), indicating that this haplogroup arose in Asia and rose in frequency in New Guinea via local expansion. In addition, all NRY and mtDNA haplogroups that are frequent in Polynesia (C-M208, O-M122, B4a, as well as PM and K-M9—the latter two not shown here) reveal haplotype networks indicating a bottleneck event in the genetic history of Polynesians, with one or two central haplotypes that are shared between many Polynesians from different islands (and between Polynesia and Asia/Melanesia). Genetic evidence for a bottleneck in the history of Polynesians is quite strong.

The most extreme models of Polynesian origins (i.e., the "Express train" and the "Entangled bank") are thus not supported by the combined NRY and mtDNA data, which show Asian as well as Melanesian contributions to Polynesians. However, the data are in agreement with the "Slow boat" model (as well as other intermediate models, such as the "Out-of-Taiwan" model, which do not specify the type or amount of admixture between Polynesian ancestors and Melanesians) we already proposed some years ago based on a more limited Polynesian data set (Kayser et al. 2000a). According to this model, Polynesian ancestors did originate from Eastern Asia but did not move rapidly through Melanesia; rather, they interacted with and mixed extensively with Melanesians, leaving behind their genes and incorporating many Melanesian genes before colonizing the Pacific. Moreover, this admixture was highly asymmetric in that the Melanesian admixture into Polynesian ancestors was primarily by males, as reflected in the higher contribution of Melanesian Y chromosomes than mtDNA in Polynesia, whereas the Polynesian ancestor admixture into Melanesians was primarily female, as reflected by the higher contribution of Asian mtDNAs than Y chromosomes in New Guinea (Kayser et al. 2006). Intriguingly, the languages in Polynesia track with the maternal history of Polynesians: Polynesians speak Austronesian languages, which most likely originated in Taiwan (Blust 1999). Another recent study, involving a discrepancy between the mtDNA and NRY relationships of groups in the South Caspian (Nasidze et al. 2006), also finds that the language of the groups tracks with the mtDNA, rather than the NRY, relationships (although in this case language and mtDNA came from the local and not the migrating population, whereas in the case of Polynesia, mtDNA and language came from the migrating and not the local population). It remains to be seen how widespread this phenomenon is, but in any event, these cases illustrate the unique ability of comparative mtDNA and NRY studies to provide insights into the paternal versus maternal history of various cultural aspects of human populations.

mtDNA AND NRY VARIATION IN FORENSICS AND GENEALOGY STUDIES

Due to their unique properties, as discussed above, especially their uniparental inheritance pattern, mtDNA and NRY markers are also used in forensics and genealogical studies for female and male lineage identification, respectively. In genealogical studies, they are applied to trace back the male or female ancestors of an individual of interest. Often, because surnames are inherited through the paternal line and the majority of human societies are patrilineal in descent, NRY genetic evidence is combined with surname information to trace male individual ancestry in genealogical studies. Y-STRs in particular are extremely popular in genealogical studies, and a number of companies offer Y-STR typing as a commercial service for genealogical purposes. However, mutational events, which occur at rates as high as on average 2–3 mutations in every thousand generations per single marker for the most widely used Y-STRs (Kayser et al. 2000b; Dupuy et al. 2004; Gusmao et al. 2005), as well as hidden nonbiological paternity (and, more rarely, maternity), can cause conflicts between genealogical information from family records and genetic data, especially for deep genealogies going back many generations (Kayser et al. 2007). A number of prominent cases of human identification through DNA-based family studies, using contemporary descendants or in combination with ancient DNA analyses, have demonstrated the power of mtDNA and NRY markers in genealogical studies; examples include the identification of the skeletons of Czar Nicholas II of Russia and his family (Gill et al. 1994; Ivanov et al. 1996), and the likely paternity of Thomas Jefferson of Eston Hemmings Jefferson, son of Jefferson's slave Sally Hemmings (Foster et al. 1998).

Autosomal STRs are normally preferred in forensics for human identification purposes, since they provide essentially individual identification if analyzed in a reasonable number (usually 10–15 STRs), and there are large, established STR-based criminal databases for providing suspect profiles. However, there are special cases where mtDNA and NRY markers can reveal useful information while autosomal STRs are not informative. NRY markers (i.e., Y-STRs) are used for male lineage identification in rape cases where the mixture of female victim DNA and male perpetrator

DNA in a case sample can cause difficulties in revealing the autosomal STR profile of the male perpetrator. Y-STRs can be detected successfully even when sperm are absent (i.e., in cases where oligospermic or azoospermic males are involved) by the presence of male epithelial cells (Betz et al. 2001). Autosomal STR analysis can be especially problematic in sexual assault cases with multiple male perpetrators, but those cases can be solved using Y-STRs which can differentiate between male lineages. NRY markers are also preferred over autosomal markers in DNA-based dragnets to trace unknown suspects due to the paternal relationships (and hence, identical or nearly identical Y-STR profiles) of men often seen in restricted geographic regions (Dettlaff-Kakol and Pawlowski 2002). NRY markers are also useful for paternity testing of male offspring in so-called deficiency cases, where the putative father is not available for DNA testing. If neither parent of the deceased putative father is available for DNA testing, the paternity cannot usually be established with high certainty using autosomal STRs, but it can be established by use of Y-STRs from any of the paternal male biological relatives of the deceased alleged father.

It has been advocated that multicopy Y-STRs (such as DYS385a/b) should be particularly useful in forensic casework, since they often show high diversity values because of combining multiple polymorphic Y-STR loci (Redd et al. 2002). However, they should be avoided in cases where the number of perpetrators involved is under question, such as in multiple rape cases (Butler et al. 2005). An additional concern is that some of the multicopy Y-STRs are located in duplicated Y chromosomal regions involved in fertility problems (Bosch and Jobling 2003), and their profiles can thus provide information on the fertility status of a man, which is unwanted (and in some countries illegal) information.

Due to its multicopy property, mtDNA often is the only source of DNA for human identification from remains such as bones or burned corpses, which is especially relevant in cases of mass disasters (Budowle et al. 2005). mtDNA is also useful for establishing the maternal relationships within a pedigree, as mentioned above. Because many mtDNA and NRY haplogroups show restricted geographic distributions, another recent use of mtDNA and NRY markers is to trace the geographic origin or genetic ancestry of the donor of a case sample. In principle, such information could assist law enforcement agencies in tracing suspects in cases where no suspects are known (Jobling and Gill 2004). However, for the latter purpose, mtDNA and NRY markers should be combined with ancestry-informative autosomal markers (Lao et al. 2006), to avoid misinterpretation of the uniparental genetic information in cases where the individual has a sex-biased genetic admixture history.

Finally, it should be emphasized that the use of NRY and mtDNA markers in forensic cases of nonexclusions represents the identification of the paternal and maternal lineage that the perpetrator belongs to, and not an actual individual identification (i.e., as can be obtained in principle using autosomal STRs). Due to the nonrecombining and thus uniparental inheritance of mtDNA and NRY-DNA, all maternal relatives of a given individual carry the same mtDNA genome, whereas all paternal male relatives of a given man carry the same Y chromosome (NRY). Consequently, if NRY and mtDNA analyses are used in a forensic case, e.g., because of limited success with autosomal STR markers, further investigation is necessary to determine whether the suspect committed the crime, as opposed to any of the maternal (mtDNA evidence) or paternal male (NRY evidence) relatives.

ACKNOWLEDGMENTS

We thank Richard Cordaux for assistance with the networks, Knut Finstermeir for assistance with figures, and all of our students and colleagues who have contributed to the studies described in this chapter.

REFERENCES

Anderson S., Bankier A.T., Barrell B.G., de Bruijn M.H.L., Coulson, A.R., Drouin J., Eperon I.C., Nierlich D.P., Roe B.A., Sanger F., et al. 1981. Sequence and organization of the human mitochondrial genome. *Nature* **290:** 457–465.

Awadalla P., Eyre-Walker A., and Smith J.M. 1999. Linkage disequilibrium and recombination in hominid mitochondrial DNA. *Science* **286:** 2524–2525.

Bamshad M.J., Watkins W.S., Dixon M.E., Jorde L.B., Rao B.B., Naidu J.M., Prasad B.V.R., Rasanayagam A., and Hammer M.F. 1998. Female gene flow stratifies Hindu castes. *Nature* **395:** 651–652.

Bandelt H.J., Quintana-Murci L., Salas A., and Macaulay V. 2002. The fingerprint of phantom mutations in mitochondrial DNA data. *Am. J. Hum. Genet.* **71:** 1150–1160.

Barbujani G., Vernesi C., Caramelli D., Castri L., Lalueza-Fox C., and Bertorelle G. 2004. Etruscan artifacts: Much ado about nothing. *Am. J. Hum. Genet.* **75:** 923–927.

Bellwood P. 2004. Colin Renfrew's emerging synthesis: Farming, languages and genes as viewed from the Antipodes. In *Traces of ancestry: Studies in honour of Colin Renfrew* (ed. M. Jones), pp. 31–39. McDonald Institute, Cambridge, United Kingdom.

Bendall K.E. and Sykes B.C. 1995. Length heteroplasmy in the first hypervariable segment of the human mtDNA control region. *Am. J. Hum. Genet.* **57:** 248–256.

Betz A., Bassler G., Dietl G., Steil X., Weyermann G., and Pflug W. 2001. DYS STR analysis with epithelial cells in a rape case. *Forensic Sci. Int.* **118:** 126–130.

Blust R. 1999. Subgrouping, circularity and extinction: Some issues in Austronesian comparative linguistics. *Symp. Ser. Inst. Linguistics Acad. Sinica* **1:** 31–94.

Bosch E. and Jobling M.A. 2003. Duplications of the *AZFa* region of the human Y chromosome are mediated by homologous recombination between HERVs and are compatible with male fertility. *Hum. Mol. Genet.* **12:** 341–347.

Budowle B., Bieber F.R., and Eisenberg A.J. 2005. Forensic aspects of mass disasters: Strategic considerations for DNA-based human identification. *Leg. Med.* **7:** 230–243.

Butler J.M., Decker, A.E., Kline, M.C., and Vallone, P.M. 2005. Chromosomal duplications along the Y-chromosome and their potential impact on Y-STR interpretation. *J. Forensic Sci.* **50:** 853–859.

Cann R.L., Stoneking M., and Wilson A.C. 1987. Mitochondrial DNA and human evolution. *Nature* **325:** 31–36.

Casanova M., Leroy P., Boucekkine C., Weissenbach J., Bishop C., Fellous M., Purrello M., Fiori G., and Siniscalco M. 1985. A human Y-linked DNA polymorphism and its potential for estimating genetic and evolutionary distance. *Science* **230:** 1403–1406.

de Knijff P. 2000. Messages through bottlenecks: On the combined use of slow and fast evolving polymorphic markers on the human Y chromosome. *Am. J. Hum. Genet.* **67:** 1055–1061.

Dettlaff-Kakol A. and Pawlowski R. 2002. First Polish DNA "manhunt"—An application of Y-chromosome STRs. *Int. J. Leg. Med.* **116:** 289–291.

Diamond J. 1988. Express train to Polynesia. *Nature* **336:** 307–308.

Dupuy B.M., Stenersen M., Egeland T., and Olaisen B. 2004. Y-chromosomal microsatellite mutation rates: Differences in mutation rate between and within loci. *Hum. Mutat.* **23:** 117–124.

Excoffier L. 1990. Evolution of human mitochondrial DNA: Evidence for departure from a pure neutral model of populations at equilibrium. *J. Mol. Evol.* **30:** 125–139.

Forster P. 2003. To err is human. *Ann. Hum. Genet.* **67:** 2–4.

Foster E.A., Jobling M.A., Taylor P.G., Donnelly P., de Knijff P., Mieremet R., Zerjal T., and Tyler-Smith C. 1998. Jefferson fathered slave's last child. *Nature* **396:** 27–28.

Giles R.E., Blanc H., Cann H.M., and Wallace D.C. 1980. Maternal inheritance of human mitochondrial DNA. *Proc. Natl. Acad. Sci.* **77:** 6715–6719.

Gill P., Ivanov P.L., Kimpton C., Piercy R., Benson N., Tully G., Evett I., Hagelberg E., and Sullivan K. 1994. Identification of the remains of the Romanov family by DNA analysis. *Nat. Genet.* **6:** 130–135.

Green R.C. 1991. The Lapita cultural complex: Current evidence and proposed models. *Bull. Indo-Pacific Prehist. Assoc.* **11:** 295–305.

Green R.E., Krause J., Ptak S.E., Briggs A.W., Ronan M.T., Simons J.F., Du L., Egholm M., Rothberg J.M., Paunovic M., and Paabo S. 2006. Analysis of one million base pairs of Neanderthal DNA. *Nature* **444:** 330–336.

Groube L.M., Chappell J., Muke J., and Price D. 1986. A 40,000 year-old human occupation site at Huon Peninsula, Papua New Guinea. *Nature* **324:** 453–455.

Gusmao L., Sanchez-Diz P., Calafell F., Martin P., Alonso C.A., Alvarez-Fernandez F., Alves C., Borjas-Fajardo L., Bozzo W.R., Bravo M.L., et al. 2005. Mutation rates at Y chromosome specific microsatellites. *Hum. Mutat.* **26:** 520–528.

Hage P. 1998. Was Proto-Oceanic society matrilineal? *J. Polynesian Soc.* **107:** 365–379.

Hagelberg E., Goldman N., Liò P., Whelan S., Schiefenhövel W., Clegg J.B., and Bowden D.K. 1999. Evidence for mitochondrial DNA recombination in a human population of island Melanesia. *Proc. R. Soc. Lond. B Biol. Sci.* **266:** 485–492.

———. 2000. Evidence for mitochondrial DNA recombination in a human population of island Melanesia: Correction. *Proc. R. Soc. Lond. B Biol. Sci.* **267:** 1595–1596.

Ingman M., Kaessmann H., Paabo S., and Gyllensten U. 2000. Mitochondrial genome variation and the origin of modern humans. *Nature* **408:** 708–713.

Ivanov P.L., Wadhams M.J., Roby R.K., Holland M.M., Weedn V.W., and Parsons T.J. 1996. Mitochondrial DNA sequence heteroplasmy in the Grand Duke of Russia Georgij Romanov establishes the authenticity of the remains of Tsar Nicholas II. *Nat. Genet.* **12:** 417–420.

Jobling M.A. and Gill P. 2004. Encoded evidence: DNA in forensic analysis. *Nat. Rev. Genet.* **5:** 739–751.

Jobling M.A. and Tyler-Smith C. 2003. The human Y chromosome: An evolutionary marker comes of age. *Nat. Rev. Genet.* **4:** 598–612.

Johnson M.J., Wallace D.C., Ferris S.D., Rattazzi M.C., and Cavalli-Sforza L.L. 1983. Radiation of human mitochondria DNA types analyzed by restriction endonuclease cleavage patterns. *J. Mol. Evol.* **19:** 255–271.

Kayser M., Vermeulen M., Knoblauch H., Schuster H., Krawczak M., and Roewer L. 2007. Relating two deep-rooted pedigrees from Central Germany by high-resolution Y-STR haplotyping. *Forensic Sci. Int. Genet.* **1:** 125–128.

Kayser M., Brauer S., Weiss G., Underhill P.A., Roewer L., Schiefenhovel W., and Stoneking M. 2000a. Melanesian origin of Polynesian Y chromosomes. *Curr. Biol.* **10:** 1237–1246.

Kayser M., Brauer S., Weiss G., Schiefenhovel W., Underhill P., Shen P., Oefner P., Tommaseo-Ponzetta M., and Stoneking M. 2003. Reduced Y-chromosome, but not mitochondrial DNA, diversity in human populations from West New Guinea. *Am. J. Hum. Genet.* **72:** 281–302.

Kayser M., Brauer S., Cordaux R., Casto A., Lao O., Zhivotovsky L.A., Moyse-Faurie C., Rutledge R.B., Schiefenhoevel W., Gil D., et al. 2006. Melanesian and Asian origins of Polynesians: mtDNA and Y chromosome gradients across the Pacific. *Mol. Biol. Evol.* **23**: 2234–2244.

Kayser M., Caglia A., Corach D., Fretwell N., Gehrig C., Graziosi G., Heidorn F., Herrmann S., Herzog B., Hidding M., et al. 1997. Evaluation of Y-chromosomal STRs: A multicenter study. *Int. J. Leg. Med.* **110**: 125–133, 141–149.

Kayser M., Kittler R., Erler A., Hedman M., Lee A.C., Mohyuddin A., Mehdi S.Q., Rosser Z., Stoneking M., Jobling M.A., et al. 2004. A comprehensive survey of human Y-chromosomal microsatellites. *Am. J. Hum. Genet.* **74**: 1183–1197.

Kayser M., Krawczak M., Excoffier L., Dieltjes P., Corach D., Pascali V., Gehrig C., Bernini L.F., Jespersen J., Bakker E., et al. 2001. An extensive analysis of Y-chromosomal microsatellite haplotypes in globally dispersed human populations. *Am. J. Hum. Genet.* **68**: 990–1018.

Kayser M., Lao O., Anslinger K., Augustin C., Bargel G., Edelmann J., Elias S., Heinrich M., Henke J., Henke L., et al. 2005. Significant genetic differentiation between Poland and Germany follows present-day political borders, as revealed by Y-chromosome analysis. *Hum. Genet.* **117**: 428–443.

Kayser M., Roewer L., Hedman M., Henke L., Henke J., Brauer S., Kruger C., Krawczak M., Nagy M., Dobosz T., et al. 2000b. Characteristics and frequency of germline mutations at microsatellite loci from the human Y chromosome, as revealed by direct observation in father/son pairs. *Am. J. Hum. Genet.* **66**: 1580–1588.

Kirch P.V. 2000. *On the road of the winds: An archaeological history of the Pacific Islands before European contact.* University of California Press, Berkeley.

Kittler R., Erler A., Brauer S., Stoneking M., and Kayser M. 2003. Apparent intrachromosomal exchange on the human Y chromosome explained by population history. *Eur. J. Hum. Genet.* **11**: 304–314.

Kivisild T., Shen P., Wall D.P., Do B., Sung R., Davis K., Passarino G., Underhill P.A., Scharfe C., Torroni A., et al. 2006. The role of selection in the evolution of human mitochondrial genomes. *Genetics* **172**: 373–387.

Kraytsberg Y., Schwartz M., Brown T.A., Ebralidse K., Kunz W.S., Clayton D.A., Vissing J., and Khrapko K. 2004. Recombination of human mitochondrial DNA. *Science* **304**: 981.

Krings M., Stone A., Schmitz R.W., Krainitzki H., Stoneking M., and Paabo S. 1997. Neandertal DNA sequences and the origin of modern humans. *Cell* **90**: 19–30.

Kumar S., Hedrick P., Dowling T., and Stoneking M. 2000. Questioning evidence for recombination in human mitochondrial DNA. *Science* **288**: 1931.

Lao O., van Duijn K., Kersbergen P., de Knijff P., and Kayser M. 2006. Proportioning whole-genome single-nucleotide-polymorphism diversity for the identification of geographic population structure and genetic ancestry. *Am. J. Hum. Genet.* **78**: 680–690.

Macpherson J.M., Ramachandran S., Diamond L., and Feldman M.W. 2004. Demographic estimates from Y chromosome microsatellite polymorphisms: Analysis of a worldwide sample. *Hum. Genomics* **1**: 345–354.

Mishmar D., Ruiz-Pesini E., Golik P., Macaulay V., Clark A.G., Hosseini S., Brandon M., Easley K., Chen E., Brown M.D., et al. 2003. Natural selection shaped regional mtDNA variation in humans. *Proc. Natl. Acad. Sci.* **100**: 171–176.

Nachman M., Brown W., Stoneking M., and Aquadro C. 1996. Non-neutral mitochondrial DNA variation in humans and chimpanzees. *Genetics* **142**: 953–963.

Nasidze I., Quinque D., Rahmani M., Alemohamad S.A., and Stoneking M. 2006. Concomitant replacement of language and mtDNA in South Caspian populations of Iran. *Curr. Biol.* **16**: 668–673.

Nasidze I., Quinque D., Dupanloup I., Cordaux R., Kokshunova L., and Stoneking M. 2005. Genetic evidence for the Mongolian ancestry of Kalmyks. *Am. J. Phys. Anthropol.* **128**: 846–854.

Nasidze I., Ling E.Y., Quinque D., Dupanloup I., Cordaux R., Rychkov S., Naumova O., Zhukova O., Sarraf-Zadegan N., Naderi G.A., et al. 2004. Mitochondrial DNA and Y-chromosome variation in the Caucasus. *Ann. Hum. Genet.* **68**: 205–221.

Noonan J.P., Coop G., Kudaravalli S., Smith D., Krause J., Alessi J., Chen F., Platt D., Paabo S., Pritchard J.K., and Rubin E.M. 2006. Sequencing and analysis of Neanderthal genomic DNA. *Science* **314**: 1113–1118.

Oota H., Settheetham-Ishida W., Tiwawech D., Ishida T., and Stoneking M. 2001. Human mtDNA and Y-chromosome variation is correlated with matrilocal versus patrilocal residence. *Nat. Genet.* **29**: 20–21.

Pakendorf B., Novgorodov I.N., Osakovskij V.L., Danilova A.P., Protod'jakonov A.P., and Stoneking M. 2006. Investigating the effects of prehistoric migrations in Siberia: Genetic variation and the origins of Yakuts. *Hum. Genet.* **120**: 334–353.

Pakendorf B. and Stoneking M. 2005. Mitochondrial DNA and human evolution. *Annu. Rev. Genomics Hum. Genet.* **6**: 165–183.

Piganeau G. and Eyre-Walker A. 2004. A reanalysis of the indirect evidence for recombination in human mitochondrial DNA. *Heredity* **92**: 282–288.

Redd A.J., Agellon A.B., Kearney V.A., Contreras V.A., Karafet T., Park H., de Knijff P., Butler J.M., and Hammer M.F. 2002. Forensic value of 14 novel STRs on the human Y chromosome. *Forensic Sci. Int.* **130**: 97–111.

Roberts R.G., Jones R., and Smith M.A. 1990. Thermoluminescence dating of a 50,000-year-old human occupation site in northern Australia. *Nature* **345**: 153–156.

Roewer L., Arnemann J., Spurr N.K., Grzeschik K.-H., and Epplen J.T. 1992. Simple repeat sequences on the human Y chromosome are equally polymorphic as their autosomal counterparts. *Hum. Genet.* **89**: 389–394.

Roewer L., Croucher P.J., Willuweit S., Lu T.T., Kayser M., Lessig R., de Knijff P., Jobling M.A., Tyler-Smith C., and Krawczak M. 2005. Signature of recent historical events in the European Y-chromosomal STR haplotype distribution. *Hum. Genet.* **116**: 279–291.

Rogers A.R. and Harpending H. 1992. Population growth makes waves in the distribution of pairwise genetic differences. *Mol. Biol. Evol.* **9**: 552–569.

Rosser Z.H., Zerjal T., Hurles M.E., Adojaan M., Alavantic D., Amorim A., Amos W., Armenteros M., Arroyo E., Barbujani G., et al. 2000. Y-chromosomal diversity in Europe is clinal and influenced primarily by geography, rather than by language. *Am. J. Hum. Genet.* **67**: 1526–1543.

Ruiz-Pesini E., Mishmar D., Brandon M., Procaccio V., and Wallace D.C. 2004. Effects of purifying and adaptive selection on regional variation in human mtDNA. *Science* **303**: 223–226.

Salas A., Carracedo A., Macaulay V., Richards M., and Bandelt H.J. 2005. A practical guide to mitochondrial DNA error prevention in clinical, forensic, and population genetics. *Biochem. Biophys. Res. Commun.* **335**: 891–899.

Schwartz M. and Vissing J. 2002. Paternal inheritance of mitochondrial DNA. *N. Engl. J. Med.* **347**: 576–580.

Seielstad M., Minch E., and Cavalli-Sforza L. 1998. Genetic evidence for a higher female migration rate in humans. *Nat. Genet.* **20**: 278–280.

Seielstad M., Yuldasheva N., Singh N., Underhill P., Oefner P., Shen P., and Wells R.S. 2003. A novel Y-chromosome variant puts an upper limit on the timing of first entry into the Americas. *Am. J. Hum. Genet.* **73:** 700–705.

Semino O., Passarino G., Oefner P.J., Lin A.A., Arbuzova S., Beckman L.E., De Benedictis G., Francalacci P., Kouvatsi A., Limborska S., et al. 2000. The genetic legacy of Paleolithic *Homo sapiens sapiens* in extant Europeans: A Y chromosome perspective. *Science* **290:** 1155–1159.

Serre D., Langaney A., Chech M., Teschler-Nicola M., Paunovic M., Mennecier P., Hofreiter M., Possnert G., and Paabo S. 2004. No evidence of Neandertal mtDNA contribution to early modern humans. *PLoS Biol.* **2:** E57.

Skaletsky H., Kuroda-Kawaguchi T., Minx P.J., Cordum H.S., Hillier L., Brown L.G., Repping S., Pyntikova T., Ali J., Bieri T., et al. 2003. The male-specific region of the human Y chromosome is a mosaic of discrete sequence classes. *Nature* **423:** 825–837.

Sun C., Kong Q.P., and Zhang Y.P. 2006. The role of climate in human mitochondrial DNA evolution: A reappraisal. *Genomics* **89:** 338–342.

Terrell J.E. 1988. History as a family tree, history as an entangled bank: Constructing images and interpretations of prehistory in the South Pacific. *Antiquity* **62:** 642–657.

Thangaraj K., Chaubey G., Kivisild T., Reddy A.G., Singh V.K., Rasalkar A.A., and Singh L. 2005. Reconstructing the origin of Andaman Islanders. *Science* **308:** 996.

Thomson R., Pritchard J.K., Shen P., Oefner P.J., and Feldman M.W. 2000. Recent common ancestry of human Y chromosomes: Evidence from DNA sequence data. *Proc. Natl. Acad. Sci.* **97:** 7360–7365.

Torroni A., Schurr T.G., Yang C.-C., Szathmary E.J.E., Williams R.C., Schanfield M.S., Troup G.A., Knowler W.C., Lawrence D.N., Weiss K.M., and Wallace D.C. 1992. Native American mitochondrial DNA analysis indicates that the Amerind and Nadene populations were founded by two independent migrations. *Genetics* **130:** 153–162.

Trejaut J.A., Kivisild T., Loo J.H., Lee C.L., He C.L., Hsu C.J., Lee Z.Y., and Lin M. 2005. Traces of archaic mitochondrial lineages persist in Austronesian-speaking Formosan populations. *PLoS Biol.* **3:** e247.

Underhill P.A., Shen P., Lin A.A., Jin L., Passarino G., Yang W.H., Kauffman E., Bonne-Tamir B., Bertranpetit J., Francalacci P., et al. 2000. Y chromosome sequence variation and the history of human populations. *Nat. Genet.* **26:** 358–361.

Vigilant L., Stoneking M., Harpending H., Hawkes K., and Wilson A.C. 1991. African populations and the evolution of human mitochondrial DNA. *Science* **253:** 1503–1507.

Wall J.D. and Hammer M.F. 2006. Archaic admixture in the human genome. *Curr. Opin. Genet. Dev.* **16:** 606–610.

Wilder J.A., Kingan S.B., Mobasher Z., Pilkington M.M., and Hammer M.F. 2004. Global patterns of human mitochondrial DNA and Y-chromosome structure are not influenced by higher migration rates of females versus males. *Nat. Genet.* **36:** 1122–1125.

Y Chromosome Consortium. 2002. A nomenclature system for the tree of human Y-chromosomal binary haplogroups. *Genome Res.* **12:** 339–348.

Zerjal T., Dashnyam B., Pandya A., Kayser M., Roewer L., Santos F.R., Schiefenhovel W., Fretwell N., Jobling M.A., Harihara S., et al. 1997. Genetic relationships of Asians and Northern Europeans, revealed by Y-chromosomal DNA analysis. *Am. J. Hum. Genet.* **60:** 1174–1183.

Zerjal T., Xue Y., Bertorelle G., Wells R.S., Bao W., Zhu S., Qamar R., Ayub Q., Mohyuddin A., Fu S., et al. 2003. The genetic legacy of the Mongols. *Am. J. Hum. Genet.* **72:** 717–721.

34 | Forensic DNA Testing

John M. Butler

Biochemical Science Division, National Institute of Standards and Technology, Gaithersburg, Maryland 20899-8311

INTRODUCTION

One of the most widely recognized applications of DNA testing—and the measurement of genetic variation—involves human identification, thanks in large measure to the popularity of TV programs such as *CSI: Crime Scene Investigation*. In the real world, beyond the hype of Hollywood, forensic DNA typing has been used to solve countless crimes—both implicating the guilty and exonerating the innocent.

A number of other applications also use the same technology and genetic markers. In the United States, almost a million samples are run each year for parentage testing, primarily to identify the father for child support purposes (AABB 2005). DNA testing is used by the U.S. military to identify casualties of war to prevent there ever being another "Unknown Soldier." Human remains from natural or man-made disasters or terrorist attacks can be identified using DNA testing of biological relatives or direct reference samples (Whitaker et al. 1995; Biesecker et al. 2005; NIJ 2006). National DNA databases containing millions of genetic profiles from convicted criminals are used in the U.S., the United Kingdom, and many other nations to link crimes and their perpetrators and to solve crimes in ways that were not possible only a few years ago (Gill 2002).

Attempts to address historical questions, such as the possibility of Thomas Jefferson fathering a slave (Foster et al. 1998), or identifying the remains of the Romanov family (Gill et al. 1994), are also made with the same forensic DNA testing techniques. In addition, genetic genealogy has become a popular tool to reconnect family pedigrees where the paper trail grows thin (Brown 2002).

A

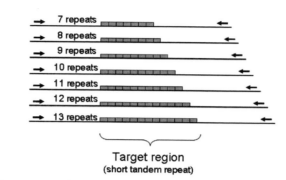

TCCCAAGCTCTTCCTCTTCCCTAGATCAATACAGAC
AGAAGACAGGTG**GATAGATAGATAGATAGATAGAT**
AGATAGATAGATAGATAGATAGATATCATTGAAAG
ACAAAACAGAGATGGATGATAGATACATGCTTACAG
ATGCACAC

= 12 GATA repeats ("12")

B

7 repeats
8 repeats
9 repeats
10 repeats
11 repeats
12 repeats
13 repeats

Target region
(short tandem repeat)

FIGURE 34-1. (A) Example sequence of an STR marker with underlined nucleotides representing the PCR primer binding sites and the bolded region showing the 12 GATA tetranucleotide repeats. (B) Schematic representation of possible alleles ranging from 7 to 13 repeats in length with arrows illustrating the PCR primer positions.

Over the past two decades, techniques for measuring genetic variation at the molecular level have advanced rapidly. Alec Jeffrey's 1984 discovery of minisatellites and hypervariable tandem repeats (Jeffrey et al. 1985) is generally considered the birth of modern forensic DNA testing, although the multilocus variable number of tandem repeat (VNTR) probe method (Gill et al. 1985) has been replaced over the past decade or so by short tandem repeat (STR) typing (NIJ 2000; Butler 2005).

STRs, sometimes referred to as microsatellites or simple sequence repeats (SSRs), are accordion-like stretches of DNA containing core repeat units of two to seven nucleotides in length that are tandemly repeated from approximately a half dozen to several dozen times (Fig. 34-1). Although the human genome contains thousands upon thousands of STR markers, only a small core set of loci have been selected for use in forensic DNA testing (Butler 2006). Commercial kits (Table 34-1) have been created to examine these core STR loci (Table 34-2). Millions of these

TABLE 34-1. Some Widely Used Commercial Kits for Autosomal STR Markers

Kit name	Source	STR loci included (amel = sex-typing marker amelogenin)	Power of discrimination[a]
PowerPlex 16	Promega	CSF1PO, FGA, TH01, TPOX, VWA, D3S1358, D5S818, D7S820, D8S1179, D13S317, amel, D16S539, D18S51, D21S11, Penta D, Penta E	1 in 3.4 x 10^17
Profiler Plus	Applied Biosystems	FGA, VWA, D3S1358, D5S818, D7S820, D8S1179, D13S317, D18S51, D21S11, amel	1 in 9.4 x 10^10
COfiler	Applied Biosystems	CSF1PO, TH01, TPOX, D3S1358, D7S820, D16S539, amel	1 in 1.3 x 10^6
SGM Plus	Applied Biosystems	FGA, TH01, VWA, D3S1358, D8S1179, D16S539, D18S51, D21S11, D2S1338, D19S433, amel	1 in 4.8 x 10^12
Identifiler	Applied Biosystems	CSF1PO, FGA, TH01, TPOX, VWA, D3S1358, D5S818, D7S820, D8S1179, D13S317, D16S539, D18S51, D21S11, D2S1338, D19S433, amel	1 in 2.5 x 10^17

[a]Based on random match probabilities for each STR locus found in Table 34-2.

TABLE 34-2. Characteristics of Common Autosomal STR Markers

STR marker	Chromosomal location	Repeat motif	Allele range[a]	PCR product size (kit)	RMP[b]
CSF1PO	5q33.1	TAGA	6–15	305–342 bp (Identifiler)	0.112
FGA	4q31.3	CTTT	17–51.2	215–355 bp (Identifiler)	0.036
TH01	11p15.5	TCAT	4–13.3	163–202 bp (Identifiler)	0.081
TPOX	2p25.3	GAAT	6–13	222–250 bp (Identifiler)	0.195
VWA	12p13.31	[TCTG] [TCTA]	11–24	155–207 bp (Identifiler)	0.062
D3S1358	3p21.31	[TCTG] [TCTA]	12–19	112–140 bp (Identifiler)	0.075
D5S818	5q23.2	AGAT	7–16	134–172 bp (Identifiler)	0.158
D7S820	7q21.11	GATA	6–15	255–291 bp (Identifiler)	0.065
D8S1179	8q24.13	[TCTA] [TCTG]	8–19	123–170 bp (Identifiler)	0.067
D13S317	13q31.1	TATC	8–15	217–245 bp (Identifiler)	0.085
D16S539	16q24.1	GATA	5–15	252–292 bp (Identifiler)	0.089
D18S51	18q21.33	AGAA	7–27	262–345 bp (Identifiler)	0.028
D21S11	21q21.1	[TCTA] [TCTG]	24–38	185–239 bp (Identifiler)	0.039
D2S1338	2q35	[TGCC] [TTCC]	15–28	307–359 bp (Identifiler)	0.027
D19S433	19q12	AAGG	9–17.2	102–135 bp (Identifiler)	0.087
Penta D	21q22.3	AAAGA	2.2–17	376–449 bp (PP16)	0.059
Penta E	15q26.2	AAAGA	5–24	379–474 bp (PP16)	0.030
Amelogenin (sex-typing)	Xp22.22 Yp11.2	not applicable		X = 107 bp Y = 113 bp	

[a]Ranges are calculated from kit allelic ladders and do not represent the full range of alleles observed in world populations. (See Butler 2006 for a more complete allele listing of these STR loci.)

[b]RMP: Random match probability for Caucasian individuals. The 13 CODIS loci data are from FBI data reported in NIJ (2000). D2S1138 and D19S433 RMP data are from the Identifiler Kit User's Manual from Applied Biosystems. The Penta D and Penta E information comes from PowerPlex 16 (PP16) population data available from Promega Corporation.

genetic variation assays are run each year by government, university, and private laboratories performing DNA databasing, forensic casework, or parentage testing.

OVERVIEW OF FORENSIC DNA TESTING

The genome of each individual is inherited from his or her parents and, due to recombination and chromosome segregation during meiosis, is unique, with the notable exception of identical twins. However, because it is prohibitive of both cost and time to examine the entire genome of an individual, only small subsets are probed for genetic variation in order to differentiate between individuals during human identity or forensic DNA testing. Thus, statistical probabilities of a random match are used, based on expected frequencies of various alleles observed in the DNA profile.

Because information gathered during the DNA process must hold up in a court of law, a chain of custody is demonstrated through tracking the sample as it passes each step in the analytical process. In addition, only validated testing methods may be used to ensure reliable and reproducible results. It is also important to realize that current standard DNA tests do not look at genes, but rather at the variation occurring in the "junk" regions within introns or between genes. Thus, little or no information is provided regarding race, ethnicity, or such aspects of phenotype as eye color, height, or hair color with a standard forensic DNA test.

STR typing involves the use of the polymerase chain reaction (PCR) to recover information from small amounts of material. The relatively small PCR product sizes of 100–500 bp generated with STR testing are generally compatible with degraded DNA that may be present due to environmental insults on the evidentiary biological material found at a crime scene. PCR amplification of multiple STR loci simultaneously, or "multiplexing," is possible with different-colored fluorescent dyes and different-sized PCR products. Use of multiple loci enables a high power of discrimination in a single test without consuming much DNA (e.g., 1 ng or less of starting material). Commercially available kits simplify generation of STR profiles and provide a uniform set of core STR loci to make possible national and international sharing of criminal DNA profiles. Commercial kits are preferred over in-house assays, even though the kits are more expensive, as they help simplify and standardize procedures and remove the burden of PCR component quality control from the busy end user. The steps involved in STR typing (Fig. 34-2) include sample collection, DNA extraction, DNA quantitation, PCR amplification of multiple STR loci, STR allele separation and sizing, STR typing and profile interpretation, and a report of the statistical significance of a match (if observed). Each of these steps is described in more detail in the following sections (see also Budowle et al. 2000).

Sample Collection

Blood stained onto paper cards, or buccal swabs containing cheek cells from the inside of an individual's mouth, are commonly used for collection of DNA from reference samples, such as a suspect or a biological relative. Crime scene stains may come from cuttings of clothing or vaginal swabs in the case of rape victims.

DNA Extraction

To be accessible to PCR primers and other reagents, DNA must be removed from its cellular environment and put into solution. The two conventional DNA extraction methods employed by forensic laboratories are organic extraction and Chelex extraction. Organic extraction involves serial dilution of sodium dodecylsulfate and proteinase K to break down the cell walls, followed by the addition of a phenol–chloroform mixture to separate the proteins from the DNA molecules. With centrifugation, proteins and other cellular debris can be pulled away from the aqueous phase that contains double-stranded DNA. Samples are often concentrated using Centricon or Microcon dialysis (Comey et al. 1994).

Chelex uses a chelating ion exchange resin suspension that is directly added to the sample (Walsh et al. 1991). Most protocols involve adding a bloodstain to a 5% Chelex suspension and boil-

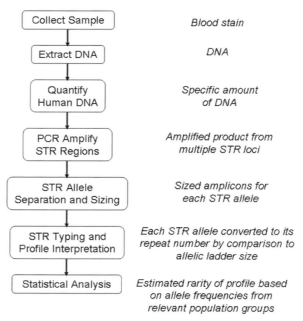

FIGURE 34-2. Summary of steps involved in STR typing with example outcome of each step.

ing for several minutes to break open the cells and release the DNA. Following centrifugation, the supernatant containing the denatured DNA molecules is removed for postextraction processing.

Alternatively, many reference samples consist of dried blood or saliva on FTA paper (Whatman) that is prepared for PCR by removing a portion of the sample as a paper punch. Following several chemical washes to remove PCR-inhibiting compounds, the cleaned FTA card punch with its bound DNA is added directly to the PCR amplification reaction (Vanek et al. 2001). In addition, more automated procedures involving silica-based solid-phase extraction methods such as QIAamp spin columns (Qiagen) and DNA IQ magnetic beads (Promega) are gaining popularity within the forensic DNA community.

When ejaculation occurs in a sexual assault, vaginal swabs contain a mixture of DNA from the female victim and from the male perpetrator. A process known as differential extraction is often used on sexual assault evidence to physically separate sperm and vaginal epithelial cells prior to breaking the cells open to release the DNA (see Chapter 3 in Butler 2005). Differential extraction involves the use of dithiothreitol (DTT) to lyse the sperm cells following removal of the female epithelial cells (Gill et al. 1985).

DNA Quantitation

When DNA is extracted from biological evidence recovered at crime scenes, nonhuman sources such as bacterial, plant, or animal DNA may be coextracted. Because only the human DNA will be targeted during PCR amplification, it is valuable to determine the specific amount of human DNA recovered. Quantitative PCR (qPCR) methods for DNA quantitation have become popular within the forensic community due to their sensitivity, human specificity, and dynamic range. A commonly used qPCR assay is the Quantifiler kit (Applied Biosystems).

Multiplex PCR amplification using STR kits is very sensitive and works best with a fairly narrow range of input DNA. Accurate quantitation of input DNA helps to produce well-balanced and on-scale STR profiles. At levels higher than ~2 ng, the fluorescence detection system can be overwhelmed, causing numerous artifacts that make data interpretation more challenging. Too much DNA can cause split peaks due to incomplete adenylation or off-scale peaks that produce bleed-through between dye colors. At levels much below 200 pg, stochastic amplification may result in locus or allelic imbalance or even dropout. Adjustment of DNA concentrations is usually made on the basis of quantitation results, so that the amount of DNA template added to the PCR reaction is in the "sweet spot" of ~0.5–1.5 ng that enables well-balanced and on-scale STR profiles to be generated.

PCR Amplification

Commercial STR kits have greatly simplified PCR amplification by providing premixed primers and a standard master mix containing the enzyme buffers and dNTPs. Use of kits, although more expensive, frees the end user from the burden of quality control of PCR reaction components, particularly when more than 30 oligonucleotide primers are present in the multiplex amplification, such as with the PowerPlex 16 (Krenke et al. 2002) and Identifiler kits (Collins et al. 2004). AmpliTaq Gold DNA Polymerase (Applied Biosystems) is regularly used to enable a specific, hot-start reaction. Typical amplification protocols involve ~10 minutes at 95°C to activate the TaqGold polymerase, followed by 28–30 cycles of denaturation at 94°C, primer annealing at 59°C or 60°C, and extension at 72°C.

STR Allele Separation and Sizing

Following PCR amplification, the overall length of the STR amplicon is measured to determine the number of repeats in each allele in the DNA profile. This length measurement is made via a size-based separation involving gel or capillary electrophoresis. Each STR amplicon has been fluorescently labeled during PCR, since either the forward or reverse locus-specific primer contains a fluorescent dye. Thus, by recording the dye color and migration time of each DNA fragment relative to an internal size standard, the size for each STR allele may be determined following its separation from other STR alleles. Commonly used instruments for STR allele separation and sizing include the ABI 310 and ABI 3100 Genetic Analyzers (Butler et al. 2004).

STR Typing and Profile Interpretation

Commercial STR kits come with allelic ladders, which are a mixture of common alleles characterized for the number of repeat units via DNA sequencing. Through comparison of the allelic ladder allele sizes (run with the same internal size standard), each sample allele may be converted from its size in base pairs to its number of repeat units (Fig. 34-3). The allelic ladder permits calibration of individual alleles to their repeat size, which enables database comparisons with data collected from laboratories with different electrophoretic conditions. The final DNA profile is the string of genotypes for the various STR loci analyzed. Two example profiles are listed below, each containing the genotyping results from the 13 core STR loci commonly used in the forensic DNA community:

Profile 1: 16,17-17,18-21,22-12,14-28,30-14,16-12,13-11,14-9,9-9,11-6,6-8,8-10,10
Profile 2: 14,17-17,17-21,22-13,13-29,33.2-14,14-12,12-11,13-9,10-12,14-6,7-8,11-12,12

The profiles contain two alleles (separated by a comma) for each locus (separated by dashes). The order of the STR loci in this example is D3S1358, VWA, FGA, D8S1179, D21S11, D18S51, D5S818, D13S317, D7S820, D16S539, TH01, TPOX, and CSF1PO (Table 34-2) (for more information on each locus, see Butler 2005, 2006). An electropherogram for Profile 2 is shown in Figure 34-4.

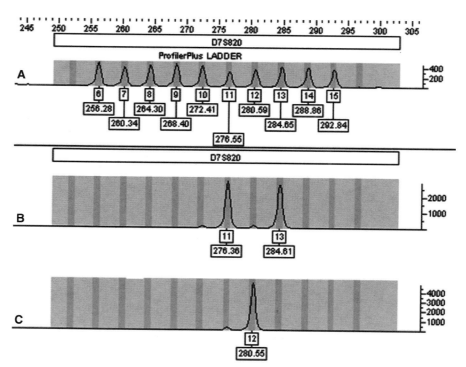

FIGURE 34-3. (A) Allelic ladder sizes for the STR locus D7S820 shown with two samples: (B) heterozygous 11,13 and (C) homozygous 12,12.

A number of biological and instrumental artifacts must often be sorted through to generate a complete and accurate STR profile (SWGDAM 2000; see also Chapters 6 and 15 in Butler 2005). Biological artifacts include stutter products, split peaks from incomplete adenylation, triallelic patterns, and variant alleles containing mutations in the repeat or flanking regions that cause an allele to be "off-ladder." Instrumental artifacts arise from voltage spikes, dye blobs, and bleed-through between dye colors. Sample mixtures can also complicate matters, as it may be difficult to fully decipher the various components present in the STR profiles of two or more individuals.

Statistical Analysis

Once a DNA profile has been generated for a particular sample, it must be considered in the context of the case at hand. DNA results are of no value in and of themselves—they must always be compared to a reference sample. Forensic casework evidence is compared to a suspect or (if no suspect exists) to a DNA database of previously convicted offenders. DNA profiles from the remains of missing persons or mass disaster victims are compared to biological relatives or direct reference samples such as DNA recovered from toothbrushes. Remains from military casualties are compared to direct reference bloodstains created when an individual enters the military. The DNA profile from an alleged father in a paternity test is compared to that of a child (and possibly the child's mother in order to identify the obligate alleles that would need to come from the father). Figure 34-5 illustrates parentage test results from a father, a mother, and three children. Mendelian inheritance of the STR alleles can be seen in the children's genotypes. For example, child #1 inherits the "12" allele from her mother and the "14" allele from her father.

There are three possible conclusions from DNA profile comparisons: (1) a match or inclusion, (2) a mismatch or exclusion, and (3) inconclusive results if not enough data are available. If a

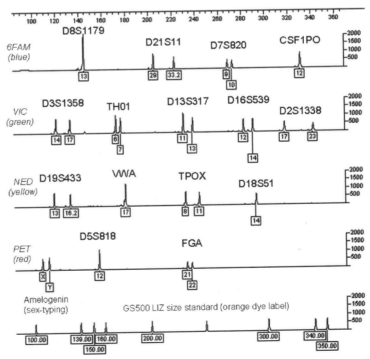

FIGURE 34-4. Electropherogram of STR typing result from the Identifiler kit matching Profile 2 example in the text. Genotypes are listed below the peaks for each STR locus.

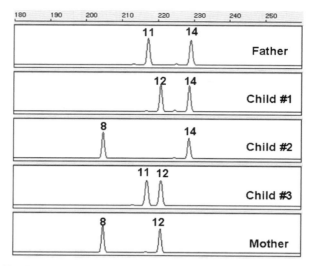

FIGURE 34-5. STR profiles for the D13S317 locus from a family illustrating Mendelian inheritance of alleles present in the three children.

match is observed, then an estimate of the rarity of the particular profile is typically given. In other words, how significant is the match observed? This profile rarity estimate, sometimes referred to as the random match probability (RMP), is based on allele frequencies from relevant population groups. The allele frequency information is gathered by examining a random set of usually 100 or more individuals (who have self-declared their ethnicity) and counting the number of times each allele is observed at an STR locus. The genotype frequencies for each locus are then estimated based on assumptions of Hardy-Weinberg equilibrium along with corrections to account for possible subpopulation structure. Typically, forensic laboratories utilize a computer program called PopStats that is supplied by the FBI Laboratory with the Combined DNA Index System (CODIS) software to perform their RMP calculations. U.S. Caucasian, African-American, and Hispanic population groups are most commonly included in reports based on allele frequencies recorded in a large population study organized by the FBI (Budowle et al. 2001). It is important to note that use of allele frequencies from different population studies can result in slightly different RMPs. For example, Profile 1 above is estimated to have a RMP of 1 in 837 trillion using allele frequencies from one U.S. Caucasian population study (Butler et al. 2003a) and a RMP of 1 in 2.46 quadrillion from another Caucasian data set (Budowle et al. 2001).

Time Considerations

The sample processing steps outlined above generally take 1–2 days but can be accomplished in as little as about 5 hours, with most of the time being spent in PCR thermal cycling. STRs have been typed in less than 1 hour using rapid cycling methods (Belgrader et al. 1998). However, rapid cycling does not work well with large multiplex amplification reactions. With a single 16-capillary ABI 3100, one individual can easily generate data from more than 100 extracted DNA samples in a single day. Robotic pipetting workstations are being introduced in many forensic laboratories in order to automate batch processing of samples (see Figs. 34-6 and 34-7). Some labs have sample tracking capabilities with bar-coded tubes and plates and laboratory information management systems (LIMS) to aid chain-of-custody documentation. A major bottleneck in the whole process is data review. Expert system software is beginning to be implemented in many labs to speed data review and STR profile interpretation. See Table 34-3 for a summary of forensic DNA testing resources.

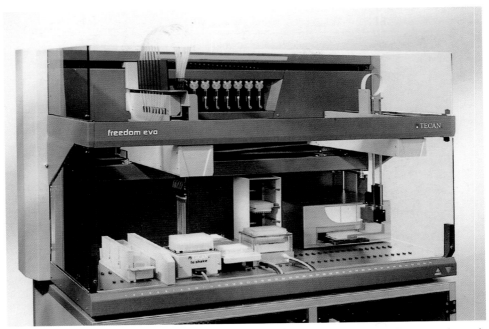

FIGURE 34-6. Robotic liquid handler for performing the manual steps of STR typing. The image shows the automated Freedom EVO Forensics workstation with enhanced features for sample preparation and analysis in the laboratory. (Courtesy of Tecan Schweiz AG.)

Costs for STR Typing

Reagents for commercial STR kits using standard reaction volumes of 50 μl or 25 μl can cost a user about $30 to $35 per sample. Several groups have demonstrated that reliable STR results can routinely be obtained with small-volume PCR, thereby reducing the cost of these kit reagents. The Royal Canadian Mounted Police (LeClair et al. 2003) and the North Louisiana Crime Lab

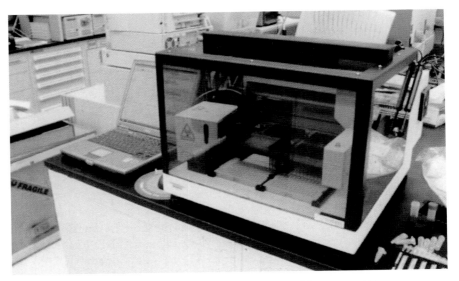

FIGURE 34-7. Picture of Corbett robot in the laboratory at NIST.

TABLE 34-3. Electronic Resources for DNA Testing

Databases	OmniPop program to calculate RMP
	Yfiler database to calculate haplotype frequency
Software	FSS-i3 expert system software for automated STR typing
	GeneMapperID v3.2
	Sequencer for mtDNA sequence analysis
Web sites for software used in match probability calculations	**Canadian Random Match Calculator** (http://www.csfs.ca/pplus/profiler.htm) enables calculation of STR profiles for the 13 U.S. core STR loci amplified by the Profiler Plus and COfiler kits sold by Applied Biosystems. This program enables comparison of results from limited FBI and Canadian collected allele frequencies.
	European Network of Forensic Science Institutes DNA Working Group STR Population Database (http://www.str-base.org/index.php) uses 5699 samples from 24 European populations to make match probability calculation on STR profiles containing the 10 STR loci present in the SGM Plus kit (Applied Biosystems) that is widely used in Europe.
	OmniPop is an Excel-based program developed by a forensic scientist, Brian Burritt, of the San Diego Police Department. It calculates an STR profile's frequency using allele frequencies from 202 published databases. The program is freely available at http://www.cstl.nist.gov/biotech/strbase/populationdata.htm.

(Gaines et al. 2002) have published 5-μl PCR protocols with commercial STR kits. Labor remains the most expensive part of the process. A contract laboratory typically charges several hundred dollars per sample for paternity testing or forensic casework performed by STR typing.

Quality Assurance in Forensic DNA Testing

The results of forensic DNA testing influence courts to make decisions regarding guilt or innocence, which in turn affects the liberty of defendants. Today, it is rare that the science underlying forensic DNA testing is challenged in court. Rather, the performance of the laboratory and the analysts involved is typically reviewed to demonstrate reliability in their DNA testing efforts.

There are five interlocking pieces in demonstrating reliability and maintaining quality assurance in DNA testing results. First, quality assurance standards have been issued by the FBI Director (FBI 2000a,b), and guidelines on topics such as validation and training have been established by the Scientific Working Group on DNA Analysis Methods (SWGDAM). These standards and guidelines set the mark and govern the operations of forensic DNA laboratories (see Table 34-4). The FBI Laboratory-sponsored Technical Working Group on DNA Analysis Methods (TWGDAM) began in November 1988 to address quality assurance issues in the process of forensic DNA testing. In 1998, the name was changed to SWGDAM. SWGDAM meets twice a year, usually in January and July, at the FBI Academy in Quantico, Virginia. It consists of 40–50 DNA technical leaders and scientists from forensic laboratories around the United States who meet in subcommittees to develop guidelines on various topics. These guidelines are then published in *Forensic Science Communications*, which is available on the FBI.gov Web site. (Early TWGDAM materials were published in the *Crime Laboratory Digest*.) The DNA Advisory Board (DAB) was a 13–voting member, congressionally mandated body that existed from 1995 to 2000 to issue standards for the forensic DNA community. Since 2000, SWGDAM has operated as the group responsible for offering recommendations to the forensic community within the United States. The FBI Laboratory oversees audits of all laboratories participating in the Combined DNA Index System (CODIS).

Second, laboratories become accredited (and are re-accredited on a regular basis), which in effect is a statement that the laboratory has developed standard operating procedures (SOPs) in accordance with the quality assurance standards and follows these SOPs in its daily opera-

TABLE 34-4. Standards and Guidelines Established by SWGDAM

TWGDAM guidelines	Kearney et al. (1989) Guidelines for a quality assurance program for DNA restriction fragment length polymorphism analysis. *Crime Lab Digest* 16: 40–59.
	Kearney et al. (1991) Guidelines for a quality assurance program for DNA analysis. *Crime Lab Digest* 18: 44–75.
	Budowle et al. (1995) Guidelines for a quality assurance program for DNA analysis. *Crime Lab Digest* 22: 20–43.
DAB standards	http://www.fbi.gov/hq/lab/fsc/backissu/july2000/codis2a.htm Quality Assurance Standards for Forensic DNA Testing Laboratories (issued by the DNA Advisory Board July 1998)
	http://www.fbi.gov/hq/lab/fsc/backissu/july2000/codis1a.htm Quality Assurance Standards for Convicted Offender DNA Databasing Laboratories (issued by the DNA Advisory Board April 1999)
	http://www.fbi.gov/hq/lab/fsc/backissu/july2000/dnastat.htm Statistical and Population Genetics Issues Affecting the Evaluation of the Frequency of Occurrence of DNA Profiles Calculated from Pertinent Population Database(s) (issued by the DNA Advisory Board February 2000)
SWGDAM guidelines	http://www.fbi.gov/hq/lab/fsc/backissu/april2003/swgdambylaws.htm Bylaws of the Scientific Working Group on DNA Analysis Methods
	http://www.fbi.gov/hq/lab/fsc/backissu/april2003/swgdammitodna.htm Guidelines for Mitochondrial DNA (mtDNA) Nucleotide Sequence Interpretation
	http://www.fbi.gov/hq/lab/fsc/backissu/april2003/swgdamsafety.htm Guidance Document for Implementing Health and Safety Programs in DNA Laboratories
	http://www.fbi.gov/hq/lab/fsc/backissu/oct2001/kzinski.htm Training Guidelines
	http://www.fbi.gov/hq/lab/fsc/backissu/july2000/strig.htm Short Tandem Repeat (STR) Interpretation Guidelines
	http://www.fbi.gov/hq/lab/fsc/backissu/july2004/standards/2004_03_standards02.htm Revised Validation Guidelines
	http://www.fbi.gov/hq/lab/fsc/backissu/july2004/standards/2004_03_standards03.htm Report on the Current Activities of the Scientific Working Group on DNA Analysis Methods Y-STR Subcommittee (recommended core Y-STR loci)
Audit document used in review of all forensic DNA testing laboratories	http://www.fbi.gov/filelink.html?file=/hq/lab/fsc/backissu/july2004/pdfs/seubert.pdf Quality Assurance Audit for Forensic DNA and Convicted Offender DNA Databasing Laboratories (issued by the FBI Laboratory July 2004)

tions. Crime laboratory accreditation is performed by the American Society of Crime Laboratory Directors-Laboratory Accreditation Board (ASCLD-LAB). Contract laboratories conducting DNA databasing rather than casework may be certified by the National Forensic Science Technology Center (NFSTC). Third, proficiency testing is performed on a semiannual basis to demonstrate that an individual analyst is capable of performing reliable testing when following laboratory SOPs. Fourth, equipment calibration and protocol performance are demonstrated through running certified reference materials, such as National Institute of Standards and Technology (NIST) Standard Reference Materials (SRMs). Fifth, inspections and laboratory audits are performed on an annual or biannual basis to verify that the quality assurance standards are being adhered to, including analyst proficiency testing and instrument/protocol calibration.

DATA ISSUES

Biological Artifacts

Some common biological artifacts observed when evaluating STR data include stutter products, incomplete adenylation, variant alleles, and triallelic patterns (see Chapter 6 in Butler 2005).

Stutter products arise from strand slippage during the PCR amplification process of tandem repeats. For tetranucleotide repeat loci, stutter products may be observed 4 bp shorter than the true allele and are usually less than 10% of the allele's height. Incomplete adenylation results in split peaks—with each allele possessing a "–A" and a "+A" product. The "–A" peak is the full-length PCR product, and the "+A" peak results from the DNA polymerase producing an over-hanging base. It is worth noting that not all polymerases add this extra nucleotide, which is highly dependent on the 3' end of the PCR product that may be defined by the 5' end of the reverse primer (see Table 6.1 in Butler 2005). Variant alleles result from an insertion or deletion of a nucleotide either within or near the repeat region. Triallelic patterns are most often observed in TPOX and D18S51 and probably result from a duplication of the primer binding sites on one of the two chromosomes being amplified. NIST maintains an Internet database on STR markers used in human identity testing known as STRBase (http://www.cstl.nist.gov/biotech/strbase). STRBase catalogs variant alleles as well as triallelic patterns that have been observed by forensic DNA scientists around the world.

Degraded DNA Profiles

When DNA degrades and breaks into smaller pieces, larger PCR products are not generated as efficiently, since their template regions are more likely to be fragmented. Signal loss is observed as PCR product size increases, because the number of full-length, intact molecules around a STR locus is reduced when the DNA sample has been damaged. Moving primers closer to the repeat region to produce mini-STR systems can result in improved amplification with degraded DNA samples (Butler et al. 2003a).

Mixtures

Samples containing biological mixtures from two or more individuals can be observed in forensic casework, especially with sexual assault evidence. Mixtures with DNA in near equal proportions can be fairly easily discerned by the presence of more than two alleles at multiple loci, as shown in Figure 34-8. Deciphering the full STR profiles of all donors can be challenging and is often performed with the aid of a known profile such as that of the victim. Tetra- and pentanucleotide markers are preferred in forensic arenas, due to their lower tendency to generate stutter products that can confound interpretation of mixed DNA profiles (see Chapter 7 in Butler 2005).

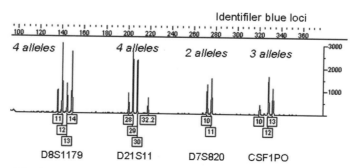

FIGURE 34-8. Plot of the four Identifiler blue labeled STR loci observed in analysis of a mixed DNA sample. Note the presence of more than two alleles at three out of the four loci shown, suggesting that more than one individual contributed to the DNA profile.

OTHER FORENSIC DNA TESTING TECHNIQUES

Y Chromosome STR Testing

The Y chromosome is found only in males and, therefore, Y-STR testing can enable recovery of male-specific information from samples containing a mixture of male and female DNA, even when the female DNA is present in greater than 1000-fold excess of the male component (see Chapter 9 in Butler 2005). Several hundred STRs have been identified on the Y chromosome, but only a dozen or so have been settled upon for use in human identity testing (Butler 2003). However, these Y-STRs are all linked on the nonrecombining region of the Y chromosome, making them less able to differentiate two unrelated individuals than a set of autosomal STR markers and unable to distinguish fathers from sons, or brothers from one another (except in the case of a mutation).

Mitochondrial DNA

In highly degraded samples or material with limited DNA content such as telogen hair shafts, forensic scientists often turn to mitochondrial DNA (mtDNA) analysis (see Chapter 10 in Butler 2005). mtDNA is present in cells at a much higher copy number than the nuclear DNA from which STRs are amplified. Whereas there are two copies of each STR locus in a cell, hundreds of copies of mtDNA exist. In humans, mtDNA is inherited only from an individual's mother and thus is not unique, as siblings, aunts, and other maternal relatives will possess the same mtDNA sequence, provided that no mutations have occurred. Two hypervariable regions within the mtDNA control region are usually sequenced, and differences in the 610 bp examined (out of 16,568 bp usually present in the mtDNA genome) are noted relative to a standard sequence known as the revised Cambridge reference sequence (see Fig. 34-9) (Anderson et al. 1981; Andrews et al. 1999). Analysis of mtDNA sequence information is more labor-intensive, time-consuming, and expensive than STR typing, but it often produces a result when none is possible with conventional STR typing methods.

SNPs for Estimating Ethnicity and Phenotypic Characteristics

Single-nucleotide polymorphisms (SNPs) will likely supplement rather than supplant current STR typing systems (Gill et al. 2004). Because SNPs have lower mutation rates relative to STRs (10^{-8} vs. 10^{-3}), SNPs are more likely to become "fixed" in a particular population of genetic variants and thus be useful for estimating ethnicity based on minor genetic differences. Efforts are under way to distinguish hair color, eye color, and other phenotypic traits using SNPs (see Chapter 8 in Butler 2005). Most of these assays are still in various stages of research and far from routine use in human identity testing.

SUMMARY

STRs have been, and will continue to be, the primary method for measuring genetic variation in human identity testing for the foreseeable future (NIJ 2000). A set of less than two dozen core STR loci is entrenched in forensic and paternity testing laboratory protocols (Butler 2006). These core STR loci are used to generate millions of DNA profiles each year for human identity testing purposes.

ACKNOWLEDGMENTS

This work was supported in part by the National Institute of Justice through Interagency Agreement 2003-IJ-R-029 with the NIST Office of Law Enforcement Standards. The assistance

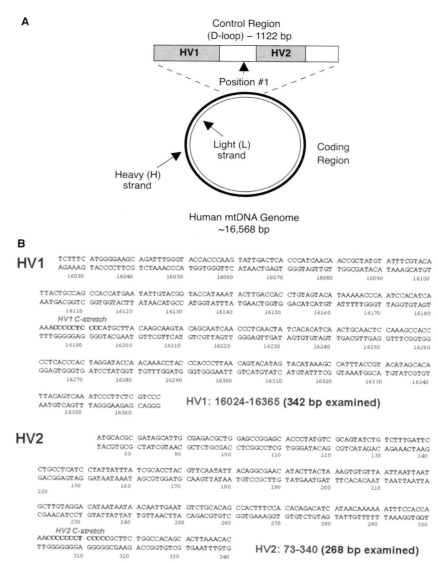

A

Control Region
(D-loop) – 1122 bp

HV1   HV2

Position #1

Light (L) strand

Heavy (H) strand

Coding Region

Human mtDNA Genome
~16,568 bp

B

HV1

```
        TCTTTC ATGGGGAAGC AGATTTGGGT ACCACCCAAG TATTGACTCA CCCATCAACA ACCGCTATGT ATTTCGTACA
        AGAAAG TACCCCTTCG TCTAAACCCA TGGTGGGTTC ATAACTGAGT GGGTAGTTGT TGGCGATACA TAAAGCATGT
        16030      16040       16050       16060       16070       16080       16090       16100

        TTACTGCCAG CCACCATGAA TATTGTACGG TACCATAAAT ACTTGACCAC CTGTAGTACA TAAAAACCCA ATCCACATCA
        AATGACGGTC GGTGGTACTT ATAACATGCC ATGGTATTTA TGAACTGGTG GACATCATGT ATTTTTGGGT TAGGTGTAGT
        16110      16120       16130       16140       16150       16160       16170       16180
        HV1 C-stretch
        AAACCCCCTC CCCATGCTTA CAAGCAAGTA CAGCAATCAA CCCTCAACTA TCACACATCA ACTGCAACTC CAAAGCCACC
        TTTGGGGGAG GGGTACGAAT GTTCGTTCAT GTCGTTAGTT GGGAGTTGAT AGTGTGTAGT TGACGTTGAG GTTTCGGTGG
        16190      16200       16210       16220       16230       16240       16250       16260

        CCTCACCCAC TAGGATACCA ACAAACCTAC CCACCCTTAA CAGTACATAG TACATAAAGC CATTTACCGT ACATAGCACA
        GGAGTGGGTG ATCCTATGGT TGTTTGGATG GGTGGGAATT GTCATGTATC ATGTATTTCG GTAAATGGCA TGTATCGTGT
        16270      16280       16290       16300       16310       16320       16330       16340

        TTACAGTCAA ATCCCTTCTC GTCCC
        AATGTCAGTT TAGGGAAGAG CAGGG            HV1: 16024-16365 (342 bp examined)
        16350      16360
```

HV2

```
                  ATGCACGC GATAGCATTG CGAGACGCTG GAGCCGGAGC ACCCTATGTC GCAGTATCTG TCTTTGATTC
                  TACGTGCG CTATCGTAAC GCTCTGCGAC CTCGGCCTCG TGGGATACAG CGTCATAGAC AGAAACTAAG
                  80        90         100        110        120        130        140

        CTGCCTCATC CTATTATTTA TCGCACCTAC GTTCAATATT ACAGGCGAAC ATACTTACTA AAGTGTGTTA ATTAATTAAT
        GACGGAGTAG GATAATAAAT AGCGTGGATG CAAGTTATAA TGTCCGCTTG TATGAATGAT TTCACACAAT TAATTAATTA
        220    150        160        170        180        190        200        210

        GCTTGTAGGA CATAATAATA ACAATTGAAT GTCTGCACAG CCACTTTCCA CACAGACATC ATAACAAAAA ATTTCCACCA
        CGAACATCCT GTATTATTAT TGTTAACTTA CAGACGTGTC GGTGAAAGGT GTGTCTGTAG TATTGTTTTT TAAAGGTGGT
        230    240        250        260        270        280        290        300
        HV2 C-stretch
        AACCCCCCCT CCCCCGCTTC TGGCCACAGC ACTTAAACAC
        TTGGGGGGGA GGGGGCGAAG ACCGGTGTCG TGAATTTGTG            HV2: 73-340 (268 bp examined)
        310        320        330        340
```

FIGURE 34-9. (A) Schematic of the human mtDNA genome showing relative positions of the coding and control regions along with the two hypervariable (HV) portions commonly examined in forensic DNA testing. (B) Annotated sequence of the 610 bp present in the revised Cambridge reference sequence for human mtDNA HV1 and HV2. Sample results are typically reported as differences (position and nucleotide change) from the reference sample.

of NIST human identity project team members Amy Decker, Dave Duewer, Becky Hill, Margaret Kline, Jan Redman, and Peter Vallone is greatly appreciated. Points of view are those of the author and do not necessarily represent the position of the U.S. Department of Justice. Certain commercial equipment, instruments, and materials are identified in order to specify experimental procedures as completely as possible. In no case does such identification imply a recommendation or endorsement by the National Institute of Standards and Technology, nor does it imply that any of the materials, instruments, or equipment identified are necessarily the best available for the purpose.

REFERENCES

AABB (American Association of Blood Banks). 2005. Annual report summary for testing in 2004 prepared by the Relationship Testing Program Unit. Available at http://www.aabb.org/Documents/Accreditation/Parentage_Testing_Accreditation_Program/rtannrpt04.pdf

Anderson S., Bankier A.T., Barrell B.G., de Bruijn M.H., Coulson A.R., Drouin J., Eperon I.C., Nierlich D.P., Roe B.A., Sanger F., et al. 1981. Sequence and organization of the human mitochondrial genome. *Nature* **290:** 457–465.

Andrews R.M., Kubacka I., Chinnery P.F., Lightowlers R.N., Turnbull D.M., and Howell N. 1999. Reanalysis and revision of the Cambridge reference sequence for human mitochondrial DNA. *Nature Genet.* **23:** 147.

Belgrader P., Smith J.K., Weedn V.W., and Northrup M.A. 1998. Rapid PCR for identity testing using a battery-powered miniature thermal cycler. *J. Forensic Sci.* **43:** 315–319.

Biesecker L.G., Bailey-Wilson J.E., Ballantyne J., Baum H., Bieber F.R., Brenner C., Budowle B., Butler J.M., Carmody G., Conneally P.M., et al. 2005. DNA identifications after the 9/11 World Trade Center attack. *Science* **310:** 1122–1123.

Brown K. 2002. Tangled roots? Genetics meets genealogy. *Science* **295:** 1634–1635.

Budowle B., Smith J., Moretti T., and DiZinno J. 2000. *DNA typing protocols: Molecular biology and forensic analysis.* Eaton Publishing, Natick, Massachusetts.

Budowle B., Shea B., Niezgoda S., and Chakraborty R. 2001. CODIS STR loci data from 41 sample populations. *J. Forensic Sci.* **46:** 453–489.

Butler J.M. 2003. Recent developments in Y-short tandem repeat and Y-single nucleotide polymorphism analysis. *Forensic Sci. Rev.* **15:** 91–111.

———. 2005. *Forensic DNA typing: Biology, technology, and genetics of STR markers.* Elsevier, New York.

———. 2006. Genetics and genomics of core STR loci used in human identity testing. *J. Forensic Sci.* **51:** 253–265.

Butler J.M., Shen Y., and McCord B.R. 2003b. The development of reduced size STR amplicons as tools for analysis of degraded DNA. *J. Forensic Sci.* **48:** 1054–1064.

Butler J.M., Buel E., Crivellente F., and McCord B.R. 2004. Forensic DNA typing by capillary electrophoresis: Using the ABI Prism 310 and 3100 Genetic Analyzers for STR analysis. *Electrophoresis* **25:** 1397–1412.

Butler J.M., Schoske R., Vallone P.M., Redman J.W., and Kline M.C. 2003a. Allele frequencies for 15 autosomal STR loci on U.S. Caucasian, African American, and Hispanic populations. *J. Forensic Sci.* **48:** 908–911.

Collins P.J., Hennessy L.K., Leibelt C.S., Roby R.K., Reeder D.J., and Foxall P.A. 2004. Developmental validation of a single-tube amplification of the 13 CODIS STR loci, D2S1338, D19S433, and amelogenin: The AmpFlSTR Identifiler PCR Amplification Kit. *J. Forensic Sci.* **49:** 1265–1277.

Comey C.T., Koons B.W., Presley K.W., Smerick J.B., Sobieralski C.A., Stanley D.M., and Baechtel F.S. 1994. DNA extraction strategies for amplified fragment length polymorphism analysis. *J. Forensic Sci.* **39:** 1254–1269.

FBI (Federal Bureau of Investigation). 2000a. Quality assurance standards for forensic DNA testing laboratories. *Forensic Sci. Commun.* **2:** no. 3 (July 2000A). Available online: http://www.fbi.gov/hq/fsc/backissu/july2000/codispre. htm.

FBI (Federal Bureau of Investigation). 2000b. Quality assurance standards for convicted offender DNA databasing laboratories. *Forensic Sci. Commun.* **2:** no. 3 (July 2000B). Available online: http://www.fbi.gov/hq/lab/fsc/backissu/july2000/codispre.htm.

Foster E.A., Jobling M.A., Taylor P.G., Donnelly P., de Knijff P., Mieremet R., Zerjal T., and Tyler-Smith C. 1998. Jefferson fathered slave's last child. *Nature* **396:** 27–28.

Gaines M.L., Wojtkiewicz P.W., Valentine J.A., and Brown C.L. 2002. Reduced volume PCR amplification reactions using the AmpFlSTR Profiler Plus kit. *J. Forensic Sci.* **47:** 1224–1237.

Gill P. 2002. Role of short tandem repeat DNA in forensic casework in the UK—Past, present, and future perspectives. *BioTechniques* **32:** 366–372.

Gill P., Jeffreys A.J., and Werrett D.J. 1985. Forensic application of DNA 'fingerprints'. *Nature* **318:** 577–579.

Gill P., Werrett D.J., Budowle B., and Guerrieri R. 2004. An assessment of whether SNPs will replace STRs in national DNA databases: Joint considerations of the DNA working group of the European Network of Forensic Science Institutes (ENFSI) and the Scientific Working Group on DNA Analysis Methods (SWGDAM). *Sci. Justice* **44:** 51–53.

Gill P., Ivanov P.L., Kimpton C., Piercy R., Benson N., Tully G., Evett I., Hagelberg E., and Sullivan K. 1994. Identification of the remains of the Romanov family by DNA analysis. *Nat. Genet.* **6:** 130–135.

Jeffreys A.J., Wilson V., and Thein S.L. 1985. Hypervariable 'minisatellite' regions in human DNA. *Nature* **314:** 67–73.

Krenke B.E., Tereba A., Anderson S.J., Buel E., Culhane S., Finis C.J., Tomsey C.S., Zachetti J.M., Masibay A., Rabbach D.R., et al. 2002. Validation of a 16-locus fluorescent multiplex system. *J. Forensic Sci.* **47:** 773–785.

Leclair B., Sgueglia J.B., Wojtowicz P.C., Juston A.C., Fregeau C.J., and Fourney R.M. 2003. STR DNA typing: Increased sensitivity and efficient sample consumption using reduced PCR reaction volumes. *J. Forensic Sci.* **48:** 1001–1013.

NIJ (National Institute of Justice). 2000. The future of forensic DNA testing: Predictions of the Research and Development Working Group of the National Commission on the Future of DNA Evidence. Washington, D.C. http://www.ojp.usdoj.gov/nij/pubs-sum/183697.htm

———.2006. Lessons learned from 9/11: DNA identification in mass fatality incidents. Washington, D.C. NCJ214781. Available at http://massfatality.dna.gov

SWGDAM (Scientific Working Group on DNA Analysis Methods). 2000. Short tandem repeat (STR) interpretation guidelines. *Forensic Sci. Commun.* 2(3): on-line at http://www.fbi.gov/hq/lab/fsc/backissu/july2000/strig.htm

Vanek D., Hradil R., and Budowle B. 2001. Czech population data on 10 short tandem repeat loci of SGM Plus STR system kit using DNA purified in FTA cards. *Forensic Sci. Int.* **119:** 107–108.

Walsh P.S., Metzger D.A., and Higuchi R. 1991. Chelex 100 as a medium for simple extraction of DNA for PCR-based typing from forensic material. *BioTechniques* **10:** 506–513.

Whitaker J.P., Clayton T.M., Urquhart A.J., Millican E.S., Downes T.J., Kimpton C.P., and Gill P. 1995. Short tandem repeat typing of bodies from a mass disaster: High success rate and characteristic amplification patterns in highly degraded samples. *BioTechniques* **18:** 670–677.

35 The Human Genome: What Lies Ahead

Michael P. Weiner[1] and J. Claiborne Stephens[2]

[1]RainDance Technologies, Guilford, Connecticut 06437; [2]Motif BioSciences Inc., New York, New York 10017

ADVANCES IN TECHNOLOGY

Most molecular biology technologies change rather rapidly. The methods we used ten years ago are much different from those we use today. In many cases, these changes represent improvements

that make current methods faster, cheaper, or more automated. Occasionally, however, changes arise that are the technological equivalent to punctuated equilibrium. The history of DNA sequencing technology and the major improvements to it can be used to illustrate this point. In 1977, Fred Sanger at The Medical Research Council in the United Kingdom invented a method to sequence DNA using a chain-terminating method. Consisting of a DNA polymerase, radioactive labels, chain-terminating dideoxynucleotides, and polyacrylamide gel electrophoresis, this method gained widespread acceptance around the world. Nearly ten years later, Leroy Hood and colleagues (1986) invented a machine to automate certain laborious parts of the Sanger-sequencing method (and by using fluorescent dyes, eliminated the need for radioactivity). The first commercially successful DNA sequencing instrument based on the Cal Tech design appeared a year later (1987). With these instruments available, researchers could sequence small bacterial genomes and rather simple organisms (yeast, fly, and worm), but the labor costs needed to sequence a human genome were still too high and our collective genome had to wait until a different technology could be implemented. That technology arrived eleven years later, in May 1998, with the introduction of the first 96-lane capillary electrophoresis-based DNA sequencer. As a result of this technology and its associated reduced costs, we were able to complete the draft human genome sequence three years later (2001). It is now late 2007, and the next-generation sequencers have been available for several years. The first of these sequencers, introduced in 2004 by 454 Life Sciences and this year by Solexa, uses massively parallel sequence-by-synthesis methods to decipher hundreds of mega- to giga-bases in a single run.

What does the evolution of sequencing technology have to do with the subject of this manual, genome variation? The answer is that it is becoming more and more apparent that both the sequencing and genome variation technologies are beginning to merge. The cost of whole-genome sequencing is closely approaching the cost of a whole-genome SNP analysis. The current technologies will probably not get us there, but it is exciting to know that we are, at most, only one generation away.

SEQUENCING APPLICATIONS AND IMPLICATIONS

The Human Genome Project provided, after years of cooperative efforts, the first revelation of the human genome sequence, which was a mosaic of many different genomes. As we have just discussed, the ultrahigh-throughput sequencing platforms now yield hundreds of millions of base pairs in a single run, thereby greatly reducing the time frame. It is notable also that these approaches provide high individual read accuracy with extremely low error rate (typically less than one in 20,000). Encouraged by these promising results, two companies have initiated projects to sequence an individual genome. 454 Life Sciences has announced the preliminary genomic sequence of a 78-year-old white male, in fact, James Watson. The selection of Watson is curiously satisfying—he is, of course, known for his role in the discovery of the structure of DNA, the Watson of the Watson-Crick DNA double helix, and was the first Head of the Human Genome Project. In a related effort, Illumina has undertaken sequencing the genome of an individual from the Yoruba, an ethnic group in Nigeria, selected in consequence of findings from The International HapMap Project. The HapMap Project has continued and even extended the spirit of the Human Genome Project with its studies of genetic diversity among four populations: the Yoruba people of Ibadan, Nigeria; the Han Chinese in Beijing; the Japanese in Tokyo; and the resource of the French Centre d'Etude du Polymorphisme Humaine (CEPH). These haplotype studies indicate that the Yoruba show a greater degree or more complete representation of human genetic diversity than the other populations—clearly, genomic sequences from this group will be of great interest in comparison to the reference groups. It is significant that samples for these variation studies were collected according to approved protocols developed by appropriate ethics committees and with community engagement to address specific concerns and obtain informed consent (see Chapter 1 for a more thorough discussion of these procedures and requirements). There are, however, remaining ethical issues. Privacy is protected as the samples are not associated with personal identifiers, but potential associations or implications to do with race or ethnic groups, which we discuss in greater detail below, are of concern.

These sequencing and genomic variation studies represent the first attempts to explore the relationships between an individual's genome and his or her biological traits and predispositions. The expectation is that improvements in sequencing technologies in the near future may make a genome analysis as routine as a blood test. There are clear and promising advantages here, especially on the medical front—the ability, for example, to screen for mutations associated with various forms of cancer. Such genomic information could, in principle, help a physician to individualize (and thereby optimize) treatment to ensure better health or medical outcomes. However, there are concerns as well about the misunderstanding or, worse, the misuse of this information. Our ability to generate genomic data exerts tremendous pressure on all aspects of medical infrastructure to keep up with the burgeoning amount of data. Every stakeholder in the healthcare network—patient, physician, payor, insurance company, regulators, educators, etc.—wants to have "validated" genetic tests available, yet the current infrastructure is unlikely to be able to keep pace with the availability of data. How will society, and in particular, those agencies required to approve and regulate genetic testing, keep up with the likely availability and demand?

CHANGES IN BIOLOGICAL DIVERSITY

We have addressed issues related to genetic variation in the human population and considered some likely future prospects. However, if we extend our discussion to other species that share our planet, it is clear that we are now witnessing some rapid (and, in some cases, distressing) changes in variation within the plant and animal kingdoms.

In the cult-favorite science fiction movie *Silent Running*, all that remains of the biodiversity of a deforested Earth is several off-world spaceships containing self-sustaining biospheres. In time, the mission of these ships is deemed unnecessary and the biospheres themselves are jettisoned into space. The film was released in the early 1970s, at a time when ecology and the environment were at the forefront of the public consciousness. Deforestation of the Amazon basin and increasing pollution affecting the viability of the coral reefs were of immense public concern.

Thirty years later, at the time of this writing, the reduction of Earth's biodiversity is again at the forefront of public awareness. Concern for retaining our biodiversity has begun to transcend national boundaries. With foresight, as a response to this concern, Norway has constructed a "Doomsday vault" in the Arctic as a repository for plant seeds in the event of a major Earth cataclysm. It is becoming increasingly clear from a number of studies that human activities are contributing to an increase in the temperature of the planet and that this rising temperature is promoting a reduction in biodiversity. But global warming is only one of several causes of a loss of biodiversity. An unintended consequence of our increasingly intensive agricultural methods is the homogenization of high-yielding foodstuff and livestock, causing a loss of genomic diversity.

Several chapters in this manual describe both animal and plant species that are endangered, now extinct, or nearly extinct. We have come to a unique point in history where we are learning intimate details about the genetic makeup of organisms that we are causing to become endangered and extinct. Who knows what we lose in our understanding of ourselves when a species is lost? As we begin to know more about, and to appreciate, the genetic diversity of the organisms in the world around us, we must strive even harder to ensure their survival.

Appendix: Cautions

GENERAL CAUTIONS

Please note that the Cautions Appendix in this manual is not exhaustive. Readers should always consult individual manufacturers and other resources for current and specific product information. Chemicals and other materials discussed in text sections are not identified by the icon (!) used to indicate hazardous materials in the protocols. However, they may be hazardous to the user without special handling. Please consult your local safety office or the manufacturer's safety guidelines for further information.

The following general cautions should always be observed.

- **Become completely familiar with the properties of substances used** before beginning the procedure.

- **The absence of a warning** does not necessarily mean that the material is safe, because information may not always be complete or available.

- **If exposed** to toxic substances, contact your local safety office immediately for instructions.

- **Use proper disposal procedures** for all chemical, biological, and radioactive waste.

- **For specific guidelines on appropriate gloves**, consult your local safety office.

- **Handle concentrated acids and bases** with great care. Wear goggles and appropriate gloves. A face shield should be worn when handling large quantities.

   Do not mix strong acids with organic solvents because they may react. Sulfuric acid and nitric acid especially may react highly exothermically and cause fires and explosions.

   Do not mix strong bases with halogenated solvent as they may form reactive carbenes, which can lead to explosions.

- **Handle and store pressurized gas containers** with caution because they may contain flammable, toxic, or corrosive gases; asphyxiants; or oxidizers. For proper procedures, consult the Material Safety Data Sheet that must be provided by your vendor.

- **Never pipette** solutions using mouth suction. This method is not sterile and can be dangerous. Always use a pipette aid or bulb.

- **Keep halogenated and nonhalogenated** solvents separately (e.g., mixing chloroform and acetone can cause unexpected reactions in the presence of bases). Halogenated solvents are organic solvents such as chloroform, dichloromethane, trichlorotrifluoroethane, and dichloroethane. Some nonhalogenated solvents are pentane, heptane, ethanol, methanol, benzene, toluene, *N,N*-dimethylformamide (DMF), dimethyl sulfoxide (DMSO), and acetonitrile.

- **Laser radiation**, visible or invisible, can cause severe damage to the eyes and skin. Take proper precautions to prevent exposure to direct and reflected beams. Always follow manufacturer's safety guidelines and consult your local safety office. See caution below for more detailed information.

- **Flash lamps**, due to their light intensity, can be harmful to the eyes. They also may explode on occasion. Wear appropriate eye protection and follow the manufacturer's guidelines.

- **Photographic fixatives, developers, and photoresists** also contain chemicals that can be harmful. Handle them with care and follow manufacturer's directions.

- **Power supplies and electrophoresis equipment** pose serious fire hazard and electrical shock hazards if not used properly.

- **Microwave ovens and autoclaves** in the lab require certain precautions. Accidents have occurred involving their use (e.g., to melt agar or Bacto-agar stored in bottles, or to sterilize). If the screw top is not completely removed and there is not enough space for the steam to vent, the bottles can explode and cause severe injury when the containers are removed from the microwave or autoclave. Always completely remove bottle caps before microwaving or autoclaving. An alternative method for routine agarose gels that do not require sterile agar is to weigh out the agar and place the solution in a flask.

- **Ultrasonicators** use high-frequency sound waves (16–100 kHz) for cell disruption and other purposes. This "ultrasound," conducted through air, does not pose a direct hazard to humans, but the associated high volumes of audible sound can cause a variety of effects, including headache, nausea, and tinnitus. Direct contact of the body with high-intensity ultrasound (not medical imaging equipment) should be avoided. Use appropriate ear protection and display signs on the door(s) of laboratories where the units are used.

- **Use extreme caution when handling cutting devices** such as microtome blades, scalpels, razor blades, or needles. Microtome blades are extremely sharp! Use care when sectioning. If you are not familiar with their use, have someone demonstrate proper procedures. For proper disposal, use the "sharps" disposal container in your lab. Discard used needles *unshielded*, with the syringe still attached. This prevents injuries (and possible infections; see Biological Safety) while manipulating used needles; many accidents occur while trying to replace the needle shield. Injuries may also be caused by broken Pasteur pipettes, coverslips, or slides.

GENERAL PROPERTIES OF COMMON CHEMICALS

The hazardous materials list can be summarized in the following categories.

- Inorganic acids, such as hydrochloric, sulfuric, nitric, or phosphoric, are colorless liquids with stinging vapors. Avoid spills on skin or clothing. Spills should be diluted with large amounts of water. The concentrated forms of these acids can destroy paper, textiles, and skin, as well as cause serious injury to the eyes.

- Inorganic bases such as sodium hydroxide are white solids that dissolve in water and under heat development. Concentrated solutions will slowly dissolve skin and even fingernails.

- Salts of heavy metals are usually colored, powdered solids that dissolve in water. Many of them are potent enzyme inhibitors and therefore toxic to humans and to the environment (e.g., fish and algae).

- Most organic solvents are flammable volatile liquids. Avoid breathing the vapors, which can cause nausea or dizziness. Also avoid skin contact.

- Other organic compounds, including organosulfur compounds such as mercaptoethanol or organic amines, can have very unpleasant odors. Others are highly reactive and should be handled with appropriate care.

- If improperly handled, dyes and their solutions can stain not only your sample, but also your skin and clothing. Some of them are also mutagenic (e.g., ethidium bromide), carcinogenic, and toxic.

- Nearly all names ending with "ase" (e.g., catalase, β-glucuronidase, or zymolyase) refer to enzymes. There are also other enzymes with nonsystematic names like pepsin. Many of them are provided by manufacturers in preparations containing buffering substances, etc. Be aware of the individual properties of materials contained in these substances.

- Toxic compounds are often used to manipulate cells. They can be dangerous and should be handled appropriately.

- Be aware that several of the compounds listed have not been thoroughly studied with respect to their toxicological properties. Handle each chemical with the appropriate respect. Although the toxic effects of a compound can be quantified (e.g., LD_{50} values), this is not possible for carcinogens or mutagens where one single exposure can have an effect. Be aware that dangers related to a given compound may also depend on its physical state (fine powder vs. large crystals/diethylether vs. glycerol/dry ice vs. carbon dioxide under pressure in a gas bomb). Anticipate under which circumstances during an experiment exposure is most likely to occur and how best to protect yourself and your environment.

HAZARDOUS MATERIALS

Note: In general, proprietary materials are not listed here. Kits and other commercial items, as well as most anesthetics, dyes, fixatives, and stains, are also not included. Anesthetics require special care. Follow the manufacturer's safety guidelines that accompany these products.

Acetic acid (concentrated) must be handled with great care. It may be harmful by inhalation, ingestion, or skin absorption. Wear appropriate gloves and goggles. Use in a chemical fume hood.

Acetone causes eye and skin irritation and is irritating to mucous membranes and upper respiratory tract. Do not breathe the vapors. It is also extremely flammable. Wear appropriate gloves and safety glasses. Keep away from heat, sparks, and open flame.

Ammonium chloride, NH_4Cl, may be harmful by inhalation, ingestion, or skin absorption. Wear appropriate gloves and safety glasses and use in a chemical fume hood.

Ammonium sulfate, $(NH_4)_2SO_4$, may be harmful by inhalation, ingestion, or skin absorption. Wear appropriate gloves and safety glasses.

Cetyltrimethylammonium bromide (CTAB) is toxic and an irritant and may be harmful by inhalation, ingestion, or skin absorption. Wear appropriate gloves and safety glasses. Avoid breathing the dust.

$CHCl_3$, *see* **Chloroform**

Chloroform, $CHCl_3$, is irritating to the skin, eyes, mucous membranes, and respiratory tract. It is a carcinogen and may damage the liver and kidneys. It is also volatile. Avoid breathing the vapors. Wear appropriate gloves and safety glasses and always use in a chemical fume hood.

CTAB, *see* **Cetyltrimethylammonium bromide**

DEPC, *see* **Diethyl pyrocarbonate**

Diethyl pyrocarbonate (DEPC) is a potent protein denaturant and is a suspected carcinogen. Aim bottle away from you when opening it; internal pressure can lead to splattering. Wear appropriate gloves, safety goggles, and lab coat, and use in a chemical fume hood.

Dithiothreitol (DTT) is a strong reducing agent that emits a foul odor. It may be harmful by inhalation, ingestion, or skin absorption. When working with the solid form or highly concentrated stocks, wear appropriate gloves and safety glasses and use in a chemical fume hood.

DTT, *see* **Dithiothreitol**

Ethidium bromide is a powerful mutagen and is toxic. Consult the local institutional safety officer for specific handling and disposal procedures. Avoid breathing the dust. Wear appropriate gloves when working with solutions that contain this dye.

Formamide is teratogenic. The vapor is irritating to the eyes, skin, mucous membranes, and upper respiratory tract. It may be harmful by inhalation, ingestion, or skin absorption. Wear appropriate gloves and safety glasses and always use a chemical fume hood when working with concentrated solutions of formamide. Keep working solutions covered as much as possible.

$HOCH_2CH_2SH$, *see* **β-Mercaptoethanol**

IAA, *see* **Isoamyl alcohol**

Isoamyl alcohol (IAA) may be harmful by inhalation, ingestion, or skin absorption, and presents a risk of serious damage to the eyes. Wear appropriate gloves and safety goggles. Keep away from heat, sparks, and open flame.

Isopropanol is flammable and irritating. It may be harmful by inhalation, ingestion, or skin absorption. Wear appropriate gloves and safety glasses. Do not breathe the vapor. Keep away from heat, sparks, and open flame.

KOH, *see* **Potassium hydroxide**

β-Mercaptoethanol (2-Mercaptoethanol), $HOCH_2CH_2SH$, may be fatal if inhaled or absorbed through the skin and is harmful if ingested. High concentrations are extremely destructive to the mucous membranes, upper respiratory tract, skin, and eyes. β-Mercaptoethanol has a very foul odor. Wear appropriate gloves and safety glasses and always use in a chemical fume hood.

NaOAc, *see* **Sodium acetate**

NaOH, *see* **Sodium hydroxide**

NH_4Cl, *see* **Ammonium chloride**

$(NH_4)_2SO_4$, *see* **Ammonium sulfate**

Octanol is highly flammable. Keep away from heat, sparks, and open flame. It may be harmful by inhalation, ingestion, or skin absorption. Wear appropriate gloves and safety goggles.

Phenol is extremely toxic, highly corrosive, and can cause severe burns. It may be harmful by inhalation, ingestion, or skin absorption. Wear appropriate gloves, goggles, protective clothing, and always use in a chemical fume hood. Rinse any areas of skin that come in contact with phenol with a large volume of water and wash with soap and water; do not use ethanol!

Potassium hydroxide (KOH) and KOH/methanol are highly toxic and may be fatal if swallowed. They may be harmful by inhalation, ingestion, or skin absorption. Solutions are corrosive and can cause severe burns. They should be handled with great care. Wear appropriate gloves and safety goggles.

Proteinase K is an irritant and may be harmful by inhalation, ingestion, or skin absorption. Wear appropriate gloves and safety glasses.

Radioactive substances: When planning an experiment that involves the use of radioactivity, include the physicochemical properties of the isotope (half-life, emission type, and energy), the chemical form of the radioactivity, its radioactive concentration (specific activity), total amount, and its chemical concentration. Order and use only as much as really needed. Always wear appropriate gloves, lab coat, and safety goggles when handling radioactive material. **X-rays** and **gamma rays** are elec-

tromagnetic waves of very short wavelengths either generated by technical devices or emitted by radioactive materials. They may be emitted isotropically from the source or may be focused into a beam. Their potential dangers depend on the time period of exposure, the intensity experienced, and the wavelengths used. Be aware that appropriate shielding is usually of lead or other similar material. The thickness of the shielding is determined by the energy(s) of the X-rays or gamma rays. Consult the local safety office for further guidance in the appropriate use and disposal of radioactive materials. Always monitor thoroughly after using radioisotopes. A convenient calculator to perform routine radioactivity calculations can be found at: http://graphpad.com/quickcalcs/index.cfm

RNase A is an irritant and may be harmful by inhalation, ingestion, or skin absorption. Wear appropriate gloves and safety glasses. Do not breathe the dust.

SDS, *see* **Sodium dodecyl sulfate**

Sodium acetate (NaOAc), *see* **Acetic acid**

Sodium dodecyl sulfate (SDS) is toxic, an irritant, and poses a risk of severe damage to the eyes. It may be harmful by inhalation, ingestion, or skin absorption. Wear appropriate gloves and safety goggles. Do not breathe the dust.

Sodium hydroxide (NaOH) and solutions containing NaOH are highly toxic and caustic and should be handled with great care. Wear appropriate gloves and a face mask. All other concentrated bases should be handled in a similar manner.

Triton X-100 causes severe eye irritation and burns. It may be harmful by inhalation, ingestion, or skin absorption. Wear appropriate gloves and safety goggles. Do not breathe the vapor.

Xylene is flammable and may be narcotic at high concentrations. It may be harmful by inhalation, ingestion, or skin absorption. Wear appropriate gloves and safety glasses and use only in a chemical fume hood. Keep away from heat, sparks, and open flame.

Index

461